MW01040637

VALUE OF INFORMATION IN THE EARTH SCIENCES

Integrating Spatial Modeling and Decision Analysis

Value of information (VOI) is a concept in decision theory for analyzing the value of obtaining additional information to solve a problem. Gathering the right kind and the right amount of information is crucial for any decision-making process. Already commonly used in medicine, economics, and finance, VOI is becoming increasingly popular with Earth scientists.

This book presents a unified framework for assessing the value of potential data gathering schemes by integrating spatial modeling and decision analysis, with a focus on the Earth sciences. The authors discuss and compare the value of imperfect versus perfect information and the value of total versus partial information, where only subsets of the data are acquired. Concepts are illustrated using a suite of quantitative tools from decision analysis, such as decision trees and influence diagrams, as well as models for continuous and discrete dependent spatial variables, including Bayesian networks, Markov random fields, Gaussian processes, and multiple-point geostatistics. Numerous examples are used to illustrate the applicability of VOI to topics such as energy, geophysics, geology, mining, and environmental science. Real datasets and MATLAB codes are also provided as online supplementary material.

Unique in its scope, this book is of interest to students, researchers, and industry professionals in the Earth and environmental sciences who use applied statistics and decision analysis techniques, and particularly to those working in petroleum, mining, and environmental geoscience.

Jo Eidsvik is Professor of Statistics at the Norwegian University of Science and Technology (NTNU), Norway, and has previous industry work experience from the Norwegian Defense Research Establishment and Statoil.

Tapan Mukerji is Associate Professor (Research) in the Department of Energy Resources Engineering and the Department of Geophysics at Stanford University. He is a co-author of *The Rock Physics Handbook* and *Quantitative Seismic Interpretation.*

Debarun Bhattacharjya is a research staff member in the Cognitive Computing Research Group at the IBM T. J. Watson Research Center, New York. He applies his expertise in decision analysis toward developing algorithms and tools for both research and consulting purposes.

VALUE OF INFORMATION IN THE EARTH SCIENCES

Integrating Spatial Modeling and Decision Analysis

JO EIDSVIK, TAPAN MUKERJI, AND DEBARUN BHATTACHARJYA

CAMBRIDGE
UNIVERSITY PRESS

CAMBRIDGE
UNIVERSITY PRESS

University Printing House, Cambridge CB2 8BS, United Kingdom

Cambridge University Press is part of the University of Cambridge.

It furthers the University's mission by disseminating knowledge in the pursuit of education, learning and research at the highest international levels of excellence.

www.cambridge.org
Information on this title: www.cambridge.org/9781107040267

First published 2015

Printed in the United Kingdom by TJ International Ltd. Padstow Cornwall

A catalogue record for this publication is available from the British Library

Library of Congress Cataloguing in Publication Data
Eidsvik, Jo, 1973–
Value of information in the earth sciences : integrating spatial modeling and decision analysis / Jo Eidsvik, Tapan Mukerji, and Debarun Bhattacharjya.
pages cm
Includes bibliographical references and index.
ISBN 978-1-107-04026-7 (hbk)
1. Earth sciences–Information services. 2. Geochemical prospecting. 3. Communication in science. I. Mukerji, Tapan, 1965– II. Bhattacharjya, Debarun. III. Title.
QE26.3.E35 2015
550.01'156–dc23 2015022572

ISBN 978-1-107-04026-7 Hardback

Additional resources for this publication at www.cambridge.org/eidsvik

Contents

Preface

This book is a result of our collaboration over the past decade on addressing problems related to the value of information (VOI) in Earth sciences applications by building links between statistics, geosciences, and decision analysis. We believe that such an interdisciplinary approach will become increasingly essential for the careful stewardship of our natural resources.

Decisions related to the Earth's natural resources are often consequential, and making these decisions under uncertainty is a ubiquitous challenge. Since there is a lot at stake, it may be worthwhile for the decision maker to obtain more information before the decision is actually made – i.e., before an irrevocable allocation of resources. When faced with uncertainty, gathering the right kind and right amount of information is crucial. Today, geo-coded data are commonly purchased, processed, and interpreted to provide information about uncertain variables, such as the spatial distribution of trees in a forest, the amount of oil or gas in the subsurface, the level of groundwater in an aquifer, or the mineral content in a mine. A crucial question to answer is: how much information should one purchase, and at what price? This question is related to the well-established concept of VOI. Additional information may help to reduce the uncertainty, but if the information has no impact on the decision, then purchasing it is not economic.

A key characteristic of applying the decision theoretic notion of VOI to the Earth sciences that makes it different from other applications is the spatial aspect: spatial uncertainty, spatially distributed information, and spatial decisions. The decision theoretic formalism provides a consistent basis for relating statistical models of spatial phenomena to the decisions. This connection facilitates decision making by providing clarity of action and also fosters innovative approaches for designing spatial information-gathering schemes.

The book presents a unified framework for VOI analysis based on statistical concepts, geological and geophysical modeling, and decision analysis. Often in the Earth sciences, information is sensed remotely – for example, from geophysical surveys that provide indirect and imperfect knowledge about the spatially varying surface. How valuable is this imperfect information? We study the comparison of various kinds of practical schemes by considering the value of imperfect versus perfect information and the value of total versus partial information, where only subsets of the possible data are acquired – for example, a sparse versus a spatially dense survey.

We focus on areas of our own expertise for the benefit and interest of others with similar scientific backgrounds. Throughout the book, we will discuss and reference the work of others, but we do not aim to provide an exhaustive summary of what has been done on applying VOI to other fields, such as in medicine. Applications from the Earth sciences are highlighted, and we describe the practical use of our methods and tools via a number of illustrative examples and hands-on exercises so that readers can learn the concepts by applying them.

Even though this is a specialized book, we aim to reach a diverse group of readers. The primary intended readers include scientists, engineers, graduate students, and professionals who use applied statistics and decision theoretic models in the quantitative Earth sciences. We believe that the topics will be of interest to researchers and industry professionals in different fields of the Earth sciences: energy resources, mining, groundwater, and environmental sciences. It will also be of interest to applied statisticians and decision analysts. We hope that this book will be a practitioner's guide.

The book requires some background in basic probability and statistics and mathematical calculus, as well as an interest in Earth sciences applications. Although it is not essential, it helps to know basic multivariate analysis and decision analysis or optimization. The reader must be open to learning unfamiliar topics and be able to appreciate the added value obtained from the multidisciplinary approach. If more background knowledge is needed for a particular topic, readers can consult some of the references suggested in the bibliographic notes at the end of each chapter. The chapters define the concepts using mathematics, but without going into too much detail. Additional mathematical details about the most important models and methods used in the book are provided in the appendix. The last chapter contains exercises and larger projects including data. On the website (srb.stanford.edu/VOI), we provide more background for these examples, including data and computer code.

We hope that you find this book useful!

Acknowledgments

We would like to acknowledge the contribution of several friends, colleagues, and collaborators who have helped us in our endeavors toward the preparation and completion of this book. This includes our co-authors on various papers pertaining to value of information (VOI) in the Earth sciences: Jef Caers, Geetartha Dutta, Maren Drange-Espeland, Steinar Ellefmo, Evangelos Evangelou, Ragnar Hauge, Ketil Hokstad, Rosemary Knight, Marie Lilleborge, Gabriele Martinelli, Sara Rekstad, Javad Rezaie, Richard Sinding-Larsen, and Whitney Trainor-Guitton. We received valuable feedback on preliminary chapter drafts from Lea Deleris, Jeffrey Keisler, Gabriele Martinelli, Sameer Parakh, Deepanshu Kumar, Javad Rezaie, Ingelin Steinsland, and Ulrich Theune. We are very grateful to those who helped design and prepare creative artwork for the book: Hessam Moussavinik, Ryan Nilsen, Mamta Parakh, and Sachin Premasuthan. We are indebted to Henning Omre for laying the foundation for our NTNU–Stanford collaboration, as well as Gary Mavko and John Weyant for their encouragement on these research topics, including supporting Debarun Bhattacharjya for a few quarters during his graduate studies at Stanford University. Gary and Henning in particular planted the seeds that grew into this collaboration. Ross Shachter and Ron Howard played an important role in revealing the power of VOI. We acknowledge the long-term support from the members of the Uncertainty in Reservoir Evaluation (URE) at NTNU, the Stanford Center for Reservoir Forecasting (SCRF), and the Stanford Rock Physics and Borehole Geophysics (SRB) consortia. The Department of Energy Resources Engineering at Stanford University hosted Jo Eidsvik during his sabbatical, allowing us to finally finish the manuscript. Finally, we thank our families for their unwavering support.

1
Introduction

Figure 1.1 data crunching times
numbers pulsing 'round the globe:
how much is enough?

Carl Sagan famously wrote about the "pale blue dot" that all of us share: "That's home. That's us ... there is nowhere else, at least in the near future, to which our species could migrate. Like it or not, for the moment the Earth is where we make our stand" (Sagan 1994, p. 8). More than 20 years after Sagan's famous book, there are plans for human settlement on Mars. But Earth is still home, and on this rock revolving around a golden sun, people everywhere constantly use information to make decisions about utilizing, managing, and sustaining our valuable natural resources. How can we quantitatively analyze and evaluate different information sources for supporting decisions in the Earth sciences?

1.1 What is the value of information?

Making decisions in the Earth sciences can be challenging. There is often significant uncertainty pertinent to the decision – for instance, the availability and spatial distribution of the resources under consideration. Moreover, there could be a lot at stake, as investments may be considerable and there may be huge financial losses or adverse environmental consequences. Petroleum exploration and production, mining, agriculture, and forestry are examples of domains where large-scale efforts are typical. Similarly, endeavors in domains such as conservation biology, ecology, groundwater management, and climate sciences strive to sustain and better manage natural resources and affect several stakeholders.

A unique aspect of the complexity of decisions in the Earth sciences is the inherent spatial variability. The subsurface formations have been forged through millions of years of coupled geological, physical, chemical, and biological processes, resulting in spatial trends and dependence between rock properties at different spatial locations. The subsurface properties have spatial dependence, but at the same time they are very heterogeneous, variable, and uncertain. The decision maker's characterization of uncertainties is then best represented by spatial statistics, because properties at a particular geographic location cannot be treated independently of those at other locations. Reliable information that resolves some of the uncertain properties at one location could therefore go a long way toward improving the overall quality of decisions.

The past few years have seen a tremendous surge of interest in "big data." This has largely been driven by the development in electronics, telecommunications, computer science, online commerce, social media, and our ability to automatically acquire and store data. While this may be a subject of great current interest in popular culture, the Earth sciences have arguably been dealing with big data for a while now. The aspect of volume – the first of the five "Vs" of big data – has certainly been well represented; examples include large geophysical surveys, especially in exploration reflection seismology, and weather and atmospheric data from remote sensing satellites. The utilization and sustenance of the Earth's resources involve multidisciplinary work that can entail acquiring, processing, modeling, and interpreting copious amounts of a variety of data types – the second of the five "Vs." As an example, typical analysis of basin and petroleum systems includes geophysical seismic data, well logs, geochemical analysis, information from biostratigraphy and paleoclimate studies, structural geology, the study of depositional environments, and core analysis. The other "Vs" include velocity – data acquired at a rapid rate (e.g., continuous streams of data from remote sensing sensors) – and veracity – whether the data are accurate and trustworthy. Eventually, the goal is to make better decisions. This is where the last of the five "Vs" of big data becomes important: **value**.

What is the value of the data and how much data are enough? Information almost always comes at a price, so when is the information worth its price? At the very core of this book lies the decision theoretic notion of **value of information** (henceforth referred to as VOI), which we use to evaluate and analyze various sources of data. The power of analyzing information sources using VOI is that: (i) it allows the decision maker to perform a reasonable evaluation before the information is purchased and therefore revealed and (ii) if the decision maker can model value using monetary units, then VOI is also in monetary units. These capabilities make VOI an extremely practical tool that addresses real problems in the real world.

Figure 1.2 demonstrates what we refer to as the "pyramid of conditions" that makes information valuable. Although all technical details are postponed until later in the book,

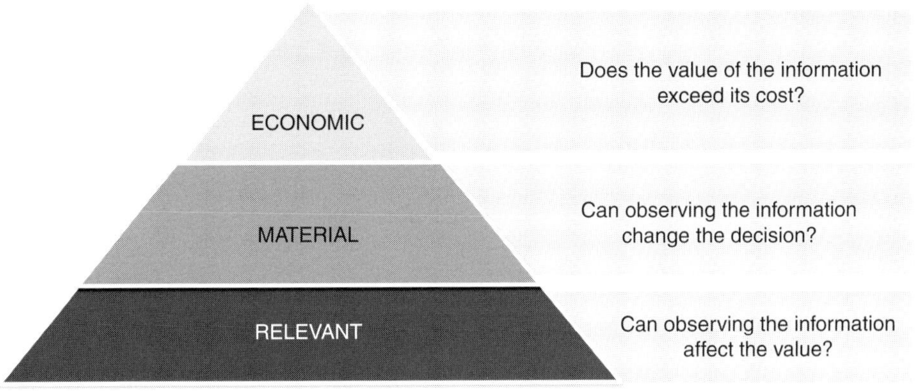

Figure 1.2 The pyramid of conditions specifies that information-gathering schemes should be economic, material, and relevant.

we feel that the figure captures the essence of VOI. The condition at the bottom of the pyramid specifies that information must be **relevant** to the value from the decision, so observing the information can impact the decision maker's beliefs about the key uncertainties. As an (extreme) example, while making a decision about whether you should take your umbrella to work, information about what your friend ate for breakfast hardly seems relevant to whether it will rain! The condition at the middle level of the pyramid specifies that information must be **material**, in the sense that observing it should have the capability to change your decision. If you enjoy both the rain and the sun and would rather not carry an umbrella, then listening to a weather forecast hardly seems worthwhile – you will choose to leave the umbrella at home, regardless of what the forecast has to say. Information must have the potential to allow the decision maker to flexibly adapt and differ from what he or she would have otherwise done. Finally, the condition at the very top of the pyramid is that information must be **economic** – the price of the information must be less than its value. The three requirements are shown as a pyramid because higher conditions cannot be satisfied unless those lower in the pyramid are satisfied. If an information source is not relevant, it cannot be material; if it is not material, it cannot be economic.

The reader may well ask: what is the catch? What do I need to do to harness the wonderful capabilities of VOI? The only catch is that the power of such a practical tool requires some modeling sophistication – it requires understanding and characterizing how the various pieces of the puzzle fit together, and for applications in the Earth sciences, this can often require an interdisciplinary effort. In this book, we recommend a four-stage workflow for using VOI to support information-gathering decisions in the Earth Sciences, as indicated in Figure 1.3:

1. To start with, the decision maker should **frame the underlying decision situation** to understand how the potential information would be used. What are the questions the decision maker is trying to address?

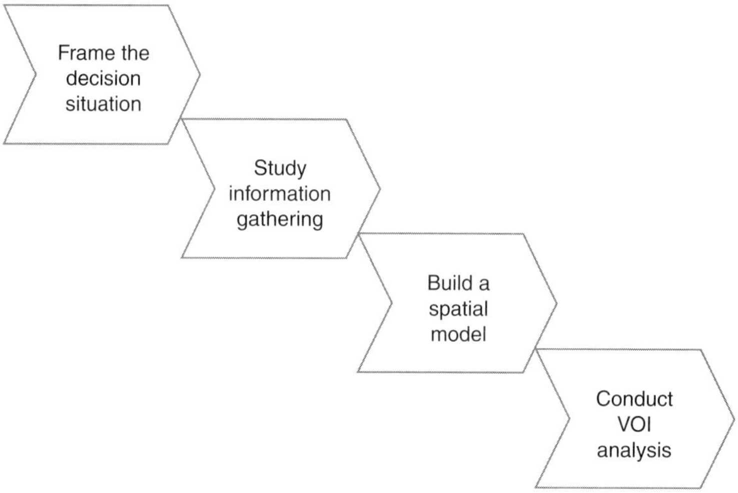

Figure 1.3 A workflow for value of information analysis consists of four steps.

2. Next, the decision maker should identify and **study the potential information-gathering schemes**. How is the information relevant to the underlying decision situation? How might the information affect the decision?
3. In the Earth sciences, spatial dependence is an important aspect of most problems. Therefore, it is often essential to **build a spatial model** that is a reasonable approximation to the real-world domain. This model captures how the various uncertainties, such as geological properties, are connected to each other and to the relevant data.
4. Finally, now that all the pieces are in place, **VOI analysis can be conducted** to address various issues of interest. Information sources of varying reliability and price can be compared, and the decision maker can proactively identify schemes that will increase their value for the decision situation.

As we will later show, the VOI for a data-gathering scheme is computed from comparing the values with and without the data. It is important to note that the underlying decision situation informed by the data plays an explicit part in the calculations. VOI analysis is useful for comparing different schemes. Since the VOI is computed before the data are actually revealed, the value with data must include some kind of averaging over the possible data sets. We highlight how such VOI analysis can be conducted in spatial contexts in the following section.

1.2 Motivating examples from the Earth sciences

Figure 1.4 shows a map view of data from a proposed mining project. The spatial distribution of the oxide grade is highly variable and uncertain. The map displays locations where oxide grade data have been acquired (black). Two types of data have been collected here, at different levels of accuracy. The lower-accuracy data set is obtained with a handheld

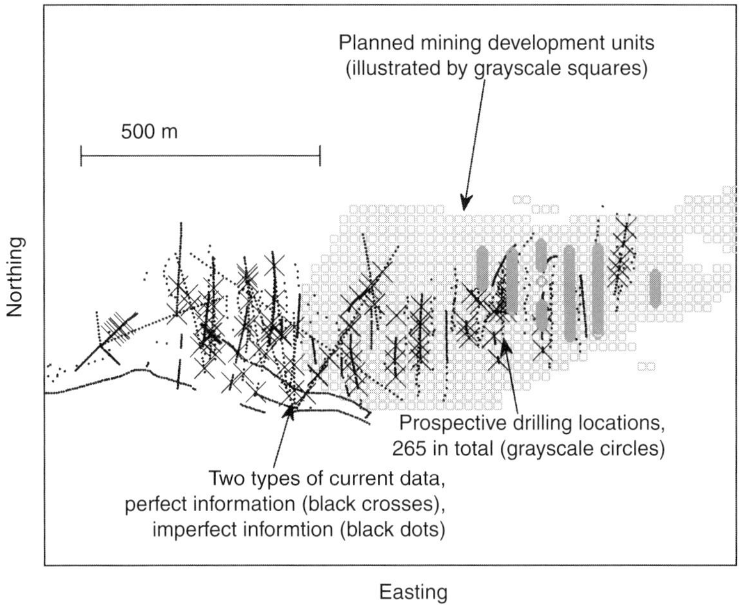

Figure 1.4 Map view of borehole measurement locations and potential mining locations for an oxide resource. The mining company can drill more boreholes before making the mining decision.

meter and requires almost no time for processing (marked with black dots in the display), while the higher-accuracy data set is acquired by taking a core plug from the borehole to the laboratory for careful scanning (marked with black crosses in the display). Based on these observations and expert geological knowledge about the ore, the question is: would mining be profitable? The company has made a careful plan for how the mining would be done (illustrated by gray squared units in the display), but it is difficult to make a decision under uncertainty due to the spatial distribution of oxide. Before making the decision, would mining company can collect more data. Potential locations for additional measurements are defined (marked with gray circles in the display). The data, of course, come at a price for drilling and processing. Are these data worth gathering? And, if yes, which type of data should we acquire – the low-accuracy data or high-accuracy data, or a combination of both? VOI analysis can be used to answer these questions by embedding the decision situation in a spatial modeling framework.

Figure 1.5 shows a network consisting of 38 nodes. The 13 nodes numbered with prefix "P" represent petroleum prospects, and the 25 bottom nodes of the network (illustrated by grayscale circles in the display) are segments of the prospects. This graph, a Bayesian network, is constructed to capture the relationships arising from the geological mechanisms within basin and petroleum systems, and the edges indicate physical connections between geological attributes at the prospects and segments. There is uncertainty about the presence of oil and gas, and the network has an associated probabilistic model that describes these

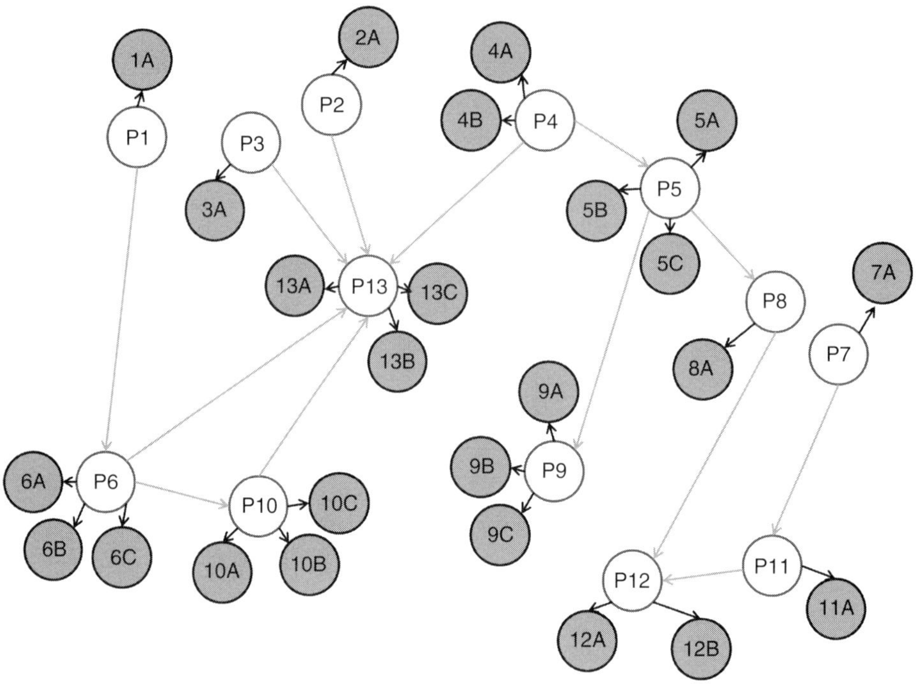

Figure 1.5 Network illustrating 13 oil field prospects and 25 segments where a petroleum company can drill exploration wells before making decisions about prospect development.

uncertainties. A petroleum company can use this model to evaluate their decisions about petroleum field development. Before pursuing expensive development decisions, would it be worthwhile to invest in some exploration wells? An exploration well landed at one segment would reveal the presence or absence of oil at that segment, resolving the uncertainty at that node, and because of the common geological mechanisms indicated by the network edges, the information at the exploration segment would also provide some information about the other segments. If exploration wells will be drilled, where should the company place them? VOI analysis is useful for evaluating such data-gathering schemes.

For the petroleum application, let us zoom in to a finer granularity. A common decision situation is whether to drill production wells at specified reservoir units or to avoid drilling. Before going through with the expensive drilling operations, it could be useful to do a careful subsurface characterization using geophysical data. Seismic and electromagnetic data can be useful in reservoir characterization, improving the prediction of reservoir variables such as lithology, porosity, and saturation. The seismic data undergo processing, which typically provides seismic amplitude information merged in a stack over all angles of the data into one entity (post-stack amplitudes). Would it be worthwhile to invert and interpret the pre-stack data to additionally obtain seismic amplitudes as a

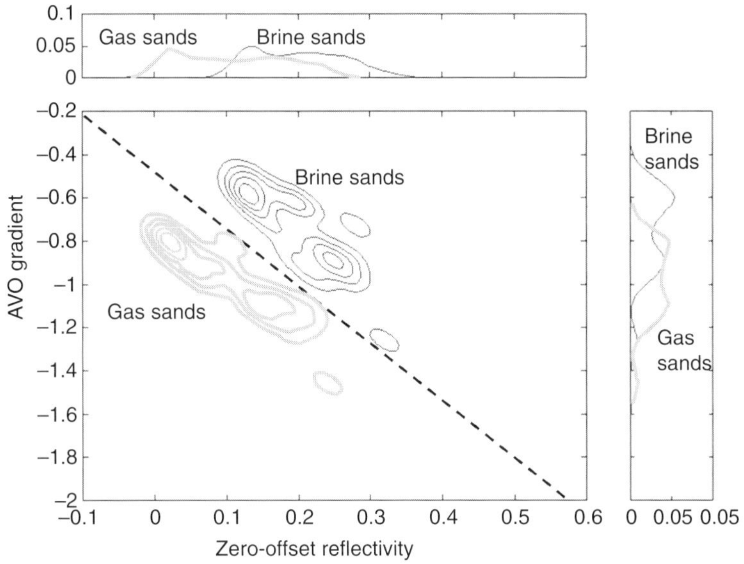

Figure 1.6 Seismic amplitude data are informative of the reservoir rock type and saturation. The contours indicate non-parametric probability density functions (pdfs) of the likely response of the seismic amplitude data for brine sands (black) and gas sands (thick, light gray curves). The marginal pdfs are shown on the right and at the top. The dashed black lines indicate classification boundaries for the two classes.

function of the incidence angle? And what processing accuracy is required for significantly improving the prediction of reservoir rocks and saturations?

Figure 1.6 shows contours of the modeled seismic amplitude responses for two reservoir facies classes (gas sands and brine sands). The contours are representative of bivariate distribution of the expected response for the seismic zero-offset attribute (first axis) and the amplitude-versus-angle attribute (second axis), given any of the two facies classes. These distributions can be assessed from well logs and rock physics and seismic models, but the uncertainty (or the spread and overlap of the distributions) depends on the underlying rock and fluid properties, their natural variability, and the accuracy of the seismic processing scheme. As shown by the marginal distribution on the top panel, the gas sands in this particular reservoir have generally lower zero-offset amplitudes, while the brine sands have somewhat higher amplitudes. However, there is a lot of overlap, and the classification based only on zero-offset amplitudes has a high misclassification error (~25%). If, on the other hand, the classification is based on both the zero-offset amplitude as well as the amplitude-versus-angle attribute, the bivariate classification of the two classes is much better, with a very small misclassification error (~1%). Is it worth purchasing the pre-stack attribute? How would a better classification impact the decision? Perhaps it may be better to get more accurate post-stack amplitudes and reduce the overlap in the marginal distributions of the post-stack amplitudes – or maybe purchase

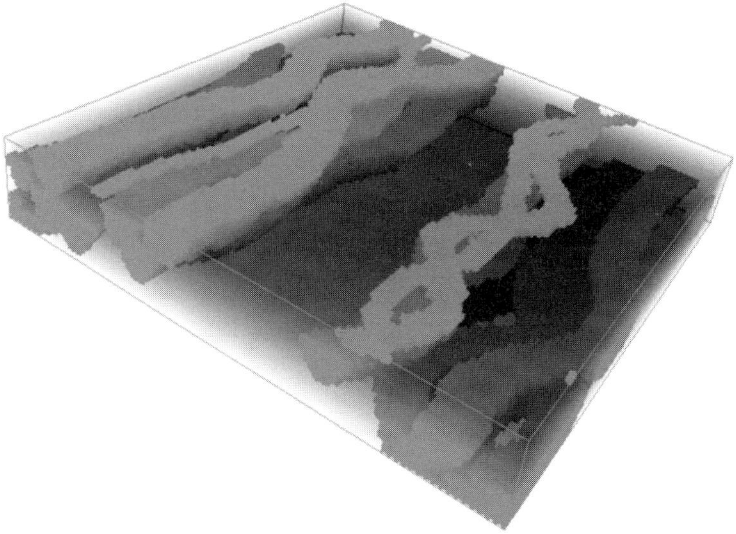

Figure 1.7 Three-dimensional view of one geostatistical realization of uncertain sand channel facies in a groundwater reservoir.

electromagnetic data. Which of the different information-gathering options is most valuable for the decision maker?

VOI analysis can play an important role in better managing groundwater, a valuable natural resource. Will purchasing geophysical electromagnetic measurements help a groundwater manager make better decisions about recharge? As an example, consider the case where groundwater is pumped out from a coastal fluvial aquifer and is used for agriculture. Seawater from the coast starts intruding due to excessive groundwater pumping, leading to increased water salinity and a decrease in usability for crops. Artificial groundwater recharge by pumping freshwater into the subsurface is considered as a way to mitigate salt water intrusion. Where should the recharge site be located? The subsurface distributions of the high-permeability sand channels and low-permeability shales, which impact the groundwater flow and the effectiveness of recharge in the aquifer, are uncertain. Geophysical measurements may be valuable for better characterizing the subsurface and thus help make informed decisions about the selection of possible recharge locations. Assessing the value of the data before actually acquiring them requires modeling the uncertain subsurface channel geometries, accounting for multiple possible scenarios, and conducting flow simulations and geophysical simulations using Monte Carlo computations.

Figure 1.7 shows one realization of subsurface channel geometries generated using multiple-point geostatistical methods. Many such realizations need to be generated to model the subsurface facies uncertainties, and one must simulate their geophysical signatures, as well as the effects of flow and recharge under different recharge alternatives available to the decision maker. The important question is whether the value of the geophysical electromagnetic data is more than the price.

We will return to these examples as well as others in the following chapters. They illustrate current-day applications in the Earth sciences where VOI analysis can be useful.

1.3 Contributions of this book

The main contribution of this book is the blending of decision analysis and spatial statistical concepts to support decisions around information gathering in the Earth sciences. Compared with the traditional use of VOI, we stress the spatial aspects of the statistical model, the alternatives, and the potential information-gathering schemes; many applications of VOI analysis in other disciplines do not need to contend with complexities arising from spatial dependence – a feature that is pervasive in the Earth sciences. Furthermore, we go well beyond the traditional and simplistic high-level models for VOI analysis that have been common for decades and work toward building more realistic models at a finer granularity. We feel that decision makers in the Earth sciences can benefit greatly from our models and methods by deploying them to analyze real-world information schemes. With burgeoning computational capabilities, efficient algorithms, and the ever-increasing availability of data, we sense that more sophisticated models will become increasingly popular and will greatly improve the quality of decisions.

In this book, we refer to decision situations that are typical in the Earth sciences as **spatial decision situations**, building a taxonomy of models based on the decision flexibility – i.e., whether there are a large or small number of alternatives and whether the decision maker's value function is coupled or decoupled, i.e. how complex it is to reasonably compute the decision maker's value from his or her decision situation. We compare various sorts of information-gathering schemes, categorizing them based on whether they provide perfect or imperfect information and whether they are partial or total schemes. We also categorize models based on assumptions around whether decision situations and information-gathering schemes are static or sequential.

We advocate the use of VOI analysis for evaluating data for spatial decision situations. There are several information measures, such as entropy, that are more popular in spatial applications. Unlike VOI, however, these measures only address aspects of the relevancy of the experiment without addressing the monetary gain in value and the ability to make better decisions. VOI is most useful in situations where data are rather expensive. When data come for free, there is not much point in evaluating them – but keep in mind that there is often much effort spent in processing vast amounts of inexpensive data.

Every day, people make important decisions about the development of the Earth's resources. At least as often, decision makers contemplate whether gathering more information will assist them in their difficult decision-making processes. In the future, our demands may change; there may be a drive for recovering unconventional resources in petroleum, we may explore subsea mining, or we may head toward renewable sources such as solar or wind energy. Similarly, environmental challenges will change, and new sustainability questions will arise. What will not change is the advantage of being able to frame decision situations, build useful models for the spatial variables, and evaluate possible information-gathering schemes for improved decision making.

1.4 Organization

This book is organized into seven chapters, including this introduction, as well as an appendix. Chapter 2 introduces basic probability and statistics while providing the fundamental notation for the book. We introduce some specific models, methods, and examples that are used throughout the book. Chapter 3 introduces decision analysis. We focus primarily on the concepts required to understand and appreciate the decision theoretic notion of VOI. Chapter 4 describes spatial statistics models. We motivate spatial modeling through several illustrative examples that are used for VOI analysis in subsequent chapters. These three topics – probability and statistics, decision analysis, and spatial modeling – lay the foundation for the subsequent formulation of methods and tools for VOI in Earth sciences applications. Each of these three topics is, of course, extensive with vast amounts of dedicated literature. The aim is not to cover them comprehensively; rather, the goal in Chapters 2, 3, and 4 is to provide an overview and lay the groundwork for the following chapters.

Chapter 5 integrates concepts from previous chapters. We define spatial decision situations and demonstrate VOI analysis for information-gathering schemes for various categories of spatial decision situations. We describe a taxonomy based on the different combinations of opportunities for spatial decision making and information gathering. Several examples are used to illustrate the concepts.

Chapter 6 provides a number of real-world examples of VOI analysis. We demonstrate applications using data from petroleum, mining, and groundwater applications. For petroleum exploration and development, information-gathering schemes include exploration wells, seismic data, or electromagnetic data. For mine development and safety, various kinds of borehole information could provide valuable information. In the hydrology example, geophysical electromagnetic data could be used to better characterize the subsurface and thus help make better decisions about groundwater recharge.

Chapter 7 contains a number of exercises and hands-on projects. On the book website (srb.stanford.edu/voi), we provide further information about these examples, including various data sets, Netica project examples, and a collection of MATLAB m-files to help readers reproduce many of the results described in the book. The code will also be useful for the hands-on projects.

In every chapter, we end with bibliographic notes, where we present our views on connections with relevant literature as well as some references. Mathematical details about the most important models used in the book are provided in the appendix.

A number of examples recur throughout the book. Table 1.1 summarizes these examples in terms of key assumptions, application domains, and the relevant sections of the book.

1.5 Intended audience and prerequisites

This book is primarily intended for practitioners, professionals, and graduate and advanced undergraduate students in domains associated with the Earth sciences. It may be used as a supplementary text for a class on spatial statistics or decision making in the Earth sciences

Table 1.1. *Summary of the main examples running throughout the book*

	Decision	Information	Model	Sections
Treasure island – the pirate example	Dig for treasure or not	Metal detector experiment	Binary variables	2.3, 3.2, 3.4
Gotta get myself connected – Bayesian network example	Develop petroleum reservoir prospects or not	Exploration wells	Bayesian network model	2.3, 5.7, 6.2, 7.2
Never break the chain – Markov chain example	Develop prospects or not	Exploration wells	Markov chain model	2.3–4, 5.4, 7.1
For whom the bell tolls – Gaussian projects example	Invest in projects or not	Reports of profits for projects	Bivariate Gaussian model	2.3, 3.2–4, 7.1
I love rock and ore – mining oxide grade example	Start mining or not	X-ray of cores in boreholes	Gaussian spatial regression model	2.4, 4.4, 6.4, 7.4
Risky business – petroleum prospect risking example	Develop petroleum reservoir prospects or not	Exploration wells	Empirical forward modeling	2.5, 6.2, 7.2
Time after time – time-lapse seismic example	Perform seismic monitoring or not	(No VOI analysis)	Graphical model	3.3, 7.1
MacKenna's gold – oil and gold example	(Lottery, not a decision situation)	(No VOI analysis)	Graphical model	3.3, 7.1
Norwegian wood – forestry example	Harvest forest units or not	Surveys of the forest	Gaussian random field	4.4, 5.3–4, 5.9, 7.1
We will rock you – rock hazard example	Add support in mining tunnels or not	Rock joint data in boreholes	Spatial generalized linear model	4.5, 6.4, 7.4
Black gold in a white plight – reservoir characterization example	Drill at reservoir units or not	Seismic data	Markov random field	4.6, 6.3, 7.3
Go with the flow – petroleum simulation example	Develop petroleum reservoir or not	Interpreted seismic data	Multiple-point geostatistics and flow simulator	4.7, 5.5, 7.1
The tree amigos – conservation biology example	Establish conservation sites or not	Surveying of spatial domain	Markov random field	5.4, 5.7, 5.9, 7.1

Table 1.1 (*cont.*)

	Decision	Information	Model	Sections
Frozen – hydropower example	Optimal water level behind dam	Snow depth data	Transformed Gaussian random field	5.6, 7.1
Basin street blues – basin modeling example	Develop reservoir prospects or not	Exploration wells	Basin modeling and Bayesian networks	6.2, 7.2
Reservoir dogs – seismic and electromagnetic data example	Drill at petroleum reservoir units or not	Seismic and/or electromagnetic data	Hierarchical Gaussian spatial model	6.3, 7.3
Salt water wells in my eyes – groundwater management example	Perform recharge for groundwater management	Electromagnetic data	Multiple geological scenario models	6.5, 7.5

or can be a main text for a course on the VOI in the Earth sciences. The exercises, hands-on projects, and data sets should make it conducive for classroom use.

The book requires a background in basic statistics and probability, mathematical calculus, and some understanding of multivariate concepts such as vectors and matrices. Knowledge of multivariate statistics is a bonus, but not essential, as some of the basics are presented here. Similarly, knowledge of basic optimization or decision theory is a bonus, but not essential. Readers familiar with probability and statistics will find Chapter 2 rather straightforward, but we recommend skimming this chapter anyway to get familiar with the notation and the running examples. Similarly, readers with a background in decision analysis would recognize most concepts from Chapter 3. Readers not trained in multivariate statistics, decision analysis, or spatial statistics should be aware that Chapters 2–4 are merely brief overviews. The interested reader should be willing to learn the unfamiliar topics and to appreciate the benefits gained from taking a multidisciplinary approach. If further background knowledge is required for any topic, the reader may wish to consult some of the texts suggested in the bibliographic notes at the end of each chapter.

If you are a statistician, you will likely benefit from learning about decision analysis concepts (Chapter 3) and possibly also spatial statistics concepts (Chapter 4) before going on to integrate the concepts presented in the core of the book (Chapters 5–7). If your expertise is decision analysis, you could use this book to learn about spatial statistics concepts (Chapter 4) and bring this together with your decision analysis knowledge to study VOI for the Earth sciences (Chapters 5–7). If your background pertains to the geosciences, you will likely benefit from learning about some basic statistical tools (Chapter 2) and decision analysis (Chapter 3), but you may choose to go quicker through spatial statistics (Chapter 4) before understanding spatial decision situations and information-gathering schemes (Chapters 5–7). Our intent is that the book should be readable for people from diverse backgrounds. The book builds on multidisciplinary ideas, and we try to appreciate that in our presentation.

1.6 Bibliographic notes

At the end of each chapter in the book, we point to further reading material or discuss some current trends pertaining to the topics that were presented. Here we discuss connections to other books related to VOI in the Earth sciences.

The book by Newendorp and Schuyler (2013) is the third edition of a classic text that has become a standard reference. It covers many important aspects of decision and risk analysis, with special attention paid to investment decisions in petroleum exploration. There is an extensive section with examples on VOI analysis. Compared to our book, Newendorp and Schulyer (2013) focus more on high-level decision analysis, economics, policy, and risk and less on quantitative spatial models of the underlying geologic uncertainties. In addition to petroleum exploration, we also cover a broader range of applications in the Earth sciences.

Bratvold and Begg (2010) cover decision making for petroleum applications, but they do not phrase the VOI in the context of spatial decision situations – i.e., where there is a multivariate statistical model for probabilistically dependent variables and data. We argue that the spatial and multivariate modeling aspect is crucial for Earth sciences problems.

The book by Caers (2011) has a good discussion on VOI applied to the Earth sciences without going into extensive mathematics. Our presentation contains more advanced VOI modeling techniques while trying to present the mathematical representations in a friendly way.

There are, of course, several books on subtopics of our book such as spatial statistics, graphical models, and decision analysis. Some books on information theory and experimental design are also pertinent and will be discussed as appropriate in the following chapters.

2
Statistical models and methods

Figure 2.1 will you strike black gold?
or else be left high and dry?
do you feel lucky?

In this chapter, we provide a fit-for-purpose introduction to probability and statistics, with an emphasis on models and concepts used later in the book. This chapter is not meant to be a comprehensive review of statistical concepts. We focus instead on presenting selected models and methods and setting the notation that will be used throughout this book.

We start with some background discussion on uncertainty quantification and information gathering and motivating the use of data sets. Section 2.2 defines random variables and probability distributions for univariate and multivariate settings. Section 2.3 covers conditional probability and Bayes' rule. Here we emphasize the applicability of graphs and hierarchical models, which will be particularly important in the book. Selected statistical inference methods are discussed in Section 2.4, and Monte Carlo methods are reviewed in Section 2.5. In Section 2.6, we provide bibliographic notes.

2.1 Uncertainty quantification, information gathering, and data examples

Data acquisition, modeling, interpretation, and analysis rely on qualitative and quantitative assessments. In most Earth sciences applications, there is insufficient evidence to use deterministic rules, and probabilistic assessments are more realistic. Decision making under uncertainty requires a probabilistic representation of the **distinction of interest** and the relevant data variables. The distinction of interest is an uncertainty that is fundamentally important to the decision maker. (The rationale behind this specific term will be revealed in the next chapter.) We motivate probabilistic modeling via data examples from the Earth sciences. We will use many of these data sets later in the book for decision analysis, spatial modeling, and value of information (VOI) analysis.

A useful first step in statistical model building and any kind of exploratory data analysis is to create graphical displays and compute various summary statistics. However, it is even more important to clearly define the question that needs to be addressed. Only then, along with experience, can we start to study relevant displays and summary statistics and build appropriate probabilistic models for uncertainty assessment and VOI calculations.

The most important summary statistics of data are the empirical mean, or sample average, and the empirical standard deviation. The average is representative of the "center" of the data, while the empirical standard deviation is a measure of the "variability" in the data. There are many other ways to quantify the central tendency and the variation. One can, for instance, gain insight by sorting the data from smallest to largest. From this sorted list, we can pick the median, defined as the data variable in the middle, and assess percentiles in the data. An empirical percentile has a fraction α of the variables smaller than itself.

In the Earth sciences, attributes are often multivariate – for instance, the subsurface is characterized by porosity, permeability, fluid saturations, rock or soil types, and so on. Moreover, these variables vary across spatial coordinates, and some attributes also vary in time. The statistical modeling assumptions should ideally incorporate multivariate trends. As a result, the univariate summary statistics are only one particular aspect of understanding components of the underlying phenomena and the probabilistic assessment. Bivariate summaries such as the empirical correlation may also be insightful. Similarly, we can analyze the conditional properties of a variable by extracting its values only when another variable is at a specified level. For instance, we may be interested in the variability of porosity in a stratigraphic layer or in the distribution of elastic properties in the oil column.

In each discipline, certain standards for visualizing data have emerged, and experts are familiar with interpreting and analyzing data in this form. Nevertheless, the use of histograms or cross-plotting is common at various stages of model building. We illustrate common visualization methods for some of the data examples used in the book. This discussion motivates methods for combining data with physical knowledge to build probabilistic models for uncertainty assessments.

Consider the data in Figure 2.2 from a vertical well log in the North Sea. The well is drilled from the sea bed to about 3000 m below sea level. The goal of such expensive wells is to explore or produce hydrocarbons. To improve the reservoir-specific understanding, several data are acquired every 15 cm (half foot) of the well path. We plot gamma-ray (which is a measure of the radioactivity; units are in API) and P-wave velocity (measured in m/s) versus depth (second axis). The data are displayed for two stratigraphic layers (Sequence 1 (solid line) and Sequence 2 (dots)), which are clearly separated in depth. There are 591 measurements in Sequence 1 and 722 measurements in Sequence 2. Such plots of physical variables as a function of depth are useful for visualizing well logs. A trained log analyst uses well log data (many others in addition to gamma ray and velocity) to interpret the properties of the subsurface at the location where the well has been drilled. The analysis builds on experience from modeling and interpreting such plots. For

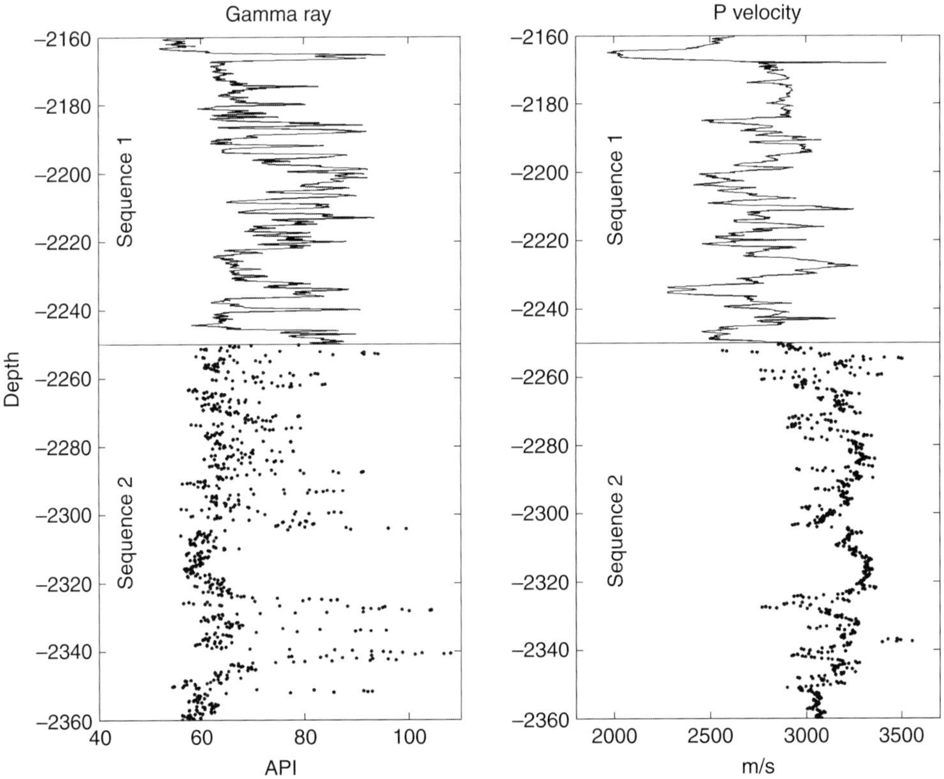

Figure 2.2 Well log data of gamma ray (left) and P-wave velocity (right) from a North Sea reservoir. The display shows two geological sequences separated in depth. The top stratigraphic layer is plotted as a solid line, while the bottom one is plotted as dots. The well log measurements are provided every 15 cm.

instance, velocity would typically increase with depth, but deviations occur as a function of rock types, saturation, porosity, and so on. The gamma-ray measurements are indicative of shaliness in the rock, because sand tends to contain less radioactive material than shale. This particular data set is from Eidsvik et al. (2004b). See also Avseth et al. (2005) for further details.

Figure 2.3 shows histograms that are made by binning the well log data in groups, organized from smallest to largest. Such displays indicate the range of the data. The histogram for the gamma-ray variable has a bimodal (or trimodal) shape in Sequence 1, while it appears skewed in Sequence 2. The empirical mean value of the gamma ray is 73 API in Sequence 1 and 67 API in Sequence 2. The empirical standard deviations are 9.5 and 9.8 in Sequences 1 and 2, respectively. The empirical means and standard deviations (in parentheses) for P-wave velocity are 2740 m/s (220) in Sequence 1 and 3150 m/s (140) in Sequence 2. Findings from such histograms are useful for building realistic models based on rock physics and statistical assessments.

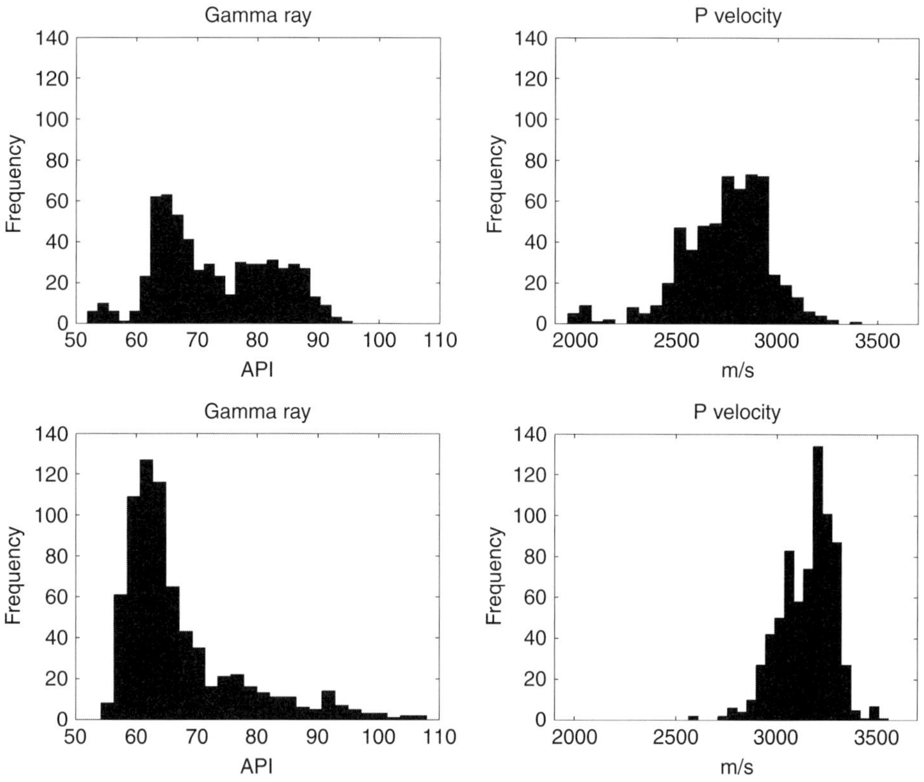

Figure 2.3 Histogram of well data. The top stratigraphic layer is in the upper row, while the deeper stratigraphic layer is at the bottom. Gamma ray (left) and P-wave velocity (right).

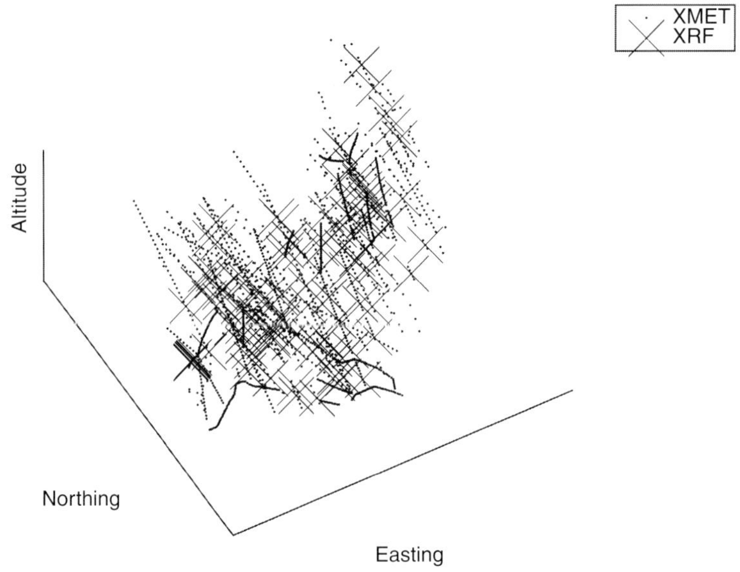

Figure 2.4 Oxide grade data observed along boreholes in a mine. Locations where only the handheld data are collected (dots) and locations where both handheld and laboratory measurements are collected (crosses).

In many spatial applications, we gain knowledge by plotting the data in map views. Figure 2.4 shows the three dimensional locations (north, east, and altitude) of oxide grade measurements in a mineral resource. We see the boreholes lining up in the map view. Most of the data are acquired by a handheld device applied to the rock core samples (dots), while some samples are transported to the laboratory for extracting more accurate data (crosses). Note that this illustration tells us nothing about the actual grade measurements – it only indicates the locations where they are made.

Figure 2.5 shows a smoothed and normalized histogram of the oxide grade data. The top display shows the data acquired with the handheld device (1871 in total). The bottom display shows the data acquired in the laboratory (103 in total). These data appear to have multiple modes caused by the underlying geology of the resource. It is helpful to incorporate such additional explanatory variables, if available, since this would allow for more realistic modeling and reliable prediction of the oxide grades. In this case, geologists have classified the mining resource in three mineralization classes. The class varies spatially across the resource. For the laboratory data, the averages in the three classes are 0.7, 2.4, and 4.2. The data are analyzed further in Eidsvik and Ellefmo (2013).

Seismic data are often visualized as a function of the spatial acquisition locations. Figure 2.6 shows the seismic amplitude response at the top of the reservoir horizon in a North Sea oil field. The top display represents the zero-offset amplitude of P waves reflected back at normal incidence (nominally vertical) from the subsurface horizon. The bottom display is representative of the change in reflection amplitudes as a function of incidence angle or source-receiver offset. It is often called the amplitude-versus-offset

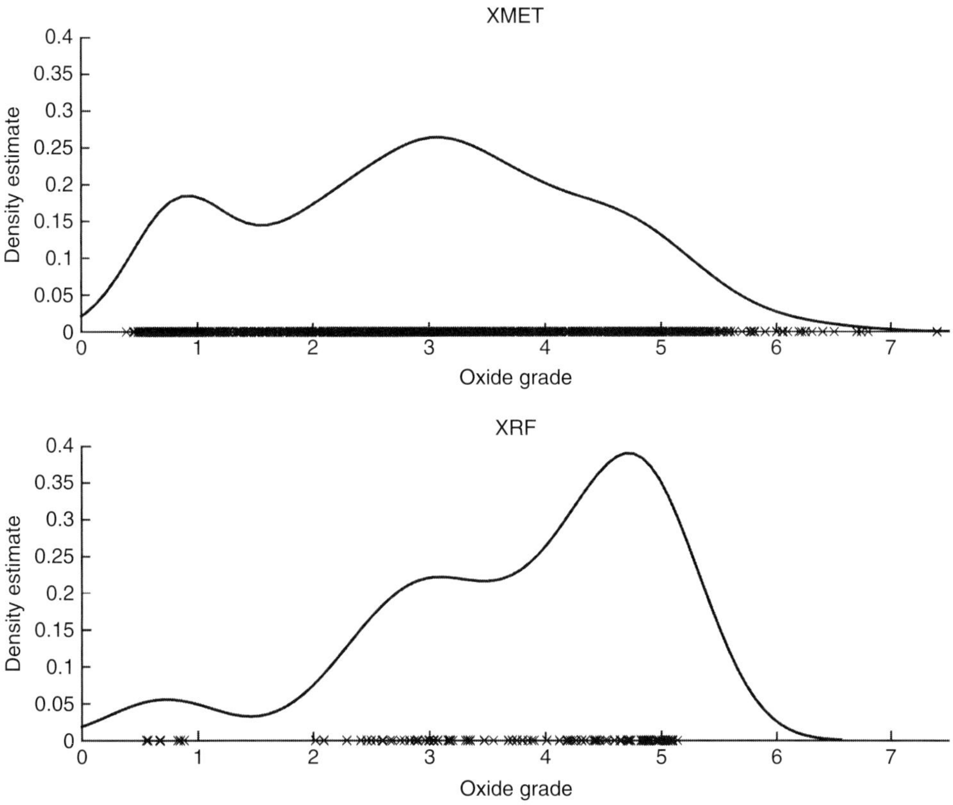

Figure 2.5 Smoothed histogram of the oxide grade observations made in a mine. X-ray fluorescence (XRF) are the laboratory data; XRF meter (XMET) are the data acquired by the handheld meter.

(AVO) gradient in geophysical terminology. The crossline (first axis) and inline (second axis) coordinates are defined by the acquisition geometry. Detailed processing of the seismic reflectivity data gives the amplitudes at the top reservoir. In this display, the zero-offset amplitudes and the AVO gradient are indicated by the grayscale color of the grid cells. The data set is discussed in detail in Avseth et al. (2005).

In Figure 2.7, we cross-plot the two seismic attributes. There is clearly a negative correlation. At locations where the intercept (zero-offset reflectivity) attribute is high, the AVO attribute (gradient) tends to be small, and vice versa. This is also visible in the images shown in Figure 2.6. For instance, near crossline 350 and inline 175, the zero-offset response is rather high (0.05–0.1), while the AVO shows a large negative gradient (about −0.3). In fact, these values of reflectivity could be the lucrative ones in this example, because they indicate sandstones.

The oxide grade measurements and the seismic amplitude data examples consist of observations made by instruments in the field. It has also become routine to generate data

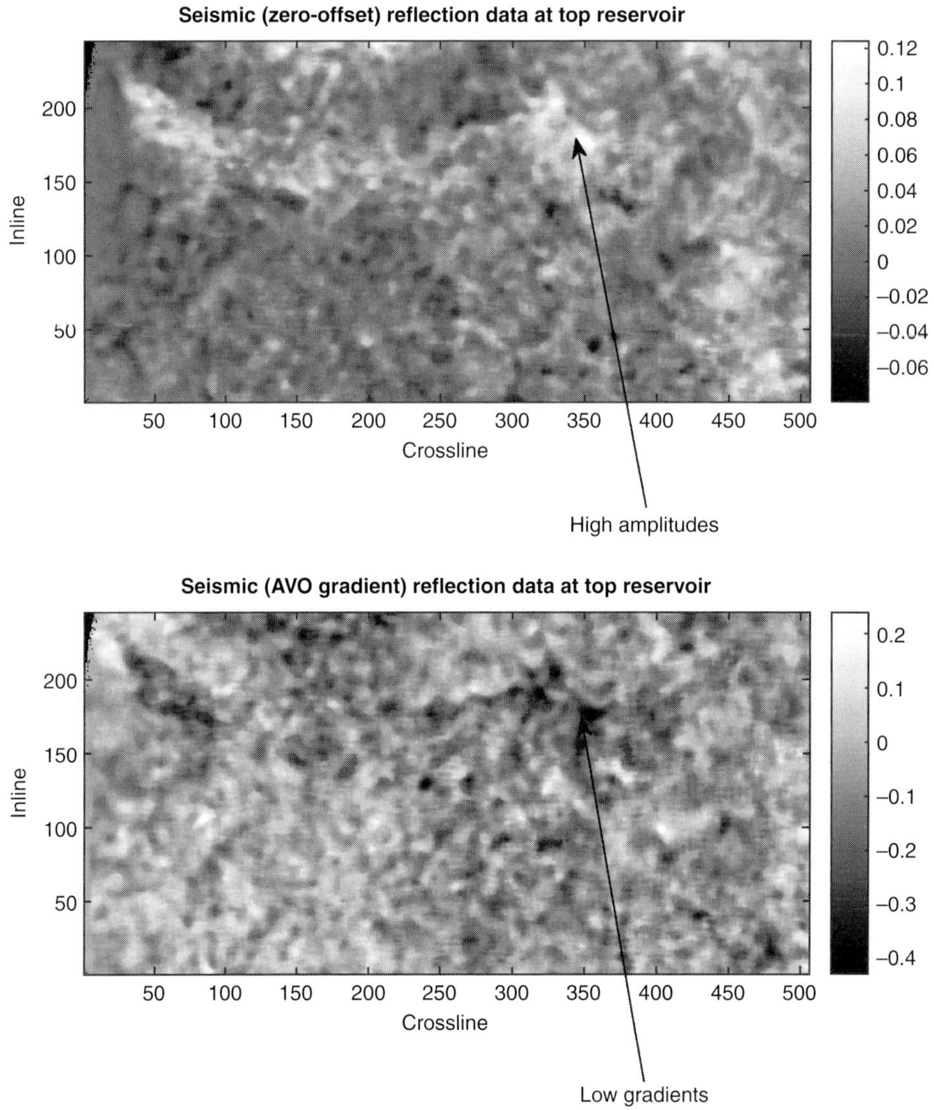

Figure 2.6 Seismic amplitude data at top reservoir level in a North Sea oil reservoir. The top plot shows amplitudes for seismic waves going straight up and down (zero-offset or intercept), while the bottom plot shows the amplitude-versus-offset (AVO) gradient. The offset refers to the incidence angle of the seismic wave. The crossline and inline coordinates are given by the sailing direction of seismic data acquisition. The arrow indicates amplitude attributes that may indicate sands.

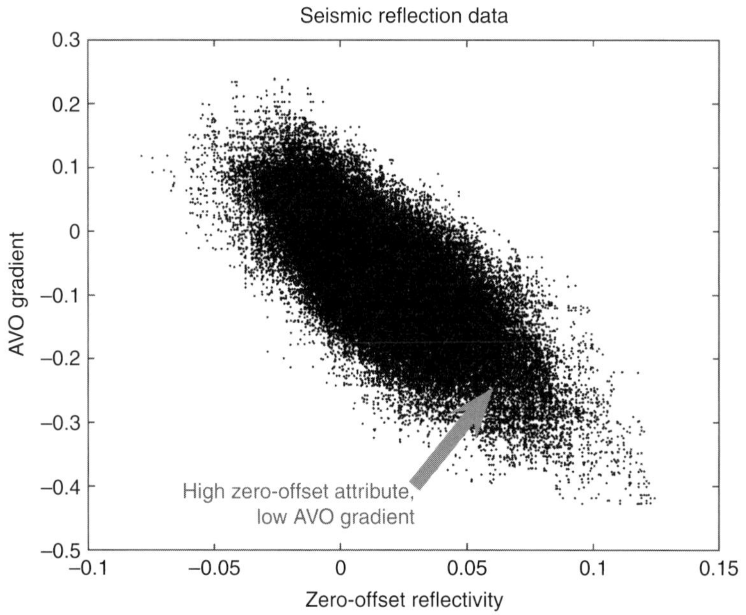

Figure 2.7 Cross-plot of seismic amplitude data at top reservoir level in a North Sea oil reservoir. Zero-offset amplitude is on the first axis. Amplitude versus offset (AVO) is on the second axis. There is a negative correlation between these two seismic amplitude variables.

by computer experiments. These "data" are not acquired on location but rather through a simulator that is set to run for various initial conditions or parameter settings. We present one example of multiple basin modeling scenarios. The basin and petroleum system is simulated using numerical finite-element computational software for the geological processes over time, with different initial conditions and input forcing parameters. These may include heat flow, rock properties such as total organic carbon content, degree of faulting, etc. The simulated data are ensembles of accumulated hydrocarbon volumes at geological prospects.

Figure 2.8 shows a diagram for propagating geological uncertainty in basin and petroleum system modeling. Martinelli et al. (2013b) focused on some selected prospects and varied the most important input parameters at the assumed high/low or high/medium/low levels. For each design of the experiment, they used the output from the simulator to model the uncertainty and interactions in the generation, expulsion, outflow, and accumulation of oil and gas.

In this design of experiment, the porosity is set to low (0) and high (1) levels; the heat flow to expected cold (0), medium (1), and high (2) values; a fault variable to closed (0) and open (1); and the total organic carbon content to the expected high (0) and low (1) levels.

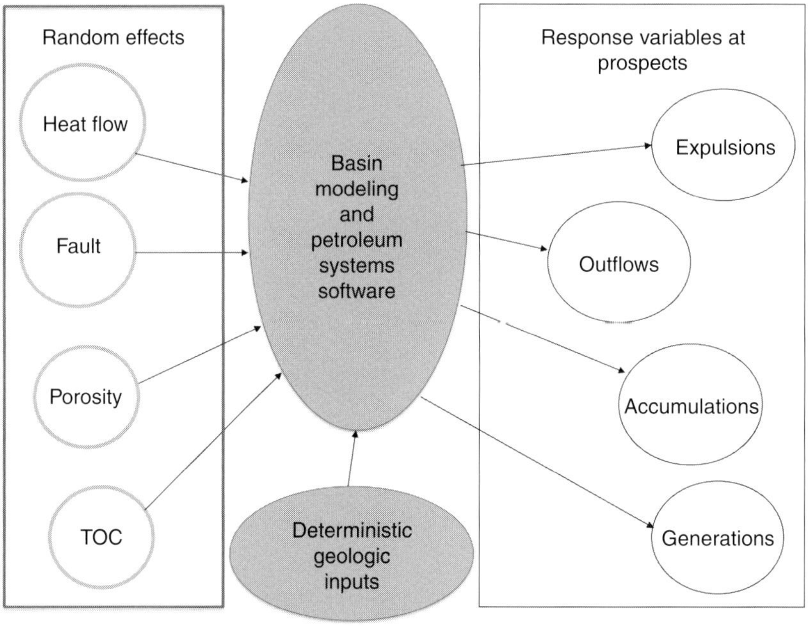

Figure 2.8 Schematic view of a method for generating basin modeling scenarios or "data." By varying input parameters, we study the variability in important response variables or data outputs. The method relies on numerical solutions of the mathematical equations for the migration of hydrocarbons over geological time. Some inputs to this simulator may be fixed, while others are random.

We summarize the data or results of the experiments in the 24 rows of Table 2.1. The table shows the accumulations of oil in the two biggest prospects denoted Top East (TE) and Bottom East (BE) due to the geographical location of the prospects. The controlled experiment could be exchanged by a random sampling strategy (risking) over variables in the input space, in which case one would obtain a realization of the output variables for each run.

In summary, the computation of some summary statistics, along with the visualization of data, is helpful for preliminary exploratory data analysis, which is then typically followed by modeling the important elements of the phenomenon to address specific questions. In the next few sections, we present probability models that are useful for capturing trends, variability, and the dependency structure in Earth sciences data.

2.2 Notation and probability models

A statistical model is defined by a sample space, random variables, and a probability distribution. Here we introduce some basic assumptions and requirements for a valid statistical model. Univariate probability distributions are presented first, but in most applications it becomes important to work in the multivariate context.

Table 2.1. *Summary of inputs and outputs in a design of experiment used in basin and petroleum systems modeling. TOC is total organic carbon. The outputs are oil (in millions barrels) in the Top East (TE) prospect and the Bottom East (BE) prospect*

Run	Porosity	Heat Flow	Fault	TOC	Oil in TE	Oil in BE
1	0	0	0	0	580	340
2	1	0	0	0	247	90
3	0	1	0	0	776	172
4	1	1	0	0	220	35
5	0	2	0	0	736	2
6	1	2	0	0	212	1
7	0	0	1	0	537	343
8	1	0	1	0	247	91
9	0	1	1	0	773	167
10	1	1	1	0	218	35
11	0	2	1	0	731	5
12	1	2	1	0	207	7
13	0	0	0	1	265	213
14	1	0	0	1	218	95
15	0	1	0	1	659	106
16	1	1	0	1	218	38
17	0	2	0	1	528	1
18	1	2	0	1	206	5
19	0	0	1	1	265	213
20	1	0	1	1	218	95
21	0	1	1	1	661	84
22	1	1	1	1	218	37
23	0	2	1	1	527	3
24	1	2	1	1	206	8

2.2.1 Univariate probability distributions

We denote a univariate distinction of interest by x. This random variable x is defined on its sample space – i.e., $x \in \Omega$ – where the sample space Ω denotes all possible outcomes for x. Note that some textbooks in statistics distinguish between the random variable using a capital letter and the outcome or realization with a lowercase letter. We use lowercase for both and try to clarify in the text whenever there may be ambiguity. The sample space Ω is very important for the specification of the statistical model, and it plays a critical role in framing the situation. Once the sample space is understood, we can start contemplating a probability model for a phenomenon. The different sample spaces go hand in hand with the common probability distribution models. In particular, there is a clear distinction between discrete and continuous random variables.

The discrete case entails a random variable taking outcomes on a finite or countable domain. When the random variable must be either 0 or 1, we write $x \in \{0,1\}$. This occurs for binary experiments characterized by success or failure, such as a petroleum exploration well that is either dry or contains hydrocarbons. In other experiments – for instance, a Geiger counter for the number of radioactive events – the possible outcomes are integer valued but may not have a natural upper bound on the sample space. For convenience, one often uses the countable sample space defined on all non-negative integers: $\Omega = \{0,1,2,3,\ldots\}$.

The continuous case entails a random variable taking outcomes on a part of the real line. For instance, rock fall in a tunnel can occur somewhere between its beginning (set to 0) and end (set to 1), so $\Omega = (0,1)$. The annual profit of a company could be negative or positive, with no natural bounds, and it is convenient to set $\Omega = (-\infty, \infty)$.

There is an endless discussion in the sciences about the use of discrete versus continuous variables. The discussion typically relates to the underlying physical mechanisms and the scale of a problem. For instance, the analyst may choose to model profits as discrete, since money cannot be worth anything less than a cent. On the other hand, if one considers a larger and more interesting scale, one may model money as a continuous variable. A continuous view often makes it easier to move between different scales. At the same time, history indicates that scientists have been successful by categorizing and classifying phenomena. This discrete view of the world has been important for communication and for focusing attention. In the Earth sciences, soil or rock types are sometimes described by elementary constituents, but it has also been useful to work at a larger-scale classification of discrete lithofacies, depositional systems, and geological scenarios. No matter whether variables are modeled as discrete or continuous, every task requires careful thought about the sample space, and in every context this should relate to what one can practically evaluate and use for decision making.

We denote the probability density function (**pdf**) of a continuous random variable $x \in \Omega$ by $p(x)$. For a discrete sample space, the probability specification $p(x)$, $x \in \Omega$, is sometimes called a probability mass function. For short, we will use pdf for both the discrete and continuous case, but we will next clarify distinctions between the two.

In the **discrete** case, the pdf or probability mass function $p(x)$ assigns a value to any outcome $x \in \Omega$ under the constraints that

$$p(x) \geq 0, \quad x \in \Omega, \quad \sum_{x \in \Omega} p(x) = 1. \tag{2.1}$$

To compute the probability of an event, defined as a subset of the sample space, we sum over all relevant outcomes. Notably, the cumulative distribution function is defined by the set of all smaller outcomes – i.e., $F(x) = \sum_{y \leq x} p(y)$, which starts at 0 and increases monotonically to 1.

For a **continuous** random variable, the pdf or density function $p(x)$ must satisfy

$$p(x) \geq 0, \quad x \in \Omega, \quad \int_{\Omega} p(x)dx = 1. \tag{2.2}$$

The cumulative distribution function is defined by $F(x) = \int_{y<x} p(y)dy$.

Given a pdf defined as in Equation (2.1) or (2.2), we can derive any feature of the probabilistic model. Two important concepts are the theoretical **mean** and **variance**, defined by

$$\begin{aligned} E(x) &= \mu = \int_{x\in\Omega} xp(x)\,dx, \\ \mathrm{Var}(x) &= \sigma^2 = \int_{x\in\Omega} (x-\mu)^2 p(x)\,dx. \end{aligned} \tag{2.3}$$

They are the first and second (central) moments of $p(x)$. With a discrete sample space, the integrals become sums. The theoretical **standard deviation** is the square root of the variance. As illustrated in Section 2.1, the theoretical mean and standard deviation have empirical counterparts computed from data. The empirical situation can be interpreted as a data-driven discrete pdf assigning equal weights to each observation. If one collects data from a known probability model, the empirical mean and variance will converge to their theoretical counterparts as the number of data increases. The problem, of course, is that we observe some data and have to use that as a basis for finding a useful probability model.

Another important summary of a probability distribution is the **entropy** (or disorder) given by

$$\mathrm{Ent}(x) = -\int_{x\in\Omega} \log\big(p(x)\big)p(x)dx, \tag{2.4}$$

where the integral is replaced by a sum over the sample space for the discrete setting. The entropy is large when there is notable uncertainty.

We briefly present some common **discrete univariate pdfs**. A key element for many discrete distributions is the indicator variable, which is 1 if an event occurs (success) and 0 otherwise (failure). We assign a probability p of success. If we perform a fixed number of independent trials with this success probability p, the resulting random number of successes x follows the **binomial distribution**. The number of trials until the first success has a **geometric distribution**. The **multinomial distribution** is a natural extension of the binomial, where the number of outcomes for every trial is larger than two. When the number of trials in the binomial experiment goes to infinity while the probability of success goes to 0, the resulting random number of successes x is naturally described by the **Poisson distribution**.

The most common **continuous univariate pdf** is the Gaussian or normal distribution. A Gaussian pdf is characterized by a quadratic function of x in the exponent, resulting in a bell-shaped form. The **Gaussian distribution** with theoretical mean μ and variance σ^2 has pdf

$$p(x) = N(\mu,\sigma^2) = \frac{1}{\sqrt{2\pi\sigma^2}}\exp\left(-\frac{(x-\mu)^2}{2\sigma^2}\right), \quad -\infty < x < \infty. \tag{2.5}$$

The quantiles of the Gaussian distribution are defined by symmetric distances from the mean. Many distributions can be derived from the Gaussian pdf. For instance, the **chi-square distribution** is defined by the squared Gaussian, and the **log-normal** is the exponential of a Gaussian variable. Various kinds of truncation or damping of Gaussian variables impose skew distributions.

Another commonly used continuous distribution is the **power law or Pareto** pdf. It has, for instance, been applied to describe the well-known Richter law for magnitudes of earthquakes, prospect size (Kaufman 1993), fault size (Borgos et al. 2002), and many other properties (Newman 2005). A characteristic of the power law pdf is that most outcomes are very small, but some are very large, and it is thus skewed upward with a heavy tail. The pdf is defined by

$$p(x) = \frac{(\beta-1)}{\alpha}\left(\frac{\alpha}{x}\right)^{\beta}, x > \alpha > 0, \beta > 1. \tag{2.6}$$

The cumulative distribution function is $F(x) = 1-(\alpha/x)^{\beta-1}$, and the probability of exceeding a particular threshold c is given by $p(x > c) = (\alpha/c)^{\beta-1}$. The latter probability is sometimes referred to as the exceedance or survival rate. To verify if available data follow a power law, it is common to plot the sorted logarithms of data against the log empirical exceedance probability. Such a display should show a linear decline.

Figure 2.9 shows percentiles and sorted data generated from a power law distribution (top) and on the log-scale (bottom). In the bottom display, the log exceedance probability on the second axis declines with slope 2. This is because we generated this data set using $\beta = 3$ ($\alpha = 0.1$), and we have

$$\log(p(x > c)) = (\beta-1)\log(\alpha) - (\beta-1)\log(c). \tag{2.7}$$

Suppose we predict a new observation from this distribution – say, the volume of an ore body. A very small outcome would be likely. The mean is 0.2, but more than 70% of the data are below 0.2. There is only a 4% chance of obtaining data larger than 0.5.

The discrete and continuous pdfs discussed above are parametric. They are functional representations of the stochastic variable x parameterized by a few model parameters, such as μ, σ in the Gaussian distribution, or α and β in the power law model. A non-parametric model avoids enforcing such parametric assumptions. Instead, it relies on the empirical representation through data. If we have B data or realizations of variables $x^1,\dots,x^B$, a non-parametric model imposes a pdf $p(x) = \frac{1}{B}\sum_{i=1}^{B} I(x = x^i)$ – i.e., a discrete pdf valid only at the sample values. It is common to impose smoothing via kernels around the samples. This gives $p(x) = \frac{1}{B}\sum_{i=1}^{B} \kappa(x; x^i)$, where the non-negative kernel $\kappa(x; x^i)$ is centered at x^i and decays as the variable x gets further away from x^i. We require $\int \kappa(x; x^i)dx = 1$ to maintain a valid pdf. An example is $\kappa(x; x^i) = N(x^i, \tau^2)$, which denotes a Gaussian distribution centered at each realization with constant standard deviation τ. One could further add flexibility to the sample weights, extending beyond uniform weighting, or one could use some sort of clustering of the B realizations. The selection of kernels has been discussed a lot in the literature – see,

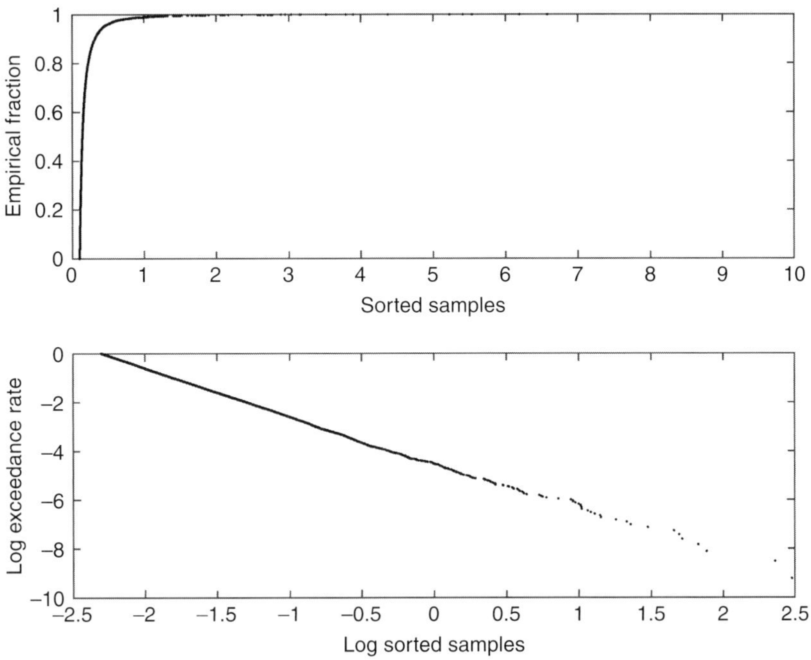

Figure 2.9 Illustration of the power law distribution. Top: sorted samples (first axis) versus the empirical cumulative distribution function (second axis). Bottom: logarithm of sorted samples (first axis) versus the empirical exceedance probability. The data were generated from a power law distribution with parameters $\beta = 3$ and $\alpha = 0.1$.

e.g., Silverman (1988). Simply speaking, kernel representations can be viewed as smoothed histograms. Figure 2.5 was constructed using a kernel for the oxide grade data.

2.2.2 *Multivariate probability distributions*

The random variable $\boldsymbol{x} = (x_1,\ldots,x_n)$ is a vector of size n in sample space $\Omega = \Omega_1 \times \ldots \times \Omega_n$. We let $p(\boldsymbol{x})$ denote the joint pdf of $\boldsymbol{x}$. The **discrete multivariate pdf** $p(\boldsymbol{x})$ naturally extends the univariate version in Equation (2.1). We require

$$p(\boldsymbol{x}) \geq 0, \quad \boldsymbol{x} \in \Omega, \quad \sum_{x_1 \in \Omega_1} \ldots \sum_{x_n \in \Omega_n} p(\boldsymbol{x}) = 1. \tag{2.8}$$

The **joint probability** $p(x_1 = k_1,\ldots,x_n = k_n)$ indicates that the first variable equals k_1, the second k_2, and so on until x_n equals k_n. This joint pdf defines simultaneous properties – i.e., the way the n variables interact in a probabilistic setting. There is a probability for every joint outcome.

We similarly extend Equation (2.2) for a valid **continuous multivariate pdf**.

$$p(\boldsymbol{x}) \geq 0, \quad \boldsymbol{x} \in \Omega, \quad \int_{x_1 \in \Omega_1} \cdots \int_{x_n \in \Omega_n} p(\boldsymbol{x}) dx_1 \ldots dx_n = 1. \tag{2.9}$$

The interpretation of this **joint continuous pdf** is that probabilities of simultaneous events are volumes under the pdf surface for the specified subset of the sample space.

When the variables x_i, $i = 1,\ldots,n$ are **independent**, the joint pdf is defined by the product of n univariate distributions – i.e., $p(\boldsymbol{x}) = \prod_{i=1}^{n} p_i(x_i)$, where $p_i(x_i)$ is the pdf of the i-th variable. If, in addition, the univariate pdfs are equal, $p_i(x_i) = p(x_i)$, we have the common – though often misused – assumption of independent and identically distributed variables. These assumptions simplify modeling and analysis, but more realistic situations usually involve non-identical components and **dependence** between the elements of $\boldsymbol{x}$.

From the joint distribution, we can retrieve the **marginal pdf** of single variables or blocks of variables. The marginal pdf of x_i is derived by integrating (or summing, in the discrete case) out all other variables from the joint distribution:

$$p_i(x_i) = \int_{\Omega_1} \cdots \int_{\Omega_{i-1}} \int_{\Omega_{i+1}} \cdots \int_{\Omega_n} p(\boldsymbol{x}) dx_1 \ldots dx_{i-1} dx_{i+1} \ldots dx_n. \tag{2.10}$$

Let $\boldsymbol{x}_{-i}$ denote all variables in $\boldsymbol{x}$, except x_i, then the marginal is $p_i(x_i) = \int p(\boldsymbol{x}) d\boldsymbol{x}_{-i}$, where the integral is henceforth assumed to be computed over the defined sample space. When there is no confusion, we will simply set $p_i(x_i) = p(x_i)$. The marginal for a subset of variables or block variable $\boldsymbol{x}_{\mathbb{K}}$, where we have split $\boldsymbol{x} = (\boldsymbol{x}_{\mathbb{K}}, \boldsymbol{x}_{\mathbb{L}})$, is $p(\boldsymbol{x}_{\mathbb{K}}) = \int p(\boldsymbol{x}) d\boldsymbol{x}_{\mathbb{L}}$.

The most common multivariate continuous distribution is the Gaussian distribution. This model is discussed at length in textbooks – e.g., Anderson (2003) and Johnson and Wichern (2007). (See Appendix A.1 for more details.) The **multivariate Gaussian pdf** is defined by

$$p(\boldsymbol{x}) = N(\boldsymbol{\mu}, \boldsymbol{\Sigma}) = \frac{1}{(2\pi)^{n/2}} |\boldsymbol{\Sigma}|^{-1/2} \exp\left(-\frac{1}{2}(\boldsymbol{x} - \boldsymbol{\mu})^t \boldsymbol{\Sigma}^{-1} (\boldsymbol{x} - \boldsymbol{\mu}) \right). \tag{2.11}$$

The formula clearly generalizes the univariate situation in Equation (2.5). In the multivariate setting, the contours of the pdf are ellipsoids, centered at the parameter $\boldsymbol{\mu}$, with axes defined by the quadratic form including the $n \times n$ matrix $\boldsymbol{\Sigma}$.

Given a joint pdf, we can compute any feature of the multivariate probabilistic model. The theoretical mean and variance–covariance are extensions of Equation (2.3) defined by

$$\begin{aligned} E(\boldsymbol{x}) &= \boldsymbol{\mu} = \int \boldsymbol{x} p(\boldsymbol{x}) d\boldsymbol{x}, \\ \text{Var}(\boldsymbol{x}) &= \boldsymbol{\Sigma} = \int (\boldsymbol{x} - \boldsymbol{\mu})(\boldsymbol{x} - \boldsymbol{\mu})^t p(\boldsymbol{x}) dx. \end{aligned} \tag{2.12}$$

The integrals become sums for a discrete sample space. In this multivariate situation, the mean is a vector of size n, while the variance–covariance is an $n \times n$ matrix with marginal variances $\sigma_i^2 = \Sigma_{ii}$, $i = 1,\ldots,n$ along the diagonal and covariance terms $Cov(x_i, x_j) = \Sigma_{ij}$ on the off-diagonal. The theoretical correlation is $Corr(x_i, x_j) = \Sigma_{ij}/(\sigma_i \sigma_j)$. The Gaussian pdf in Equation (2.11) is explicitly parameterized by the mean $\boldsymbol{\mu}$ and the covariance matrix $\boldsymbol{\Sigma}$.

The entropy is a scalar variable defined similar to Equation (2.4). For the multivariate model,

$$\text{Ent}(\boldsymbol{x}) = -\int \log(p(\boldsymbol{x}))\, p(\boldsymbol{x})\, d\boldsymbol{x}. \tag{2.13}$$

For the Gaussian distribution, the entropy is $\text{Ent}(\boldsymbol{x}) = \frac{n}{2}(1 + \log(2\pi)) + \frac{1}{2}\log|\boldsymbol{\Sigma}|$.

Figure 2.10 shows the Gaussian pdf in $n = 2$ dimensions under different parameterizations. In all plots, the mean vector is fixed to $\boldsymbol{\mu} = (1,1)$, while the covariance matrix $\boldsymbol{\Sigma}$ varies in the displays. The top left plot has an identity covariance matrix. The two variables x_1 and x_2 are independent in this case; therefore, the joint pdf decouples to the product of marginal pdfs for x_1 and x_2. The other plots show variables with larger variance 1.5^2 and have dependence between x_1 and x_2. The top right (and bottom left) display illustrates positive (and

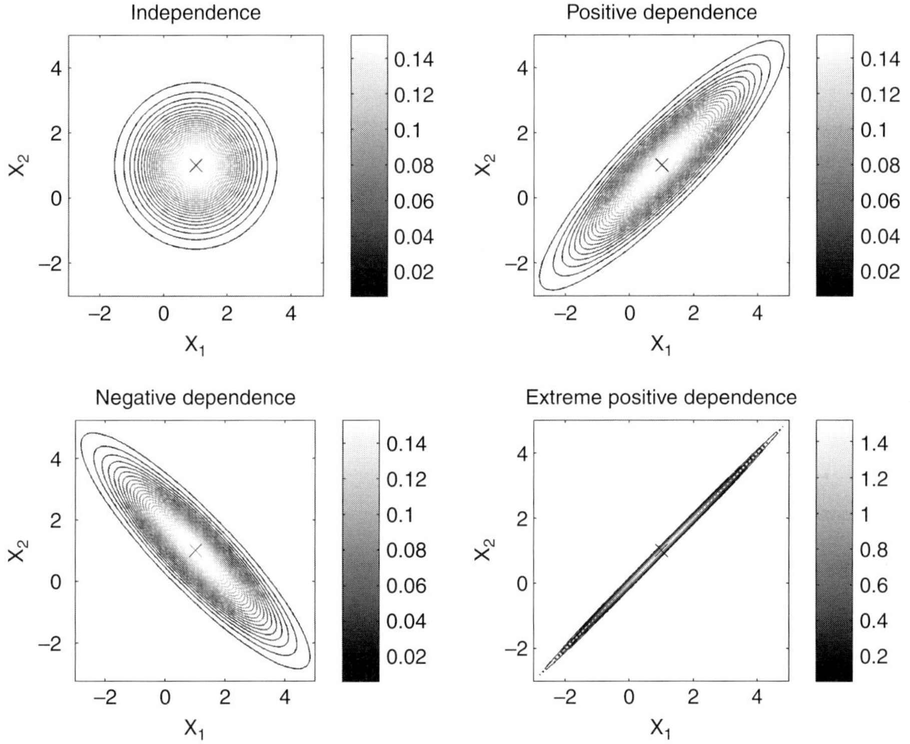

Figure 2.10 Gaussian distribution in two dimensions. All displays have the mean set to 1 for both variables. The variances are equal for the two variables but vary between the displays. The correlation also varies. Top left: independent variables with identity covariance matrix. Top right: positive dependence (correlation 0.74) in the two variables with variance 1.5^2. Bottom left: negative dependence (correlation −0.74) in the two variables with variance 1.5^2. Bottom right: correlation is set to 0.999 and variance is 1.5^2.

negative) correlation $\text{Corr}(x_1, x_2) = 0.74$. This entails $|\boldsymbol{\Sigma}| = 1$, similar to the independent case, and the entropy is $\text{Ent}(\boldsymbol{x}) = 2.84$ in all these cases. The positive off-diagonal entry (top right display) gives a tendency for variables to interact – i.e., they are often simultaneously large or small. With negative correlation –0.74 in the bottom left display, the two variables tend to occur as opposites. The fourth display (bottom right) has an extreme correlation of 0.999, so the density contours appear almost like a line. Knowing one variable then provides very precise information about the other.

In Figure 2.11, we show the bivariate Gaussian distribution fitted to the seismic reflection data discussed in Section 2.1. The white line superimposed on the data is the 95th percentile ellipse based on the empirical mean and covariance matrix from the data (see Section 2.4 for parameter estimation). There is some non-symmetry outside the indicated ellipse, indicating that a Gaussian pdf may not be sufficient to capture the data distribution.

There are numerous applications with both discrete and continuous components. The Gaussian mixture model combines $d \geq 2$ Gaussian components. In the seismic example discussed previously, three different facies classes were identified from well log analysis. The facies classes span different rock types and saturation: oil sand, brine sand, and shale ($d = 3$). Let the bivariate seismic reflection attributes at one location be $\boldsymbol{x} = (x_1, x_2)$. We define pdf

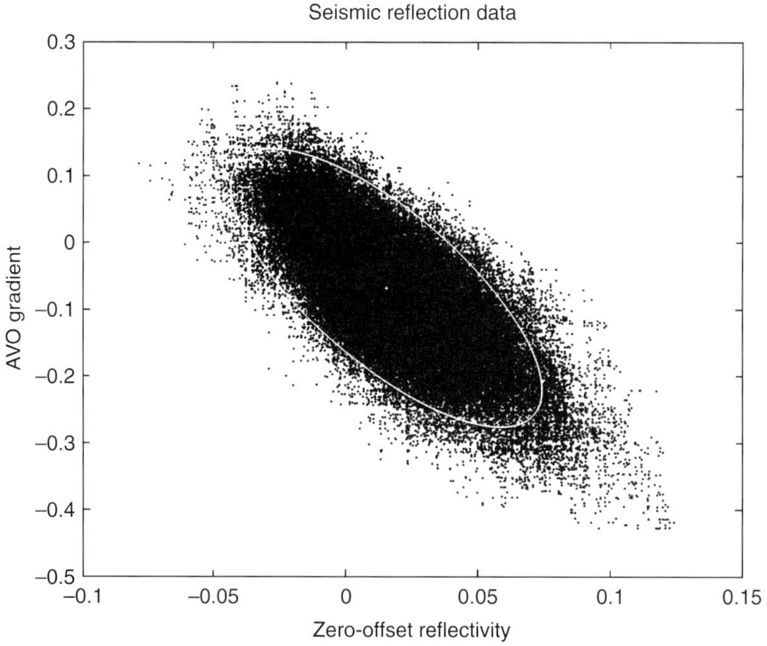

Figure 2.11 Gaussian approximation for the seismic amplitude data example. The zero-offset attribute is on the first axis, and the amplitude-versus-offset (AVO) attribute is on the second axis. The Gaussian approximation is displayed in white, indicating the 95th percentile ellipse.

$$p(\boldsymbol{x}) = \sum_{j=1}^{d} \frac{\alpha_j}{(2\pi)} |\boldsymbol{\Sigma}_j|^{-1/2} \exp\left(-\frac{1}{2} (\boldsymbol{x} - \boldsymbol{\mu}_j)^t \boldsymbol{\Sigma}_j^{-1} (\boldsymbol{x} - \boldsymbol{\mu}_j) \right), \tag{2.14}$$

where $\sum_{j=1}^{d} \alpha_j = 1$ and $0 < \alpha_j$ for all classes $j = 1, \ldots, d$. For each class j, the model allows for a separate mean vector $\boldsymbol{\mu}_j$ and a positive definite covariance matrix $\boldsymbol{\Sigma}_j$.

Figure 2.12 shows the three components of a bivariate Gaussian mixture distribution fitted from well data and plotted with the seismic reflection data. The three mean levels for oil sand, brine sand, and shale are different due to rock physics relations. In this illustration, the covariance matrix for each rock type is the same. The Gaussian mixture model is likely more suitable than the Gaussian pdf in this example. The mixture model is based on physical properties of various facies types and represents their tendency to have different seismic responses.

The mixture model in Equation (2.14) is often used in classification and clustering (Hastie et al. 2009). For instance, if an analyst wants to classify an observation $\boldsymbol{x}$ into the underlying category (rock type in our example), one can compute distances to every class center, scaled with the associated covariance:

$$\text{distance}_j = (\boldsymbol{x} - \boldsymbol{\mu}_j)^t \boldsymbol{\Sigma}_j^{-1} (\boldsymbol{x} - \boldsymbol{\mu}_j), \quad j = 1, \ldots, d. \tag{2.15}$$

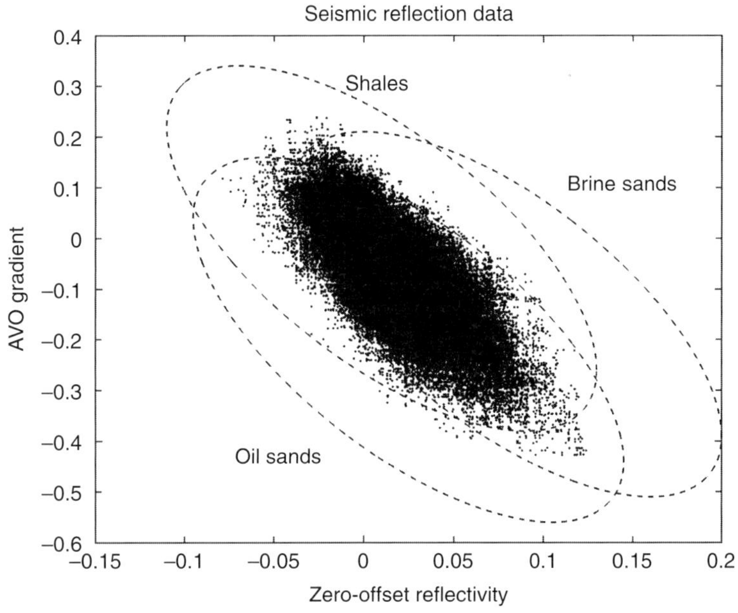

Figure 2.12 Gaussian mixture model for the seismic amplitude data example. The zero-offset attribute is on the first axis, and the amplitude-versus-offset (AVO) attribute is on the second axis. The Gaussian mixture is estimated from well log data and indicated by the 95th percentile ellipse for three facies classes (oil sand, brine sand, and shale).

The analyst classifies the observation by choosing the class index with the smallest distance.

We end this section with a discussion about the transformation of variables. Let $\boldsymbol{y}$ be another random variable defined by $\boldsymbol{y} = \boldsymbol{F}\boldsymbol{x} + \boldsymbol{b}$ for fixed matrix $\boldsymbol{F}$ and vector $\boldsymbol{b}$. Then, we have mean $E(\boldsymbol{y}) = \boldsymbol{F}E(\boldsymbol{x}) + \boldsymbol{b} = \boldsymbol{F}\boldsymbol{\mu} + \boldsymbol{b}$ and variance $\text{Var}(\boldsymbol{y}) = \boldsymbol{F}\text{Var}(\boldsymbol{x})\boldsymbol{F}^t = \boldsymbol{F}\boldsymbol{\Sigma}\boldsymbol{F}^t$. The covariance of $\boldsymbol{y}$ and $\boldsymbol{x}$ is $\text{Cov}(\boldsymbol{y}, \boldsymbol{x}) = \text{Cov}(\boldsymbol{F}\boldsymbol{x} + \boldsymbol{b}, \boldsymbol{x}) = \boldsymbol{F}\boldsymbol{\Sigma}$. A more complicated situation occurs with a non-linear relationship $\boldsymbol{y} = \boldsymbol{f}(\boldsymbol{x})$. If the non-linearity is not too severe, one can approximate the mean and variance of $\boldsymbol{y}$ by linearization, $\boldsymbol{y} \approx \boldsymbol{f}(\boldsymbol{\mu}) + \frac{d\boldsymbol{f}(\boldsymbol{\mu})}{d\boldsymbol{x}}(\boldsymbol{x} - \boldsymbol{\mu})$, where $\boldsymbol{\mu}$ is chosen as the linearization point, in which case $E(\boldsymbol{y}) \approx \boldsymbol{f}(\boldsymbol{\mu})$. In general, the mean of a non-linear function is not the same as the function of the mean. The pdf of a general transformation is only rarely computable. To find the pdf, one must first find an appropriate inverse transformation and then compute the Jacobian of this inverse. More generally, one can apply Monte Carlo simulation to approximate the distribution of a transformation (see Section 2.5 and Appendix B).

Like we discussed for the univariate case, the vector $\boldsymbol{x}$ could, in some cases, be represented by data or realizations $\boldsymbol{x}^1, \boldsymbol{x}^2, \ldots, \boldsymbol{x}^B$. These could be replicated data or the results of some mechanism (say, a transformation or a differential equation) over many input realizations, leading to an ensemble representation of propagated variables. Instead of assuming a parametric model for this distinction of interest, one could use non-parametric methods. The simplest such approach assumes equally likely realizations, with weights $1/B$ for each outcome. The generalization to a continuous representation involves kernels centered at each realization or obtained from clustering.

2.3 Conditional probability, graphical models, and Bayes' rule

The notion of marginal pdfs and joint pdfs was defined in the previous section. We extend these ideas to conditional probability models and Bayes' rule. Conditional probabilities are very important for VOI analysis, because conditioning on observations from information could lead to improved decisions, and VOI lets the decision maker gauge the average improvement.

2.3.1 Conditional probability

One can derive conditional pdfs from the joint distribution $p(\boldsymbol{x})$ of variables $\boldsymbol{x} = (x_1, \ldots, x_n)$. We fix one or more variables in the vector $\boldsymbol{x}$ and study how this information about a subset of variables affects beliefs about the other variables. In the situation with independence, the conditioning has no effect on the others. When there is dependence, the outcome of observed variables changes the pdf of the non-observed variables.

The **conditional pdf** of variables in block $\mathbb{K}$ given the variables in block $\mathbb{L}$, where $\boldsymbol{x} = (\boldsymbol{x}_{\mathbb{K}}, \boldsymbol{x}_{\mathbb{L}})$, is defined by

$$p(\boldsymbol{x}_{\mathbb{K}} \mid \boldsymbol{x}_{\mathbb{L}}) = \frac{p(\boldsymbol{x})}{p(\boldsymbol{x}_{\mathbb{L}})} = \frac{p(\boldsymbol{x})}{\sum_{x_{\mathbb{K}}} p(\boldsymbol{x})}. \tag{2.16}$$

One can interpret this conditional probability as follows: the numerator defines the proportion of joint occurrences of $\boldsymbol{x}_{\mathbb{K}}$ and $\boldsymbol{x}_{\mathbb{L}}$, while the denominator scales it with the proportion of $\boldsymbol{x}_{\mathbb{L}}$ events alone. If block $\mathbb{L}$ is very informative of variables in block $\mathbb{K}$, joint occurrences account for a large part of the times $\boldsymbol{x}_{\mathbb{L}}$ happens, and the conditional pdf should be close to 1, possibly different from the unconditional pdf of $\boldsymbol{x}_{\mathbb{K}}$.

To obtain marginal pdfs within the subset $\mathbb{K}$, still given $\boldsymbol{x}_{\mathbb{L}}$, the other variables must be summed out from the conditional $p(\boldsymbol{x}_{\mathbb{K}} \mid \boldsymbol{x}_{\mathbb{L}})$:

$$p(x_{\mathbb{K},i} \mid \boldsymbol{x}_{\mathbb{L}}) = \sum_{x_{\mathbb{K},1}} \cdots \sum_{x_{\mathbb{K},i-1}} \sum_{x_{\mathbb{K},i+1}} \cdots \sum_{x_{\mathbb{K},n_{\mathbb{K}}}} p(\boldsymbol{x}_{\mathbb{K}} \mid \boldsymbol{x}_{\mathbb{L}}), \tag{2.17}$$

where $n_{\mathbb{K}}$ is the number of variables in subset $\mathbb{K}$, and the sums become integrals in the continuous situation.

Note that the definition of the conditional pdf holds for any subset $\mathbb{K}$ and $\mathbb{L}$. For instance, if we condition on one variable x_i, we have

$$p(\boldsymbol{x}_{-i} \mid x_i) = \frac{p(\boldsymbol{x})}{p(x_i)}. \tag{2.18}$$

The full conditional for x_i– i.e., given all other variables $\boldsymbol{x}_{-i}$ – is

$$p(x_i \mid \boldsymbol{x}_{-i}) = \frac{p(\boldsymbol{x})}{p(\boldsymbol{x}_{-i})}. \tag{2.19}$$

One can also rearrange this conditional formula and write out the joint distribution for $\boldsymbol{x}$ as a product of a marginal and a conditional:

$$p(\boldsymbol{x}) = p(\boldsymbol{x}_{\mathbb{L}})\, p(\boldsymbol{x}_{\mathbb{K}} \mid \boldsymbol{x}_{\mathbb{L}}) = p(\boldsymbol{x}_{\mathbb{K}})\, p(\boldsymbol{x}_{\mathbb{L}} \mid \boldsymbol{x}_{\mathbb{K}}). \tag{2.20}$$

With independence, the conditioning has no effect, and Equation (2.20) simplifies to $p(\boldsymbol{x}) = p(\boldsymbol{x}_{\mathbb{K}})\, p(\boldsymbol{x}_{\mathbb{L}})$, which is rarely applicable in practice.

More generally, one can **factorize** the joint distribution in a stepwise manner as a product of a marginal and conditionals with iterative conditioning:

$$p(\boldsymbol{x}) = p(x_1)\, p(x_2 \mid x_1)\, p(x_3 \mid x_2, x_1) \ldots p(x_n \mid x_{n-1}, \ldots, x_1). \tag{2.21}$$

In this equation, we started with the marginal for variable x_1 and ended with the full conditional for x_n, but we could just as well start with the marginal for x_n and end up with the full conditional for x_1: any order is possible. We can arrange the variables (or blocks of variables) in the most convenient way, depending on the situation. In Earth sciences applications, there is often an important structure that can be used to frame the model and arrange the variables: physical forward models, geological mechanisms, etc.

The definition of properties such as expectation, variance, entropy, and so on can be generalized to the conditional setting by plugging in the conditional pdf in expressions like Equations (2.12) and (2.13). For instance, given the outcome of x_i, the conditional mean and covariance of the remaining variables become

$$\begin{aligned} E(\boldsymbol{x}_{-i} \mid x_i) &= \int \boldsymbol{x}_{-i} p(\boldsymbol{x}_{-i} \mid x_i) d\boldsymbol{x}_{-i}, \\ \text{Var}(\boldsymbol{x}_{-i} \mid x_i) &= \int (\boldsymbol{x}_{-i} - E(\boldsymbol{x}_{-i} \mid x_i))(\boldsymbol{x}_{-i} - E(\boldsymbol{x}_{-i} \mid x_i))^t p(\boldsymbol{x}_{-i} \mid x_i) d\boldsymbol{x}_{-i}. \end{aligned} \tag{2.22}$$

The conditional entropy is defined by

$$\text{Ent}(\boldsymbol{x}_{-i} \mid x_i) = -\int \log(p(\boldsymbol{x}_{-i} \mid x_i)) p(\boldsymbol{x}_{-i} \mid x_i) d\boldsymbol{x}_{-i}. \tag{2.23}$$

Pitfalls: understanding marginals and conditionals

The bivariate Gaussian distribution in Section 2.2 was illustrated using various dependence structures. There is an important difference between conditional and marginal expressions when the model has positive or negative correlation. When the two variables are independent (top left display in Figure 2.10), the conditional mean and variance of x_2 are the same as in the marginal distribution, and $p(x_2 \mid x_1) = p(x_2) = N(1,1^2)$. With dependence, the conditional mean clearly depends on what we observe for x_1: it is either pulled toward the observation (positive correlation) or pushed away (negative correlation). The conditional variance is much smaller than in the marginal (unconditional) distribution, but it does not depend on the observation itself for the Gaussian distribution. Suppose we observe that $x_1 = 3$ and there is positive dependence, $\text{Corr}(x_1, x_2) = 0.74$ (as in the top right display in Figure 2.10). The conditional distribution is $p(x_2 \mid x_1 = 3) = N(2.79, 0.67^2)$, which is shifted and tighter than the marginal $p(x_2) = N(1, 1.5^2)$. The formulas for the conditional mean and variance of the Gaussian distribution will be presented in more detail later.

The conditional entropy is reduced the most in the dependent case, where the uncertainty about x_2 is reduced by knowing the outcome of x_1. By the factorizing formula, $p(\boldsymbol{x}) = p(\boldsymbol{x}_{-i} \mid x_i) p(x_i)$, and by the definition of entropy via the log pdf in Equation (2.13), $\text{Ent}(\boldsymbol{x}) = \text{Ent}(\boldsymbol{x}_{-i} \mid x_i) + \text{Ent}(x_i)$. For the independent case and the one with $\text{Corr}(x_1, x_2) = 0.74$, the joint entropy $\text{Ent}(\boldsymbol{x})$ is the same, and this means that the reduction in entropy equals $\text{Ent}(x_1)$. For the independent case, $\text{Ent}(x_1) = 1.42$, while we have $\text{Ent}(x_1) = 1.82$ for the dependent case.

2.3.2 Graphical models

The modeling of dependent stochastic variables is often not straightforward. We have seen how the Gaussian and Gaussian mixture models define joint distributions, but they may not always provide the flexibility required for real-world phenomena. It turns out that it is convenient to define a joint model via marginal pdfs and conditional pdfs as in the factorization formula in Equation (2.21). In fact, the Gaussian mixture is a simple example of such a model: first, there is a discrete rock type with its marginal probability (defined as α_j

in Equation (2.14)); then, there is a Gaussian distribution for the seismic data, conditional on the rock type (defined by $\boldsymbol{\mu}_j$ and $\boldsymbol{\Sigma}_j$ in Equation (2.14)).

One popular way of building such models is via graphical formulations. This approach uses a modular construction, where pieces of the model are connected locally. The split-and-conquer formulation eases the modeling phase and incorporates sparseness in the model formulation. Full conditioning is rarely necessary; instead, the conditioning is only over a subset of variables. Of course, care is needed to study whether the sparseness induced by the graph is justified for the problem at hand. An in-depth discussion of graphical models is provided in textbooks such as Cowell et al. (2007) and Koller and Friedman (2009).

Let us motivate the use of graphs to visualize conditional modeling formulations. Figure 2.13 shows three possible graphical models for the joint probability distribution of two variables of interest: pore fluid saturation and shale volume (VShale). They are denoted by nodes named "Saturation" and "VShale," respectively. The joint distribution of the two nodes can be represented by any of the three diagrams. The first two diagrams show two different ways of factoring the joint probability distribution. The third diagram has no edge between the variables. This shows that the saturation and the volume of shale are independent. Thus, in such graphs, the absence of arrows indicates conditional independence. In this case, hardly any petrophysicist would claim that the saturation and the VShale are independent, so the third model would generally be inaccurate. Of the other two diagrams, the second diagram may be easier to assess. Given that we know VShale, it is easier to assess the conditional probability distribution of the saturation. For instance, we would expect a higher brine saturation if VShale were large, but not necessarily vice

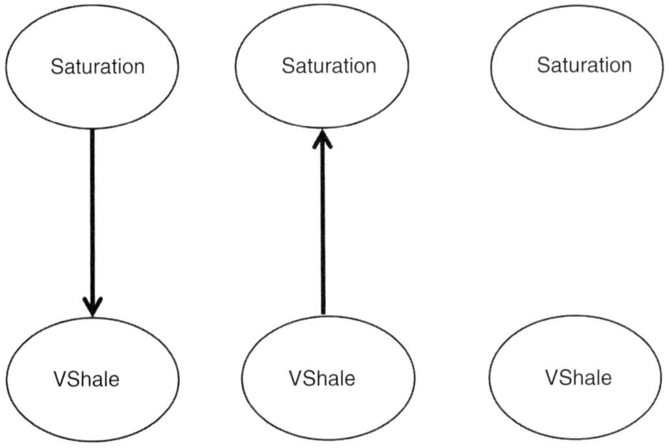

Figure 2.13 Joint probability distribution of saturation and volume of shale represented as a graphical model. The direction of the edge (or arc or arrow) indicates how the geophysical model is established. The rightmost display without any edge indicates that the two variables are independent.

versa. The second diagram may therefore be the most convenient representation for the joint distribution.

Such graphical models are referred to by various names, such as **Bayesian networks** (BNs) or Bayes nets or belief networks or relevance diagrams. They are useful models for building a valid joint probability model and reasoning under uncertainty using conditional statements. A probabilistic graphical model consists of **nodes** (representing random variables), such as VShale and saturation mentioned earlier, and **edges** that connect the nodes together (conditional statements). BNs are directed graphs – i.e., there is a sense of direction on all edges in the graph. An edge that indicates direction from one node to another is an arc or, synonymously, an arrow. If there is an arc emanating from a particular node i to another node j, node i is known as a parent. Equivalently, j is a child of i. The parents for node i are denoted $\text{pa}(i)$.

In a BN, the arcs between nodes are conditional in the sense that every node is conditioned on its parents. In other words, every node is associated with a conditional pdf, which specifies the probability for each possible value of the node conditioned on each possible value of its parents. For the case with discrete sample space, this is typically represented in the form of a matrix known as a **conditional probability table**, but other representations are also possible.

Figure 2.14 (top) illustrates a BN for reservoir variables in a North Sea reservoir. Here, the reservoir variables of interest are rock type or facies, pore fluid filling, porosity, density, and P- and S-wave velocity. There are different types of data or explanatory variables available: well log data and seismic amplitude data. The edges illustrate the conditional dependencies between the variables and the data, motivated from geophysical principles. For instance, the saturation does not influence the porosity, and it affects the S-wave velocity only through the density variable. The rock density depends on the facies (mineralogy), porosity, and pore fluid saturation. Similarly, once we know the elastic parameters (density and P- and S-wave velocity), the seismic data are independent of the other reservoir variables. Note that the graph shows the forward modeling of variables. When we condition on the data, variables like porosity and saturation will no longer be independent.

In this example, the facies and saturation variables could be multivariate distinctions of interest, and they could vary across spatial coordinates. (See Chapter 4.) Since the nodes could represent vector variables, separate nodes could consist of their own sub-networks. The split in nodes (and sub-nodes) is useful to communicate the model definitions to practitioners. This modular feature of graphs makes it easy to expand on smaller parts of the graph, possibly without changing the entire modeling framework.

Figure 2.14 (bottom) shows a BN for reservoir prospects in a region of the North Sea. The root nodes represent the source rock variable at geological kitchens, while the leaf nodes in the network represent potential drilling locations called prospects and segments. These are connected via migration paths indicated by edges. In this example, the migration paths are set by experts. This is the qualitative part of the network. The conditional independence structure is clear: once the prospect nodes (indicated by "P") are known, the bottom node segments do not depend on any other variables. The conditional pdfs of nodes given the

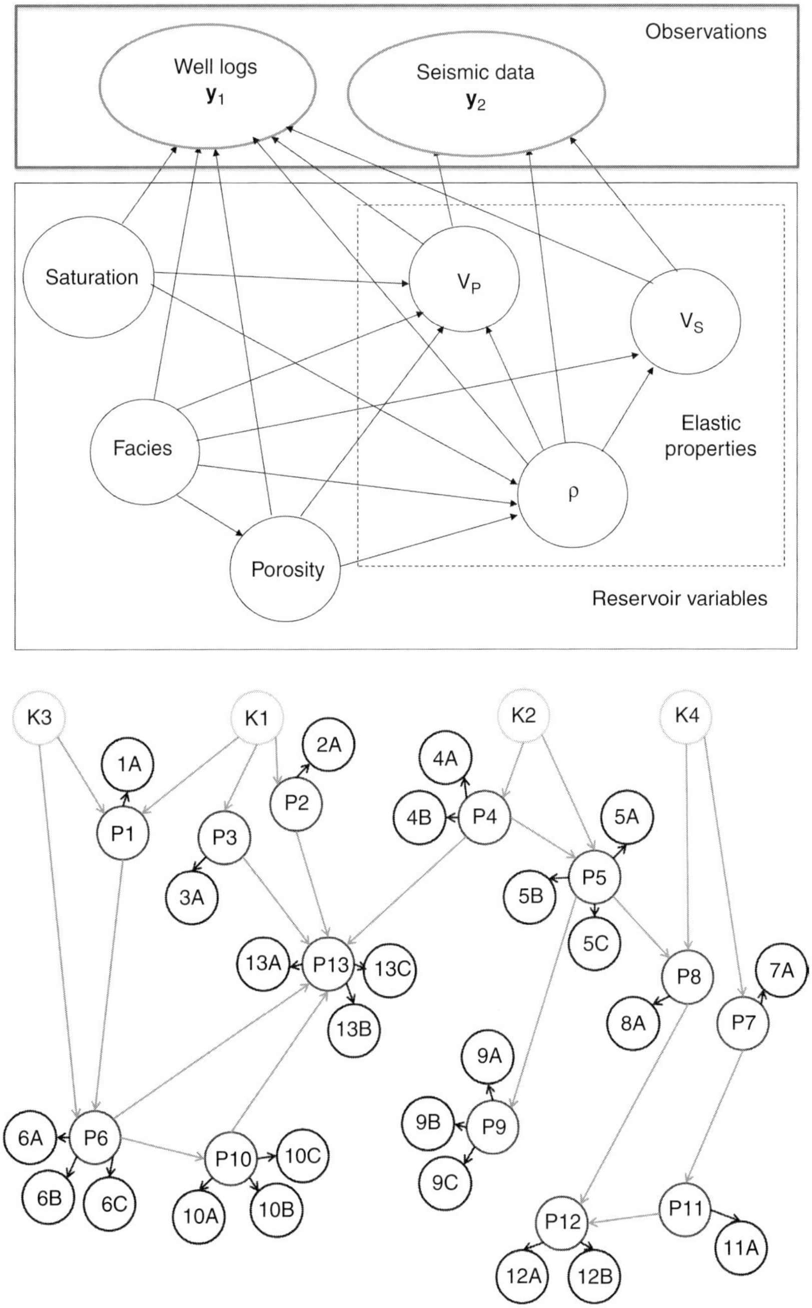

Figure 2.14 Top: Bayesian network for reservoir variables and observations. The data include well observations and seismic data. The elastic properties are P and S velocity and density ρ. Bottom: Bayesian network for 25 prospect segments in the North Sea. The segments are connected by edges indicating hydrocarbon migration paths.

outcome at parent nodes constitute the quantitative part of the model. For this case study, the conditional pdfs were constructed based on geological assumptions of gas and oil expulsion. This particular BN was presented and analyzed in Martinelli et al. (2011).

To illustrate graphical models in further detail, let us consider the two small graphs displayed in Figure 2.15. We will use these two graphs in examples described later. In the top display, Nodes 2 and 3 are prospects where we consider oil exploration. Node 1 represents a geological feature, which we cannot observe directly. The edges going from Nodes 1 to 2 and Nodes 1 to 3 mimic a causal geological mechanism via conditional probability statements. There are no direct edges between Nodes 2 and 3. Thus, the random variables x_2 and x_3 are conditionally independent given the outcome of variable x_1. We consider binary variables $x_i \in \{0,1\}$ and $i = 1,2,3$ and assume that outcome 1 corresponds to success (such as from an exploration well discovering oil), while outcome 0 corresponds to failure (the exploration well is dry).

The graph in Figure 2.15 (bottom) has edges only between the nearest adjacent nodes. This graph illustrates a lag-1 or first-order Markov chain model. In the simplest setting, the variables may be binary, $x_i \in \{0,1\}$, $i = 1,\ldots,n$, with stationary conditional probabilities $p(x_{i+1} = l \mid x_i = k) = P(k,l)$. These one-step probabilities can be organized as a Markov transition matrix as shown next:

$$P = \begin{pmatrix} P(0,0) & P(0,1) \\ P(1,0) & P(1,1) \end{pmatrix}. \tag{2.24}$$

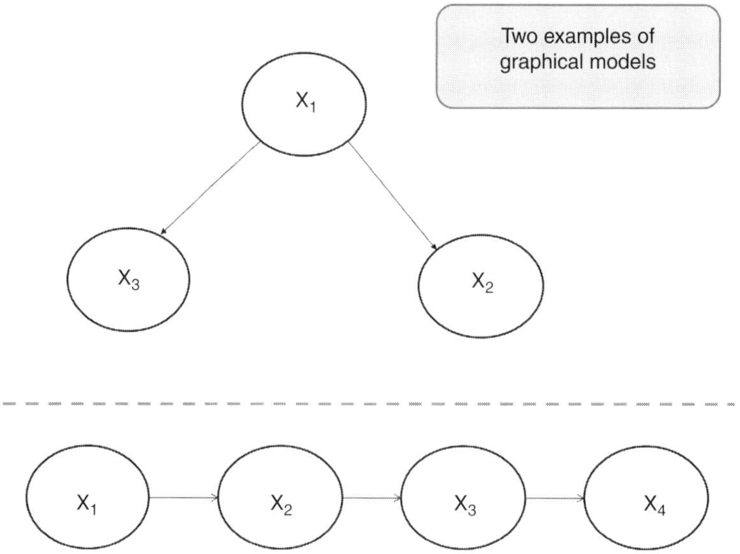

Figure 2.15 Bayesian networks are characterized by a set of nodes and directed edges between the nodes. This figure shows two examples. Top: variables at Nodes 2 and 3 are conditionally independent given the outcomes at Node 1. Below: the situation defines a first-order Markov chain model where each node only connects with the nearest nodes.

Naturally, the row sums must equal 1, so as to cover the sample space with a valid pdf. The marginal pdf for x_1 must also be specified to complete the model formulation. Further background on Markov chains is provided in Appendix A.3.

Conditional independence or sparsity makes graphs useful not only from a modeling perspective but also in the way probabilities are computed. A common practical task is to condition on data variables in the graph, as the analyst may be interested in the effect that the data have on the variables of interest. Due to dependence in the statistical model, the data or evidence will propagate to other nodes. The main computations required for conditional pdfs are marginalization over various subsets of node variables, as in Equation (2.16).

Darwiche (2009) gives an excellent description of the different types of algorithms used for exact and approximate inference using BNs. Exact inference algorithms include the variable elimination algorithm and the junction tree algorithm. These work by successively removing variables or factors to compute the probabilities of interest. Many real-world network models are complex, and exact inference may be very slow. This has led to the development of many approximate inference methods. One widely used class of algorithms is based on iterative belief propagation or loopy belief propagation. As described in Darwiche (2009), iterative belief propagation can be conceptually described in two ways: (i) as an optimization method searching for an approximate probability distribution that minimizes an appropriate statistical distance (the Kullback–Leibler divergence) to the true distribution induced by the BN or, alternatively (ii) as casting the approximate inference problem into a problem of exact inference but on an approximate, edge-deleted network. Another large class of algorithms for approximate inference is based on stochastic sampling, which simulates the network by drawing Monte Carlo samples according to the probability tables for each variable. (See Section 2.5.) The network is simulated multiple times to generate a large set of realizations from which the desired probabilities of events are assessed based on their frequency of occurrence in the simulated realizations. Stochastic sampling algorithms are relatively simple compared to the other algorithms, but a large number of samples need to be generated, especially for events of interest that have low probabilities. (See also Appendices A.5 and B.)

2.3.3 Bayesian updating from data

Let $\boldsymbol{x}$ denote the distinction of interest. One can collect data $\boldsymbol{y}$, which are informative of this distinction of interest. In the Earth sciences, the distinction of interest might be the amount and spatial distribution of oil or gas in a petroleum reservoir, the content of minerals in a mine, the level of groundwater, and so on. Observations that can be acquired to improve knowledge about these distinctions of interest include boreholes with cores or logging measurement tools, seismic data, electromagnetic signals, etc. Some data can be considered perfect – i.e., they are directly informative of the distinction of interest. Other data are measured with noise and are imperfect. Data can be partial, in the situation where only a few of the multivariate variables are observed, or there could be a total test where each variable is observed. Borehole data are an example of partial perfect testing, where certain variables are measured exactly, in which case $\boldsymbol{y} = \boldsymbol{x}_{\mathbb{K}}$, where $\mathbb{K}$ indicates a subset of the variables. We will return to categorizing the various types of information in Chapter 3.

The model for data $\boldsymbol{y}$ can be represented by a conditional pdf, known as the **likelihood function** $p(\boldsymbol{y} \mid \boldsymbol{x})$. The pdf $p(\boldsymbol{x})$ is a **prior model** for the distinction of interest, specified before the data $\boldsymbol{y}$ are available. The **posterior model** for $\boldsymbol{x}$ is obtained by conditioning on the data and is denoted $p(\boldsymbol{x} \mid \boldsymbol{y})$. From the laws of conditional probability in Equation (2.16)

$$p(\boldsymbol{x} \mid \boldsymbol{y}) = \frac{p(\boldsymbol{x}, \boldsymbol{y})}{p(\boldsymbol{y})} = \frac{p(\boldsymbol{x})p(\boldsymbol{y} \mid \boldsymbol{x})}{p(\boldsymbol{y})} \propto p(\boldsymbol{x})p(\boldsymbol{y} \mid \boldsymbol{x}). \tag{2.25}$$

Here, the numerator consists of the prior model $p(\boldsymbol{x})$ and the likelihood model $p(\boldsymbol{y} \mid \boldsymbol{x})$. The denominator is the marginal pdf of the data $p(\boldsymbol{y})$, which is not a function of $\boldsymbol{x}$, and we hence use the proportionality sign in Equation (2.25) to obtain the last expression. In some situations, one may be interested in computing the marginal likelihood in the denominator, which is also sometimes referred to as the pre-posterior model. This can be obtained by marginalizing over $\boldsymbol{x}$ in the joint pdf

$$p(\boldsymbol{y}) = \int p(\boldsymbol{x}, \boldsymbol{y}) d\boldsymbol{x} = \int p(\boldsymbol{x})p(\boldsymbol{y} \mid \boldsymbol{x}) d\boldsymbol{x}, \tag{2.26}$$

or using sums for the discrete situation.

Pitfalls: likelihood and marginal likelihood

Note that the marginal likelihood $p(\boldsymbol{y})$ is usually more difficult to represent than the likelihood $p(\boldsymbol{y} \mid \boldsymbol{x})$. The marginal likelihood is a result of averaging over all possible outcomes for the distinction of interest, while the likelihood tells us the distribution of the response variable given that the distinction of interest is at a specified level. Suppose the distinction of interest is oil saturation in a rock and that the observations are physical laboratory experiments. Geophysicists can use experience to predict the response given the oil content, since there is a physical relation. This can be summarized in the likelihood $p(\boldsymbol{y} \mid \boldsymbol{x})$. The interpretation of $p(\boldsymbol{y})$ is based on marginalizing over all oil saturations.

The equation for posterior pdf $p(\boldsymbol{x} \mid \boldsymbol{y})$ in Equation (2.25) is often referred to as **Bayes' rule**. One uses Bayes' rule to construct the posterior from the prior and the likelihood. Although the model specifies $p(\boldsymbol{x})$ and $p(\boldsymbol{y} \mid \boldsymbol{x})$ – i.e., goes forward – from $\boldsymbol{x}$ to $\boldsymbol{y}$, Bayes' rule allows for going the opposite way from $\boldsymbol{y}$ to $\boldsymbol{x}$. This kind of analysis is tagged as Bayesian inversion. The situation is summarized in Figure 2.16.

It is important to appreciate the broad applicability of merging a priori knowledge with new observations. If there is momentous prior knowledge, the observations will not add

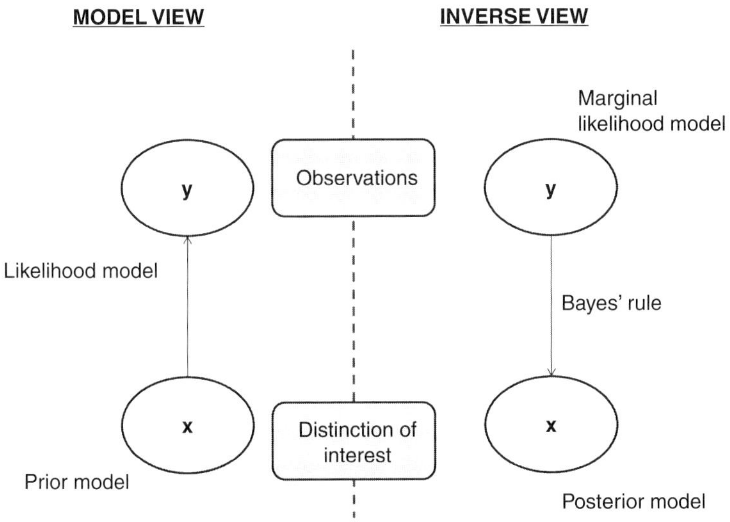

Figure 2.16 Illustration of Bayesian modeling (left) and Bayesian inversion (right). The modeling goes from the distinction of interest $\boldsymbol{x}$ to the data $\boldsymbol{y}$. The inversion uses the data $\boldsymbol{y}$ to get a posterior understanding of $\boldsymbol{x}$.

much information. Also, if the likelihood is almost the same for every outcome of the distinction of interest, there is not much difference between the prior and posterior. The posterior can be viewed as defining a trade-off between prior knowledge and the information content in the data.

The method of choice for the assessment of posterior pdf $p(\boldsymbol{x} \mid \boldsymbol{y})$ depends on the prior pdf, the likelihood model, and how the two combine in the product. Some posterior models can be analyzed analytically or by straightforward sums for marginalization, while others require more complicated numerical or approximate sampling tools. (See Appendices A and B.)

Discussion: Bayesian inversion and fully Bayesian modeling

The inverse probability interpretation of Bayes' rule is commonly used in applications. This has shown to be very useful for assimilating data to update a priori understanding of a distinction of interest. Note, however, that some authors interpret "Bayesian statistics" as a competitor to "frequentist statistics." Fully Bayesian modeling principles entail priors on all statistical modeling parameters. We will briefly discuss Bayesian versus frequentist issues in Section 2.4.

2.3.4 Examples

Treasure Island: The pirate example

Keywords: *pirate example, binary outcomes, Bayes' rule, false positives*

Suppose that a pirate is uncertain about whether a treasure is present or not. This distinction of interest is a binary variable denoted $x \in \{0,1\}$. Suppose further that "success" is a rare event. The probability of success is $p(x=1)=1/100=0.01$. The pirate can carry out some experiment $y \in \{0,1\}$, which is informative of the distinction of interest. For instance, the pirate may purchase a device such as a metal detector and conduct experiments to study whether the treasure is present or absent. We assume that the likelihood model for the test result is $p(y=0 \mid x=0)=p(y=1 \mid x=1)=0.95$. Thus, the data are fairly accurate, failing only 5 out of 100 times. Bayes' rule can be used to compute the posterior probability of a success given that the test result is positive:

$$\begin{aligned} p(x=1 \mid y=1) &= \frac{p(y=1 \mid x=1)p(x=1)}{p(y=1 \mid x=0)p(x=0)+p(y=1 \mid x=1)p(x=1)} \\ &= \frac{0.95 \times 0.01}{0.05 \times 0.99 + 0.95 \times 0.01} \approx 0.16 = 16/100. \end{aligned} \tag{2.27}$$

Perhaps surprisingly, the probability of success is still quite small even though the test result is positive. The reason is the very small a priori probability of a success and the chance of false positives (a successful test when the treasure is not present). The likelihood is not accurate enough to compensate for the small a priori probability of success.

The example highlights the fact that an uncertain test indicating treasure may still not yield a high chance of actually finding the treasure. For a petroleum company searching for oil in areas with a small probability of oil, this means that even with quite accurate measurement techniques, there is a high probability of failure. The result further explains the problem with false positives in medicine, with frustration and disbelief among patients erroneously diagnosed with a rare disease because of a positive test.

In Figure 2.17, we plot the success probability (second axis) given a positive test. Here, the first axis indicates the probability of false positives. We see that the conditional probability of success increases with higher test accuracy, but the test must be very good to get a conditional probability exceeding 0.5.

Gotta get myself connected: Bayesian network example

Keywords: *Bayesian network, exploration wells, graph, conditional independence, Bayes' rule, evidence propagation*

We use the graph in Figure 2.15 (top) to illustrate evidence propagation. As we described earlier, suppose that the node variables in Figure 2.15 (top) represent the presence or absence of hydrocarbons. We consider the situation of drilling an exploration well at Prospect Node 2. This information would result in posterior probabilities at Prospect Node 3.

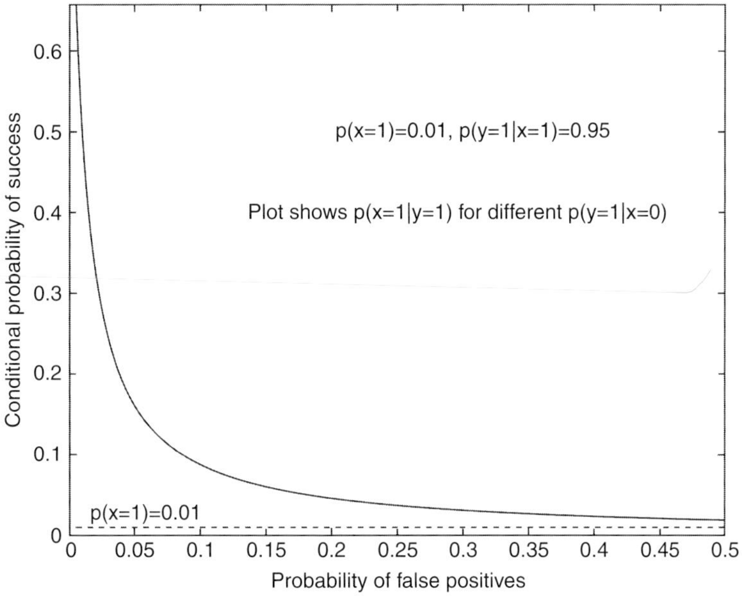

Figure 2.17 The conditional probability of a success given a positive test. The success probability is plotted as a function of the false discovery probability on the first axis.

Let the marginal probability be $p(x_1 = 1) = 0.2$, while the conditional probabilities are $p(x_2 = 1 \mid x_1 = 1) = 0.5$, $p(x_3 = 1 \mid x_1 = 1) = 0.5$, $p(x_2 = 1 \mid x_1 = 0) = 0$, and $p(x_3 = 1 \mid x_1 = 0) = 0$. The model is now completely specified, because the top node is assigned a marginal pdf, and all edges have a conditional pdf. The marginal distribution of the two prospect nodes is available by summing out the outcome at Node 1. We have:

$$p(x_3 = 1) = \sum_{y=0}^{1} p(x_3 = 1 \mid x_1 = y) p(x_1 = y) = 0.5 \cdot 0.2 + 0 \cdot 0.8 = 0.1. \qquad (2.28)$$

By symmetry, the marginal pdf at Node 2 is the same.

Recall the restrictions imposed when the outcome at Node 1 is dry. The geological mechanism prevents oil from entering Nodes 2 and 3 if Node 1 is dry. The effect of conditioning is notable in this situation: when the decision maker drills at Node 2, he or she hopes to find oil and start exploiting this resource. But the evidence obtained by drilling Node 2 is also valuable for learning the chances of finding oil at Node 3, since the two are dependent through the common parent Node 1.

Assume that the decision maker drills at Node 2 and discovers oil ($x_2 = 1$). Therefore, $x_1 = 1$; otherwise, Node 2 could not have contained oil. Then, the conditional probability at Node 3 becomes

$$p(x_3 = 1 \mid x_2 = 1) = p(x_3 = 1 \mid x_1 = 1) = 0.5. \qquad (2.29)$$

This is a 5-fold increase in probability compared with the marginal pdf. The situation with a dry outcome in Node 2 is a little harder to compute, because this can happen both when Node 1 is dry or has oil. We have to marginalize over both these outcomes for the parent node:

$$p(x_1 = 1 \mid x_2 = 0) = \frac{p(x_2 = 0 \mid x_1 = 1)p(x_1 = 1)}{\sum_{y=0}^{1} p(x_2 = 0 \mid x_1 = y)p(x_1 = y)}, \\ = \frac{0.5 \cdot 0.2}{0.5 \cdot 0.2 + 1 \cdot 0.8} = 0.11, \tag{2.30}$$

$$p(x_3 = 1 \mid x_2 = 0) = \sum_{y=0}^{1} p(x_3 = 1 \mid x_1 = y)p(x_1 = y \mid x_2 = 0), \\ = 0 \cdot 0.89 + 0.5 \cdot 0.11 = 0.055, \tag{2.31}$$

which is a clear reduction of the marginal probability for Node 3.

For larger-size BNs, the marginalization operations are not that straightforward to perform, but there are several well-known algorithms for that purpose, such as the junction tree algorithm discussed in Section 2.3.2 (and Appendix A.5).

Never break the chain: Markov chain example

Keywords: *Markov chain, dependent binary variables, conditional independence, Bayes' rule*

For the Markov chain model in Figure 2.15 (bottom), assume that we can obtain information for one of the variables – say, $x_i = k$. We are interested in propagating this evidence to the other variables. Of course, at Node i, we are now certain of the outcome since this is perfect information. We can update the probabilities at the neighboring nodes given the evidence $x_i = k$ based on the Markov transition probabilities in Equation (2.24). Unless the transition probabilities $P(k,l) = 0.5$ for all k, l, corresponding to independence between the variables, there will be a change from prior marginal probability $p(x_j)$ to posterior marginal probability $p(x_j \mid x_i = k)$. The computation of $p(x_j \mid x_i = k)$ is an exercise in the repeated use of Bayes' formula. (See Appendix A.3.)

Figure 2.18 illustrates evidence propagation for a chain graph with $n = 50$ nodes after observing $x_{20} = 1$. The difference between the four displays is the transition probability matrix. Figure 2.18 (top left) is the independent case with $P(0,0) = P(1,1) = 0.5$. Figure 2.18 (top right) is based on $P(0,0) = P(1,1) = 0.9$, imposing some continuity in the classes. Figure 2.18 (bottom left) has a larger probability of staying in the successful state – i.e., $P(0,0) = 0.5$ versus $P(1,1) = 0.9$. Figure 2.18 (bottom right) shows a non-stationary chain, not dissimilar from the failure propagating network discussed previously, with $P(0,0) = 1$ and $P(1,1) = 0.9$. Here, 0 is a state that is referred to as an absorbing state, because there is no chance of moving to other states.

The displays show varying widths of evidence propagation away from the evidence node. For the independent case, the observation at Node 20 carries no information about the states at other nodes. In Figure 2.18 (bottom left), there is a short window of evidence

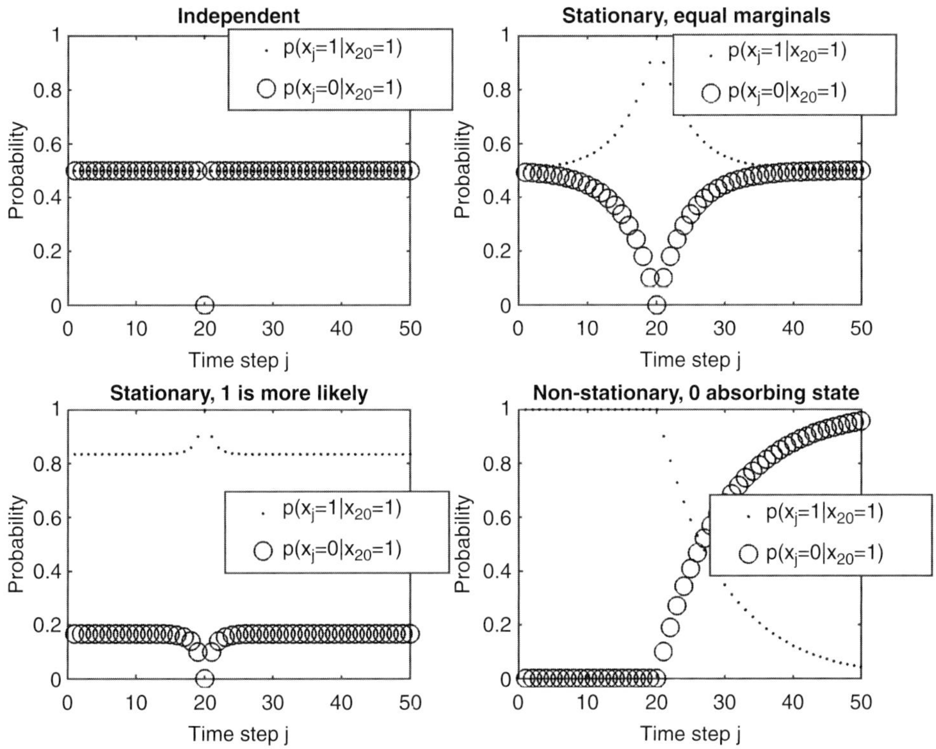

Figure 2.18 Evidence propagation in the Markov chain model. The four displays show the conditional probabilities when we obtain perfect information about a success at Node 20. The displays are for various Markov transition matrices: Independence (top left), equal diagonal entries of 0.9 (top right), larger diagonal for Class 1 (bottom left), and an absorbing failure state (bottom right).

propagation (about five indices), since the prior stationary probabilities are rather close to 0 and 1 ($p(x_i = 1) = 0.83$). For the absorbing situation, the success at Node 20 tells us that all previous nodes must be successes too. The success probability for the subsequent states dies out exponentially with factor $P(1,1) = 0.9$.

For whom the bell tolls: Gaussian projects example

Keywords: *Gaussian projects example, bivariate Gaussian distribution, correlation, linear Gaussian model, perfect information, partial information, imperfect information, total information*

The linear Gaussian model can be written as

$$\boldsymbol{y} = \boldsymbol{F}\boldsymbol{x} + \boldsymbol{\varepsilon}, \quad \boldsymbol{\varepsilon} \sim N(\boldsymbol{0}, \boldsymbol{T}), \tag{2.32}$$

where $\boldsymbol{x} = (x_1, \ldots, x_n)$ is the unknown variable of interest, $\boldsymbol{y} = (y_1, \ldots, y_m)$ is the response variable or data, and the $m \times n$ matrix $\boldsymbol{F}$ has fixed entries. The structure in this matrix is defined by the data-gathering mechanism. It may consist of 0s and 1s as in the factorial design of experiments or known explanatory variables in usual linear regression analysis, but note that variable $\boldsymbol{x}$ is random in Equation (2.32).

Suppose that the distinctions of interest are projects which are a priori Gaussian with a known mean vector and covariance matrix – i.e., $p(\boldsymbol{x}) = N(\boldsymbol{\mu},\boldsymbol{\Sigma})$. In the Earth sciences setting, this distinction of interest can represent the profits from reservoirs or mining units. Using Equation (2.32), the likelihood is $p(\boldsymbol{y} \mid \boldsymbol{x}) \sim N(\boldsymbol{Fx},\boldsymbol{T})$. The joint model for the distinction of interest and the data is Gaussian (see Appendix A.1):

$$p(\boldsymbol{x},\boldsymbol{y}) = N\left(\begin{bmatrix} \boldsymbol{\mu} \\ \boldsymbol{F\mu} \end{bmatrix}, \begin{bmatrix} \boldsymbol{\Sigma} & \boldsymbol{\Sigma F}^t \\ \boldsymbol{F\Sigma} & \boldsymbol{F\Sigma F}^t + \boldsymbol{T} \end{bmatrix}\right). \tag{2.33}$$

We get a Gaussian posterior solution defined by the conditional mean and covariance

$$\begin{aligned} \boldsymbol{\Sigma}_{x|y} &= \boldsymbol{\Sigma} - \boldsymbol{\Sigma F}^t \left(\boldsymbol{F\Sigma F}^t + \boldsymbol{T}\right)^{-1} \boldsymbol{F\Sigma}, \\ \boldsymbol{\mu}_{x|y} &= \boldsymbol{\mu} + \boldsymbol{\Sigma F}^t \left(\boldsymbol{F\Sigma F}^t + \boldsymbol{T}\right)^{-1} \left(\boldsymbol{y} - \boldsymbol{F\mu}\right). \end{aligned} \tag{2.34}$$

The matrix–vector expressions in Equation (2.34) are derived in Appendix A.1.

Let us consider two projects ($n = 2$) and set the prior mean profits of projects and the prior covariance matrix as follows:

$$\boldsymbol{\mu} = \begin{pmatrix} 0 \\ 0 \end{pmatrix}, \qquad \boldsymbol{\Sigma} = \begin{bmatrix} 1 & \rho \\ \rho & 1 \end{bmatrix}. \tag{2.35}$$

Suppose that the data are of size $m = 2$ or $m = 1$: with univariate data, we assume that there is perfect information about one project, while the bivariate data provide imperfect information acquired about both projects. The design matrix $\boldsymbol{F}$ determines which project(s) are observed. For perfect information about the first project, we set $\boldsymbol{F} = \begin{bmatrix} 1 & 0 \end{bmatrix}$ and no noise in the univariate observation – i.e., $T = 0$. For imperfect information about both projects, we set $\boldsymbol{F} = \boldsymbol{I}_2$ and $\boldsymbol{T} = \tau^2 \boldsymbol{I}_2$, assuming conditional independence between the two observations and equal uncertainty or noise standard deviation τ. Graphically, this can be displayed by edges from each project to its information source. The two distinctions of interest can be connected in either direction because their dependence can be defined either way.

A perfect observation on Project 1 leaves no posterior uncertainty on this project. The conditional variance of Project 2 is obtained by Equation (2.34), resulting in $\sigma^2_{x_2|y} = 1 - \rho^2$. With imperfect information on both projects, the posterior covariance matrix is

$$\boldsymbol{\Sigma}_{x|y} = \boldsymbol{\Sigma} - \boldsymbol{\Sigma}\left(\boldsymbol{\Sigma} + \tau^2 \boldsymbol{I}\right)^{-1} \boldsymbol{\Sigma} = \begin{bmatrix} 1 - \dfrac{(1+\tau^2)(1+\rho^2) - 2\rho^2}{\left((1+\tau^2)^2 - \rho^2\right)} & \rho - \dfrac{2\rho(1+\tau^2) - \rho(1+\rho^2)}{\left((1+\tau^2)^2 - \rho^2\right)} \\ \rho - \dfrac{2\rho(1+\tau^2) - \rho(1+\rho^2)}{\left((1+\tau^2)^2 - \rho^2\right)} & 1 - \dfrac{(1+\tau^2)(1+\rho^2) - 2\rho^2}{\left((1+\tau^2)^2 - \rho^2\right)} \end{bmatrix}, \tag{2.36}$$

and the integrated variance is the trace of this matrix.

The integrated variance reduction can now be studied for both information-gathering schemes. Figure 2.19 illustrates this posterior uncertainty as a function of the correlation

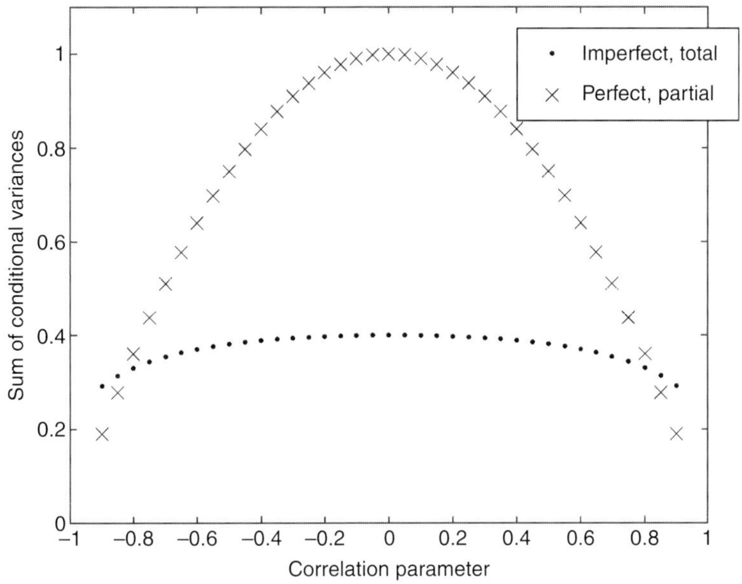

Figure 2.19 Conditional variances given imperfect and perfect information. The second axis is the sum of the posterior variances at the two projects. The first axis is the correlation between the two projects.

ρ between projects. For the case with imperfect total information, we set noise standard deviation $\tau = 0.5$. When the absolute correlation is very large, the scheme obtaining perfect information at one project gives the smallest posterior uncertainty. The sum of conditional variances is equal for a correlation parameter equal to about 0.8.

2.4 Inference of model parameters

Any statistical model has a number of parameters that must be specified in the context of an application. These parameters may include the probabilities for discrete outcomes, mean values, variances, or correlations in the prior distribution for the distinction of interest or the noise level of data in the likelihood model. They are often referred to as nuisance parameters because they are typically not of primary interest. Our main interest usually lies in making predictions about the distinction of interest and then making decisions based on those predictions. However, these higher-level activities usually require the analyst to specify the nuisance parameters.

With parametric modeling assumptions, the number of unknown parameters is usually small – say, around 1–10. We estimate the small number of nuisance parameters from the currently available data, either some perfect information $\boldsymbol{x}$ or imperfect information $\boldsymbol{y}$. After a general discussion, the focus in this section will be on maximum likelihood estimation (MLE), which is widely applicable for parametric models and exhibits desirable statistical properties.

In some applications, the underlying phenomenon is well understood, and this can be used to assign parameter values directly, without the need for complicated estimation techniques. Model parameters may also be interpreted as a summary of the data, typically the mean values, the uncertainty, the heterogeneity between groups, and so on. For instance, for the power law example in Section 2.2, we created a log-log plot of data. Such empirical summary statistics lead to moment estimators, and we can use moments to specify parameters in a model or as initial guesses for MLE.

Trial and error is also commonly used to specify statistical model parameters. One can check parameter sensitivity by predicting over a hold-out set of data for different values of input parameters. This approach can be formalized by cross-validation. Suppose that we have B data or realizations $\boldsymbol{y}^b$, $b = 1,\ldots,B$. The leave-one-out distribution for the i-th component of realization $\boldsymbol{y}^b$, given all other instances $\boldsymbol{y}^{-b} = \{\boldsymbol{y}^c; c = 1,\ldots,B, c \neq b\}$, is denoted $p(y_i^b \mid \boldsymbol{y}^{-b}, \boldsymbol{y}_{-i}^b; \boldsymbol{\theta})$. Cross-validation uses some measure of misfit between this predictive distribution and the actual data y_i^b. A common measure is the prediction mean square error. This kind of cross-validation is used substantially in model validation, especially for non-parametric models such as tuning the kernel widths of Gaussian mixtures.

Many Earth sciences applications involve physical or geological mechanisms that can be framed within parametric assumptions. Optimization methods are used to find the model parameters that best fit the data in some sense. Note, however, that this approach requires a model, and we thus assume a known distribution or mechanism. Care should be used when fitting a parametric model, and it is good practice to check if the distribution assumptions hold by studying if the empirical distribution of data looks similar to the statistics expected from the assumed distribution. A common approach is to study percentiles of the assumed distribution with fitted parameters versus the empirical percentiles of the data. This is often visualized and referred to as a quantile–quantile or QQ plot.

2.4.1 Maximum likelihood estimation

This discussion of likelihood will be fairly general. We provide more background about parameter estimation for the specific models used in this book in Appendix A.

The objective of MLE is to find the parameter that most likely generated the data under the current modeling assumptions. MLE aims to optimize $p(\boldsymbol{x};\boldsymbol{\theta}) = L(\boldsymbol{\theta})$ as a function of the unknown nuisance parameter $\boldsymbol{\theta}$ for the case of perfect information about the distinction of interest $\boldsymbol{x}$. If we have imperfect information, the marginal likelihood is $p(\boldsymbol{y}) = p(\boldsymbol{y};\boldsymbol{\theta}) = \int p(\boldsymbol{y},\boldsymbol{x};\boldsymbol{\theta})d\boldsymbol{x} = L(\boldsymbol{\theta})$, where the joint distribution in the integrand is constructed by a prior model for the distinction of interest and a likelihood model for the data – i.e., $p(\boldsymbol{y},\boldsymbol{x};\boldsymbol{\theta}) = p(\boldsymbol{x};\boldsymbol{\theta})p(\boldsymbol{y} \mid \boldsymbol{x};\boldsymbol{\theta})$. The case with imperfect information is usually more difficult than the case with perfect information. The preferred situation is to have both perfect and imperfect information. In the Earth sciences, there are often auxiliary data available (either $\boldsymbol{x}$ or $\boldsymbol{y}$ or both) for specifying important model parameters.

> **Discussion: fully Bayesian versus frequentist analysis**
>
> An alternative to the maximum likelihood principle is the fully Bayesian framework, which assigns a prior $p(\boldsymbol{\theta})$ on the nuisance parameters, and where parameter specification is based on the posterior of the model parameters denoted $p(\boldsymbol{\theta}|\,\mathbf{y})$.

The MLE of the parameters is obtained by maximizing $L(\boldsymbol{\theta})$ or, equivalently, the **log-likelihood function** $l(\boldsymbol{\theta}) = \log L(\boldsymbol{\theta})$, with respect to the parameter $\boldsymbol{\theta}$. We have

$$\hat{\boldsymbol{\theta}} = \text{argmax}_{\theta}\left\{l(\boldsymbol{\theta})\right\}, \qquad \frac{dl(\hat{\theta})}{d\boldsymbol{\theta}} = \mathbf{0}, \tag{2.37}$$

where the derivative $dl/d\boldsymbol{\theta}$ is also known as the score.

Several optimization algorithms are used in practice to compute the MLE. In some situations, there are analytical solutions to Equation (2.37), but in others, complicated numerical routines are required. One such numerical optimization algorithm is the Newton–Raphson method defined by a starting value $\boldsymbol{\theta}^0$ and updates for $b = 1, 2...$ according to

$$\boldsymbol{\theta}^b = \boldsymbol{\theta}^{b-1} - \frac{d^2 l\left(\boldsymbol{\theta}^{b-1}\right)}{d\boldsymbol{\theta}^2}^{-1} \cdot \frac{dl\left(\boldsymbol{\theta}^{b-1}\right)}{d\boldsymbol{\theta}}. \tag{2.38}$$

The final output $\hat{\boldsymbol{\theta}} = \boldsymbol{\theta}^b$, obtained after some iterations of Equation (2.38), is the MLE of $\boldsymbol{\theta}$. Figure 2.20 illustrates the Newton–Raphson method in practice. In most cases, when the

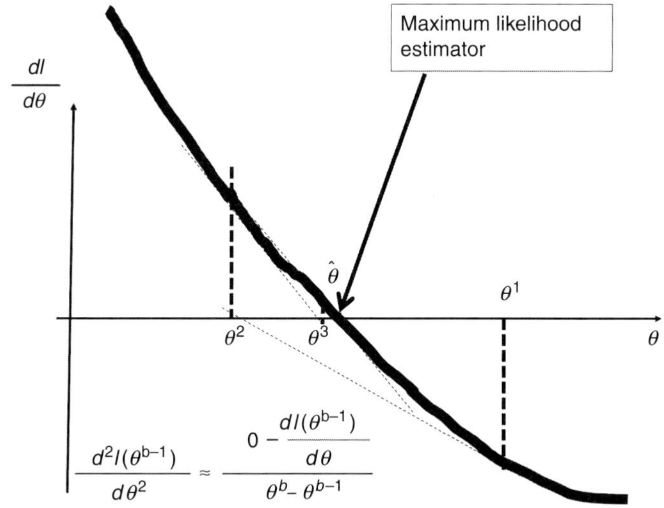

Figure 2.20 Maximum likelihood estimation, illustrated by the Newton–Raphson method for optimizing the likelihood function of a one-dimensional model parameter.

starting value is not too far from the optimum value, convergence is achieved within 5–10 iterations. Note, however, that one can experience overshooting and divergence with this method, and sometimes more robust optimization schemes are required.

Optimization of the likelihood requires derivatives with respect to the model parameters. These derivatives can be computed analytically for several models. The second derivative (Hessian) can be quite unbalanced because of the randomness induced by the data. A numerically more stable expression is obtained by taking the expected value of the random observations – i.e., $E\left(d^2 l/d\boldsymbol{\theta}^2\right)^{-1}$. The optimization method obtained from using this in Equation (2.38) is known as Fisher scoring.

Asymptotically, as the size of the data approaches infinity, the MLE is unbiased – i.e., $E\left(\hat{\boldsymbol{\theta}}\right) = \boldsymbol{\theta}$, where the expectation is taken over the data distribution. Moreover, the estimator is Gaussian distributed with variance obtained from the negative inverse Hessian $-E\left(d^2 l/d\boldsymbol{\theta}^2\right)^{-1}$. These desirable properties have ensured that the MLE is popular in statistics.

Inserting the MLE into the prior pdf, we get $p(\boldsymbol{x}) = p\left(\boldsymbol{x};\hat{\boldsymbol{\theta}}\right)$, and similarly for the likelihood of data, $p(\boldsymbol{y} \mid \boldsymbol{x}) = p\left(\boldsymbol{y} \mid \boldsymbol{x};\hat{\boldsymbol{\theta}}\right)$. In this way, the frequentist view treats these nuisance parameters as fixed but unknown. However, one might say that this expression underestimates the variability since the estimate $\hat{\boldsymbol{\theta}}$ is uncertain. There are methods to include some of this uncertainty, such as bootstrapping (Efron and Tibshirani 1993).

2.4.2 Examples

I love rock and ore: mining oxide grade example

Keywords: *oxide grade example, mining, X-ray data, linear Gaussian model, least-squares estimation, maximum likelihood estimation*

An oxide ore grade data set was shown in Section 2.1. Two data types are available at 103 of the locations in this mine. X-ray fluorescence (XRF) data consist of extensive X-ray measurements of core samples in the laboratory. XRF meter (XMET) data consist of measurements obtained with a handheld meter at the mining site. The XRF data are treated as perfect information about the grade, whereas the XMET data are assumed to be noisy or imperfect observations of the true grade. In addition, there is geological knowledge about the mineralization level, which is used as a covariate.

We will use a subset of the data to illustrate the estimation of nuisance parameters. A more thorough presentation is discussed in Chapter 4 for spatial modeling and in Chapter 6 for the value of XRF and XMET information analysis. XMET data at the 103 locations with both data types are denoted $\boldsymbol{y} = \left(y_1,\ldots,y_{103}\right)$, and XRF data are denoted $\boldsymbol{x} = \left(x_1,\ldots,x_{103}\right)$. We assume a Gaussian likelihood for the XMET data, $y_i = x_i + N\left(0,\tau^2\right), i = 1,\ldots,103$. Moreover, the oxide grade depends on a known mineralization covariate $h_i \in \{1,2,3\}$ in a regression model $x_i = \beta_0 + \beta_1 h_i + N\left(0,\sigma^2\right)$, where the regression parameters β_0 and β_1 are fixed but unknown. For short, we write this as: $\boldsymbol{x} = \boldsymbol{H\beta} + N\left(\boldsymbol{0},\sigma^2\boldsymbol{I}\right)$, where the 103 × 2 matrix $\boldsymbol{H}$

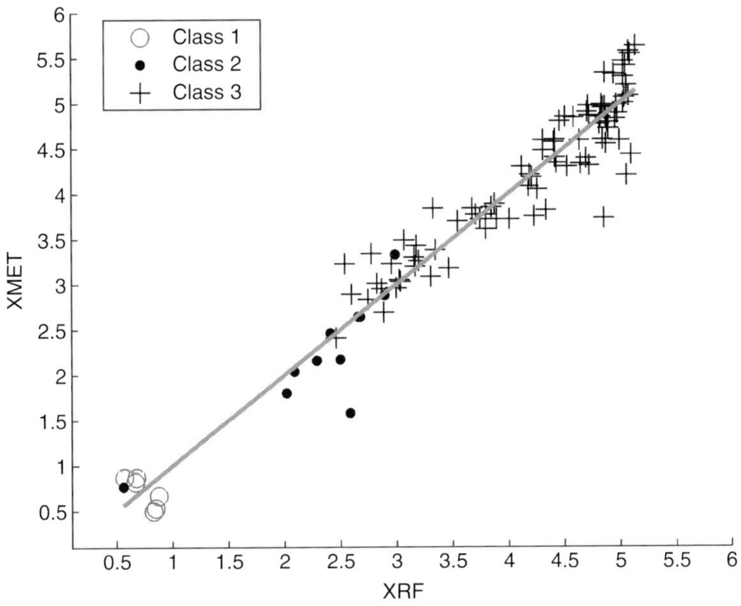

Figure 2.21 Parameter estimation in a regression model. The XMET data are assumed to provide imperfect information about the oxide grade, while the XRF data provide perfect information about the oxide grade. There are three geological classes used as covariates in the regression model for the oxide grade.

contains the known covariate classes and where the 2 × 1 vector of regression parameters is $\boldsymbol{\beta} = (\beta_0, \beta_1)$.

Since both data and the actual distinction of interest are available at the 103 locations, we can separately estimate parameters in the likelihood $p(\boldsymbol{y} \mid \boldsymbol{x}; \tau)$ and the prior $p(\boldsymbol{x}; \sigma, \boldsymbol{\beta})$. Here, we analytically compute the MLE from the Gaussian expressions.

The MLE (and the least-squares estimate) for the regression parameter is

$$\hat{\boldsymbol{\beta}} = \left(\boldsymbol{H}^t \boldsymbol{H}\right)^{-1} \boldsymbol{H}^t \boldsymbol{x}. \tag{2.39}$$

The associated MLE for the variance in the grade is defined from the residuals in the regression fit:

$$\hat{\sigma}^2 = \frac{1}{103}\left(\boldsymbol{x} - \boldsymbol{H}\hat{\boldsymbol{\beta}}\right)^t \left(\boldsymbol{x} - \boldsymbol{H}\hat{\boldsymbol{\beta}}\right). \tag{2.40}$$

The likelihood variance is estimated from the differences in the XMET and XRF data:

$$\hat{\tau}^2 = \frac{1}{103}(\boldsymbol{y} - \boldsymbol{x})^t (\boldsymbol{y} - \boldsymbol{x}) = \frac{1}{103}\sum_{i=1}^{103}(y_i - x_i)^2. \tag{2.41}$$

A cross-plot of the XRF and XMET oxide grade measurements is shown in Figure 2.21. Note that the grade measurements get larger with the increased mineralization covariate.

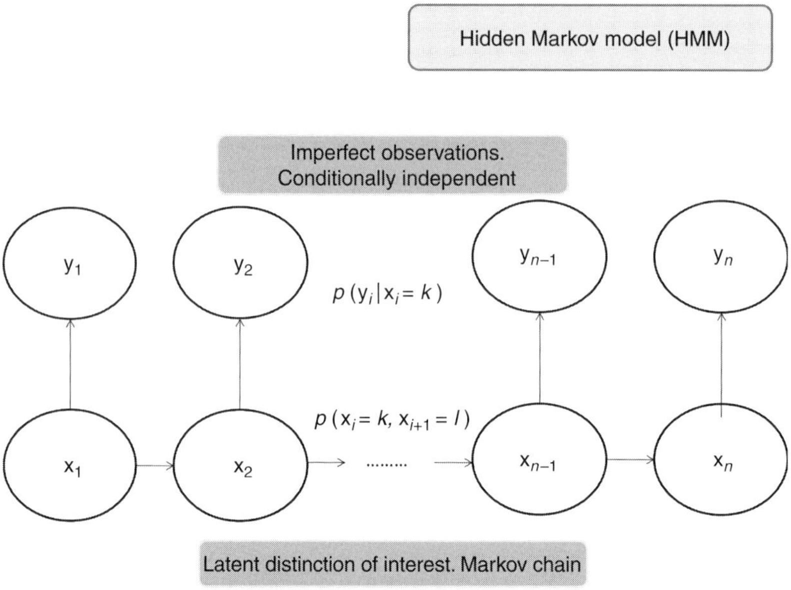

Figure 2.22 Hidden Markov model where the latent variable is a first-order Markov chain model and the data are conditionally independent given the latent variable.

The variability in the XMET measurements from the straight line is used to estimate the likelihood noise. We obtain measurement noise variance $\hat{\tau}^2 = 0.34^2$, considering only that the XRF data give regression parameters $\hat{\boldsymbol{\beta}} = (-1.1, 1.8)'$ and grade variance $\hat{\sigma}^2 = 0.75^2$.

Never break the chain: Markov chain example

Keywords: *Markov chain, dependent binary variables, conditional independence, hidden Markov chain, maximum likelihood estimation*

Let us now consider the hidden Markov model (HMM) – see, e.g., MacDonald and Zucchini (1997) – which is an extension of the Markov chain model with conditionally independent data (see Appendix A.3).

The graph for this model is illustrated in Figure 2.22. The (latent) distinction of interest has stationary first-order transition probabilities $P(k,l)$ between subsequent discrete states $x_i = k$ and $x_{i+1} = l$. We assume two possible states and equal transition rates $P(0,0) = P(1,1) = p$. The data are conditionally independent given the variable of interest, and the Gaussian likelihood is defined by $p(y_i \mid 0) = N(-1, \tau^2)$ and $p(y_i \mid 1) = N(1, \tau^2)$.

Using the factorization in Equation (2.21), the marginal likelihood for the model parameters $\boldsymbol{\theta} = (p, \tau^2)$ is the product of all sequential probabilities:

$$L(\boldsymbol{\theta}) = p(y_1, \ldots, y_n; \boldsymbol{\theta}) = \prod_i p(y_i \mid y_1, \ldots, y_{i-1}; \boldsymbol{\theta}). \tag{2.42}$$

The required probabilities can be evaluated by recursive marginalization of the discrete latent distinction of interest (see Appendix A.3).

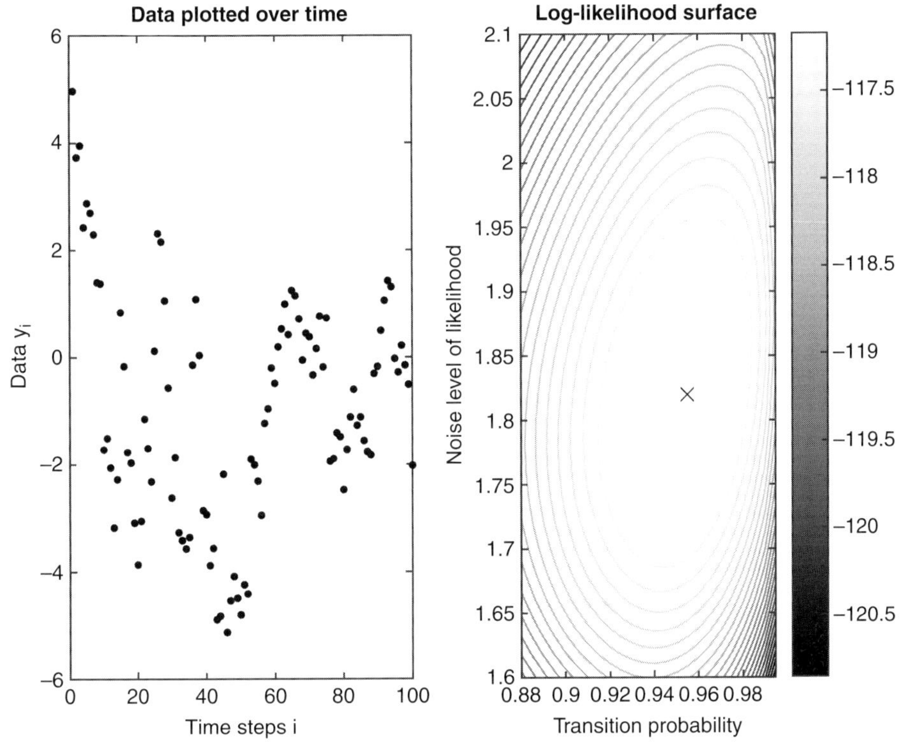

Figure 2.23 Example of numerical calculation of maximum likelihood for a hidden Markov chain model. Data (left) and log-likelihood surface (right) for the Markov transition probability (first axis) and the likelihood noise standard deviation (second axis).

We generate data of size $n = 100$. The parameters are $p = 0.95$ and the noise variance is $\tau^2 = 2^2$. In the pdf of the initial state x_1, the two states are equally likely. The data are shown in Figure 2.23 (left). In Figure 2.23 (right), we display the likelihood surface, which shows that these parameters are estimated reasonably well by the data.

2.5 Monte Carlo methods and other approximation techniques

Monte Carlo methods are widely used for various types of sampling, simulation, optimization, and integration. In this section, we will focus on stochastic simulation and solving integrals by sampling from the probability models and evaluating functional expressions. The integration approximations will be relevant for the VOI analysis described in later sections. Recent books covering Monte Carlo simulation include Liu (2001), Gamerman and Lopes (2006), and Rubinstein and Kroese (2007).

2.5.1 Analysis by simulation

In Section 2.2, we discussed the mean, variance, and pdf of a transformation. When the function or transformation $f(x)$ is non-linear, analytical solutions are rarely available.

Linearization results in bias – i.e., $E(f(\boldsymbol{x})) \neq f(\boldsymbol{\mu})$ – and a variance approximation would over- or underestimate the uncertainty. The full distribution of the transformation is even more difficult to represent. We can study the pdf of $f(\boldsymbol{x})$ directly for only some special situations. A useful solution technique is **stochastic simulation**.

Algorithm: a Monte Carlo method for uncertainty propagation

1. Sample $\boldsymbol{x}^1,\ldots,\boldsymbol{x}^B$ from pdf $p(\boldsymbol{x})$.
2. Evaluate or simulate $f(\boldsymbol{x}^1),\ldots,f(\boldsymbol{x}^B)$.

The B outputs represent the distribution of the function $f(\boldsymbol{x})$, and we can compute the sample mean, variances, or percentiles. Such statistics converge to the theoretical counterparts when $B \to \infty$.

This approach is extremely useful for uncertainty quantification in complicated systems. For instance, in the basin modeling example shown in Section 2.1, we can sample input geological variables and use established software for basin and petroleum systems modeling to simulate the output accumulations, which are representative of the mean values and the uncertainties in volumes. The random sampling approach is a viable alternative to the controlled design of experiments in many cases. In a design of experiment setup, Step 1 in the algorithm is performed in a deterministic way. One then sets the elements in the input space in a structured way, typically defined by high and low levels, and then studies the associated responses in Step 2.

Simulation is often used for preliminary sensitivity analysis. This can be helpful for studying the effect of single variables or interactions of variables. Promising recent approaches use efficient dimension reduction such as basis representations or multidimensional scaling to effectively explore the input space. This has been very useful for experiments where the function $f(\boldsymbol{x})$ is very complicated and computationally demanding. It can also be combined with clustering techniques or various adaptive designs of experiments. Simulation is also used to falsify modeling assumptions in prior models $p(\boldsymbol{x})$ or some fixed input conditions involved in the function $f(\boldsymbol{x})$. Falsification may include response variables that are unphysical in some sense or responses that are totally inconsistent with observations $\boldsymbol{y}$. The latter aspect relates to Bayesian inversion.

2.5.2 *Solving integrals*

As we will discover in later chapters, several of the computations required for VOI analysis are integrals (continuous sample space) or sums (discrete sample space). For instance, a decision situation would likely involve the expected value $\int xp(x)dx$, or the decision maker would perform some optimal choice under uncertainty, involving integral expressions such as $\int \max\{0,x\}p(x)dx$. In this section, we discuss approaches for approximating such integrals or sums. Expressions that cannot be solved analytically or numerically can be handled by Monte Carlo methods.

Consider a general integral expression

$$I = E\left(f(\boldsymbol{x})\right) = \int f(\boldsymbol{x}) p(\boldsymbol{x}) d\boldsymbol{x}, \tag{2.43}$$

where $p(\boldsymbol{x})$ is a pdf on the domain of $\boldsymbol{x}$ and where $f(\boldsymbol{x})$ is a function that we can evaluate. If $f(\boldsymbol{x}) = x_i$, the integral in Equation (2.43) is the expectation of variable x_i; if $f(\boldsymbol{x}) = I(x_i > 0)$, we compute the probability that variable x_i is positive. Analytical or numerical approximations depend heavily on the function $f(\boldsymbol{x})$. If available, they can be very fast and accurate, but for some situations they may yield poor approximations. Monte Carlo methods are far less sensitive to the function $f(\boldsymbol{x})$.

Analytical approximations of integrals are largely based on Gaussian approximations or quadratic expansions. The **Laplace approximation** uses both of these tools. It can be presented in various ways. Consider a rewritten version of Equation (2.43):

$$I = E\left(f(\boldsymbol{x})\right) = \frac{\int f(\boldsymbol{x}) p(\boldsymbol{x}) d\boldsymbol{x}}{\int p(\boldsymbol{x}) d\boldsymbol{x}}. \tag{2.44}$$

We can fit a quadratic form to the exponent of $b(\boldsymbol{x}) = \log f(\boldsymbol{x}) + \log p(\boldsymbol{x})$ at the mode of this expression – i.e., $\tilde{\boldsymbol{x}} = \arg\max_x \{b(\boldsymbol{x})\}$. The same is done for the denominator, which means fitting a Gaussian approximation to $p(\boldsymbol{x})$ at the mode $\hat{\boldsymbol{x}} = \arg\max_x \{p(\boldsymbol{x})\}$. The quadratic expansions allow Gaussian completion of the integrals. Since the first derivatives are $\mathbf{0}$ at the mode, we only need to account for the second derivatives in the (co)variance part of the Gaussian pdf. The Laplace approximation of the integral in Equation (2.44) becomes

$$\hat{I}_{\text{Laplace}} = \exp\left(b(\tilde{\boldsymbol{x}})\right) \frac{\sqrt{\left|\frac{d^2 \log p(\hat{\boldsymbol{x}})}{d\boldsymbol{x}^2}\right|}}{\sqrt{\left|\frac{d^2 b(\tilde{\boldsymbol{x}})}{d\boldsymbol{x}^2}\right|}}. \tag{2.45}$$

This Laplace approximation can be very useful for models dominated by Gaussian components and weakly non-linear functions $f(\boldsymbol{x})$.

Numerical approximations of integrals define a grid of evaluation nodes in the support of the integrand. Direct approaches assume a constant, linear, or quadratic form of the integrand between the evaluation nodes. For lower-dimension integrals, these traditional numerical approaches tend to be extremely accurate, but for high-dimension integrals, the numerical approximations are difficult to compute because the number of evaluation points gets very large. Useful guidelines for the placement of evaluation points in related contexts have been worked out, such as central composite designs or more general response surface methodologies, but it is not straightforward to achieve reliable results in a high dimension.

We now turn to Monte Carlo methods for solving integrals like Equation (2.43). As mentioned earlier, Monte Carlo approximations are often the method of choice in difficult problems (complex $f(\boldsymbol{x})$ or $p(\boldsymbol{x})$) and in high dimensions (large n) because they are more generally applicable than analytical and numerical methods.

The **Monte Carlo approximation** is based on generating samples from the pdf $p(\boldsymbol{x})$ and computing function averages.

Algorithm: the Monte Carlo approximation of an integral

1. Draw samples $\boldsymbol{x}^1,\ldots,\boldsymbol{x}^B$ from $p(\boldsymbol{x})$.
2. Evaluate the function for each sample and average $\hat{I} = \frac{1}{B}\sum_{b=1}^{B} f(\boldsymbol{x}^b)$.

Recall that the function $f(\boldsymbol{x})$ represents whatever we would like to approximate. It could be a linear function (mean value), a quadratic function (variance), an indicator (a percentile), etc. Notably, we can approximate different integrals or statistics from the same samples. The Monte Carlo approximation is easy to implement once we can sample from $p(\boldsymbol{x})$. Here, $E(f(\boldsymbol{x}^b)) = \int f(\boldsymbol{x})p(\boldsymbol{x})d\boldsymbol{x} = I$ for all samples $b = 1,\ldots, B$, and the Monte Carlo approximation is an unbiased estimator of the integral. The quality of the Monte Carlo approximation of the integral I depends on the probability distribution $p(\boldsymbol{x})$ relative to the function $f(\boldsymbol{x})$. In this way, the variability of the Monte Carlo estimator depends on the integral expression that we wish to estimate. If we attempt to estimate a rare event such as a tail probability, we often need several Monte Carlo samples B to get a sufficiently small relative Monte Carlo error. When we estimate the mean, fewer samples are required for small error bounds.

We study various approximation methods for the evaluation of a logistic integral

$$I = \int_{-\infty}^{\infty} \frac{\exp(x)}{1+\exp(x)} \frac{1}{\sqrt{2\pi}\sigma} \exp\left(-\frac{(x-\mu)^2}{2\sigma^2}\right) dx. \tag{2.46}$$

This integral appears, for instance, when modeling the presence or absence of an asset, success or failure of events, etc. The probability of success is $p = \exp(x)/(1+\exp(x))$, and the integral computes the expected value of this probability. Generalized linear models (GLMs) (McCullagh and Nelder 1989; Dobson and Barnett 2008) stress the modeling of the logistic variable x, typically having a linear form depending on the explanatory variables (see Appendix A.2).

We solve this logistic integral by three methods:

1. Numerical approximation using a dense grid of evaluation points (Num).
2. Analytical expansion of the integrand around the mode (Laplace method: LA).
3. Monte Carlo approximation based on sampling from the Gaussian distribution (MC).

The numerical solution is done on a dense grid of 10 000 evaluation points for this example, and this is considered the exact solution. For Monte Carlo sampling, we use

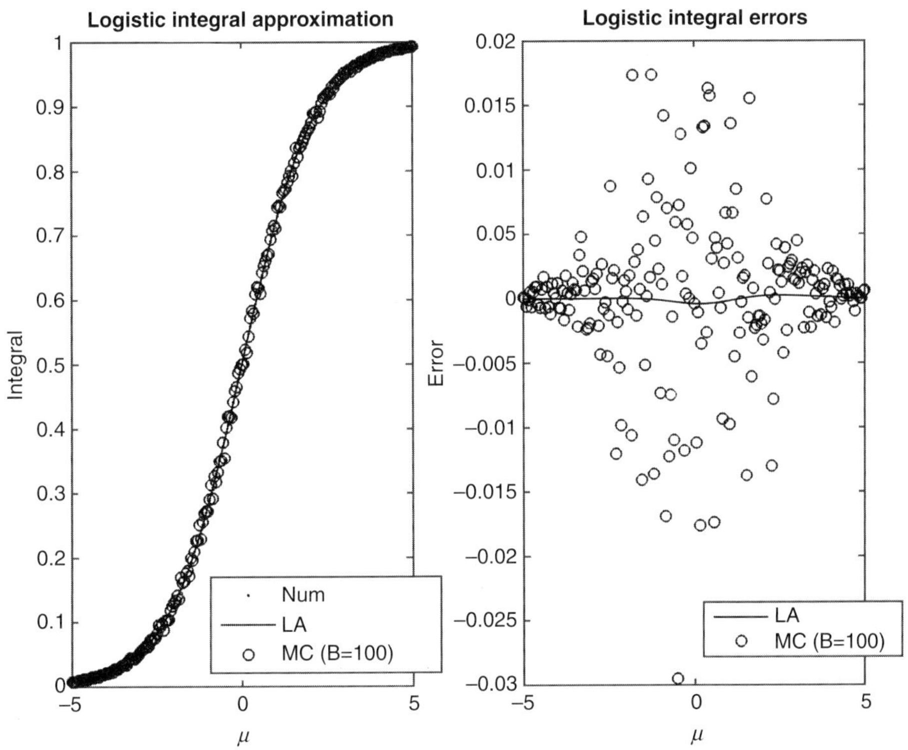

Figure 2.24 Left: logistic integral for different approximations (numerical solution, Laplace approximation, and Monte Carlo with $B = 100$ realizations) and for varying levels of the Gaussian mean. Right: comparison of Laplace and Monte Carlo approximation against the numerical reference solution.

$B = 100$ samples from the Gaussian pdf and take the average of the logistic function over the samples.

Figure 2.24 shows the results for a range of mean values μ and standard deviation $\sigma = 1/2$. All approximations deliver accurate results in this example, as we can hardly discriminate the three curves in the left display. When we look at the errors compared with the dense numerical approximation in the right display, we notice that the Monte Carlo approach has the largest errors near zero, where the success probability is about 0.5. The Laplace approximation performs well for most mean values.

Several tricks are possible for reducing the variability in the Monte Carlo approximation. For instance, one can apply importance sampling to reduce the Monte Carlo variability. This technique generates samples $\boldsymbol{x}^1,\ldots,\boldsymbol{x}^B$ from another pdf $q(\boldsymbol{x})$, called the proposal distribution, rather than the true model $p(\boldsymbol{x})$. The Monte Carlo estimator is a weighted sum

$$\hat{I} = \sum_{b=1}^{B} f\left(\boldsymbol{x}^b\right) w\left(\boldsymbol{x}^b\right), \qquad w\left(\boldsymbol{x}^b\right) = \frac{p\left(\boldsymbol{x}^b\right)/q\left(\boldsymbol{x}^b\right)}{\sum_{c=1}^{B} p\left(\boldsymbol{x}^c\right)/q\left(\boldsymbol{x}^c\right)}. \tag{2.47}$$

This integral estimator is unbiased and will converge to the true value as the sample size increases under mild regularity conditions on the proposal distribution $q(\boldsymbol{x})$. The importance sampling estimator in Equation (2.47) could have smaller variance than the straightforward Monte Carlo approximation. The idea is that the proposal distribution highlights the important variables for the particular case in which we are interested. In this way, the best proposal distribution $q(\boldsymbol{x})$ changes with the case-specific goals via the function $f(\boldsymbol{x})$.

2.5.3 Sampling methods

Appendix B presents a number of methods for generating random variables. This section only provides a brief discussion of the main ideas.

There is a large variety of methods for generating random variables from known distributions. Independent uniform variables $U(0,1)$ play a critical role here. If we can generate independent uniform variables, we can then apply **transformations** to get a sample from the relevant pdf. For instance, a power law distributed variable can be generated by first sampling a uniform variable and then using the **inverse cumulative distribution function**. We can sample a Gaussian variable by first sampling uniform variables and transforming these to independent standard Gaussian variables. To sample from $N(\boldsymbol{\mu},\boldsymbol{\Sigma})$, we can first Cholesky factorize the covariance matrix, $\boldsymbol{\Sigma} = \boldsymbol{L}\boldsymbol{L}^t$, and set $\boldsymbol{x} = \boldsymbol{\mu} + \boldsymbol{L} \cdot N(\boldsymbol{0}, \boldsymbol{I}_n)$. Similar tricks are common for a wide range of distributions. For instance, independent variables can be used in a stepwise propagation – say, by using conditional probabilities like that of a graph or network. However, not all pdfs are straightforward to sample from, and we will briefly discuss accept/reject methods and iterative methods.

Rejection sampling generates samples from the proposal distribution pdf $q(\boldsymbol{x})$ which is usually easier to sample from. Samples are subsequently either rejected or accepted as samples from the true pdf $p(\boldsymbol{x})$. The number of accepted samples depends on the similarity between the proposal mechanism $q(\boldsymbol{x})$ and the target density $p(\boldsymbol{x})$. Although rejection sampling is an exact method, it is not always easily applicable, especially in high dimensions.

Approximate Bayesian computation relies on ideas similar to rejection sampling. Here, many samples $\boldsymbol{x}^1,\ldots,\boldsymbol{x}^B$ are generated from the prior model $p(\boldsymbol{x})$. Next, the samples are forward-propogated to the data space, and we accept samples that are in some sense close to the data $\boldsymbol{y}$. Approximate Bayesian computing is looser in its acceptance criteria than rejection sampling. In fact, it avoids the full specification of a likelihood model, which would be required for correct rejection sampling. Instead, it relies on a similarity measure between the simulated data and the observations. This algorithm was popularized in biological applications (Beaumont 2002) where the function model is very complex but amenable to evaluation or simulation. The same setting with a complicated forward model holds in many Earth sciences applications.

Markov chain Monte Carlo sampling is a method for generating dependent samples from a probability model $p(\boldsymbol{x})$. Unlike rejection sampling, which proposes independent variables in each trial, the Markov chain Monte Carlo algorithm uses the current sample to construct a new sample. This iterative construction is based on a Markov transition matrix (or kernel in continuous state spaces), depending on the current state. There are two famous Markov chain Monte Carlo algorithms: the Metropolis–Hastings sampler and the Gibbs sampler.

2.5.4 Example

Risky business: petroleum prospect risking example

Keywords: *risking, petroleum prospects, transformed Gaussian variables, Monte Carlo sampling, approximate Bayesian computing*

Let us consider a situation not dissimilar to the basin modeling setup described in Section 2.1. In Figure 2.8, we studied different variables affecting the volumes of hydrocarbons at geologically defined prospects. A design of experiment was run to study the effect of prospect volumes (Table 2.1). An alternative to the design of experiment in this case is Monte Carlo risking, which entails sampling variables in the input space and evaluating the output variables of interest for each Monte Carlo input sample.

In this example, we assume that the key uncertainties in prospect volumes are the thickness and porosity of the reservoir zone in an oil field. The thickness is influenced by the geological history of the basin (erosion events, sea level over geologic time, etc.). Similarly, the porosity depends on the geological deposition environment over time. We assume a simplified formulation where porosity is assumed to be a logistic Gaussian variable (similar to the probability in Equation (2.46)) but bounded within the interval $(0.15, 0.4)$. We can sample by first generating a Gaussian variable and then taking the logistic transform. The thickness is assumed to be power law distributed. We can simulate a power law variable by first simulating a Gaussian variable, then computing the cumulative distribution function, and finally taking the inverse power law. The volumes are obtained by multiplying the uncertain thickness and porosity variables and a fixed scalar term.

Suppose further that there are $n = 100$ prospects with some common geological mechanisms inducing correlation. This pairwise correlation is assumed to be the same for all prospects, defined by a parameter ρ. In our formulation, it is straightforward to model the correlation for the Gaussian variables. We can thus simulate Gaussian correlated variables, transform these to the correct logistic porosity variables and power law variables, and multiply to get the volumes. This construction is similar to what is typically done for copulas (Joe 2014). In addition to the pairwise correlation between prospects, the transformed porosity and thickness variables at every prospect have a 2×2 covariance matrix $\boldsymbol{\Sigma}_0$. The modeled covariance matrix for basic variables at the prospects is then

Table 2.2. *Illustration of porosity, thickness, and volumes from the prior (top) and posterior (bottom) distribution. The prior and posterior pdfs are explored by sampling*

	5th Percentile	25th Percentile	Median	75th Percentile	95th Percentile
Prior, porosity	0.17	0.22	0.27	0.33	0.38
Prior, thickness	5.2	5.8	7.1	10.1	22.1
Prior, total volume	15	18	20	24	43
	5th Percentile	25th Percentile	Median	75th Percentile	95th Percentile
Posterior, porosity	0.18	0.21	0.24	0.27	0.32
Posterior, thickness	7.4	8.6	9.9	11.6	14.4
Posterior, total volume	19	22	25	27	32

$$\boldsymbol{\Sigma} = \text{kron}\left(\boldsymbol{\Sigma}_0, \begin{bmatrix} 1 & \rho & .. & \rho \\ \rho & 1 & \rho & .. \\ \rho & .. & 1 & \rho \\ \rho & .. & \rho & 1 \end{bmatrix}\right), \tag{2.48}$$

where the Kronecker product enforces separability of the correlation between porosity and thickness between prospects. The between-prospect correlation (ρ) is set very high here, and there is some positive correlation between the porosity and thickness variates ($\boldsymbol{\Sigma}_0$).

The Monte Carlo representation of the volumes requires three steps: (i) generate random samples from the specified Gaussian distribution, (ii) transform variables to obtain samples from the logistic porosity and power law distribution for thickness, and (iii) compute the volumes by multiplication.

Table 2.2 (top) shows the results of $B = 10\,000$ prior samples of the porosity and thickness. The table presents percentiles of porosity and thickness for Prospect 1, as well as the total volume for all prospects (sum over all $n = 100$ prospects). The percentiles are obtained after sorting the samples from smallest to largest and picking the appropriate fraction.

Assume that there is perfect information about the thickness of Prospect n. We know that this is 10 m. Posterior samples can be obtained by approximate Bayesian computing, where samples closer than a threshold to the observed 10-m thickness are selected. The results of posterior sampling, based on a quadratic loss criterion, are presented in the bottom rows of Table 2.2. Information about the thickness at Prospect n propagates to all variables because of the dependence in the model. Both the thickness and the porosity distribution at Prospect 1 are clearly narrower than for the prior, but the effect of conditioning is smaller for porosity than for thickness. In the bottom row, we list the percentiles of the distribution for total volume. The posterior for total volumes has smaller uncertainty than the prior. In particular, very large volumes are less likely in the posterior.

Discussion: computing

Programming languages like MATLAB have built-in routines for generating samples from known pdfs, such as the Gaussian in this example. It is much faster to draw many samples in one operation than running through a for-loop. In MATLAB, we can sample B Gaussian vectors with common mean $\boldsymbol{\mu}$ and covariance $\boldsymbol{\Sigma}$ as follows:

$$\boldsymbol{x}_{\text{mat}} = \boldsymbol{\mu} * \mathbf{1}_B^t + \text{chol}(\boldsymbol{\Sigma})^t * \text{randn}(n, B).$$

It is also possible to use the *mvnrnd* function from the MATLAB statistics toolbox. Additional speed-up can be gained through parallel implementation.

2.6 Bibliographic notes

It is not our intention to cover probability models and statistical methods comprehensively. There are numerous textbooks on statistical models and methods – for instance, the book by Jensen et al. (2000), which is widely referenced in the Earth sciences. The appendix contains further details pertaining to the primary probability models and sampling methods we focus on for VOI analysis in this book.

Models

There are plenty of well-known univariate probability distributions (Johnson et al. 1994). Useful multivariate probability distributions are much harder to construct, but the books by Johnson et al. (1997) and Kotz et al. (2000) cover commonly used multivariate discrete and continuous distributions. The increase in data size and computing power means that the t-distribution described in introductory statistics textbooks is perhaps not that relevant anymore. The Gaussian distribution has remained extremely popular because of its ease of applicability, its properties, and its ability to handle high-dimensional situations. If not directly useful as a stand-alone model, the Gaussian distribution will continue to be a building block in more complex settings.

As we have discussed, it has become popular to construct graphical models or networks from generic and modular principles, including conditional independence. Several books on graphical models and Bayesian networks demonstrate modeling aspects, methods for inference, and a wealth of applications (Green et al. 2003; Cowell et al. 2007; Jensen and Nielsen 2007; Koller and Friedman 2009; Sucar et al. 2012). There is also plenty of software available for computing Bayesian networks – we use the Bayesian Network Toolbox (BNT) in MATLAB developed by K. Murphy (bnt.googlecode.com) and Netica (www.norsys.com/netica.html).

Estimation and sampling

The boundaries between fully Bayesian statistical approaches and frequentist methods are not as clear in practice as they used to be. Practical solutions borrow the most appropriate models and methods from different communities and a variety of tools are often required. Approximations are often necessary, yet statisticians carefully study the properties of approximations under various conditions. An excellent book on likelihood principles and Bayesian theory is the one written by Carlin and Louis (2000). Scoring rules have become popular for fitting models to ensembles or scenario-based data, where the goal is to trade model complexity for predictive power while maintaining robustness – see, e.g., Gneiting and Raftery (2007).

New types of data sets, which can be enormous in size, have encouraged new approaches for summarizing data, inferring parameters, classification, and learning. Rasmussen and Williams (2006) is a great introduction to machine learning. Hastie et al. (2009) provide an excellent overview of recent tools and general techniques for learning from data. These tools build on both parametric and non-parametric principles. We have focused mainly on parametric models here because there tend to be several parametric, physical relations in the Earth sciences. Moreover, in the Earth sciences, there are substantial auxiliary data that should be used for learning. For instance, well logs, geological information, core samples, basin modeling, etc., are used to specify model parameters at the seismic interpretation stage.

Markov chain Monte Carlo methods have been particularly popular in Bayesian statistics – see, e.g., Gamerman and Lopes (2006). Expectation maximization (EM) algorithms have similarly been very popular in frequentist statistics – see, e.g., McLachlan and Krishnan (2008). Both these iterative computational techniques use the structure of the model to efficiently sample or estimate subsets of the variables while keeping the other variables fixed. In the Bayesian literature, recent alternative techniques construct faster approximations to the posterior distributions – for instance, nested Laplace approximations (Rue et al. 2009), which rely on Gaussian approximations for subsets of the variables of interest, or variational methods (Jaakkola 2000), which assume certain factorization forms of the joint posterior model. The massive dimensions of current data sets have motivated the use of inference methods using local properties such as pseudo-likelihoods or composite models (Varin et al. 2011).

Traditionally, approaches for the design of experiments popularized statistics in some applications; these methods are still relevant. However, experimental design has become more sophisticated within the realm of computers and new data sets. A great introduction to the design of computer experiments is given in Santner et al. (2003). For the Earth sciences, computer experiments and simulations are used extensively, and the domain is likely to benefit from the recent advances in the literature.

Recent years have seen the use of many tools of multivariate statistics to analyze high-dimensional data, reduce dimensions, and visualize them in lower dimensions. These tools include principal component analysis and kernel principal component analysis, clustering, and multidimensional scaling. The books by Izenman (2008) and Hastie et al. (2009) describe many of these multivariate statistical techniques. Multidimensional scaling for dimension reduction and data visualization arose in the fields of psychology and behavioral sciences but has now been used in many disciplines, including Earth sciences. In reservoir modeling, multidimensional scaling has been used to visualize the uncertainty of very-high-dimensional spatial Earth models by computing an appropriate distance between pairs of models in an ensemble of model realizations and then using multidimensional scaling to project them to a lower dimension space. Clustering, screening of multiple geologic scenarios, and model selection can then be done in this lower dimension space. Cox and Cox (2001) and Borg and Groenen (2005) give extensive coverage of general multidimensional scaling theory and techniques. Caers (2011) and Scheidt and Caers (2009) provide applications of multidimensional scaling for visualization and uncertainty quantification in reservoir modeling.

Parallel computing will no doubt make all these techniques more efficient in the future – see, e.g., Kontoghiorghes (2005). The bottleneck at the current time appears to be user-friendly interfaces for connecting desktop or internet code to parallel computing units, but we expect such issues to be resolved in the near future.

3
Decision analysis

Figure 3.1 scuffles in my mind
traversing waters unknown
how should I decide?

We all make decisions in our personal and professional lives. Many of these decisions are trivial and fleeting, without profound significance in the broader scheme of things. Many decisions we make, however, have a meaningful impact on our lives and on society at large. Decision analysis is the study of applying the principles of decision theory to practical decision situations, providing guidance that can help us with important decisions that deserve our attention through analysis.

We review the fundamental concepts of decision analysis in this chapter. We focus on concepts that have a bearing on value of information (VOI) analysis, but the reader should hopefully be able to glean a general overview of the field. We start with a background discussion in Section 3.1, outlining key principles that lie at the core of decision analysis. Section 3.2 defines the notation and terminology for decision situations, providing language that will be required for eventually integrating decision analysis with other subjects such as spatial statistics. Section 3.3 reviews graphical models, focusing solely on the two most popular models in decision analysis: decision trees and influence diagrams. VOI is introduced in Section 3.4. Here, we highlight concepts that will be critical for understanding the forthcoming chapters. Finally, we outline some bibliographic notes in Section 3.5.

3.1 Background

Decision analysis stands on the solid foundation built over hundreds of years of thought regarding decision making under uncertainty – its fundamental notions can be traced back to some of the pioneers of probability theory – for instance, Bernoulli, Bayes, and Laplace. The term "decision analysis" was coined by Ronald Howard (1964) to define the field of study that logically evaluates available alternatives in a decision situation once the uncertainties involved and the preferences of the decision maker have been recognized. The first book that used the phrase "decision analysis" in the title is by Raiffa (1968).

Figure 3.2 shows a fundamental distinction between various fields of study pertaining to decision making, categorizing them into one of three types: (a) **descriptive** domains are those that effectively describe behavior around how people make decisions, (b) **normative** ones are those that lay down a set of norms (or axioms), and (c) **prescriptive** fields of study regarding decision making are concerned with methods and tools for helping people make better decisions. Decision analysis is a prescriptive approach with the intent of guiding a decision maker through a potentially opaque decision situation, leading him or her to clarity of action through a process of assessing his or her beliefs and preferences. It is based on the normative principles of decision theory. Prescriptive disciplines like decision analysis attempt to address and rectify cognitive as well as motivational biases that creep into a person's decision-making process, and there is a rich history of active interaction between researchers in prescriptive and descriptive domains (see von Winterfeldt and

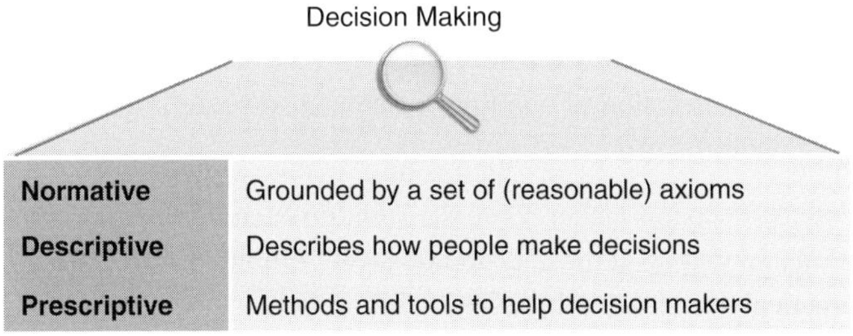

Figure 3.2 Decision making studies can be categorized into one of three types: normative, descriptive, and prescriptive.

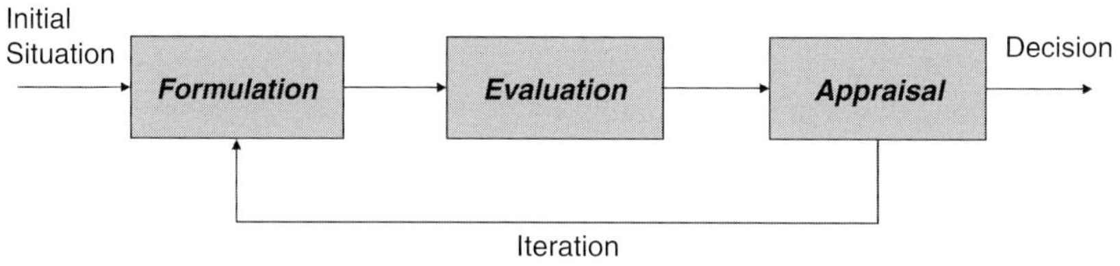

Figure 3.3 The decision analysis cycle contains three stages: formulation, evaluation, and appraisal.

Edwards 1986). We should note here that there are several authors who do not make a distinction between normative and prescriptive domains, treating them as one and the same. For research on descriptive aspects such as biases during judgment and decision making, we refer the reader to Tversky and Kahneman (1974) and Lichtenstein and Slovic (2006).

Figure 3.3 depicts the stages through which decision analysis is typically applied in practical situations. The decision analysis cycle comprises three main stages, much like other domains involving mathematical modeling: formulation, evaluation, and appraisal. First, a model is formulated by the analyst in accordance with the beliefs of the decision maker. It is then evaluated using the tools and techniques of decision analysis to arrive at a formal recommendation. In the appraisal stage, further insights are developed by exploring the implications of the formulation and evaluation stages. The model is often refined in an iterative fashion until the decision maker is comfortable with the suggested recommendation and the reasoning on which this is based, and finally the decision maker is encouraged to commit to action.

One of the key foundational takeaways in decision analysis is the distinction between **decisions and outcomes**. The quality of a decision can and should be determined before the outcome has been observed; sometimes, good decisions lead to bad outcomes and bad decisions lead to good ones. This sort of reasoning dates back at least as early as the Greek historian Herodotus, as can be gauged from his discussions about policy decisions made by the Persian kings. He notes that a decision was deemed wise or foolish based on whether the evidence at hand indicated that it was the best one to make, regardless of the consequences. Decision analysis suggests that decisions should be deemed good or bad at the time they are made.

If decisions should not be gauged by their outcomes, then how should one measure **decision quality**? Figure 3.4 presents a spider diagram that depicts six elements of decision quality. A high-quality decision is one that is deemed to score high on as many of these elements as possible:

1. **Frame:** Did the decision maker solve the right problem? Was the context of the decision situation understood at the appropriate level? Improper framing is a common error in many practical decisions. A broadening of the frame is often required to address the decision maker's true concerns.
2. **Alternatives:** Did the decision maker consider a comprehensive set of alternatives that were available? Was there an attempt to pursue and consider creative alternatives?

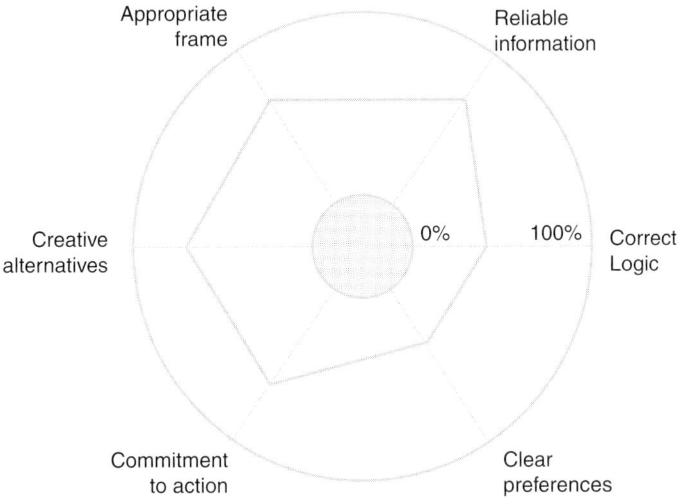

Figure 3.4 Gauging decision quality with a spider diagram consisting of six elements.

3. **Information:** Did the decision maker try to obtain and incorporate meaningful and reliable information that was available at the time of the decision?
4. **Preferences:** Did the decision maker deliberate about and incorporate all the important attributes that truly reflect his or her preferences? Did he or she incorporate his or her preferences?
5. **Logic:** Did the decision maker take a logical and reasonable approach consistent with their alternatives, information, and preferences? In decision analysis, sound reasoning occurs when the norms of decision theory are deployed.
6. **Commitment to action:** Did the decision maker commit to the alternative that was recommended by the analysis? Often in organizations, the analysis suggests one alternative, but another is chosen due to organizational issues such as motivational biases. This results in a poor decision, regardless of the effort that was put into the analysis.

Naturally, the goal of decision analysis – and in fact any prescriptive approach to making decisions – is to help the decision maker choose alternatives that lead to better outcomes than they would have otherwise attained; however, there are often uncertainties prevalent in the world that do not guarantee achieving the best outcome for any particular decision. For instance, a person who chooses to step out of the house without an umbrella in the dry season may still end up getting drenched. If the person knew about the downpour before leaving the house, he or she would have been able to make an even better decision. Modeling the ramifications of resolving an uncertainty before making a decision – and quantifying whether it is worthwhile to do so – is the key idea behind VOI, making it a powerful decision analytic tool for decision makers in any domain.

3.2 Decision situations: terminology and notation

Decision analysis attempts to guide a decision maker to clarity of action in dealing with a situation where one or more decisions are to be made, typically in the face of uncertainty. Keeney (2007) makes a distinction between two kinds of decision situations. The first kind is a decision problem, named thus because the decision is motivated by the decision maker's circumstances. Here, there is some sort of external trigger that causes the decision maker to commit to action. The second kind is a decision opportunity, where the decision maker recognizes an opportunity for making a change in their environment and thus perceives an internal trigger. The latter type of situation is one where the decision maker proactively considers making changes to their environment rather than reacting to extrinsic factors and is a crucial way through which decision makers should create substantial value for themselves.

The kinds of situations that decision analysis tackles are also studied in other related domains pertaining to prescriptive decision making; as a result, they are known by a variety of names. When viewed as an optimization problem, they are also known as maximum expected utility problems, since the objective function involves maximization of expected utility. These problems are often also referred to as sequential decision problems when they involve several decisions made over time or simply as decision making under uncertainty problems when there are random variables involved. In this book, we will use the term **decision situation** (Keeney 2007) since this appears to be the most general and encompassing term.

In this section, we introduce some basic terminology and notation around decision situations. We apply the notation to a simple decision situation, illustrating the ideas with examples.

3.2.1 Decisions, uncertainties, and values

A decision maker's decision situation involves distinctions (synonymously, variables) that are of interest for that particular situation. A distinction defines possibilities that are mutually exclusive and collectively exhaustive – i.e., exactly one of these possibilities can occur. We distinguish between two important types of distinctions – those that are directly under the control of the decision maker are **decisions**, and those that are not are **uncertainties** (or, more technically, random variables). We refer to the possibilities for decisions as alternatives or actions and the possibilities for uncertainties as states or realizations. Decisions are denoted a, and the decision maker can choose an alternative from the set of available alternatives, $a \in A$. On the other hand, uncertainties are denoted x, and they can take on any state in the sample space, $x \in \Omega$, as described in greater detail in Chapter 2. In any discussion about a particular distinction, the possibilities should pass the clarity test – i.e., all the parties involved should understand precisely how all possibilities for that distinction are defined. Just as we did in Chapter 2, we will sometimes use the phrase "distinction of interest" to refer to the primary uncertainty that is relevant in a decision situation; it should be clear from the context that the distinction being referred to is an uncertainty. Also, recall from Chapter 2 that the possibilities for a distinction may be discrete or continuous. To avoid confusion between summation and integration operations, we describe variables in most of this chapter as if they are associated with a finite set of alternatives A and discrete sample space Ω, but all the concepts extend naturally to when these sets are continuous.

Consider the pirate example from Chapter 2, where a pirate is informed about a potential treasure on an island. We will treat this as a recurring example of a decision situation in this chapter. The distinction of interest is whether the treasure is present or not, modeled as a binary random variable $x \in \{0,1\}$. The pirate must decide whether it is worthwhile to make the trip to the island and dig for treasure. The pirate believes that there is a 1% chance that treasure is present – i.e., $p(x=1)=0.01$. Using our notation, an alternative from the following set must be chosen: $A=\{0,1\}=\{\text{do not dig, dig}\}$.

It is important for the decision maker to identify the appropriate decisions in the context of the situation, as well as those uncertainties that are most relevant. Once they have been identified, the decision maker can consider the various scenarios. A **scenario** is the set consisting of a possibility for each distinction or, equivalently, an instantiation of every distinction in the decision situation. Our simple decision situation involves one uncertainty x and one decision a; therefore, there are a total of $|\Omega| \otimes |A|$ scenarios. Each scenario is associated with a **prospect**, which is how the decision maker views the future given the scenario. We prefer the word "prospect" rather than "outcome," as it signifies a beginning rather than an end of the decision maker's future.

The decision maker can associate each prospect with measures that we refer to as **values**. In this book, we assume that the value for each scenario is measured in terms of a single numeraire; specifically, we assume that this numeraire is monetary units. The advantage of using monetary units as a measure is that this permits the decision maker to consider the implication of choices in terms of a value scale of common experience – most people are very much acclimatized to using monetary units to think about value – and to use VOI computations to help with information purchase decisions. VOI techniques can also be deployed using units of value besides monetary units, but they lose the buying price interpretation that we will describe later in this chapter.

The decision maker's values for prospects in their decision situation are represented by a **value function**, denoted $v(.)$. For our simple decision situation, the value function would be applied to the domain of $|\Omega| \otimes |A|$ – i.e., $v(x,a)$. We will also use the notation $v = v_a$ to refer to the value of an alternative a, with implicit dependence on x. Note that since x is a random variable, so is its function $v(x,a)$. Table 3.1 presents an example of a value table for the pirate's decision situation. There are four scenarios in this decision situation, since there are two alternatives for the drilling decision and two states for the uncertainty around the treasure. The value table specifies how the pirate values the prospect for each scenario. In this case, the pirate believes that finding the treasure results in a profit of \$100 000, but if the treasure is absent, then it would result in a loss of \$10 000 from travel and drilling expenses. If the pirate does not dig for treasure, then there is no gain and no loss, resulting in \$0.

In our review of basic probability and statistics in Chapter 2, we discussed modeling the relationships among uncertainties using joint probability distributions. We have seen in this section thus far that decision analysis builds upon the framework of probabilistic modeling of uncertainties by also incorporating decisions and values. Note that there may also be

Table 3.1. *Value table for the pirate example. The value depends on the alternative (dig or do not dig) and the uncertain variable (presence or absence of treasure)*

Digging Decision	Presence of Treasure Uncertainty	Value (Profit)
Dig	Present	\$100 000
Dig	Absent	–\$10 000
Do not dig	Present	\$0
Do not dig	Absent	\$0

situations where there is some notion of value from uncertain prospects *without* any decision to be made. We use the term **lottery** to refer to such situations; they are also known as gambles or deals. As posed, the pirate example is a decision situation, of course – but if the pirate were compelled to dig and did not have a choice in the matter, possibly under mutiny from the crew (!), he would be faced with the following lottery:

$$v_{\text{dig}} = \begin{cases} \$100\,000 & \text{with probability } 0.01 \\ -\$10\,000 & \text{with probability } 0.99 \end{cases} \tag{3.1}$$

where v_{dig} denotes the pirate's value if he digs for treasure, and the numbers are the same as those presented previously.

Now that we have set up the terminology for the basic ideas, we can proceed to more advanced concepts.

3.2.2 Utilities and certain equivalent

A decision maker's risk preferences represent their preferences for lotteries that involve uncertain value prospects. According to the principles of decision analysis, risk is very much like beauty, in the sense that it lies in the eyes of the beholder. In other words, it is a characteristic of the decision maker, not the decision situation. If the decision maker is **risk neutral**, then there is no need to go beyond values. By definition, a risk-neutral decision maker should make decisions by maximizing the expected value. Such a decision maker would only be concerned with the means of random variables. To account for more general risk preferences, values need to be extended to **utilities**. Incorporating risk preferences requires assessing a **utility function**, denoted $u(.)$ (von Neumann and Morgenstern 1947). $u(.)$ is a function that takes units of value as input and returns units of utility (sometimes referred to as utiles).

The left side of Figure 3.5 presents examples of three utility functions that extend over the domain of values from v^0 to v^*. The utility functions are associated with different decision makers. The figure indicates that a risk-neutral decision maker has a linear utility function – i.e., of the form

$$u(v) = \alpha + \beta v, \tag{3.2}$$

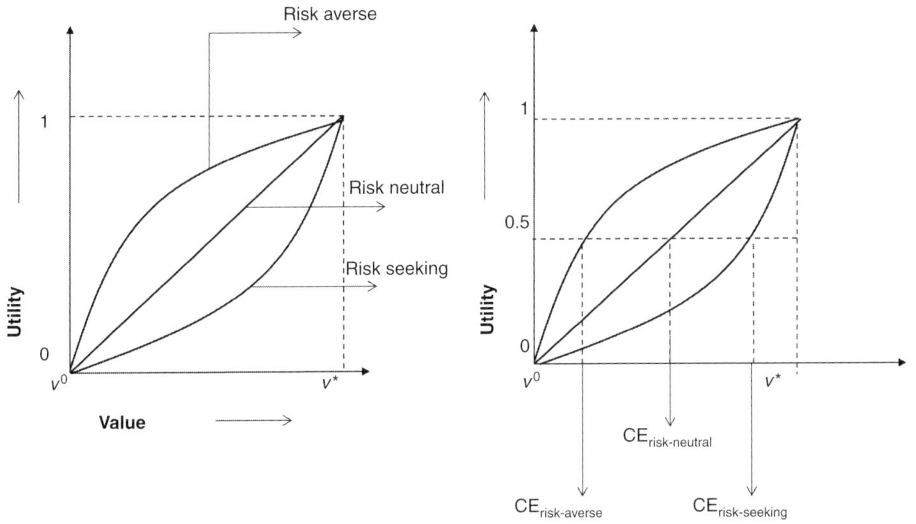

Figure 3.5 Utility function examples. Left: three utility functions: one each for a risk-neutral, risk-averse, and risk-seeking decision maker. Right: certain equivalents for the three utility functions for a lottery with an expected utility of 0.5.

where α and β are constants such that $\beta > 0$. **Risk-averse** and **risk-seeking** decision makers have concave and convex utility functions, respectively. For reasons that we will study later in this subsection, we will also pay particular attention to another family of utility functions, exponential utility functions, which are of the form

$$u(v) = \alpha + \beta e^{-\gamma v}, \tag{3.3}$$

where α and β are constants such that if $\gamma > 0$, then the decision maker is risk averse and $\beta < 0$, whereas if $\gamma < 0$, then the decision maker is risk seeking and $\beta > 0$. In the limiting condition where $\gamma = 0$, the decision maker is risk neutral and the linear utility function should be used. γ is the risk aversion coefficient – it is the only parameter of the exponential utility function, which is sometimes parameterized by the risk tolerance $1/\gamma$. The constants α and β in the utility function do not matter as long as they satisfy the required conditions, since utility functions are invariant to affine transformations (Keeney and Raiffa 1976). Therefore, whenever we deploy the exponential utility function for numerical examples, we will arbitrarily set $\alpha = 1$ and $\beta = -1$ for risk-averse decision makers, which results in the functional form $u(v) = 1 - e^{-\gamma v}$.

A utility function is a mathematical construct that is used merely to capture and incorporate the decision maker's risk preferences. It is therefore often inappropriate to use utilities for communicating the results of the decision analysis to decision makers who are more concerned about the practical implications of their actions. The **certain equivalent** is usually a more meaningful measure because it is in units of value. It is defined as the price at which the decision maker would be indifferent between selling their situation versus retaining it, and it is therefore also referred to as the personal indifference selling price for a situation. The certain equivalent is a measure of how much a situation is worth to the decision maker because they should not be willing to sell it for less than this price.

The notion of a certain equivalent applies to lotteries as well as decision situations. To introduce the concept, let us first apply it to lotteries in this subsection. Consider a person with wealth w who owns a lottery where the uncertainty is denoted x, and the value from the prospects is $v(x)$. The person's beliefs about uncertainty $x \in \Omega$ are represented by the probability distribution $p(x)$, and his or her risk preferences are characterized by the utility function $u(\cdot)$. The expected utility of the lottery is

$$E(u) = \sum_{x \in \Omega} u(v(x)+w)\,p(x) = E\big(u(v(x)+w)\big). \tag{3.4}$$

Computing the certain equivalent for this lottery is straightforward. If the certain equivalent is denoted *CE*, then the person is indifferent between selling the lottery, thereby receiving price *CE* for certain, as opposed to retaining the lottery with uncertain value prospects. The utilities for these two situations are identical by definition, because the person is indifferent, and therefore

$$\begin{aligned} &u(w+CE) = E(u) = E\big(u(v(x)+w)\big). \\ &\Rightarrow CE = u^{-1}\Big(E\big(u(v(x)+w)\big)\Big) - w. \end{aligned} \tag{3.5}$$

For a certain class of utility functions, the *CE* does not depend on the person's wealth, in which case assessments and computations are simpler because it is not necessary to incorporate the wealth. These utility functions are referred to as constant risk aversion utility functions because the risk aversion coefficient, defined as $\gamma = -\frac{u''(v)}{u'(v)}$, is constant and does not depend on the value. The linear and exponential utility functions are the only constant risk aversion utility functions. It can be shown that these utility functions (and only these) satisfy what is known as the **delta property**: if a constant value Δ were added to all the prospects in a situation with uncertain value, then the certain equivalent would increase by Δ. Formally, the utility function would have to satisfy the following property:

$$u^{-1}\Big(E\big(u(v(x)+\Delta)\big)\Big) = u^{-1}\Big(E\big(u(v(x))\big)\Big) + \Delta \tag{3.6}$$

for any uncertain value distribution $v(x)$ and any constant value Δ. Using this property in Equation (3.5), we observe that the certain equivalent in this lottery simplifies to

$$CE = u^{-1}\Big(E\big(u(v(x))\big)\Big). \tag{3.7}$$

Consider the right side of Figure 3.5, which demonstrates how to graphically estimate the certain equivalents for three utility functions for the following lottery:

$$v(x) = \begin{cases} v^* & \text{with probability } 0.5 \\ v^0 & \text{with probability } 0.5 \end{cases} \tag{3.8}$$

In this lottery, there is a 50–50% chance of receiving the best prospect v^* versus the worst prospect v^0. Since utilities can be scaled arbitrarily, we can normalize the utilities such that the best and worst prospects have utility 1 and 0, respectively – i.e., $u(v^*)=1$ and $u(v^0)=0$. The expected utility for this lottery is $E(u(v(x)))=0.5$. From Equation (3.7), the certain equivalent is computed by taking the utility inverse of the expected utility. This can be done graphically by drawing a horizontal line from the expected utility of the lottery – in this case, 0.5 – and then drawing a vertical line from the point where the horizontal line meets the utility function. Note that when the decision maker is risk neutral, he or she is indifferent between a lottery with uncertain prospects and having the expected value of the lottery for certain. In this case, the expected value of the lottery is $(v^*+v^0)/2$. The figure demonstrates that the certain equivalent for a risk-averse decision maker is less than the expected value of the lottery. On the other hand, the certain equivalent for a risk-seeking decision maker is more than the expected value of the lottery. Thus, a risk-seeking decision maker would prefer a lottery with uncertain prospects over receiving the expected value for certain, and vice versa for a risk-averse decision maker.

When utilities are scaled between 0 and 1, the utility of any intermediate prospect can be elicited by identifying an equivalent lottery where the best prospect is obtained with probability p versus the worst prospect with probability $1-p$. If the decision maker is indifferent between the intermediate prospect and the lottery, the utility of the prospect equals the expected utility of the lottery, which is p. When interpreted in this fashion, utilities are also known as **preference probabilities**.

There are a variety of well-known elicitation schemes to elicit a utility function. An elicitation scheme is composed of a sequence of questions. Standard gamble methods present the decision maker with a lottery and a certain amount and pose questions of mainly two types: a comparison (or choice) question is one where the decision maker is asked to compare the lottery with the certain amount, indicating which one is preferred. In an equivalence (or matching) question, the decision maker is asked to provide any one of the parameters pertaining to either the lottery or the certain amount such that he or she is indifferent between them. The utility curve is estimated from the decision maker's responses to the elicitation questions. Farquhar (1984) provides an excellent review of some elicitation schemes.

Discussion: multiattribute utility theory

We have described an approach to solving decision situations under uncertainty where a one-dimensional utility function is assessed over a deterministic value function. This approach makes no assumptions about utility independence (Keeney and Raiffa 1976) and is particularly suitable for our purposes since the presence of a value function enables VOI computations. The reader is referred to Matheson and Howard (1968) and Dyer and Sarin (1982) for further details about the approach. There are other approaches where a utility function is assessed directly. These techniques are often useful for multiattribute (or multicriteria) decision situations. Please see Keeney and Raiffa (1976) for details.

3.2.3 Maximizing expected utility

Consider the following simple decision situation, which is merely an abstraction of the pirate example. Suppose that a decision maker with wealth w faces a decision a involving unresponsive uncertainty x – i.e., uncertainty x is not affected by the decision. For instance, whether the pirate digs for treasure has no effect on whether the treasure is present or not. Furthermore, suppose that the decision maker will obtain value $v_a = v(x,a)$ from the decision situation if alternative a is chosen, where $v(.)$ is the value function. The decision maker's beliefs about uncertainty $x \in \Omega$ are represented by the probability distribution $p(x)$, and his or her risk preferences are characterized by the utility function $u(.)$. Again, note that v_a is a function of a random variable x; therefore, it is a random variable itself.

We now have all the necessary assumptions, terminology, and notation to determine the decision maker's optimal alternative. According to the principles of decision theory, he or she should choose the alternative that maximizes the expected utility. If the decision maker chooses alternative a, then he or she will obtain value $v_a + w$, the sum of their uncertain value from their decision situation, $v_a = v(x,a)$, and their prior wealth. The decision maker's uncertain utility is $u(v_a + w) = u(v(x,a) + w)$. According to the norms of decision theory, the **maximum expected utility** is

$$\begin{aligned} MEU &= \max_{a \in A} \left\{ \sum_{x \in \Omega} u(v_a + w) p(x) \right\} \\ &= \max_{a \in A} \left\{ E\left(u(v_a + w)\right) \right\} = E\left(u(v_{a^*} + w)\right), \end{aligned} \tag{3.9}$$

where a^* is the optimal alternative – i.e., the one that maximizes the expected utility

$$a^* = \arg\max_{a \in A} \left\{ E\left(u(v_a + w)\right) \right\}. \tag{3.10}$$

Calculations for the certain equivalent for our simple decision situation are similar to those for the lottery in the previous subsection. If the certain equivalent is denoted CE, then the decision maker is indifferent between selling the situation, thereby receiving price CE for certain, as opposed to retaining the decision situation with uncertain value prospects. The utilities for these two situations are identical because the decision maker is indifferent, and therefore

$$\begin{aligned} &u(w + CE) = MEU = E\left(u(v_{a^*} + w)\right) \\ &\Rightarrow CE = u^{-1}\left(E\left(u(v_{a^*} + w)\right)\right) - w. \end{aligned} \tag{3.11}$$

When the decision maker's utility function satisfies the delta property, we can use Equation (3.6) and simplify Equation (3.11) to

$$CE = u^{-1}\left(E\left(u(v_{a^*})\right)\right). \tag{3.12}$$

When the decision maker is risk neutral, $u(v) = \alpha + \beta v$, $\beta > 0$, the equations simplify further such that the certain equivalent equals the maximum expected value:

$$CE = E\left(v_{a^*}\right). \tag{3.13}$$

Later, we will discuss how utility functions that satisfy the delta property make VOI computations simpler and more tractable. In fact, the delta assumption is routinely made in the VOI literature as well as in practice. This is not as limiting as it may appear because: (i) the assumption is often reasonable in practical situations, particularly those where the certain equivalent is much less than the decision maker's wealth and (ii) even if this is not the case, the exponential utility function is versatile and can effectively approximate many utility functions encountered in practice (Kirkwood 2004).

Discussion: the five rules of actional thought

The reader may well ask: why should I maximize expected utility? Edwards et al. (2007) respond that "systematic and repeated violations of these principles will result in inferior long-term consequences of actions and a diminished quality of life" (p. 1). Decision analysis is a normative subject and there are, of course, many ways to axiomatize a normative subject. Howard and Abbas (2015) present the axioms of decision theory as the "five rules of actional thought." Any decision maker who agrees to these (rather reasonable) rules follows the norms and should make their decisions by maximizing expected utility:

1. **Probability rule:** The decision maker's degree of belief about the possibilities for uncertainties is represented using probabilities.
2. **Order rule:** The decision maker should be able to declare their preference order for all prospects.
3. **Equivalence rule:** For any prospect, the decision maker should be able to construct an equivalent lottery involving the best and worst prospects.
4. **Substitution rule:** The decision maker should be willing to substitute any prospect with the lottery, as constructed above.
5. **Choice rule:** Given two lotteries with uncertainty around the best and worst prospects, the decision maker should choose the lottery with the higher probability of obtaining the best prospect.

3.2.4 Examples

Treasure island: the pirate example

Keywords: *pirate example, binary outcomes, certain equivalent, risk-neutral decision maker*

We return to the pirate example that was introduced in Chapter 2 and is referred to throughout this section. As a reminder, a pirate must decide whether to make the trip to an island and dig for potential treasure or not. The pirate is uncertain about whether the treasure is present or not. This distinction of interest is modeled as a binary random variable $x \in \{0,1\}$. If the treasure is indeed present and it is dug up, the pirate expects to make a profit of \$100 000, but if it is not present, he will incur a loss of \$10 000. The pirate believes there is a 1% chance that the treasure is present.

Suppose that the pirate is risk neutral – i.e., has a linear utility function for monetary values. Since utility functions are invariant to affine transformations, we will use $u(v) = v$. Furthermore, since the linear utility function satisfies the delta property, we do not need to incorporate the pirate's wealth into the calculations.

Let us apply the notation from this subsection. Decision a is the digging decision, and uncertainty x is whether the treasure is present or not. There are two alternatives, $A = \{\text{dig, do not dig}\}$. Similarly, there are two states $\Omega = \{0,1\}$ that denote the absence and presence of the treasure, respectively. The distribution for x and the value function are as follows:

$$\begin{aligned} &x = \begin{cases} 1 & \text{with probability } 0.01 \\ 0 & \text{with probability } 0.99 \end{cases} \\ &v_{\text{dig}} = \begin{cases} \$100\,000 & \text{with probability } 0.01 \\ -\$10\,000 & \text{with probability } 0.99 \end{cases} \qquad v_{\text{do not dig}} = \$0. \end{aligned} \tag{3.14}$$

Using Equation (3.12), the certain equivalent is

$$\begin{aligned} &CE = u^{-1}\left(\max\left\{E\left(u\left(v_{\text{dig}}\right)\right), E\left(u\left(v_{\text{do not dig}}\right)\right)\right\}\right) \\ &\Rightarrow CE = u^{-1}\left(\max\left\{E\left(v_{\text{dig}}\right), E\left(v_{\text{do not dig}}\right)\right\}\right) \\ &\Rightarrow CE = u^{-1}\left(\max\left\{(100\,000 \cdot 0.01) + (-10\,000 \cdot 0.99), 0\right\}\right) \\ &\Rightarrow CE = u^{-1}(0) = \$0 \end{aligned} \tag{3.15}$$

We could also have used Equation (3.13) directly since the pirate is risk neutral. In this case, it is optimal for the pirate not to dig for the treasure, and the pirate's certain equivalent for this decision situation is \$0.

For whom the bell tolls: Gaussian projects example

Keywords: *Gaussian projects example, bivariate Gaussian distribution, certain equivalent, risk-averse decision maker*

Suppose that the distinction of interest $\boldsymbol{x} = (x_1, \ldots, x_n)$ represents uncertain profits from projects that a decision maker is considering for investment purposes. We consider two projects ($n = 2$) with random profits (Section 2.3). A bivariate Gaussian distribution $p(\boldsymbol{x}) = N(\boldsymbol{\mu}, \Sigma)$ is used for the profits. We specify the prior mean and covariance as follows:

$$\boldsymbol{\mu}=\begin{pmatrix}100\\70\end{pmatrix},\qquad \boldsymbol{\Sigma}=\begin{bmatrix}30^2 & 300\rho\\300\rho & 10^2\end{bmatrix}\text{ in units of \$, millions.}\tag{3.16}$$

The decision maker can choose exactly one of the projects. Note that in this problem, the two projects are correlated, but the correlation ρ does not matter – only the marginal distributions of the projects matter. Since $\boldsymbol{x}=(x_1,x_2)$ is bivariate Gaussian, the marginal distributions are Gaussian with $p(x_1)=N(100,30^2)$ and $p(x_2)=N(70,10^2)$ in units of \$, millions. Furthermore, assume that the decision maker has an exponential utility function with risk aversion coefficient $\gamma = 0.1$ \$, millions^{-1}.

Using Equation (3.12) with utility function $u(x)=1-e^{-\gamma x}$, the certain equivalent is

$$CE = u^{-1}\left(\max\left\{\int_{x_1}\left(1-e^{-\gamma x_1}\right)p(x_1)dx_1,\int_{x_2}\left(1-e^{-\gamma x_2}\right)p(x_2)dx_2\right\}\right).\tag{3.17}$$

We can use the following result from Howard (1971), showing that when x follows a Gaussian distribution and the decision maker has an exponential utility function, the certain equivalent for x can be computed using a simple closed-form solution (Appendix A.1):

$$CE(x)=\mu-\frac{1}{2}\gamma\sigma^2,\tag{3.18}$$

where μ and σ are the mean and standard deviation of x, respectively, and γ is the risk aversion coefficient.

Using the result in Equation (3.18),

$$\begin{aligned}CE &= \max\left\{\mu_1-\frac{1}{2}\gamma\sigma_1^{\,2},\mu_2-\frac{1}{2}\gamma\sigma_2^{\,2}\right\}=\max\left\{100-\frac{30^2}{2\cdot 10},70-\frac{10^2}{2\cdot 10}\right\}\\ &= \max\{55,65\}=\$65\text{ million.}\end{aligned}\tag{3.19}$$

It is therefore optimal for the decision maker to choose the second project, because its certain equivalent is greater than that of the first (\$65 million versus \$55 million). Due to the decision maker's risk aversion and the greater uncertainty in profits for the first project as measured by the variance, the decision maker should choose the second project even though the first project has a higher mean. A risk-neutral decision maker would of course have chosen the first project, and their certain equivalent would have been \$100 million.

3.3 Graphical models

Graphical models that represent decision situations help structure the various elements. The modular approach helps in studying the essential distinctions without getting bogged down with unnecessary and less relevant details. Graphical models are extremely useful for communication between the decision maker and the analyst, since they help identify and discover relationships among the various key distinctions. In Section 2.3, we studied graphical models for uncertainties represented via Bayesian networks. In this section, we briefly review some important graphical models that also explicitly include decisions and values of prospects.

In decision analysis, the two most popular graphical models for representing and evaluating decision situations are **decision trees** and **influence diagrams**. Decision trees are simple graphical models that are typically taught in any first course in decision analysis. In the early years of decision analysis, decision trees were so popular that they were almost synonymous with the field of decision analysis. Influence diagrams have become popular since their inception in the 1970s and 1980s, as they are excellent tools for both communication and computation. In this section, we briefly review these two graphical models.

3.3.1 Decision trees

Decision trees are graphical models that represent the conditional and informational precedence relationships between distinctions. Square (or rectangular) nodes represent decisions, and the emanating arcs (as viewed from left to right) represent the alternatives along branches. Circular (or oval) nodes represent uncertainties, and the emanating arcs are associated with the corresponding probabilities for the different possibilities. Decision trees are indeed graphical models that are "trees," in the sense that children do not have more than one parent. The leaves of the decision tree represent the scenarios of the decision situation; each scenario is associated with a value for each prospect. When the decision maker has risk preferences – i.e., is not risk neutral – each scenario is also associated with utilities that are computed from the values using the decision maker's utility function.

The solution procedure is termed "rolling back" a decision tree, which is the standard backward induction procedure in dynamic programming problems. At the circular nodes, a sum–product operation is performed where probabilities are multiplied with the respective expected utilities (or values, if risk neutral), and then they are added. At the rectangular nodes, the expected utility is maximized, and the maximizing alternative is recorded as the optimal choice for that particular situation.

Figure 3.6 presents the decision tree for the pirate example. The decision node represents the pirate's digging decision, and the only uncertainty is whether the treasure is present or not. The values for each scenario are shown at the leaves of the decision tree. Since the pirate is risk neutral, the optimal alternative is computed by maximizing the expected value. The figure graphically depicts the very computations that were performed in the Examples subsection of the previous section. It is optimal for the pirate not to dig since the expected value of digging is negative.

Decision trees are extremely intuitive and easy to understand since they are structured in a way in which decision makers naturally view their decision situation. A key advantage is that they are able to capture the asymmetries that are typical in real-world decision problems. For instance, there is an asymmetry in the pirate example – when the pirate decides

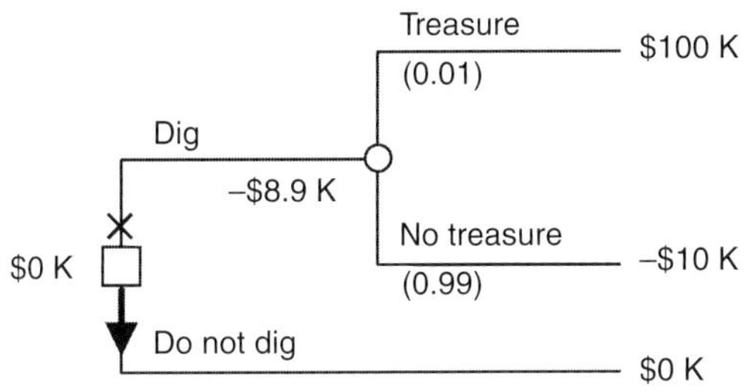

Figure 3.6 The decision tree for the pirate example. There are two alternatives: dig or do not dig. The uncertain outcome is the presence or absence of the treasure.

not to dig, the uncertainty around the presence of treasure does not affect his value. Thus, the decision tree in Figure 3.6 does not need to show all four scenarios and only shows three instead; the uncertainty around the presence of the treasure has been "collapsed" in the situation in which the pirate decides not to dig.

An important disadvantage is that decision trees become intractable for decision situations involving a large number of distinctions, as they grow exponentially with the number of distinctions. It is practically impossible to analyze decision situations with several decisions and a large number of uncertainties using trees, as the large number of leaves makes problem representation difficult. Complex situations demand the use of more sophisticated decision analysis tools. They may be more amenable for representation and solution through other graphical representations, such as influence diagrams, which we discuss in the next subsection.

Discussion: probability trees

Decision trees that represent situations with only uncertainties and no decisions are known as probability trees. They can be useful in a variety of ways. For instance, it is often easy and convenient to use them to perform probabilistic computations such as inference. For an example of Bayes' rule being performed using probability trees, the reader may look ahead to Figure 3.16. As another application of a probability tree, see Figure 3.12, which depicts a lottery for a company that owns an oil field. Here, the probability tree also depicts the values of the prospects at the leaves, just like in a decision tree.

3.3.2 *Influence diagrams*

An influence diagram captures a decision situation by depicting the relationships between decisions, uncertainties, and preferences (Howard and Matheson 1984). They capture the conditional independencies in the model and are able to represent the model as conveniently assessed by the decision maker. Influence diagrams have been particularly successful in medical expert systems as an aid in encoding expert knowledge, and they have also been applied successfully in several other domains. Influence diagrams can capture complex modeling situations due to their succinct graphical form.

Recall from Section 2.3 that in a Bayesian network, the nodes represent uncertainties, and the edges capture the conditional relationships between nodes. Every node is associated with a probability for each possible state of the node conditioned on each possible state of its parents. For discrete uncertainties, this is typically represented in the form of a conditional probability table. An influence diagram is a directed acyclic graph that can be viewed as an extension or a generalization of a Bayesian network. It has a larger variety of nodes as well as arcs, because it is intended to represent a decision maker's entire decision situation and not merely the uncertainties.

Influence diagrams have three kinds of nodes. Along with **uncertain nodes**, they also include **decision nodes** and **value nodes**. Uncertain nodes are drawn as ovals, just like in Bayesian networks, whereas decision nodes are rectangles. Different authors use different symbols for value nodes; examples of symbols include diamonds, hexagons, and octagons. We will use diamonds for value nodes in this book. A special case of an uncertain node is a deterministic node, typically drawn as a concentric oval. A deterministic node is completely determined when its parents are known. Thus, one can think of a deterministic node as an uncertain node where the conditional probability distribution for that node only involves 1s and 0s. When influence diagrams were initially conceived, they were meant to contain, at most, one value node. Influence diagrams with multiple additive or multiplicative value nodes were introduced by Tatman and Shachter (1990), and they can provide additional computational savings. In this book, we will focus on influence diagrams with a single value node.

The semantics of the edges or arcs are more involved in influence diagrams, and they take on different meanings depending on the types of nodes toward which they are directed. **Conditional arcs** are those arcs that lead into uncertain nodes, and they have the same meaning as in Bayesian networks. **Informational arcs** are those arcs directed toward decision nodes. If any node has an arc leading into a decision node, that variable is observed at the time the decision is made. Thus, the parents of a decision variable are those variables that will be observed before the decision must be made. Informational arcs are added only when these nodes are observed before the decision. **Functional arcs** are those directed toward a value node. Some authors do not distinguish between conditional and functional arcs, viewing a functional arc as a special case of a conditional arc.

Influence diagrams are not merely pictorial or graphical depictions of a decision situation. Their power lies in the fact that they have both qualitative and quantitative capabilities. The model also involves numbers that can be used for computations. Every uncertain node is associated with a conditional probability distribution, like in Bayesian networks. Every value node is associated with a value table for the value function, indicating how much the decision maker values every prospect. This might often be related to a monetary

Conditional probability table

Oil amount	Prob.	
	high	low
	0.35	0.65

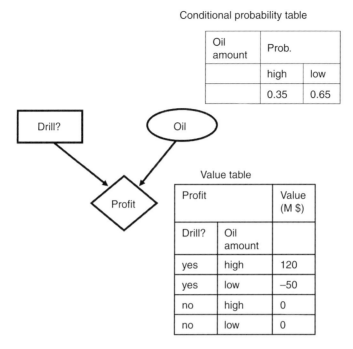

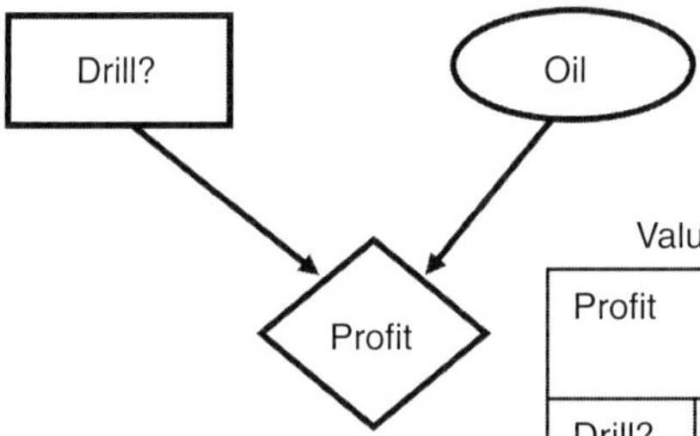

Value table

Profit		Value (M $)
Drill?	Oil amount	
yes	high	120
yes	low	–50
no	high	0
no	low	0

Figure 3.7 The influence diagram for the oil wildcatter example. There are two alternatives (drill or not) and an uncertain amount of oil.

value of the prospect, though values might also be defined otherwise. In this book, we will use monetary units of value, as our primary focus is on VOI computations.

Figure 3.7 presents an influence diagram for a classic decision situation known as the oil wildcatter example (Raiffa 1968). It is a simple influence diagram, with one node of each kind, representing the decision situation for someone who owns an oil field. In this simple version of the example, the only decision is whether to drill a well or not. The main uncertainty is the amount of oil in the field, modeled as a binary variable with the states high and low. The decision maker's value is the profit from the field, which is a function of the amount of oil and the drilling decision. Often in an influence diagram, the missing arcs reveal more than the arcs that are present. Note that there is no arc between the drilling decision "Drill?" and the uncertainty "Oil." There is no arc from the decision to the uncertainty because the decision does not affect how much oil there is underground. There is no arc in the reverse direction, from the uncertainty to the decision, because the decision maker must decide without observing the amount of oil. The figure also depicts the numbers behind the influence diagram, along with the diagram itself. The uncertain node is associated with a conditional probability table, indicating that there is a 35% chance that there is a high amount of oil. The value node is associated with a value table, indicating that drilling a well in a field with a high amount of oil returns a profit of $120 million, whereas drilling a well in a field with a low amount of oil results in a loss of $50 million.

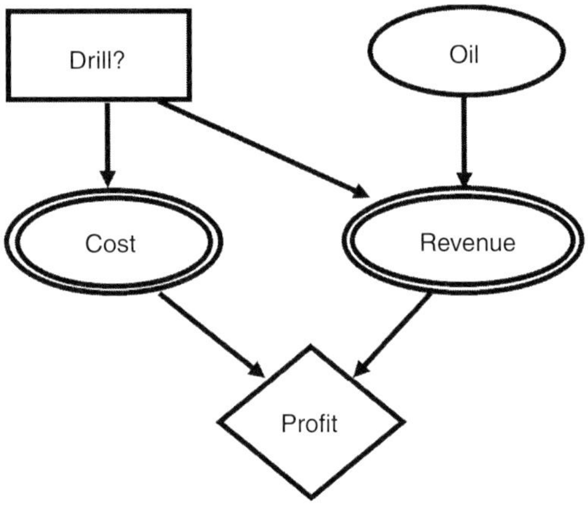

Figure 3.8 The influence diagram for the oil wildcatter example, including deterministic nodes for cost and revenue.

An influence diagram can be drawn in several ways. Consider Figure 3.8, which includes two additional (deterministic) nodes that provide more detail to the model from Figure 3.7. These nodes are depicted as concentric ovals. This influence diagram explicitly indicates that the profit is a function of the cost of the well and revenue from the oil field. Revenue is assumed to be a deterministic function of the amount of oil and the drilling decision. The uncertainty about the amount of oil propagates to the revenue, and there is uncertainty around the revenue. The cost of the well is fixed and known; therefore, cost is a deterministic function of the drilling decision. Note that an influence diagram with only deterministic nodes and no uncertain node is equivalent to a standard spreadsheet model.

As an illustration of a slightly more complex influence diagram, consider the diagram in Figure 3.9. This is an extension of the oil wildcatter example and includes seismic testing. This influence diagram has five nodes in total: two decision nodes, two uncertain nodes, and one value node. The decision maker decides whether to perform a seismic test. If he or she decides to test, a seismic report is observed before the drilling decision is made. The seismic report is assumed to only convey whether the test is positive or negative. From the conditional probability table for the report, we see that the seismic report is assumed to be stuck at negative if the test is not performed – i.e., it is uninformative. The presence of the arcs from the seismic test decision node and the report uncertainty node to the drilling decision signify that they are observed before the decision maker chooses whether to drill the well or not. The profit is a function of the cost of the test (if it is performed), the cost of drilling (if the well is drilled), and the amount of oil in the field. The reader can compare the numbers from Figures 3.7 and 3.9 and infer that the cost of the seismic test is assumed to be \$20 million. Although this influence diagram is more complicated than the previous examples, it is still highly simplified. Later in the book we present more realistic and finer-grained models for decisions involving petroleum exploration and production and seismic testing, incorporating the spatial aspects of the uncertain distribution of oil, the

Conditional probability table

Report (seismic)		Prob.	
Test?	Oil amount	positive	negative
yes	high	0.75	0.25
yes	low	0.1	0.9
no	high	0	1
no	low	0	1

Conditional probability table

Oil amount	Prob.	
	high	low
	0.35	0.65

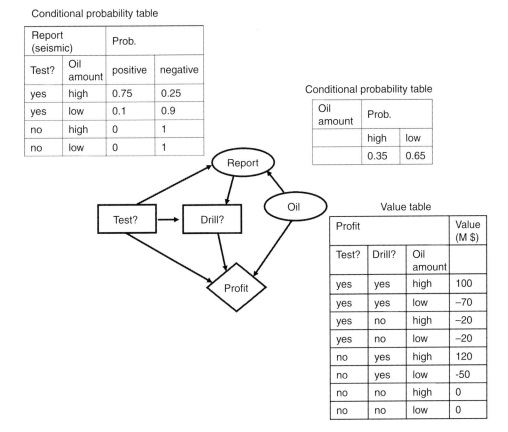

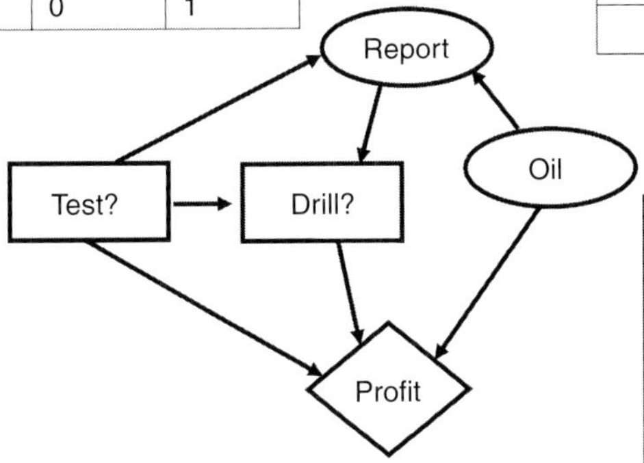

Value table

Profit			Value (M $)
Test?	Drill?	Oil amount	
yes	yes	high	100
yes	yes	low	–70
yes	no	high	–20
yes	no	low	–20
no	yes	high	120
no	yes	low	-50
no	no	high	0
no	no	low	0

Figure 3.9 The influence diagram for the oil wildcatter example with seismic testing.

spatially distributed information from seismic tests, and the choice of different locations to drill the well.

There are certain restrictions and conditions on formulating valid influence diagrams. An inconsistency arises if information known before one decision is made is not known before subsequent decisions. Therefore, to be a valid influence diagram for a rational individual, there must exist at least one "no forgetting" ordering of the decisions such that all earlier observations and decisions are observed before any later decisions are made. No forgetting is a standard assumption for influence diagrams, and often the no forgetting arcs are not explicitly represented in the diagram but are inferred implicitly. Also – of particular relevance to VOI computations – it is inconsistent to make observations that tell the decision maker anything about the decisions yet to be made, so it is required that uncertainties on which observations are made not be responsive in any way to the decisions (Heckerman and Shachter 1995). This is enforced by not allowing the nodes for such uncertainties to be descendants of the decision node. We refer to such uncertainties as unresponsive uncertainties.

Pitfalls: use the correct semantics for arcs in influence diagrams

A common mistake while formulating influence diagrams is to add arrows because we feel that the decision "depends" on certain variables. Influence diagrams have arcs with specific semantics, as we have described earlier in this section, and care must be taken to formulate the diagram correctly. Moreover, since people are usually familiar with other graphical models such as flowcharts, there is a tendency to use the arcs too casually. We have observed that although influence diagrams are intuitive diagrams for representing decision situations, there is a little bit of "unlearning" that may be required to use them correctly. Formulating influence diagrams can be an art. The decision maker or analyst will need to learn the specific semantics through examples and, once the concepts have been mastered, should be able to build influence diagrams at will.

Influence diagrams are not merely tools for representation and communication – they are also effective computational tools that evaluate and analyze decision situations. The general problem that an influence diagram solves is to determine the decisions that maximize the expected utility when the parents of each decision are observed before it is made. In other words, the goal is to find the optimal strategy. Let us first define a strategy before we discuss how the optimal one is determined.

A **strategy** is an assignment of an alternative to every scenario of observations that are made at the time of a decision for every decision in the decision situation. In other words, a strategy provides the decision maker with a plan that tells him or her what alternative to choose at any possible stage in the decision situation. For instance, consider the decision situation represented by the influence diagram in Figure 3.9. An example of a strategy is as follows: the decision maker should perform the seismic test and drill if the report is positive and refrain from drilling if the report is negative. An alternate strategy would be: the decision maker should not test and should drill the well. Yet another strategy would be: the decision maker should test and drill the well regardless of what the report declares. Intuitively, this last strategy seems far from optimal – why should the decision maker test, thereby presumably spending a non-trivial amount of money, and disregard the information from the report?

An influence diagram solution algorithm determines the optimal strategy based on the norms of decision theory. The optimal strategy is one that maximizes the expected utility, and the expected utility of the optimal strategy is known as the maximum expected utility. The certain equivalent of the decision situation can be obtained from the maximum expected utility, as described earlier. Therefore, the solution of an influence diagram reveals both the optimal strategy and the certain equivalent of the decision situation, informing the decision maker about what to do in the decision situation and what the decision situation is worth to him or her. Note that the maximum expected utility of the decision situation is an internal result that any influence diagram solution algorithm will compute in order to find the optimal strategy. Influence diagrams are also useful for performing efficient appraisal of decision situations, including sensitivity analysis and VOI analysis.

When influence diagrams were initially conceived, the solution method involved transforming them into decision trees (Howard and Matheson 1984). In the 1980s, several techniques were introduced to directly evaluate the influence diagram through a sequence of variable elimination steps without the need for converting the diagram into a decision tree network (Shachter 1986). The popular algorithms in the literature converted influence diagrams to a simpler form through a sequence of operations called value-preserving reductions. These operations maintain the feasibility of the problem, retaining the maximum expected utility at every iteration. The diagram is reduced until all that remains is a single value node so that the decision situation can be valued. The solution process takes expectation over the uncertainties and maximizes over the decisions. The algorithm terminates when all nodes except the value node have been absorbed into the graph. The optimal strategy and the maximum expected utility are computed in the process.

There have been several subsequent algorithms proposed to solve influence diagrams. Since the 1990s, various efforts have focused on converting an influence diagram into related graphical structures, such as valuation networks (Shenoy 1992), junction trees (Shachter and Peot 1992; Jensen et al. 1994; Shachter 1999), or decision circuits (Bhattacharjya and Shachter 2007) for efficient solution and analysis.

3.3.3 Examples

For whom the bell tolls: Gaussian projects example

Keywords: *Gaussian projects example, bivariate Gaussian distribution, decision tree, graphical model, risk-averse decision maker*

Decision trees are typically drawn for decision situations with distinctions of interest that have discrete sample space. For instance, Figure 3.6 presents the decision tree for the pirate example where there are two alternatives (dig or not) and where the only uncertainty has two states (treasure present or not). In this example, we remind the reader of the Gaussian projects example from Section 3.2, which includes a continuous uncertainty. Here, the decision maker can choose at most one out of two projects. The marginal distributions for the project profits are Gaussian with $p(x_1) = N(100, 30^2)$ and $p(x_2) = N(70, 10^2)$ in units of \$, millions. The decision maker has an exponential utility function with the risk aversion coefficient $\gamma = 0.1$ M\$$^{-1}$.

Figure 3.10 shows the same computations that were performed earlier, but this time they are depicted by a decision tree. The chance nodes in the tree explicitly display the continuous nature of the profits. The decision node makes a choice between the two projects and indicates that it is optimal for the decision maker to choose the second project. The certain equivalent of the second project and therefore the decision situation is \$65 million.

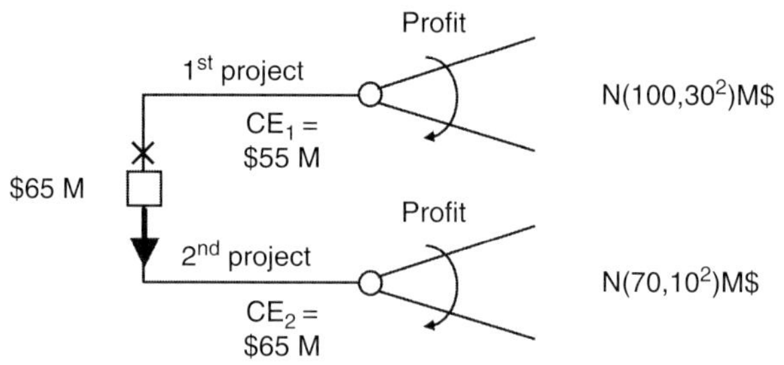

Figure 3.10 The decision tree for the Gaussian projects example. The decision maker can choose at most one of two projects with uncertain Gaussian distributed profits.

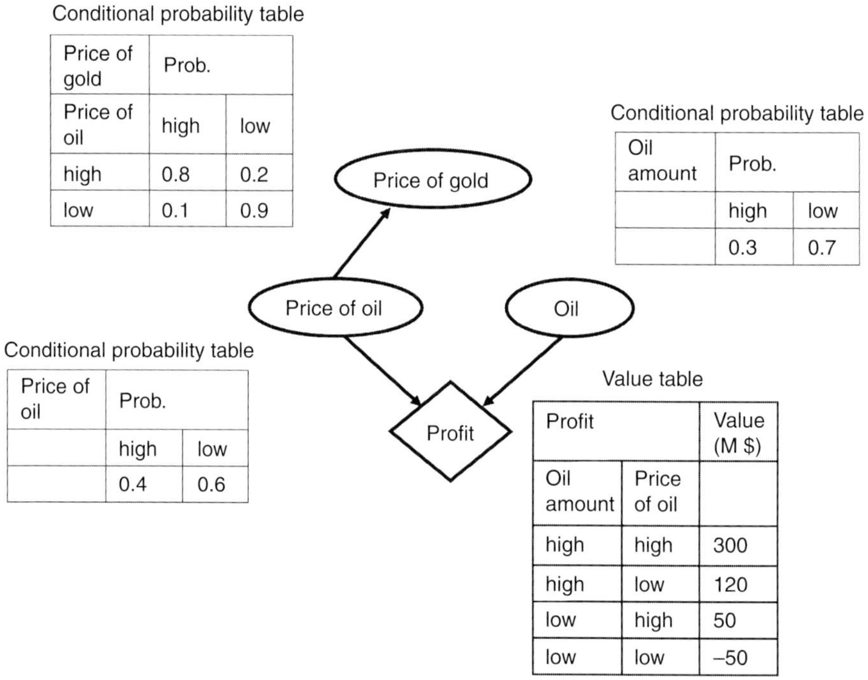

Figure 3.11 The influence diagram for the oil and gold example. Oil amount, price of oil, and price of gold are uncertain binary variables.

MacKenna's gold: oil and gold example

Keywords: *oil and gold example, graphical model, lottery, influence diagram, probability tree, risk-averse decision maker*

Let us illustrate an example of a lottery – i.e., situations that do not incorporate decisions but do include values for prospects. Consider a company that owns a producing oil field. The company is assumed to have an exponential utility function with the risk aversion coefficient $\gamma = 0.002$ M^{-1}. For the moment, all decisions pertaining to field development

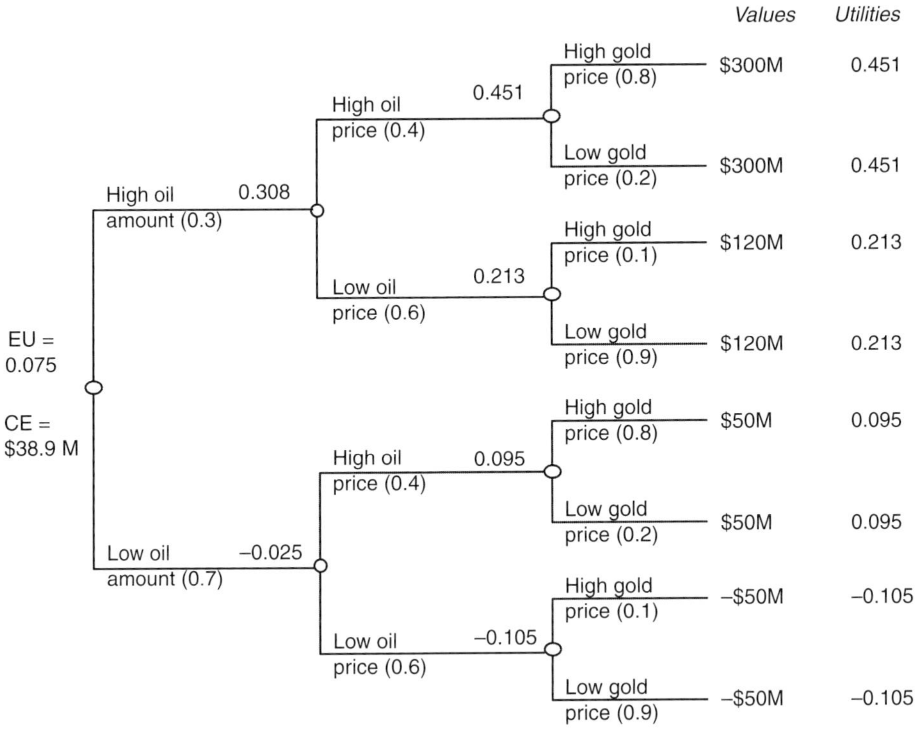

Figure 3.12 The probability tree for the oil and gold example. There are no decisions in this tree.

have been made. The company may still wish to contemplate the uncertainties involved and their implications on its profits. Figure 3.11 shows an influence diagram for the company's lottery from Bhattacharjya and Shachter (2010).

The three uncertainties are oil amount, price of oil, and price of gold – all modeled as binary random variables. The conditional probabilities for the three uncertainties are shown in the figure using conditional probability tables. What makes the diagram different from a (typical) Bayesian network is the presence of a value node, which models the company's profits as a function of the amount of oil and the price of oil. These numbers are shown in the value table associated with the node for profits, conditional on its parents.

Figure 3.12 shows the probability tree corresponding to the influence diagram in Figure 3.11. This is a probability tree – i.e., a special case of a decision tree – because there are no decision nodes in the tree. The conditional independence in this situation should be clear from looking at the appropriate branches of the tree. For instance, the amount of oil does not affect the conditional probability numbers for the price of gold in the probability trees. The price of oil and the price of gold are clearly correlated though.

The company is risk averse, so every prospect is associated with utilities that are computed from the values using the transformation $u(x) = 1 - e^{-\gamma x}$. The probability tree is used to compute the expected utility through the standard rollback procedure. Note that the notion of *maximum* expected utility is no longer pertinent, as there is no decision node that performs maximization. The certain equivalent of the situation can be computed by applying the utility inverse function to the expected utility. It is calculated to be \$38.9 million.

In the influence diagram in Figure 3.11, the price of gold clearly does not affect the company's profits, as seen from the lack of an arc into the value node. What, then, is the point of including it in the diagram? Even though this node does not currently contribute to the company's value, the company may want to include it in the model since it is relevant to the price of oil. Bhattacharjya and Shachter (2010) show how the company may wish to hedge its risk by considering lotteries that depend on the price of gold, such as through financial instruments like futures. The company can use the correlation to improve its value by being in a position to benefit from the price of gold even if the oil price becomes low and affects its revenues from the oil field.

Time after time: time-lapse seismic example

Keywords: *time-lapse example, seismic data, graphical model, influence diagram*

In the exploration and production of hydrocarbon reservoirs, decisions are complex due to several uncertainties: uncertainty in the geological properties, seismic imaging, repeatability, reservoir structure, rock and fluid properties, etc. Furthermore, several decisions need to be made over the entire lifecycle, and often it is not clear how current decisions might affect the future bottom line. In the exploration and characterization stage, more information is acquired about the reservoir structure; the reservoir rock and fluid properties; and the spatial distributions of lithology, porosity, and saturations, as well as other factors beyond the scope of this book such as oil prices and even the political scenario. Later, decisions have to be made about the technical and economic feasibility of seismically monitoring the reservoir during production.

Although several case studies show that four-dimensional (4-D) seismic tests are extremely useful – e.g., Lumley (2001) – in practice it may often be unclear whether they will actually add value to the production process for a particular reservoir. In their paper about assessing the technical risk of a 4-D project, Lumley et al. (1997) propose a scorecard method to score a particular reservoir based on reservoir and seismic properties. The scorecard provides a quantitative framework that aids in answering the question – should one perform a 4-D seismic survey? The following example, described in Bhattacharjya and Mukerji (2006), builds on the paper by Lumley et al. (1997) to model the 4-D monitoring decision problem using influence diagrams. We delve into this example in much greater detail than other examples in this chapter because it is significantly closer to a real practical application of the concepts. The reader should be aware, however, that it is still a toy problem in comparison to applications in forthcoming chapters.

Value from 4-D seismic monitoring

Economic value may be derived from 4-D seismic surveys in several ways. Time-lapse seismic images, when properly interpreted using rock physics relations, can identify bypassed oil and thus add significant reserves, helping to rejuvenate and prolong the life of the field. They may also spot potentially problematic changes like early breakthroughs and save millions of dollars by ensuring timely corrective action and adjustment of depletion plans. Other benefits include identifying inefficient sweep, optimizing injection programs and design of production strategies by monitoring injected fluid fronts, mapping reservoir compartmentalization, and identifying faults (Lumley 2001). Time-lapse seismic data are used as a tool for better management of the life of the field. The ultimate value comes from being able to detect some behavior that deviates from what was expected (Blangy et al. 2014).

The value of a 4-D survey depends on the reliability, repeatability, and overall quality of the survey, as well as on the reservoir conditions such as rock and fluid properties and reservoir temperature and pressure. If the tests are unreliable, then it may not be valuable to perform them because there is no confidence in the interpretation. However, even if the tests are reliable but the reservoir is fairly homogenous or if it is clear that it is soon to become uneconomical, then again it may not be worthwhile to spend money in trying to monitor the fluid flow.

Lumley et al. (1997) use a scorecard method to gauge the value from a 4-D seismic survey. The scorecard is divided into two basic categories: reservoir and seismic. The reservoir scorecard shows the estimated score for reservoir-related properties like dry rock bulk modulus, porosity, and fluid saturation change. A particular reservoir with a reservoir property that brings about a favorable seismic response gets a high score for that property. For instance, a typical reservoir in Indonesia with unconsolidated rocks and low bulk modulus gets a high score. This is favorable, as unconsolidated rocks with low bulk modulus are seismically more sensitive to fluid changes. The seismic scorecard indicates the score for the estimated seismic response and includes factors like image quality, resolution, and repeatability of the seismic tests. The value of the 4-D seismic data is the final score, which is the sum of the scores for all the relevant properties. Tools like scorecards and decision trees are popular and useful, but sometimes complex situations demand the use of more sophisticated decision analysis tools. The scorecard can be thought of as a special case of an influence diagram model with only deterministic nodes. Understanding the interrelationship among the reservoir factors, the seismic factors, and the associated uncertainties is a crucial element in seismic testing that may not be well captured in a scorecard. An influence diagram uses these relationships and relates the observed phenomena with the uncertain reservoir conditions to estimate the seismic response.

Influence diagrams for 4-D seismic monitoring

In our analysis for this example, we assume that an initial three-dimensional (3-D) seismic characterization has been conducted before studying the feasibility of a 4-D seismic survey. We also assume that the future prospects for the field are promising. Other information available includes well log and core data. There are certain properties that are observed from the available information. For instance, we may roughly know the average depth of the reservoir and the reservoir thickness. We may also know whether the acquisition conditions will be favorable or not before we actually perform the 4-D seismic test. These

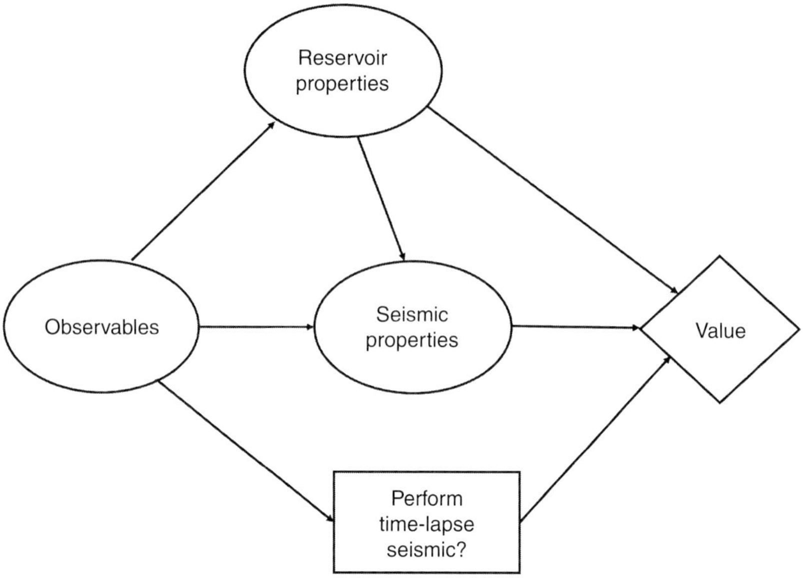

Figure 3.13 The influence diagram for the time-lapse example (collapsed nodes).

properties will be labeled as the "observable properties." Interrelationships among the observable, reservoir, and seismic properties are very important.

Figure 3.13 gives a compact graphical representation of the relationship between all the properties. In the following text we expand out all the groups to show the nodes within them. The observable properties are known before the decision is made, and they are related to the reservoir and seismic properties. The value derived from 4-D seismic testing is through some of the reservoir properties, the seismic properties, and the decision that is made. The complete influence diagram model (represented using the software Netica, as well as in MATLAB) is available on the book website and can be opened up to look at the details of the various conditional probability tables. We also use this influence diagram model for some exercises in Chapter 7.

Figure 3.14 expands the nodes for the observable, reservoir, and seismic properties. We describe each category of nodes next.

Observable property nodes

We have assumed that the shown observable properties are known from initial careful 3-D seismic characterization. Some of these properties are related to the reservoir characteristics, like the fluid and rock properties from well data. The node for spatial heterogeneity refers to our belief regarding the heterogeneity of the reservoir and has two discrete states: high and low. Here, what is taken to be observable is not the actual spatial distribution of reservoir heterogeneity but whether the spatial heterogeneity is high or low. This may be estimated from geological analysis of the depositional environment. This analysis is at a coarse level of granularity where we consider overall geological properties (high

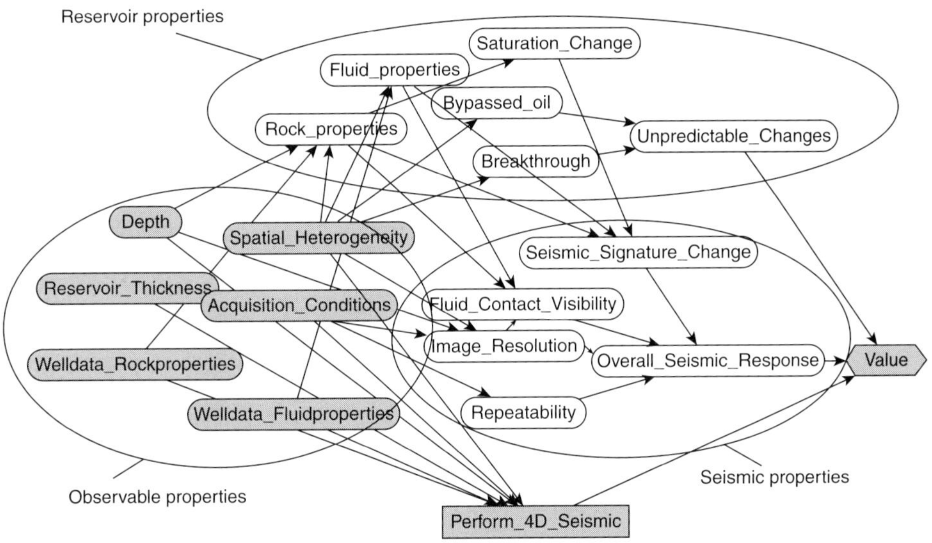

Figure 3.14 The influence diagram for the time-lapse example (expanded nodes).

or low spatial heterogeneity) without explicit spatial modeling, such as the techniques described in Chapter 4.

Reservoir property nodes

All reservoir property nodes are probabilistic. The node for unpredictable changes captures the cases where a 4-D seismic survey would be useful to monitor bypassed oil or breakthroughs due to unpredicted changes in the fluid flow behavior. The reservoir properties group also has a rock property node and a fluid property node. These depend on the well data observations of rock and fluid properties along with the level of spatial heterogeneity and depth. Observing favorable rock properties in the well makes it more likely (but not certain) that the reservoir rock properties are also favorable, more so if the spatial heterogeneity is low. Greater depth makes it more likely for the rock properties to be unfavorable for time-lapse monitoring, since at depth, rocks are usually more consolidated, elastically stiff, and less sensitive to changes in pore fluid and pressure conditions.

Seismic property nodes

The overall seismic response is considered a function of the seismic signature change, fluid contact visibility, image resolution, and repeatability of the seismic tests – again, all probabilistic nodes.

The direct parents of the value node include the decision node "Perform_4D_Seismic" (with the alternatives "true" and "false") and uncertain nodes "Unpredictable changes" (with the states "high," "low," and "none") and "Overall_Seismic_Response" (with the states "very low," "low," "medium," and "high"). The conditional probability tables are

Table 3.2. *Comparison of influence diagram results for the time-lapse example*

	Example 1	Example 2	Example 3
Observables			
Depth	Shallow	Deep	Shallow
Reservoir thickness	High	Low	Low
Rock and fluid properties	Favorable	Poor	Favorable
Acquisition	Favorable	Poor	Poor
Spatial heterogeneity	High	High	High
Value (if performed)/value (if not)	68.6/47.6	46.6/66.8	53.3/60.7
Decision	Perform	Do not/caution	Do not/caution

based on plausible relationships among the uncertainties, and the value for every outcome was assigned by comparing the *relative* benefit on a scale of 0 to 100. For instance, the joint outcome of high unpredictable changes, good seismic response, and no 4-D seismic would result in a value of 0 because of lost profits from bypassed oil. The joint outcome of high unpredictable changes, good seismic response, and true for 4-D seismic is assigned a value of 100, as is the joint outcome of no unpredictable changes, poor seismic response, and no 4-D seismic. All possible outcome combinations were enumerated in the value model of the influence diagram.

Next, we show examples of evaluating the influence diagram model for three cases. (See also Chapter 7 for hands-on exercises using this model.) Examples 1 and 2 are inspired by the "Indonesia" type and the "North Sea" type, respectively, of Lumley et al. (1997). Example 3 is a hypothetical case where the decision is not trivial because of conflicting conditions. The state of the observable nodes for each example and the results in terms of expected value are summarized in Table 3.2. The table also presents the ratio of expected value if the seismic test is performed to the expected value if it is not performed. These relative numbers are only for illustrating the methodology, and these would vary substantially depending on the particular reservoir, actual survey costs, and economic prospects. The assignments of value will also vary from one decision maker to another based on their preferences. For a risk-neutral decision maker, the optimal decision is the one with higher expected value. For the first two cases of observables listed, perhaps one may not require an influence diagram to come to conclusions; they may be fairly obvious and therefore serve as a test case for the influence diagram. The real value of the model is when there are conflicting interrelated observables and the decision is not as straightforward, as in Example 3. In this case, the rock and fluid properties are favorable, and the reservoir is not very deep, favoring the performance of a monitoring survey. On the other hand, acquisition conditions are not favorable, and the reservoir is thin – conditions that are not in favor of monitoring. However, the reservoir is highly heterogeneous and fluid flow is less predictable, conditions where a monitoring survey could be useful. Decision analytic models like influence diagrams could provide clarity of action for such complex decision situations with conflicting conditions.

Influence diagrams are useful tools for modeling, evaluating, and analyzing complex decision problems in subsurface exploration and production. Perhaps their main advantage here is that they capture the relationship between geological properties with the drivers of value. In a spreadsheet, this relationship is not explicitly represented, although the correlation between properties may be implicitly hidden in the values assigned to the properties. The scorecard can be viewed as an influence diagram with only deterministic nodes, and thus it is a special case. Influence diagrams are also useful for communicating information to experts and decision makers. The example described here is a simple one and is of course not the only way to construct an influence diagram to capture the relationships between uncertainties contributing to the value of time-lapse seismic monitoring.

3.4 Value of information

When a decision maker is faced with a decision situation, he or she is also faced with certain auxiliary decisions – i.e., decisions pertaining to some underlying decision situation. One of the most typical auxiliary decisions pertains to information gathering – it may be worthwhile for the decision maker to expend resources to acquire more information about specific uncertainties relevant to their decision situation. For instance, consider a medical decision where surgery is an alternative. Should the decision maker undergo further diagnostic tests before making the surgery decision? Although such tests would provide the medical team with a better understanding of the condition, they may come at a price, possibly through adverse side effects. It is therefore of interest to understand if performing the tests would be worthwhile. Information gathering is a crucial aspect of most decision situations, including medical, personal, and business decisions.

Decision analysis addresses issues pertaining to information gathering using the popular notion of value of information based on the construct of a **clairvoyant**, which was introduced in the early decision analysis literature. In fact, VOI is also referred to as **value of clairvoyance** (Howard 1966); in this book, we will only use the former term for the sake of consistency.

A clairvoyant can tell the decision maker how any uncertainty, past, present, or future, will be resolved, as long as: (i) the clairvoyant need not exercise any judgment in understanding what it means for the uncertainty to be resolved and (ii) the resolution does not depend upon any future action of the decision maker unless that action is specified. As far as we are aware and despite various claims by certain people, there is no evidence that clairvoyants (in the common parlance sense) actually exist in the real world. Yet, this construct is extremely useful in decision analysis. One major benefit is that it encourages clarity, ensuring that all participants involved in the decision process understand whatever is being discussed. The other benefit, which is of particular relevance to this book, is that it helps the decision maker evaluate information sources, activities, and experiments that do in fact exist. There are several potential ways for a decision maker to resolve some key uncertainties before making a decision – for instance, through weather forecasts, sensors, medical diagnostics, surveys, seismic tests, satellite images, expert consultations … the list is endless. Whether it is worthwhile for the decision maker to gather information to resolve some uncertainties is of concern in most practical decision situations.

This section is devoted entirely to explaining VOI. The reader will discover that the VOI for a particular information-gathering scheme depends on several key aspects of the decision situation, the "accuracy" of the information (we will subsequently make this notion more precise), and the decision maker's risk preferences. The literature on the subject and its applications have burgeoned appreciably since its inception in the mid twentieth century. It is now a well-known technique in the arsenal of decision analysis and has been widely applied across numerous applications for the appraisal of decision situations.

3.4.1 Definition

The VOI for an unresponsive uncertainty x and a decision a is the price at which the decision maker is indifferent between purchasing the information, therefore observing the uncertainty before making the decision, and the original decision situation without the information. The VOI is therefore the personal indifference buying price for the information – i.e., the price at which the decision maker is indifferent between having and not having this information. The word "personal" indicates that the price depends on the decision maker's preferences. The word "indifference" indicates that, at this price, the decision maker would be indifferent between receiving the information and acting without it. Therefore, a lower price should be acceptable, and the indifference price is the maximum amount of money that the decision maker should be willing to pay for the information.

Let us compute the VOI for the only uncertainty x in the simple decision situation with the only decision a from a previous section. Similar to the calculations for the certain equivalent of the original decision situation, the utilities for the two situations (with and without information) need to be equated. When the information is available, then the uncertainty is observed before making the decision. If x is observed and alternative a is chosen, then the decision maker retains wealth w and pays price P for certain and also acquires value $v(x,a)$ from the decision situation since $E(v(x,a) \mid x) = v(x,a)$. The optimal alternative if x is observed is

$$a^*(x) = \arg\max_{a \in A} \{u(v(x,a) + w - P)\}. \quad (3.20)$$

Note that the decision maker will not know how the uncertainty will resolve itself before the decision is to be made. Since x is observed with probability $p(x)$, the expected utility for the situation where information is available, denoted MEU', is

$$MEU' = \sum_x \max_{a \in A} \{u(v(x,a) + w - P)\} p(x). \quad (3.21)$$

The VOI is the price P at which the expected utilities from Equations (3.9) and (3.21) are equal – i.e.,

$$\sum_x \max_{a\in A}\{u(v(x,a)+w-P)\}p(x)=\max_{a\in A}\{E(u(v(x,a)+w))\}. \tag{3.22}$$

It can be computed by iteratively varying the price until Equation (3.22) is satisfied.

VOI computations are simpler for situations where the decision maker's utility function satisfies the delta property. The certain equivalent of the situation with information CE' can be computed from Equation (3.21) – i.e.,

$$\begin{aligned} u(CE'+w) &= \sum_x \max_{a\in A}\{u(v(x,a)+w-P)\}p(x) \\ \Rightarrow CE' &= u^{-1}\left(\sum_x \max_{a\in A}\{u(v(x,a)+w-P)\}p(x)\right)-w. \end{aligned} \tag{3.23}$$

Using the delta property from Equation (3.6), this simplifies to

$$CE'=u^{-1}\left(\sum_x \max_{a\in A}\{u(v(x,a))\}p(x)\right)-P. \tag{3.24}$$

Taking the utility inverse of both sides of the equivalence condition in Equation (3.22), we note that when the delta property holds, the VOI is the difference between the certain equivalent of the decision situation with information when it is available for free and the certain equivalent of the decision situation without information:

$$VOI(x)=u^{-1}\left(\sum_x \max_{a\in A}\{u(v(x,a))\}p(x)\right)-u^{-1}\left(\max_{a\in A}\{E(u(v(x,a)))\}\right). \tag{3.25}$$

The notation $VOI(x)$ is used to clearly specify that the VOI is being computed for uncertainty x. If the delta property is satisfied, then the VOI can be computed without the need for iterative calculations. This simplification can be extremely beneficial in practice. Since we use this assumption so extensively in the book, we will often refer to the VOI as the difference between the posterior value, which is the certain equivalent if the information is available for free, and the prior value, which is the certain equivalent without the information:

$$VOI(x)=PoV(x)-PV. \tag{3.26}$$

The reader should be aware of the more general definition in Equation (3.22).

3.4.2 *Perfect versus imperfect information*

In many real-world problems, there is a key uncertainty or distinction of interest x that has a direct impact on the decision maker's value, but it may be very difficult (or perhaps even impossible) to observe x directly. It is often possible, however, to observe a "noisy version" y of this distinction of interest. For instance, it is difficult and expensive to observe the amount of oil in a reservoir – in fact, the only way to do so is to drill many wells that can drain oil from the entire reservoir. It is more convenient and much cheaper to examine the physical properties of the reservoir through seismic tests. The distinction between information about the key uncertainty and another relevant uncertainty that is easier and cheaper to observe is referred to as that of **perfect** versus **imperfect information**. Other common examples of perfect versus imperfect information include: the dimension of a product versus gauge measurements, the condition of a patient versus his or her symptoms, the demand for a product versus market survey results, the physical location of an object versus its satellite image, etc.

In the previous subsection, we showed how to compute the value of perfect information for the general case and also for the special case where the decision maker's risk preferences are captured by a utility function that satisfies the delta property. Let us now approach computing the value of imperfect information for our simple decision situation.

To compute the value of imperfect information – i.e., the VOI for some uncertainty y, the decision maker must assess the probabilistic relationship between x and y through a joint probability distribution. A popular way to do this is to assess the **prior** distribution $p(x)$ and the **likelihood** $p(y \mid x)$. Although there are other ways to assess the joint distribution, it is often easier to assess the probabilities in the causal direction – i.e., from the cause to the effect of the cause. For instance, in medical decisions, it is typical to assess the prior distribution of the underlying disease condition and the likelihood of the symptoms given the disease. As a result, these distributions are often referred to as being in **assessed form**. Inference is typically done in the opposite direction – i.e., from effect to cause – with the help of Bayes' rule, as described in Section 2.3. As a reminder, the following equation demonstrates Bayes' rule, which manipulates the prior distribution and likelihood to compute the marginal likelihood or **pre-posterior** distribution $p(y)$ and the **posterior** distribution $p(x \mid y)$:

$$p(x \mid y) = \frac{p(x)p(y \mid x)}{p(y)} = \frac{p(x)p(y \mid x)}{\sum_x p(x)p(y \mid x)}. \tag{3.27}$$

The numbers computed as above are those in **inferred form**, since they are inferred from the numbers in assessed form; these are the numbers that are used for VOI computations.

In our decision situation, the approach to compute the VOI for an uncertainty y that is relevant to x is similar to that of computing the value of perfect information. Now, y is observed with pre-posterior probability $p(y)$; therefore, the expected utility for the situation where information on y is available, denoted MEU'', is:

$$MEU'' = \sum_y \max_{a \in A} \{E(u(v_a + w - P) \mid y)\} p(y), \tag{3.28}$$

where the posterior distribution $p(x \mid y)$ in Equation (3.27) is used to compute the conditional utility associated with value $v_a = v(x,a)$, given y. The VOI is the price P where the expected utilities from Equations (3.9) and (3.28) are equal:

$$\sum_y \max_{a \in A} \{E(u(v_a + w - P) \mid y)\} p(y) = \max_{a \in A} \{E(u(v_a + w))\}. \tag{3.29}$$

Again, this can be computed by iteratively varying the price until the equation is satisfied.

Just like in the case of perfect information from the previous subsection, VOI computations are simplified greatly for a delta property decision maker. We do not demonstrate the equations for the simplification, as they are similar to the previous subsection. The result is that the VOI is the difference between the certain equivalent of the decision situation with the imperfect information when it is available for free and the certain equivalent of the decision situation without information:

$$\begin{aligned} VOI(y) &= PoV(y) - PV \\ &= u^{-1}\left(\sum_y \max_{a\in A}\{E(u(v_a)\mid y)\}p(y)\right) - u^{-1}\left(\max_{a\in A}\{E(u(v_a))\}\right). \end{aligned} \tag{3.30}$$

3.4.3 Relevant, material, and economic information

Now that we are armed with the terminology, notation, and necessary tools to perform VOI computations, we can return to what we briefly discussed in Chapter 1. The reader can look back at Figure 1.2, which displays what we refer to as the "pyramid of conditions." These are the necessary conditions that make an experiment valuable, as per the decision analysis approach. An experiment should be:

1. Relevant to the distinction of interest – i.e., observing its results should affect the decision maker's beliefs about their value.
2. Material to the decision – i.e., there should be some experimental result for which the decision maker would change their decision.
3. Economic for the decision maker – i.e., the value should be greater than the cost.

The requirements are very much intertwined: an irrelevant experiment cannot be material, and an immaterial test cannot be economic.

Let us demonstrate the layering of conditions using VOI computations for an imperfect source of information y for our simple decision situation. Recall Equation (3.9) from earlier. Suppose that the a priori optimal alternative is a^*. Then, the prior value is

$$PV = E(u(v_{a^*} + w)). \tag{3.31}$$

If y is not relevant, then

$$\max_{a\in A}\{E(u(v_a + w - P)\mid y)\} = \max_{a\in A}\{E(u(v_a + w - P))\}, \qquad \forall y, \tag{3.32}$$

because y does not affect the random variable v_a. Since reducing all values by a constant does not change the optimal alternative

$$\max_{a\in A}\{E(u(v_a + w - P))\} = E(u(v_{a^*} + w - P)). \tag{3.33}$$

Therefore, y is not material.

If y is not material, then the optimal decision for all outcomes of y is a^*; therefore:

$$\sum_y \max_{a\in A}\left\{E\left(u\left(v_a+w-P\right)\mid y\right)\right\}p(y)=\sum_y E\left(u\left(v_{a^*}+w-P\right)\mid y\right)p(y)$$
$$=E\left(u\left(v_{a^*}+w-P\right)\right). \quad (3.34)$$

The solution to Equation (3.29) must be $P = 0$. Since the VOI is 0, the information cannot be economic at any cost.

The pyramid of conditions captures what lies at the core of the VOI approach. While observing information can reduce uncertainties, which may well occur if the information is relevant, it may not have an impact on the decision., and unless there are regulatory reasons or the decision maker is otherwise compelled to obtain the information, there is no value to purchasing immaterial information. Even if a quantitative approach to studying information gathering through VOI analysis is not possible, at the very least the decision maker would be well advised to qualitatively contemplate the pyramid of conditions as a guide to information-gathering decisions.

3.4.4 Examples

Treasure island: the pirate example

Keywords: *pirate example, binary outcomes, VOI, perfect information, imperfect information, risk-neutral decision maker*

Let us return to the pirate example, extending the problem as discussed in Section 3.2. The numbers are as before:

$$x=\begin{cases}1 & \text{with probability } 0.01\\ 0 & \text{with probability } 0.99\end{cases}$$
$$v_{\text{dig}}=\begin{cases}\$100\,000 & \text{with probability } 0.01\\ -\$10\,000 & \text{with probability } 0.99\end{cases} \quad v_{\text{do not dig}}=\$0. \quad (3.35)$$

The pirate is trying to decide whether to dig for the treasure, which he believes is present with a 1% chance. Since he is risk neutral, all computations will be performed with the linear utility function $u(v)=v$.

In Equation (3.15), we computed the value of the decision situation without any additional information. We rewrite the result, using the notation PV for prior value:

$$PV=\max_{a\in A}\left\{E\left(v(x,a)\right)\right\}=\max_{a\in A}\left\{\sum_x v(x,a)\,p(x)\right\}.$$
$$=\max\left\{0,\left(100\,000\cdot 0.01\right)+\left(-10\,000\cdot 0.99\right)\right\}=\$0 \quad (3.36)$$

The chance of success is so low that it is not worthwhile for the pirate to dig for the treasure; hence, the prior value is \$0.

Let us now compute the value of perfect information. Equation (3.26) indicates that the value of perfect information can be obtained by subtracting the prior value from the posterior value. The posterior value is the first term in Equation (3.25). Using the risk-neutral assumption:

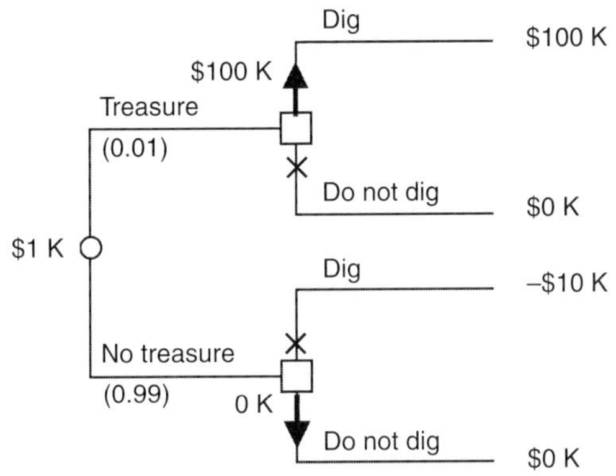

Figure 3.15 The decision tree for the pirate example with perfect information.

$$PoV(x) = \sum_x \max_{a \in A} \{v(x,a)\} p(x) \\ = \left(0.01 \cdot \max\{0, 100\,000\}\right) + \left(0.99 \cdot \max\{0, -10\,000\}\right) = \$1\,000. \tag{3.37}$$

We can perform the same computations using the decision tree in Figure 3.15, which graphically illustrates that the decision maker gets to observe the uncertainty before the decision is made in the value of the perfect information computation. In this case, if the pirate knows that the treasure is present, he will dig for treasure, whereas if he knows that there is no treasure, he will not incur the loss from digging. The value of perfect information is

$$VOI(x) = PoV(x) - PV = 1\,000 - 0 = \$1\,000. \tag{3.38}$$

The value of perfect information is the maximum the decision maker should be willing to pay to get more information about the presence of the treasure. It therefore acts as a useful upper bound for immediately removing excessively expensive information-gathering schemes from consideration.

Now suppose that a friend approaches the pirate and offers to sell him a metal detector for $500. The pirate believes that the detector has an accuracy of 95%. Should the pirate be willing to purchase the detector? The first check is that the cost is less than the value of perfect information; therefore, it *might* be worthwhile. Let us compute the value of imperfect information to confirm.

The first step is to use Bayes' rule from Equation (3.27) to compute the pre-posterior and posterior from the prior and likelihood distributions. This is shown graphically using the probability trees in Figure 3.16. On the left-hand side, we see the probability tree for numbers in assessed form. The tree indicates the prior on whether treasure is present $p(x)$ and the detector response conditional on whether the treasure is present $p(y \mid x)$. To obtain the numbers in inferred form, this probability tree needs to be "flipped" – i.e., the order of the two uncertainties should be flipped around, as shown on the right side. The flipped tree shows the pre-posterior on the detector response $p(y)$, and the posterior, whether the

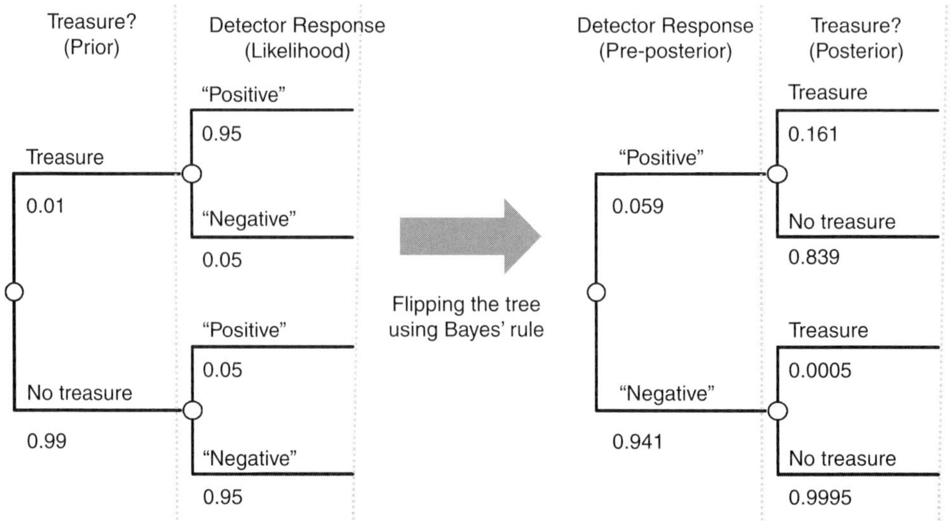

Figure 3.16 Applying Bayes' rule for flipping the probability tree. Left: the prior and likelihood model. Right: the pre-posterior and the posterior probabilities.

treasure is present conditional on the detector response, $p(x \mid y)$. A priori, the chance of success is only 1%, but if the detector suggests that the treasure is present, then the probability of finding the treasure increases to around 16%. However, if the detector does not show signs that the treasure is present, the probability decreases to 0.05%.

The posterior value from observing imperfect information can now be computed by replacing the inferred form numbers into the first term of Equation (3.26). Using the risk-neutral assumption (and using value numbers in thousands):

$$\begin{aligned} PoV(y) &= \sum_{y} \max_{a \in A} \left\{ E\left(v(x,a) \mid y\right) \right\} p(y) \\ &= \left(0.059 \cdot \max\left\{0, (100 \cdot 0.161) + (-10 \cdot 0.839)\right\}\right) \\ &+ \left(0.941 \cdot \max\left\{0, (100 \cdot 0.0005) + (-10 \cdot 0.9995)\right\}\right) \\ &= \left(0.059 \cdot \max\left\{0, 7.71\right\}\right) + \left(0.941 \cdot \max\left\{0, -9.95\right\}\right) = \$0.455 \text{ thousand.} \end{aligned} \tag{3.39}$$

Figure 3.17 shows the posterior value computations using a decision tree. We see that the pirate should dig if the detector response is positive but should refrain from digging if the response is negative. Note that if the detector response is positive, the (conditional) expected value is \$7.71 thousand. However, the pirate has no way to know for certain what the detector will declare before it is purchased. The VOI computation utilizes the pirate's beliefs and, through probabilistic inference, helps the pirate realize that there is only a chance of around 5.9% that the detector response will be positive. That is why the posterior

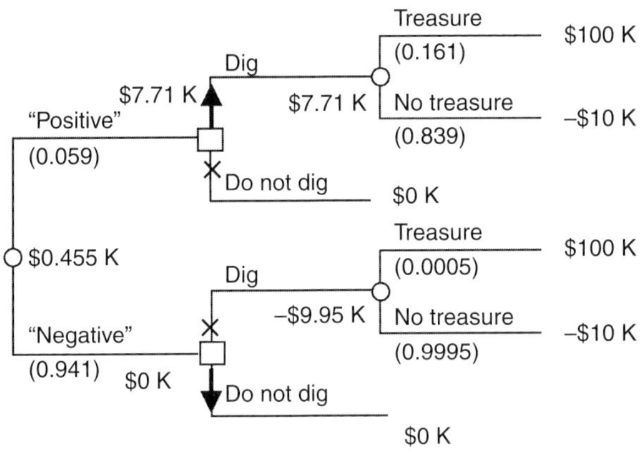

Figure 3.17 The decision tree for the pirate example with imperfect information.

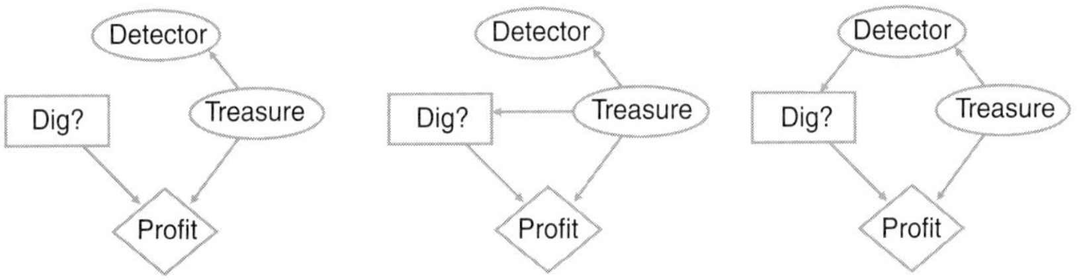

Figure 3.18 The influence diagrams for the pirate example. Left: without information; middle: with perfect information; right: with imperfect information.

value of the detector is as low as $0.455 thousand. In fact, the value of the imperfect information from the detector is

$$VOI(y) = PoV(y) - PV = 0.455 - 0 = \$0.455 \text{ thousand.} \tag{3.40}$$

Since the price of the detector ($500) is greater than its value to the pirate ($455), it is not worthwhile as is. The pirate may wish to negotiate the price with his friend and now has the necessary tools to do so. In addition, perhaps the pirate may wish to consider other detectors or sources of information that could improve his value for his decision situation.

Figure 3.18 shows influence diagrams for the various situations that we have compared for our VOI computations. On the left, we see the influence diagram for the pirate's prior decision situation. The uncertainties for the detector response and the presence of treasure are dependent, and hence there is a conditioning arc. The pirate's profit is a function of whether the treasure is present and the digging decision. The middle figure represents the case of perfect information, where there is an additional arc from the treasure uncertainty to the drilling decision. The informational arc indicates that this uncertainty is observed before the decision is made. On the right, we see the case of imperfect information; there is an informational arc from the detector response uncertainty to the drilling decision. These examples illustrate the influence diagram interpretation of VOI for an unresponsive uncertainty: the VOI can be computed by comparing the value of an influence diagram of the prior decision situation with that of an influence diagram where there is an additional informational arc from the uncertainty under consideration to the appropriate decision node.

For whom the bell tolls: Gaussian projects example

Keywords: *Gaussian projects example, bivariate Gaussian distribution, VOI, perfect information, imperfect information, partial information, total information, risk-neutral decision maker*

We return to the Gaussian projects example, this time using different numbers and assumptions. There are two projects that the decision maker can invest in; each project has uncertain profits modeled jointly as a random variable with a bivariate Gaussian distribution $p(\boldsymbol{x}) = N(\boldsymbol{\mu}, \boldsymbol{\Sigma})$. For this example, we specify the prior mean and covariance matrix as in Section 2.3 – i.e.,

$$\boldsymbol{\mu} = \begin{pmatrix} 0 \\ 0 \end{pmatrix}, \qquad \boldsymbol{\Sigma} = \begin{bmatrix} 1 & \rho \\ \rho & 1 \end{bmatrix}. \tag{3.41}$$

We will study how the VOI varies as a function of the correlation parameter ρ.

Previously in this chapter, we had assumed that the decision maker could choose exactly one project. Here, we change this assumption: instead, the decision maker is free to invest in as many projects as are profitable. Another difference is that here the decision maker is assumed to be risk neutral rather than risk averse. As a result of these assumptions, the prior value is

$$PV = \sum_{i=1}^{2} \max\{0, E(x_i)\} = \sum_{i=1}^{2} \max\{0, \mu_i\} = 0. \tag{3.42}$$

Before making the investments, the decision maker can purchase information about the projects. Such information may become available at a price by asking experts to investigate the projects, write a report, and eventually provide quantitative data that are directly informative of the profits. Depending on the accuracy of this work, such information might be either perfect or imperfect.

We consider the situation where the decision maker must select either perfect information about the profit of one project or imperfect information about the profits of both projects. The possible data are then of size $m = 2$ (imperfect information about both projects) or $m = 1$ (perfect information about one project). For bivariate imperfect information:

$$\boldsymbol{y} = \boldsymbol{x} + \boldsymbol{\varepsilon}, \qquad \boldsymbol{\varepsilon} \sim N(\boldsymbol{0}, \tau^2 \boldsymbol{I}_2), \tag{3.43}$$

where we assume equal accuracy or standard deviation τ for both projects. The other information gathering scheme is that of partial perfect information x_1. Due to the symmetry of the model, it does not matter whether we consider Project 1 or 2 for perfect information gathering.

Consider first the case with imperfect information about both projects. The marginal distribution of the data is $p(\boldsymbol{y}) = N(\boldsymbol{0}, \boldsymbol{\Sigma} + \boldsymbol{T})$. As shown in Section 2.3 and Appendix A.1, the conditional expectation of projects given data $\boldsymbol{y}$ is $\boldsymbol{\mu}_{x|y} = \boldsymbol{\mu} + \boldsymbol{\Sigma}(\boldsymbol{\Sigma} + \boldsymbol{T})^{-1}(\boldsymbol{y} - \boldsymbol{\mu})$. The posterior value in Equation (3.30) becomes

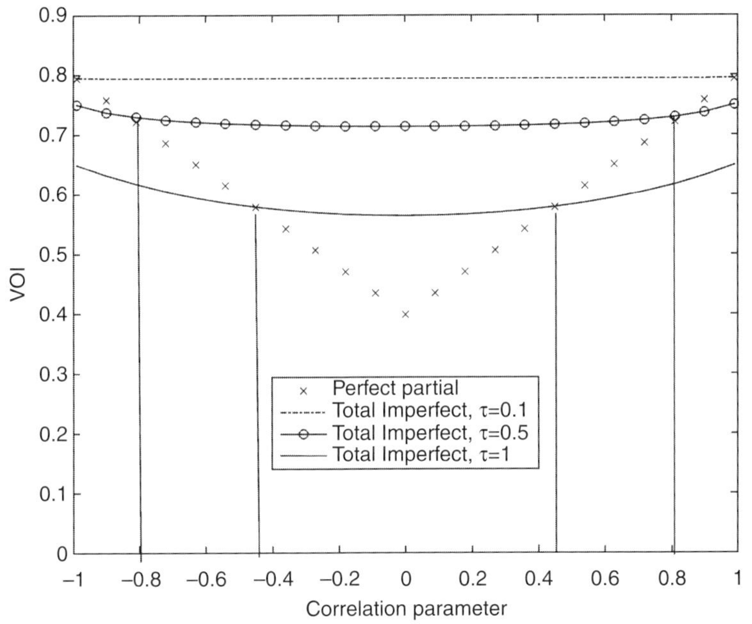

Figure 3.19 Sensitivity of VOI to the correlation parameter in the Gaussian projects example. The lines represent partial perfect testing (crosses) and total imperfect testing for various accuracies.

$$PoV(\mathbf{y}) = \sum_{i=1}^{2} \int \max\left\{0, \mu_{x_i|\mathbf{y}}\right\} p(\mathbf{y}) d\mathbf{y}. \tag{3.44}$$

With perfect information at one project, the posterior value is

$$\begin{aligned} PoV(x_1) &= \int \max\{x_1, 0\} p(x_1) dx_1 + \int \max\{E(x_2 \mid x_1), 0\} p(x_1) dx_1 \\ &= \int \max\{x_1, 0\} p(x_1) dx_1 + \int \max\{\rho x_1, 0\} p(x_1) dx_1. \end{aligned} \tag{3.45}$$

For the Gaussian model, the integral expressions in Equations (3.44) and (3.45) are available in closed form (Appendix A.1). Here, the closed-form expressions for the posterior value are

$$PoV(\mathbf{y}) = \frac{2}{\sqrt{2\pi}} \sqrt{\frac{(1+\tau^2)(1+\rho^2) - 2\rho^2}{(1+\tau^2)^2 - \rho^2}}, \qquad PoV(x_1) = \frac{(1+\rho)}{\sqrt{2\pi}}. \tag{3.46}$$

Figure 3.19 shows the VOI as a function of the correlation parameter ρ for three different levels of measurement accuracy: $\tau = 1$, $\tau = 0.5$, and $\tau = 0.1$. The VOI is larger when there is greater positive or negative dependence between the two projects. This increase with correlation is largest for the case with perfect information at one project. It is smallest when we have very accurate imperfect information about both projects.

The VOI must be compared with the price of the experiments. Assume first that the price of perfect information about one project is the same as the price of imperfect information about both projects with accuracy $\tau = 0.5$. Then, partial perfect information would be the preferred information-gathering scheme only if the correlation $|\rho| > 0.8$. If the price of perfect partial information is the same as that of imperfect information about both projects with accuracy $\tau = 1$ then the partial perfect information-gathering scheme would be preferred if $|\rho| > 0.45$.

3.5 Bibliographic notes

Decision analysis fundamentals

Miles (2007) provides an entertaining history of ideas that lead to the inception of the field of decision analysis up until the 1960s. Readers who wish to become more familiar with decision analysis fundamentals are referred to introductory textbooks such as Schlaiffer (1959), Raiffa (1968), Clemen and Reilly (1999), and Howard and Abbas (2015). We highly recommend the recent book by Howard and Abbas (2015).

Graphical models

Kirkwood (1993) discusses solving problems with large decision trees. Shachter (2007) and Bielza et al. (2010) present perspectives on modeling decision situations with influence diagrams. Although we have only presented decision trees and influence diagrams in this chapter, there are a variety of other graphical models that explicitly represent decisions in decision situations. The reader is referred to Bielza and Shenoy (1999) and Jensen and Nielsen (2007) to peruse a few of these other graphical models. As some specific examples, see the literature on sequential decision diagrams (Covaliu and Oliver 1995), valuation networks (Shenoy 1992), game trees (Shenoy 1998), and decision circuits (Bhattacharjya and Shachter 2007).

VOI fundamentals

The literature on VOI is founded upon classic work such as that by Blackwell (1953), Schlaiffer (1959), Howard (1966, 1967), and Raiffa (1968) and continues to gain popularity. Also see earlier work on VOI and sequential decisions (Miller 1975; Merkhofer 1977), properties of VOI (Hilton 1981), and value of control (Matheson 1990).

VOI for canonical problems

The hunt for analytical closed-form solutions for VOI has often led to more negative results than positive ones (Hilton 1981); as a result, many researchers have studied canonical problems – i.e., specific classes of decision problems – to understand the general effect of the parameters on information value. Among these, the two-action linear loss problem, also known as the go/no-go problem, has received particular attention in the literature (Schlaiffer 1959; Keisler 2005; Bickel 2008). Other articles about VOI for canonical problems include studies on the news vendor problem (Clemen and Winkler 1985), the portfolio problem (Keisler 2004; Bhattacharjya et al. 2013; Zan and Bickel 2013), the stopping problem (Bhattacharjya and Deleris 2014), etc. Due to the favorable mathematical properties of the Gaussian distribution, many of these decision problems involve Gaussian models.

Computational issues and application reviews

Algorithms for computing VOI using graphical models are discussed in Shachter (1999), Bhattacharjya and Shachter (2008), and Krause and Guestrin (2009). There are several articles pertaining to VOI applications – here we only list selected review articles. VOI analysis is a powerful tool for medical decision making and related domains (Yokota and Thompson 2004a, 2004b). Keisler et al. (2014) present a comprehensive review of applications related to the environment. Grayson (1960) introduced VOI to the oil and gas industry, focusing on drilling decisions. Bratvold et al. (2009) provide a perspective on the past, present, and future use of VOI in the oil and gas industry. There is extensive literature on VOI applications in the domain of artificial intelligence and expert systems – see, for instance, the early work of Heckerman et al. (1989).

4
Spatial modeling

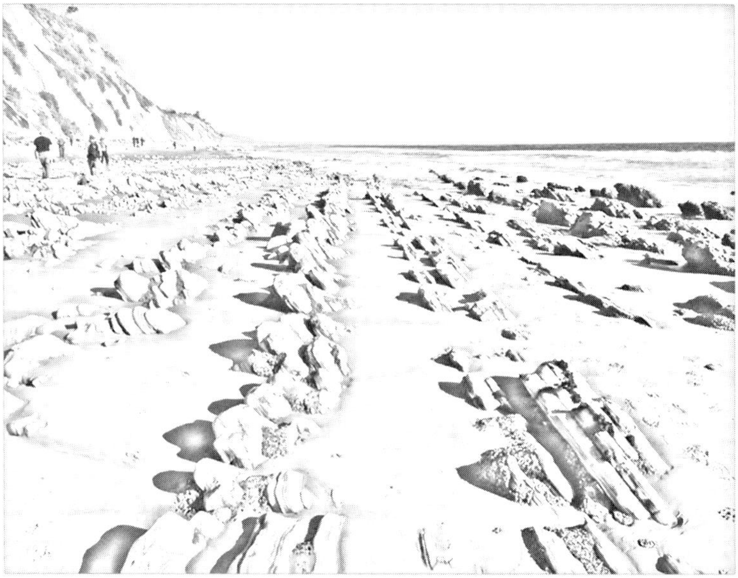

Figure 4.1 rocks along the shore
like soldiers marching in step
patterns surround us

Geoscientists and engineers who have to quantitatively characterize and forecast subsurface properties understand that there is always uncertainty in any Earth model. Geological heterogeneities contribute to interpretation uncertainty, and delineation of subsurface heterogeneity is a key factor in reliable characterization, forecasting, and decision making. These heterogeneities occur at various scales and can include spatial variations in lithology, pore fluids, clay content, porosity, permeability, etc. How do we quantitatively model these spatial uncertainties? It is important for us to discuss spatial models since our book is about spatial decision situations, data gathering at spatial locations, and value of information (VOI) analysis of such spatial data. Most of the data examples used in the current chapter will also be used later in the book for VOI analysis.

In Section 4.1, we discuss the need for stochastic modeling and uncertainty quantification in spatial modeling. We discuss random fields and spatial prediction in Sections 4.2–3. This is followed by an overview of commonly used geostatistical techniques, accounting for some of their advantages and disadvantages: Gaussian models in Section 4.4, extensions of Gaussian models in Section 4.5, discrete spatial models in Section 4.6, and multiple-point statistics in Section 4.7. The presentation is not intended to provide a complete description of models or methodologies. The final section in this chapter is a bibliography pointing out some important references in geostatistics. Further details on selected models and methods are provided in the appendix.

4.1 Goals of stochastic modeling of spatial processes

Before discussing various spatial models, it is useful to consider some of the possible goals of a modeling exercise, because the appropriateness of any particular method depends, in a large part, on the goal of the study. Stochastic methods vary considerably in their requirements for time, money, human resources, and computer hardware and software. Not all stochastic modeling studies need the latest multiple-point technique; some studies will do fine with the traditional variogram-based simulation technique or even with a robust estimation technique such as Kriging.

Some of the models and methods used in spatial process characterization are purely statistical, based on multivariate techniques for stochastic spatial correlation, interaction, templates, or patterns. Others are deterministic, based on physical models derived from established theory as well as laboratory observations. Each group of techniques can have some degree of success depending on the particular study. It is often beneficial to combine the best of each method to generate results more powerful than would be possible from purely statistical or purely deterministic techniques alone.

Spatial statistics is characterized by modeling or data analysis of variables associated with geographic locations. In Chapter 2, we described some data sets acquired and organized according to location. To motivate the current discussion, consider the illustration in Figure 4.2 with petroleum reservoir variables.

From the currently available data – say, seismic data and well log information – the petroleum company predicts the spatial reservoir variables of interest, such as the porosity and permeability variables in Figure 4.2 (top left) at any spatial location. This prediction step includes spatial modeling, not very different from that described by a graphical model in Section 2.3, but now the variables are spatially dependent. Note that the porosity variable varies smoothly in the east and north directions. Furthermore, the spatial variables tend to be interdependent; permeability typically increases with porosity, but there is often a lot of scatter in the porosity–permeability relation.

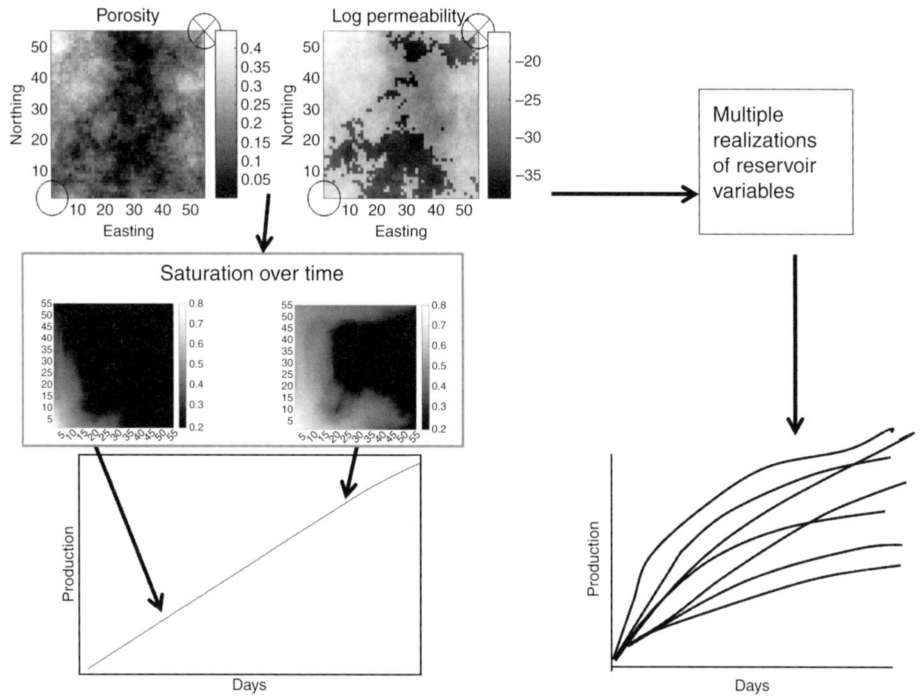

Figure 4.2 Illustration of spatial modeling. Reservoir variables such as porosity and permeability determine the fluid flow in the reservoir and the recoverable production. Rather than using one "best guess" model of the predicted reservoir variables, the uncertainty is best quantified by stochastic spatial modeling and multiple realizations of reservoir variables and their resulting production profiles.

The associated fluid flow and petroleum production from this reservoir are shown in the bottom left display of Figure 4.2. In this illustration, the petroleum company has drilled an injection well in the southwest corner and a production well in the northeast corner. The injected water assists the production, and the flow of oil and water is directed by the porosity and permeability in the reservoir. The company needs to predict the recoverable oil production, but this is not straightforward because the reservoir properties are uncertain. One possible prediction is to compute the expected value of porosity and permeability, given the currently available data, and then use these as inputs to the fluid flow calculation. Under a least-squares loss criterion, the optimal prediction is often called the Kriging surface after the famous geoscientist D. G. Krige. (The intuition and mathematical aspects of Kriging are covered in Sections 4.3 and 4.4.) This is a very common way of smoothing and visualizing spatial data, and it will continue to be a very important technique in spatial statistics. The Kriging surface or the associated variance does not, however, capture all the spatial attributes of the variable of interest. Moreover, the Kriging predictor does not necessarily result in the most reliable prediction of the production since the flow simulation (transfer function) is non-linear.

Rather than using just the expected value or some other summary of the spatial variable $\boldsymbol{x}$, it is common to generate multiple realizations $\boldsymbol{x}^1,...,\boldsymbol{x}^B$ of spatial variables. From the realizations, one can use Monte Carlo approximation to estimate integral expressions, as described in Chapter 2. For non-linear prediction, the forward-propagated realizations $f(\boldsymbol{x}^1),...,f(\boldsymbol{x}^B)$ represent the uncertainty of the prediction variable. A fluid flow forward operator f involves complicated interactions in the spatial variables, and propagating an ensemble of realizations through the flow simulator will provide the petroleum company with a much more reliable prediction of the recoverable petroleum production (Figure 4.2, right) than using just a single prediction.

Before discussing particular spatial models, we consider some of the goals of spatial modeling in more detail following the discussion of Srivastava (1994).

Stochastic modeling for visualization of spatial features

Stochastic models can serve as qualitative visual tools to catalyze better technical work. Stochastic spatial models entail random process constructions with fluctuations around a smooth description capturing the large-scale trends. Thus, stochastic models help to challenge the familiar layer-cake stratigraphy, with smoothly varying heterogeneities that are often the prevalent belief. For instance, in an application as in Figure 4.2, cross sections through stochastic models reveal much more visual complexity in the interwell region than traditional models can portray. The value of the stochastic model in this context is simply to make the experts realize that there are many possible renditions that equally honor the data; to focus the views and opinions of a wide variety of experts; and, in so doing, to point to an improved reservoir model.

Stochastic techniques are attractive not only for their ability to generate many plausible outcomes but also for their ability to produce outcomes that have a realistic level of spatial heterogeneity. Reservoir performance predictions are more accurate when based on models that reflect the actual heterogeneity of the reservoir instead of using overly simplistic models. Most traditional methods for subsurface modeling lead to models that are too smooth and continuous rather than showing the variability known to exist. Such smoothness commonly leads to biased predictions and poor development plans – actual breakthrough times of water in wells end up being much quicker than expected, for example, or sweep is not as efficient as the smooth model predicted.

Stochastic spatial modeling for assessing uncertainty and risk analysis

Stochastic modeling offers the ability to do uncertainty assessment and Monte Carlo risk analysis because the various realizations and scenarios it produces are all plausible in the sense that they honor all of the information. In such studies, hundreds, if not thousands, of alternate models are generated and processed to produce a probability distribution of possible values for some critical engineering parameters – for example, net to gross, connected pore volume, or net present value, etc. These distributions are then used to optimize decisions by minimizing some appropriate objective function.

A critical aspect of this use of stochastic modeling is the belief in some "space of uncertainty" and that the stochastic modeling technique can produce outcomes that are a fair representation of the spectrum of possibilities.

Stochastic modeling for integrating spatial information

A reason for the increasing popularity of stochastic models and methods is due to their ability to incorporate a broad range of information from different spatial data types that more conventional techniques do not accommodate. Stochastic models provide a natural framework for incorporating different information in terms of conditional probabilities and prior and posterior probabilities. Sometimes the goal of using a stochastic approach to spatial analysis may be the integration of different data types – for instance, seismic data with petrophysical data – while at the same time ensuring that the model follows some basic geologic principles. These complicated links between spatial variables and data sources can be treated consistently through a formalized stochastic model.

Spatial information gathering may carry nearly perfect information at some locations – say, well logs of reservoir variables. Due to the spatial statistical dependence, this local or partial knowledge at well locations will carry information about the surrounding sites of interest. In other situations, one may conduct a test with extensive spatial coverage, but the test only provides noisy (imperfect) information about the primary variable of interest – for example, three-dimensional (3-D) seismic attributes that may be interpreted in terms of reservoir lithologies. This results in total (spatially exhaustive) but indirect information that is available at all the sites where we aim to predict the distinction of interest. We may have variants of these situations when we collect multivariate data sets, some of which are perfect, imperfect, partial, or total and often are at different resolution scales. Integrating these data sets in a consistent manner requires spatial stochastic modeling.

4.2 Random fields, variograms, and covariance

In the broadest sense, geostatistics may be defined as a branch of statistical sciences that studies spatial phenomena and capitalizes on spatial relationships to model possible values of variables at unsampled locations (Caers 2005). In geostatistics, variables or data are modeled as random fields $y(\boldsymbol{s})$, where y denotes the attribute (e.g., porosity or seismic impedance) at a location with vector coordinates $\boldsymbol{s}$. Just as a random variable is characterized by its probability density function (pdf) (see Chapter 2), a **random field** is characterized by the set of all joint n-variate pdfs for any number n and any locations $\boldsymbol{s}_1,...,\boldsymbol{s}_n$ – i.e.,

$$p(y_1,...,y_n;\boldsymbol{s}_1,...,\boldsymbol{s}_n) = p\left(y(\boldsymbol{s}_1),...,y(\boldsymbol{s}_n)\right). \tag{4.1}$$

Recall that the random variables (or the pdf) can have a discrete valued sample space or a continuous sample space. For generality, a stronger definition of random fields via the cumulative distribution function for any n tuple is often used, but the notion of a joint pdf in Equation (4.1) for any set of variables is sufficient in our setting.

The joint uncertainty about the n values is then represented by the multivariate pdf. The random field $y(\boldsymbol{s})$ is said to be stationary if its multivariate pdf is invariant under translations of the vector $\boldsymbol{s}$. Invariance of the multivariate pdf also implies invariance of all lower-order pdfs (obtained via marginalization) and moments (expectation and variance). Choosing stationarity in a model simplifies inference (parameter estimation and prediction). However, stationarity is a property of the random field model and not of the underlying physical or geological processes. In practice, stationarity rarely holds, at least not at all scales. A common choice is to model the trends of a process and then assume a stationary residual surface of some kind. Nevertheless, the choice of which data to pool together into one stationary population is a subjective one that underlies all statistical modeling (Isaaks and Srivastava 1989; Pyrcz and Deutsch 2014).

Assume that n data are available at locations $\boldsymbol{s}_i$, $i = 1,..,n$. The **mean** and **variance**, as described in Chapter 2, are estimated by

$$\hat{\mu} = \frac{1}{n}\sum_{i=1}^{n} y(\boldsymbol{s}_i), \qquad \hat{\sigma}^2 = \frac{1}{n}\sum_{i=1}^{n}\left(y(\boldsymbol{s}_i) - \hat{\mu}\right)^2. \tag{4.2}$$

The mean and variance are summary statistics of the data that involve one datum location at a time.

The **variogram** and, equivalently, the covariance and correlogram, are two-point statistics in that they involve two data locations at a time, $\boldsymbol{s}_i + \boldsymbol{t}$ and $\boldsymbol{s}_i$, separated by distance or lag vector $\boldsymbol{t}$. The empirical variogram is estimated by

$$2\hat{\gamma}(\boldsymbol{t}) = \frac{1}{n(\boldsymbol{t})}\sum_{i}\left(y(\boldsymbol{s}_i) - y(\boldsymbol{s}_i + \boldsymbol{t})\right)^2, \tag{4.3}$$

where the sum goes over the $n(\boldsymbol{t})$ pairs of locations separated approximately by the same vector (modulus and direction) $\boldsymbol{t}$. The variogram is thus a measure of variability or dissimilarity between two y variables at a distance of $\boldsymbol{t}$.

The **covariance** is a measure of similarity and is estimated from the data as follows:

$$\begin{aligned}\hat{C}(\boldsymbol{t}) &= \frac{1}{n(\boldsymbol{t})}\sum_{i}\left(y(\boldsymbol{s}_i + \boldsymbol{t}) - \hat{\mu}\right)\left(y(\boldsymbol{s}_i) - \hat{\mu}\right) = \frac{1}{n(\boldsymbol{t})}\sum_{i} y(\boldsymbol{s}_i + \boldsymbol{t})\, y(\boldsymbol{s}_i) - \hat{\mu}^2, \\ &= \hat{C}(0) - \hat{\gamma}(\boldsymbol{t})\end{aligned} \tag{4.4}$$

where $\hat{C}(0) = \hat{\sigma}^2$ is the stationary variance. The latter equality in Equation (4.4) shows how the spatial covariance is related to the variogram for a stationary random field.

The **correlogram** standardizes the covariance to be in $(-1,1)$. It measures the linear correlation between two y variables at a spatial distance $\boldsymbol{t}$ – i.e.,

$$\hat{\rho}(\boldsymbol{t}) = \frac{\hat{C}(\boldsymbol{t})}{\hat{C}(0)} \tag{4.5}$$

Typically, the semivariogram $\hat{\gamma}(\boldsymbol{t})$ increases with the separation vector $\boldsymbol{t}$ until it stabilizes around a sill value. Correspondingly, the covariance/correlogram decreases down to its zero value $\hat{C}(\infty) = 0$; very distant data values are not correlated anymore.

Discussion: variogram or covariance

Geostatisticians have traditionally preferred the variogram over the covariance/correlogram because (i) its inference does not call for prior inference of the mean and variance and (ii) there are theoretical models such as fractals that have infinite variance; hence, no covariance is defined, and yet they may have a finite variogram. However, in recent practice it is just as common to work with the covariance and estimate the mean as a function of explanatory variables as in a regression model.

Spatial two-point correlation between two different attributes y_1 and y_2 (e.g., porosity and permeability) is described by the cross-covariance function given by:

$$\hat{C}_{12}(\boldsymbol{t}) = \frac{1}{n(\boldsymbol{t})}\sum_i \left(y_1(\boldsymbol{s}_i + \boldsymbol{t}) - \mu_1\right)\left(y_2(\boldsymbol{s}_i) - \mu_2\right), \tag{4.6}$$

where μ_1 and μ_2 are the stationary means of the two attributes.

Unless data locations are on a regular grid, tolerance of lag value (and direction) is needed to find enough pairs, $n(\boldsymbol{t})$, of data approximately at distance $\boldsymbol{t}$ apart to infer an experimental value for the variogram in Equation (4.3), covariance in Equation (4.4), and cross-covariance in Equation (4.6). The level of tolerance needed depends on the amount and spatial layout of the available data. The tolerance setting is another modeling choice that impacts the resulting experimental variograms and hence the analytical models used to fit the experimental variogram. A full angle/direction tolerance pools together all data pairs with the same distance modulus $|\boldsymbol{t}|$ irrespective of direction, and the result is then an omni-directional experimental variogram often modeled as an isotropic model. **Isotropy** – that is, invariance with direction – is then a consequence of the angle tolerance, not necessarily a physical characteristic of the underlying phenomenon.

In practice, when we have data $y(\boldsymbol{s}_1), y(\boldsymbol{s}_2), .., y(\boldsymbol{s}_n)$ available, the empirical variogram is constructed. Next, we fit a parametric model, such as those presented in Table 4.1 (or any other legitimate model) to the data. The simplest way to do this is by visual inspection to answer questions such as: does the decline with distance appear to follow an exponential trend? And is there a nugget effect? More sophisticated ways of parameter estimation will be treated in Section 4.4 in the context of Gaussian random fields.

Four common examples of spatial covariance functions are shown in Table 4.1. Many other popular spatial covariance functions are described in standard books on geostatistics – e.g., Deutsch and Journel (1992), Goovaerts, (1997), Lantuejoul, (2002), and Chiles and Delfiner (2012). In the equations in Table 4.1, τ^2 is the nugget effect, which only affects the variance; σ^2 is the variance–covariance of the spatially dependent process, so the overall variance equals $\tau^2 + \sigma^2$; and η determines the decay of the covariance function. If η is large, the covariance goes quickly to 0, while it decays more slowly for small η. For the exponential covariance function, one can parametrize the decay by the effective spatial range $3/\eta$, since $\exp(-3) \approx 0.05$, indicating that the correlation is only 0.05 at spatial distance $|\boldsymbol{t}| = 3/\eta$. The four covariance functions are displayed as a function of lag distance in Figure 4.3 (left) together with the associated variograms (Figure 4.3,

Table 4.1. *Four examples of spatial covariance functions. For all covariance models, the first term represents a nugget effect tied to uncorrelated noise or measurement noise. The latter terms include spatial correlation*

Model	Covariance
Exponential	$C(\lvert t\rvert) = \tau^2 I(\lvert t\rvert = 0) + \sigma^2 \exp(-\eta\lvert t\rvert)$
Matern 3/2	$C(\lvert t\rvert) = \tau^2 I(\lvert t\rvert = 0) + \sigma^2 (1+\eta\lvert t\rvert)\exp(-\eta\lvert t\rvert)$
Cauchy type	$C(\lvert t\rvert) = \tau^2 I(\lvert t\rvert = 0) + \sigma^2 \frac{1}{(1+\eta\lvert t\rvert)^3}$
Gaussian	$C(\lvert t\rvert) = \tau^2 I(\lvert t\rvert = 0) + \sigma^2 \exp(-\eta^2\lvert t\rvert^2)$

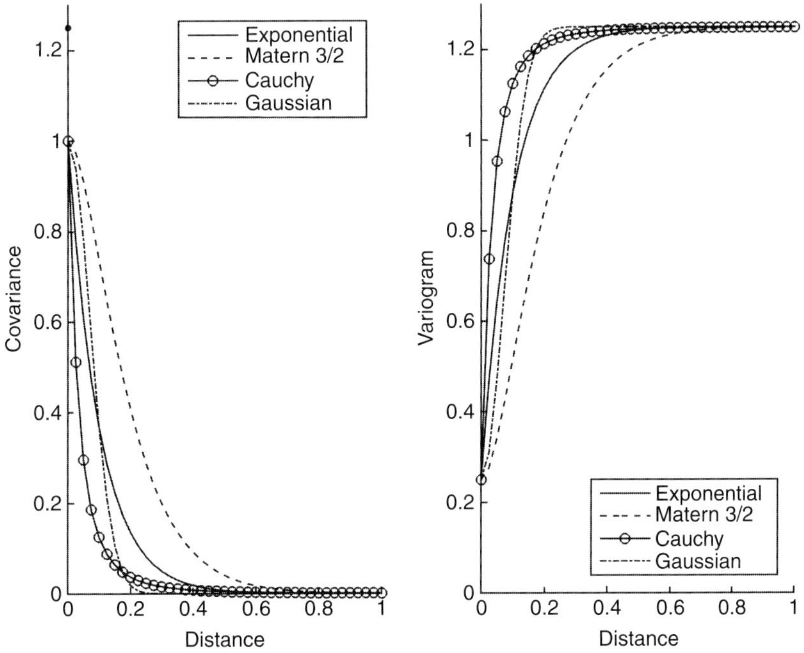

Figure 4.3 Three different covariance functions and variograms. Left: covariance functions plotted as a function of distance (first axis). Right: variogram plotted as a function of distance (first axis). The parameters are variance $\sigma^2 = 1$ for the random effect, $\tau^2 = 0.5^2$ for the nugget effect, and correlation decay $\eta = 10$.

right). The model parameters were here set to $\tau^2 = 0.5^2$, $\sigma^2 = 1^2$, and $\eta = 10$. All these four common parametric covariance functions give valid positive definite covariance matrices for n response variables at any combination of spatial locations $s_1, \ldots, s_n$.

The connection between the covariance models and the related variogram model is shown in Figure 4.3. The variogram model $\gamma(\lvert t\rvert) = C(0) - C(\lvert t\rvert)$ would approach

$\gamma(0^+) = \tau^2$ as $|t| \to 0$, and yet $\gamma(0) = 0$. When the variogram exhibits this discontinuity for very small distances, it indicates the presence of an additive white noise process. The variogram increases with larger distance $|t|$ and levels out around the effective spatial range. The asymptotic value, as $|t|$ goes to infinity, is $\gamma(\infty) = \sigma^2 + \tau^2$.

Recall that the variogram is just a summary statistic, or a theoretical second moment of a model description, but for the Gaussian distribution (Section 4.4), it plays a crucial role in the parameterization and indicates our ability to borrow information and attain predictive power in the spatial domain. For most geoapplications, we are interested in specifying parameters for spatial dependence or interaction, but it may not be the variogram directly. In Sections 4.5–7, we present a couple of non-Gaussian models in some detail.

4.3 Prediction and simulation

We now explain concepts underlying spatial prediction. As we have discussed in the previous section, in many spatial models the location-wise prediction and prediction variance may not realistically capture the spatially dependent distinctions of interest. For this reason, the simulation of spatial processes has become important and, for completeness, we briefly describe some useful ways of spatial simulation. In subsequent sections, we will focus on concepts related to the prediction and simulation methods tailored to the models for VOI analysis.

4.3.1 Spatial prediction and Kriging

One of the most important goals of modeling, and ultimately of importance for decision making, is the ability to predict an unknown response as well as assess its uncertainty. Prediction is well known in regression analysis where the model is fit from the currently available response variables and explanatory variables. When there are new explanatory variables, the goal is to use the fitted model to predict the new response. This prediction is obtained by plugging the explanatory variables into the fitted regression model. The prediction uncertainty can be computed based on the amount of data used for fitting the regression model and the noise in the model. Notably, the uncertainty also depends on the new explanatory variables. If these are near the center of the formerly collected data, there is little uncertainty, but if the explanatory variables are far from the currently available ones, there is greater uncertainty. In fact, it is unclear whether the regression model would fit well far outside the domain of former data and in general it is best to avoid extrapolation.

In the geosciences, prediction relies on explanatory variables but also relates to the dependence in the spatial variables. Observations made near a prediction site are often most valuable when predicting spatial attributes. Early applications of geostatistics in the mining industry addressed this question of predicting ore grade at a location based on measured grade values at surrounding locations. In modern applications, one includes explanatory variables as well as spatial proximity, possibly represented by a hierarchical modeling framework. Moreover, one might not only be interested in predicting the ore at a location but perhaps may also be interested in predicting the block volume or a response from a process integrated over a complicated domain, such as the fluid flow response in a reservoir. The tools we use for prediction rely very much on the situation, including both the modeling aspects and the relevant prediction tasks.

Suppose we want to predict the response at a spatial location $\mathbf{s}_0$. Denote the unknown response by $y(\mathbf{s}_0)$, and let $\mathbf{y} = (y(\mathbf{s}_1),..,y(\mathbf{s}_n))$ be the data at n other sites. Under expected square loss, the **optimal spatial predictor** of the response is the conditional expectation $E(y(\mathbf{s}_0) \mid \mathbf{y})$. To see why this holds, let us compare the conditional expectation with another predictor $g(\mathbf{y})$ for arbitrary function g. The mean square prediction error (MSPE) is

$$\begin{aligned} E\left(y(\mathbf{s}_0)-g(\mathbf{y})\right)^2 &= E\left(y(\mathbf{s}_0)-E(y(\mathbf{s}_0)\mid \mathbf{y})+E(y(\mathbf{s}_0)\mid \mathbf{y})-g(\mathbf{y})\right)^2 \\ &= E\left(y(\mathbf{s}_0)-E(y(\mathbf{s}_0)\mid \mathbf{y})\right)^2+E\left(E(y(\mathbf{s}_0)\mid \mathbf{y})-g(\mathbf{y})\right)^2 \\ &\quad +2E\left(\left(y(\mathbf{s}_0)-E(y(\mathbf{s}_0)\mid \mathbf{y})\right)\left(E(y(\mathbf{s}_0)\mid \mathbf{y})-g(\mathbf{y})\right)\right) \\ &= E\left(y(\mathbf{s}_0)-E(y(\mathbf{s}_0)\mid \mathbf{y})\right)^2+E\left(E(y(\mathbf{s}_0)\mid \mathbf{y})-g(\mathbf{y})\right)^2, \end{aligned} \tag{4.7}$$

where the expected value is taken over all data $\mathbf{y}$ as well as the random value $y(\mathbf{s}_0)$. In Equation (4.7), we first subtract and add the conditional mean and then expand the quadratic form. The cross-term cancels by double expectation – i.e., $E(E(y(\mathbf{s}_0)\mid \mathbf{y})) = E(y(\mathbf{s}_0))$. The last term in the bottom row of Equation (4.7) is 0 for $g(\mathbf{y}) = E(y(\mathbf{s}_0)\mid \mathbf{y})$, whereas it is positive for any other g function. Thus, the mean square error is smallest for the conditional expectation.

The **Kriging** predictor is the **best linear unbiased predictor** – i.e., $g(\mathbf{y}) = \sum_{i=1}^{n} \alpha_i y(\mathbf{s}_i)$, where the weights α_i, $i = 1,...,n$ are derived from minimizing the MSPE among all linear and unbiased predictors. When we assume a Gaussian model, the conditional mean $E(y(\mathbf{s}_0)\mid \mathbf{y})$ is linear in the data $\mathbf{y}$ and is hence identical to the Kriging predictor.

The Kriging equations will be derived in Section 4.4 assuming Gaussian data. They require only the mean function and the variogram (or covariance). The predictor considers not only the correlation between the unknown and each datum (taken one at a time) but also the correlation between any two data. All such correlations are two-point statistics, as provided by the variogram model. Kriging is quite versatile and lies at the origin of the success of traditional geostatistics.

Pitfalls: interpretation of Kriging results

Kriging estimates are very useful when the goal is to get a local best estimate (e.g., porosity or ore grade at a prediction site). However, maps of Kriging estimates should not be used to represent spatial uncertainty, as they were never designed to reproduce any spatial statistics. Kriging estimates at two different locations do not reflect the model covariance between these two locations. Kriging maps deliver an incorrect sense of spatial continuity. Marginal variances or two-location correlations can be derived from the conditional covariance in the Gaussian case. More generally, simulation aims at reproducing spatial continuity when a global response to the joint spatial heterogeneity is desired, such as in reservoir flow simulations. Recall that $E(f(\mathbf{y})) \neq f(E(\mathbf{y}))$, so even in the Gaussian case, where Kriging is the optimal predictor, we require function evaluations of f for samples of $\mathbf{y}$ rather than the function evaluated at the Kriging estimate.

4.3.2 Common geostatistical stochastic simulation methods

For visualization purposes, the marginal distributions at different sites may be sufficient, but for prediction and decision making, we often need to generate multiple alternative realizations from the joint distribution. This allows using the set of simulated realizations and their response (e.g., through a flow simulator) as a measure of uncertainty. There are a wealth of tools available for the spatial simulation of random processes, and this discussion only introduces some key concepts. More background can be found in references provided in the bibliography at the end of this chapter.

We first discuss some methods for unconditional and conditional simulation. The former aims to generate spatial simulations from (i) a priori distribution or (ii) a priori notions based on a training image or propagating variables using physical models (typically, differential equations).

Point (i) often requires assumptions about a parametric family of distributions. Once this distribution $p(\boldsymbol{x})$ is established, sampling can be done using the Monte Carlo techniques described in Section 2.5. The conditional simulation is from the posterior distribution $p(\boldsymbol{x} \mid \boldsymbol{y})$, which is available from the prior distribution and by assuming a likelihood model $p(\boldsymbol{y} \mid \boldsymbol{x})$ for the data. Depending on the likelihood model, the conditional simulation from the posterior distribution may be simpler or more difficult than the unconditional simulation.

Point (ii) resembles that of non-parametric modeling, where we enforce less parametric and distributional assumptions. Unconditional simulation relies on templates, objects, or patterns one would like to reproduce in the simulations. Conditioning involves comparing the simulations with data. This may be done directly or via propagating the simulations to the natural data domain.

Recall that both (i) and (ii) try to output realizations that can be used for prediction. Ultimately, prediction may involve complicated transforms of variables – say, basin modeling, fluid flow, or seismic wave propagation. One may compare unconditional predictions with conditional predictions under various data acquisition schemes. This will make it possible to evaluate the usefulness of data in our ability to predict reliably and to eventually make better decisions. The method used for sampling or simulation is very dependent on the situation; nevertheless, we will present some of the main concepts next.

Directly simulating a sample from the unconditional prior $p(\boldsymbol{x})$ or from the conditional situation with posterior $p(\boldsymbol{x} \mid \boldsymbol{y})$ is only possible in a few special and simple cases. In (low-dimensional) Gaussian models, with linear conditioning to the data, this is straightforward. It is also easy to simulate Markov chain models. (Both models are treated in depth in Appendix A.) In more complex settings, it may be possible to use rejection sampling, described in Section 2.5, where we generate realizations from another distribution $q(\boldsymbol{x})$ and then accept the likely ones as exact samples from the target prior or posterior distribution.

Unconditional plus error for the conditional: The idea here is to sample unconditionally from the prior and then perform an update step to get a conditional sample. Under Gaussian modeling assumptions, where the mean is linear in the data, this idea tends to work very well for conditional sampling. The conditional sample is obtained as a linear combination of the unconditional sample and the data. Noise is added such that the posterior sample has the correct covariance. Assume that $p(\boldsymbol{x}) = N(\boldsymbol{\mu},\boldsymbol{\Sigma})$ and $p(\boldsymbol{y} \mid \boldsymbol{x}) = N(\boldsymbol{F}\boldsymbol{x},\boldsymbol{T})$; then, the algorithm is

$$\boldsymbol{x}_{\text{uncond}} \sim N(\boldsymbol{\mu},\boldsymbol{\Sigma}), \qquad \boldsymbol{\varepsilon} \sim N(\boldsymbol{y},\boldsymbol{T})$$
$$\boldsymbol{x}_{\text{cond}} = \boldsymbol{x}_{\text{uncond}} + \boldsymbol{\Sigma}\boldsymbol{F}^t \left(\boldsymbol{F}\boldsymbol{\Sigma}\boldsymbol{F}^t + \boldsymbol{T}\right)^{-1} \left(\boldsymbol{\varepsilon} - \boldsymbol{F}\boldsymbol{x}_{\text{uncond}}\right). \tag{4.8}$$

In the case where the data provide perfect information about $\boldsymbol{x}$ (also known as hard data in traditional geostatistics jargon) at specific locations (i.e., partial perfect information), the conditional realization is updated from the unconditional using the Kriging estimate as follows: $\boldsymbol{x}_{\text{cond}} = \boldsymbol{x}_{\text{uncond}} + \hat{\boldsymbol{x}}_{\text{krig}}^{\text{data}} - \hat{\boldsymbol{x}}_{\text{krig}}^{\text{uncond}}$, where $\hat{\boldsymbol{x}}_{\text{krig}}^{\text{data}}$ is the Kriging estimate using the data, while $\hat{\boldsymbol{x}}_{\text{krig}}^{\text{uncond}}$ is the Kriging estimate using the values from the unconditional simulation at the same spatial locations as the actual data. Since the data locations remain the same, the Kriging matrix can be factored once and used for multiple conditional realizations, leading to computational efficiency. The approach requires efficient generation of samples from the unconditional distribution. This is usually done by a direct method such as spectral methods based on fast Fourier transforms or Cholesky factorization (see also Appendix A.1).

Iterative simulation algorithms like Markov chain Monte Carlo are certainly viable for sampling from the prior $p(\boldsymbol{x})$ and posterior $p(\boldsymbol{x} \mid \boldsymbol{y})$ (see Section 2.5). In this way, one can obtain dependent samples $\boldsymbol{x}^1,\ldots,\boldsymbol{x}^B$. The dependency means that subsequent samples are very similar, but each sample itself is from the right joint distribution. A challenge here is the computer time: the transient phase (burn-in) may be long, and mixing of the Markov chain may be slow.

Sequential simulation algorithms all make use of the same basic procedure that also holds for many of the direct routines for simulation (such as the Cholesky matrix factorization for Gaussian simulation). This procedure relies on the sequential way of writing a joint probability distribution – i.e.,

$$p(\boldsymbol{x}) = p(x_1)p(x_2 \mid x_1)\ldots p(x_n \mid x_{n-1},\ldots,x_1) \qquad \text{unconditional,}$$
$$p(\boldsymbol{x} \mid \boldsymbol{y}) = p(x_1 \mid \boldsymbol{y})p(x_2 \mid x_1,\boldsymbol{y})\ldots p(x_n \mid x_{n-1},\ldots,x_1,\boldsymbol{y}) \qquad \text{conditional.} \tag{4.9}$$

In a spatial setting over a defined domain of sites, one would need to sample one variable (or blocks of variables) at a time. Note that the sequential simulation algorithm must condition on previous simulations and all data. Some implementations of the sequential approach condition only locally. This may induce bias, underestimation, or other unwanted effects in the realizations, the predictions, and the decisions. In some variations of the sequential simulation workflow, the sites are visited not along a random path but along a raster path, working deterministically along columns or rows in the map. Sequential simulation with a raster path can have some advantages for unconditional simulations but can make it harder for conditioning to data ahead of the raster path (Mariethoz and Caers 2015).

Thus, the main difference between various sequential procedures, in terms of practice, is the way in which the local conditional probabilities are assessed. Under strict modeling assumptions, we may know these conditionals exactly, and then this approach becomes a direct method. Commonly, any technique that can approximate this conditional distribution is used as the basis for sequential simulation. Under multi-Gaussian assumptions, Kriging is used to assess the parameters of the conditional distribution (which is also Gaussian), and the algorithm is then called sequential Gaussian simulation.

The conditional distribution could also be obtained from training images that can provide not only two-point but also multiple-point statistics. With this approach, the conditioning to data events extends to patterns of multiple data taken together. Multiple-point statistics can simulate geologically realistic features as long as an appropriate training image is available (which may not always be the case).

The fact that the sequential approach can accommodate any technique for assessing the conditional distribution has made it a very flexible and popular technique, and many methods use the sequential principle even though their names do not contain the word "sequential." Inputs to the sequential simulation include the conditioning data that has to be honored as well as a description of the spatial continuity, either in terms of variograms and cross-variograms or in terms of a training image for multiple-point statistics. Many multiple-point simulation algorithms (see also Section 4.7) follow the sequential procedure, but instead of estimating a local conditional distribution at each pixel on the sequential path, they directly sample multiple-point patterns from the training image.

Object-based simulations or Boolean models are a family of methods working with geometric objects that have some genetic significance rather than being built up one elementary node at a time. They have been useful to generate facies and fracture models in reservoirs but have challenges in terms of conditioning. There are algorithms for conditioning object-based simulations, but they can become computationally intensive when conditioning to a large number of point data (e.g., data in many wells). The idea in object-based simulations is to realistically describe geometry such as sinuous channels, half-ellipses in cross sections, triangular wedges, lobes, disks, etc. In addition, other required specifications include the overall proportions of each shape and the distributions of the parameters describing the shape. These parameters typically include size, the anisotropy ratio, and the orientation of the long axis and rules that describe how the various shapes can be positioned relative to one another. These methods can also be considered as marked point processes, as they are based on a point process with marks (objects) attached to the points of the process. Many different Boolean simulation algorithms are described in Lantuejoul (2002) and Chiles and Delfiner (2012).

Process-based and process-mimicking methods try to capture the physics and chemistry of the geological processes that created the subsurface environment under consideration. They are motivated by the fact that it is very hard to capture all of the features of geology using purely random function models. Geology has random as well as deterministic components governed by the physics and chemistry of erosion, transportation, deposition, heat and fluid flow, and diagenesis. Process-based models numerically solve the governing partial differential equations, computationally forward-simulating the processes that gave rise to the geological architectures. Process-based models can also be run in the physical laboratory rather than numerically on the computer using tank experiments to physically forward-simulate the erosion, transportation, and deposition of sediments giving rise to various geologic architectures. Computer simulations require inputs such as initial and boundary conditions over geologic time (which could be millions of years for basin models), as well as empirical parameters governing rock and fluid properties as a function of temperature and stress. Very realistic geological models can be made with process-based simulations. These models can be conditioned to global and large-scale data but are much harder to condition to dense local point data. They are also computationally very intensive, and it can take days or weeks to run a single simulation. As an alternative to "full physics" differential equation-based simulations, process-mimicking methods attempt to create realistic geologic features using various algorithmic rules without numerically solving any differential equations. Various process-mimicking algorithms (some of them are also called event-based or surface-based methods) have been developed for different types of geological environments. These are computationally much faster to run than the full process-based models. Like the process-based simulations, the process-mimicking simulations are harder to condition to dense point data, but because they take much less time to simulate, iterative or "brute force" rejection methods might be applicable for conditioning these simulations to point data. Further discussions and references to the literature on these methods can be found in Pyrcz and Deutsch (2014) and Mariethoz and Caers (2015).

A combination of different methods is usually implemented for practical spatial modeling. The success of spatial modeling comes from the development of hybrid methods that borrow useful tools from all techniques, combining them for real-case analysis. For example, multiple-point simulation from training images might be used to simulate the facies, which are then populated with petrophysical properties using Gaussian simulation methods. The training image used in the multiple-point simulation might come from a Boolean simulation or perhaps from a process-mimicking method. The process-mimicking simulation might borrow statistics and rules from a physical laboratory simulation or from a "full physics" numerical simulation. Though all methods have shortcomings, all of the methods have been successfully used in actual case studies. One must be wary of the tendency to promote a single method as better than all others. There is no stochastic modeling method that is universally best for all possible subsurface modeling problems. As stochastic modeling becomes more accepted in the Earth sciences, and as more stochastic modeling techniques are developed, the most successful case studies will be those that view the assortment of spatial modeling methods as a tool kit rather than as a silver bullet.

4.4 Gaussian models

The Gaussian random field model is a key construction in geostatistics and is often used as a self-standing model, especially for very-high-dimensional models or very large data sizes. Moreover, it is common to apply the Gaussian model as a building block when constructing hierarchical model formulations. In this section, we discuss traditional but important aspects of the Gaussian model in spatial statistics.

4.4.1 The spatial regression model

The spatial regression model for continuous response data relies on Gaussian variables and linear association. This is possibly the most commonly applied model in spatial statistics – see, e.g., Cressie (1993), Stein (1999), and Banerjee et al. (2004). The model has historical importance related to (universal) Kriging. Moreover, the model provides a natural extension of the usual linear regression model, as it also accounts for the spatially correlated error terms. The attractive computational properties of the Gaussian model make it one of the few applicable models for massive data sets. The Gaussian assumption can often be justified by the central limit theorem, stating that sums and means of random variables converge to Gaussian variables. Thus, even though the response is not really Gaussian, these assumptions can provide useful results in many situations.

The response variable is assumed to be partially explained by (i) explanatory variables, (ii) a smooth Gaussian noise process – a Gaussian random field, and (iii) independent errors. The idea is to incorporate the spatial smoothness and, as a result, obtain more reliable estimates of the regression parameters for improved predictions. Ignoring a spatially structured noise term could give biased estimates and erroneous uncertainty bounds. The following exposition is based on a univariate response variable. At the end of the section, a more general framework is presented. We assume that the process is defined at all locations $\boldsymbol{s} \in \mathcal{D}$, where $\mathcal{D}$ denotes a continuous spatial domain in two or three dimensions. The model for the response $y(\boldsymbol{s})$ at an arbitrary site is

$$y(\boldsymbol{s}) = \boldsymbol{h}^t(\boldsymbol{s})\boldsymbol{\beta} + w(\boldsymbol{s}) + \varepsilon(\boldsymbol{s}), \tag{4.10}$$

where $\boldsymbol{\beta} = (\beta_1, \ldots, \beta_k)$ is the vector of k regression parameters and $\boldsymbol{h}(\boldsymbol{s}) = (h_1(\boldsymbol{s}), \ldots, h_k(\boldsymbol{s}))$ is the vector of k covariates at site $\boldsymbol{s}$. The residual is split into two parts: $w(\boldsymbol{s})$ and $\varepsilon(\boldsymbol{s})$. The spatially structured residual $w(\boldsymbol{s})$ provides dependence, $\mathrm{Cov}(w(\boldsymbol{s}), w(\boldsymbol{s}')) = \Sigma(\boldsymbol{s}, \boldsymbol{s}')$, capturing the effect of unobserved covariates with a spatial pattern. Statisticians often refer to the regression parameters $\boldsymbol{\beta}$ as fixed effects, while the structured Gaussian field $w(\boldsymbol{s})$ is the random effects. The non-structured spatial residual $\varepsilon(\boldsymbol{s})$ is independent white noise with

$\mathrm{Var}\left(\varepsilon(s)\right) = \tau^2$, which can be interpreted as the measurement error. The spatial regression model in Equation (4.10) can also be written as:

$$y(s) = x(s) + N\left(0, \tau^2\right), \quad E\left(x(s)\right) = \boldsymbol{h}^t(s)\boldsymbol{\beta}, \quad \mathrm{Cov}\left(x(s), x(s')\right) = \Sigma(s, s'). \quad (4.11)$$

The latent process is often of key interest to the decision maker. In Equation (4.11), this latent process $x(s)$ is imperfectly observed by $y(s)$.

Discussion: spatial model versus ordinary least squares

The basic linear regression model, which forms the starting point in many contexts of exploratory data analysis, assumes independent non-structured error terms. Here, in the spatial context, the noise process is modeled using the tools of variograms and covariance functions described previously. The covariance structure of the spatial residual $w(s)$ is thus typically characterized by a few parameters describing the scale and correlation range. Customarily, the spatial covariance $\mathrm{Cov}(w(s), w(s')) = \Sigma(s, s')$ is modeled by a stationary, isotropic process – i.e., it only depends on the absolute distance between the locations s and s' (see Table 4.1). For the independent noise process, we assume that $\varepsilon(s) \sim N\left(0, \tau^2\right)$ for all s. This was interpreted as the nugget effect in Section 4.2.

Assume that we can observe the spatial process $y(s)$ and associated covariates $h^t(s)$ at n locations $s_1, \ldots, s_n$. Under the specified assumptions, we can now write the Gaussian regression model as a **hierarchical model**. Let us denote the collection of data by length n vector $\boldsymbol{y} = \left(y(s_1), \ldots, y(s_n)\right)$; the latent random effects by $\boldsymbol{x} = \left(x(s_1), \ldots, x(s_n)\right)$; and the covariates by a size $n \times k$ matrix $\boldsymbol{H}$, where row i is $h^t(s_i)$. Then

$$p(\boldsymbol{x}) = N(\boldsymbol{H}\boldsymbol{\beta}, \boldsymbol{\Sigma}), \qquad p(\boldsymbol{y} \mid \boldsymbol{x}) = N(\boldsymbol{x}, \tau^2 \boldsymbol{I}_n), \quad (4.12)$$

and the marginal pdf of the response (integrating out the random effects) becomes

$$p(\boldsymbol{y}) = N(\boldsymbol{H}\boldsymbol{\beta}, \boldsymbol{C}), \qquad \boldsymbol{C} = \boldsymbol{C}(\boldsymbol{\theta}) = \boldsymbol{\Sigma} + \tau^2 \boldsymbol{I}_n, \quad (4.13)$$

where $\boldsymbol{C}$ is a $n \times n$ covariance matrix, and $\boldsymbol{\theta} = \left(\sigma^2, \eta, \tau^2\right)$ denotes the set of covariance parameters.

The set of observation locations is called the spatial design. A regular design has observation sites $s_1, \ldots, s_n$ on a grid of the spatial domain $\mathcal{D}$. Irregular sampling designs are, however, more common in practice. For instance, in a situation with monitoring sites for precipitation and wind or air pollution, one would tend to place the sites near roads or cities for logistical reasons. On the other hand, remotely sensed data such as satellite data or seismic data are often processed to be represented on a regular grid.

Figure 4.4 shows a realization of a Gaussian random field on the unit square. This is shown on a regular grid in the left display. An irregular sampling of the data of a much

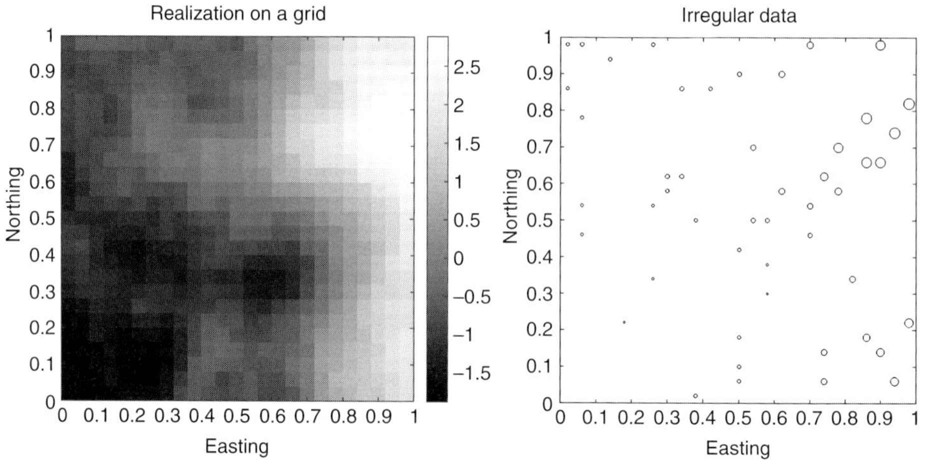

Figure 4.4 Realization of a Gaussian process in two spatial dimensions. Left: the spatial variable represented on a dense regular grid. Right: the spatial variable is represented at only 100 irregular locations.

smaller size is also illustrated in the right display. In Figure 4.4, the model is based on using the east and north coordinates as covariates, and the regression parameters are set to $\boldsymbol{\beta} = (-2,3,1)$, where the first entry corresponds to an intercept term at the origin and the next two are the spatial covariates. The covariance function is a Matern type with smoothness parameter (3/2), as in Table 4.1, and with parameters $\sigma^2 = 0.5^2$, $\eta = 9$, and $\tau^2 = 0.05^2$, which corresponds to a correlation range of about a third of the unit square. Note that there is only a small nugget effect here. In Figure 4.4, we see that the random field increases with the east (and north) coordinate with smooth variability defined by the Gaussian residual process.

We assume that the parameters $\boldsymbol{\beta}$ and $\boldsymbol{\theta}$ are fixed but unknown. These parameters must be estimated based on the data and explanatory variables. A common way of specifying the parameter values is by maximum likelihood estimation (MLE), described in Section 2.4 and Appendix A.1. The Gaussian distribution defines the log-likelihood as a function of parameters $\boldsymbol{\beta}$ and $\boldsymbol{\theta}$ – i.e.,

$$l(\boldsymbol{\theta},\boldsymbol{\beta}) = -\frac{n}{2}\log(2\pi) - \frac{1}{2}\log|\boldsymbol{C}(\boldsymbol{\theta})| - \frac{1}{2}(\boldsymbol{y} - \boldsymbol{H\beta})^t \boldsymbol{C}^{-1}(\boldsymbol{\theta})(\boldsymbol{y} - \boldsymbol{H\beta}). \tag{4.14}$$

The maximum likelihood estimates are defined by

$$(\hat{\boldsymbol{\beta}},\hat{\boldsymbol{\theta}}) = \operatorname{argmax} l(\boldsymbol{\theta},\boldsymbol{\beta}). \tag{4.15}$$

An algorithm for locating the maximum is presented in Appendix A.1. For fixed covariance parameters $\boldsymbol{\theta}$, the estimate for the regression parameters is

$$\hat{\boldsymbol{\beta}} = \hat{\boldsymbol{\beta}}(\boldsymbol{y};\boldsymbol{\theta}) = \left(\boldsymbol{H}^t\boldsymbol{C}^{-1}\boldsymbol{H}\right)^{-1}\boldsymbol{H}^t\boldsymbol{C}^{-1}\boldsymbol{y}, \qquad \operatorname{Var}\left(\hat{\boldsymbol{\beta}}\right) = \left(\boldsymbol{H}^t\boldsymbol{C}^{-1}\boldsymbol{H}\right)^{-1}. \tag{4.16}$$

4.4.2 Optimal spatial prediction: Kriging

We showed how Kriging is the method for optimal pointwise spatial prediction in the Gaussian model. Suppose that we wish to predict responses $\boldsymbol{y}_0 = (y(\boldsymbol{s}_{0,1}),\ldots,y(\boldsymbol{s}_{0,N}))$ at N prediction sites $\boldsymbol{s}_{0,j}$, $j=1,\ldots,N$, given data $\boldsymbol{y}=(y(\boldsymbol{s}_1),\ldots,y(\boldsymbol{s}_n))$ at n observation sites. The joint distribution of $\boldsymbol{y}_0$ and $\boldsymbol{y}$ is Gaussian with the following mean and covariance:

$$p(\boldsymbol{y}_0,\boldsymbol{y}) = N\left(\begin{pmatrix}\boldsymbol{H}_0\\ \boldsymbol{H}\end{pmatrix}\boldsymbol{\beta},\begin{pmatrix}\boldsymbol{C}_0 & \boldsymbol{C}_{0,*}\\ \boldsymbol{C}_{0,*}^t & \boldsymbol{C}\end{pmatrix}\right). \tag{4.17}$$

The $N\times k$ matrix $\boldsymbol{H}_0$ contains the covariates at the prediction sites. Moreover, the $N\times N$ matrix $\boldsymbol{C}_0$ is the covariance matrix for the responses at all prediction sites, while the $N\times n$ matrix $\boldsymbol{C}_{0,*}$ contains the covariances between the N variables at prediction sites and the n variables at observation sites. The covariance matrices depend on the statistical model parameters $\boldsymbol{\theta}$, but the cross-covariance matrix $\boldsymbol{C}_{0,*}$ does not depend on the measurement error variance τ^2 since we assume independent nugget effects.

Recall from Chapter 2 that the conditional pdf of $\boldsymbol{y}_0$ given $\boldsymbol{y}$ (and for fixed $\boldsymbol{\beta}$ and $\boldsymbol{\theta}$) is also Gaussian. The length N vector of conditional means or **Kriging predictions** is

$$E(\boldsymbol{y}_0 \mid \boldsymbol{y}) = \boldsymbol{H}_0\boldsymbol{\beta} + \boldsymbol{C}_{0,*}\,\boldsymbol{C}^{-1}(\boldsymbol{y}-\boldsymbol{H}\boldsymbol{\beta}), \tag{4.18}$$

and the associated $N\times N$ conditional covariance is

$$\mathrm{Var}(\boldsymbol{y}_0 \mid \boldsymbol{y}) = \boldsymbol{C}_0 - \boldsymbol{C}_{0,*}\,\boldsymbol{C}^{-1}\,\boldsymbol{C}_{0,*}^t. \tag{4.19}$$

The conditional variances are defined by the diagonal elements of this matrix. See Appendix A.1 for further details about these properties of the Gaussian pdf.

Discussion: interpreting the Kriging prediction variance

For sites that are close to other data, there is high correlation in $\boldsymbol{C}_{0,*}$, and conditioning will reduce the variances in $\boldsymbol{C}_0$ substantially. Sites that are farther from the observation sites will have larger prediction variances. The reduction of prediction variance also depends on the clustering of the observation sites according to $\boldsymbol{C}^{-1}$. Two data at almost the same location will not contribute twice the information, since the two observations will be correlated and therefore somewhat redundant. It is quite remarkable that the Kriging variances do not depend on the data – they only depend on the geographic locations of the data and the prediction site. This holds for the Gaussian situation but may not hold in non-Gaussian settings where a large (or small) observation may influence the prediction variance as well.

Based on the modeling assumptions, the prediction distribution is also Gaussian. The 5th and 95th percentiles of the standard Gaussian distribution are −1.64 and 1.64. A 90% prediction interval for the response $y(\boldsymbol{s}_{0,j})$ at the prediction site $\boldsymbol{s}_{0,j}$ is then

$$\left(E(y(\boldsymbol{s}_{0,j})\mid \boldsymbol{y}) - 1.64\sqrt{\mathrm{Var}(y(\boldsymbol{s}_{0,j})\mid \boldsymbol{y})},\ E(y(\boldsymbol{s}_{0,j})\mid \boldsymbol{y}) + 1.64\sqrt{\mathrm{Var}(y(\boldsymbol{s}_{0,j})\mid \boldsymbol{y})}\right). \tag{4.20}$$

4.4.3 Multivariate hierarchical spatial regression model

Recall that the spatial regression model can be written as a hierarchical model as in Equation (4.12) or a random effects model, with the likelihood function $p(\boldsymbol{y} \mid \boldsymbol{x})$ conditional on the latent process $\boldsymbol{x}$, and a Gaussian random field prior model. One example is a multivariate hierarchical model. Consider, for instance, the case with reservoir characterization: the (multivariate) latent distinction of interest are the porosity, permeability, and saturation variables at all reservoir grid cells. Seismic data, which provide imperfect information about the reservoir properties, consist of (multivariate) angle gathers of reflection amplitudes at each spatial grid cell. There could also be well data providing perfect information at some cells. These multivariate situations can be conveniently modeled using a hierarchical framework.

Suppose we have K latent spatial processes denoted $\boldsymbol{x} = (\boldsymbol{x}_1, \ldots, \boldsymbol{x}_K)$, each of length n, and L different response variables $\boldsymbol{y} = (\boldsymbol{y}_1, \ldots, \boldsymbol{y}_L)$ of lengths $m_1, .., m_L$. The data can be acquired at different locations for the various responses. Some may have large spatial coverage with perhaps more uncertainty (say, seismic data), while others may be sparsely sampled but carry accurate information where they are available (say, well data). The hierarchical model can easily incorporate such varying dimensions in the data – i.e., $m_k \neq m_l$, $k, l \in \{1, \ldots, L\}$.

Suppose that we assign a Gaussian prior model $p(\boldsymbol{x}) = N(\boldsymbol{\mu}, \boldsymbol{\Sigma})$ at the top of the hierarchy, where $\boldsymbol{\mu}$ is a size nK vector containing the prior mean of the latent variables at the n locations of interest, while Σ is a $nK \times nK$ covariance matrix for the K processes. The mean may, of course, include covariates as in the usual regression setting. A convenient way of modeling a multivariate spatial covariance matrix is $\mathrm{Cov}(x_k(\boldsymbol{s}), x_l(\boldsymbol{s}')) = R(\boldsymbol{s}, \boldsymbol{s}')\Sigma_0(k, l)$, where we use a separable structure in the space and multiple variable dimensions. Spatial correlation is defined by $R(\boldsymbol{s}, \boldsymbol{s}')$, while the $K \times K$ matrix $\boldsymbol{\Sigma}_0$ defines the spatially invariant covariance matrix between the K variables. The full covariance matrix is, in this way, constructed by a Kronecker product between the spatial correlation matrix and the intervariable covariance matrix.

Suppose that there is a linear Gaussian likelihood model for the multivariate spatial data – i.e., $p(\boldsymbol{y} \mid \boldsymbol{x}) = N(\boldsymbol{F}\boldsymbol{x}, \boldsymbol{T})$, where $\boldsymbol{F}$ is a $\left(\sum_{l=1}^{L} m_l\right) \times nK$ matrix containing the design of experiments or forward models of the different data, conditional on $\boldsymbol{x}$. If the forward model is a result of a local operator, this matrix can be decomposed in blocks – say, $\boldsymbol{F}_0$ – which is a $L \times K$ matrix that couples the $\boldsymbol{x}$ latent variables at a spatial location to the L response variables at the same location (if all data types are available at that location). The matrix $\boldsymbol{T}$ defines the covariance of the measurement noise. In the simplest case, the matrix is diagonal $\boldsymbol{T} = \tau^2\boldsymbol{I}$, but a more flexible structure would allow for different variances in the various response variables – for instance, there could be different diagonal terms $\tau_1^2, \tau_2^2 \ldots, \tau_L^2$ in the covariance matrix $\boldsymbol{T}$. There might also be correlation in the likelihood model for the L response variables, resulting in a block diagonal structure, or some sort of smoothing in the data acquisition scheme.

As in the previous description, the marginal likelihood becomes

$$p(\boldsymbol{y}) = \int p(\boldsymbol{y} \mid \boldsymbol{x}) p(\boldsymbol{x}) d\boldsymbol{x} = N(\boldsymbol{F}\boldsymbol{\mu}, \boldsymbol{F}\Sigma\boldsymbol{F}^t + \boldsymbol{T}). \tag{4.21}$$

The model parameters can again be estimated by likelihood maximization. In general, it can be hard to identify all parameters based on the marginal likelihood in Equation (4.21). For instance, it may be difficult to separate out parameters in $\boldsymbol{F}$ and $\boldsymbol{\Sigma}$. Alternatively, we could specify parameters from auxiliary data sources. The estimation simplifies if one has observed both $\boldsymbol{x}$ and $\boldsymbol{y}$ for similar conditions, because one can separate out the effects. Prediction of the multivariate latent variable, given all the multivariate spatial data, can again be conducted from the Gaussian posterior model:

$$p(\boldsymbol{x} \mid \boldsymbol{y}) = N(\boldsymbol{\mu} + \Sigma\boldsymbol{F}^t \left(\boldsymbol{F}\Sigma\boldsymbol{F}^t + \boldsymbol{T}\right)^{-1} (\boldsymbol{y} - \boldsymbol{F}\boldsymbol{\mu}), \Sigma - \Sigma\boldsymbol{F}^t \left(\boldsymbol{F}\Sigma\boldsymbol{F}^t + \boldsymbol{T}\right)^{-1} \boldsymbol{F}\Sigma). \tag{4.22}$$

4.4.4 Examples

Norwegian wood: forestry example

Keywords: *forestry example, Gaussian random field, spatial regression model, maximum likelihood estimation, spatial prediction, Kriging, partial information*

Let us illustrate parameter estimation and prediction for a Gaussian random field example. Consider a regular grid of size 25 × 25, where the variable on this grid is observed only imperfectly at irregular sites. The goal is to use the data to estimate regression parameters and predict on the regular grid.

The example is motivated from forestry, where the grid of cells may represent forest units that a farmer can choose to harvest. The profits of timber or wood at the cells are related to the height and volume of the trees, as well as the cost of harvesting and processing timber, and the selling price for products. For simplicity, we assume that the profits are Gaussian distributed with spatial correlation. In the current section, the goal is to estimate model parameters and predict the spatial distribution of forest profits (which depends on the spatial distribution of the trees). We return to this example to study VOI analysis in Chapter 5.

In this example, the realization to the left in Figure 4.4 is treated as the "truth" or reference spatial distribution. Would it be possible to estimate the parameters of the Gaussian model and predict the spatial profits at the 625 units, given data acquired at some sites? Two different designs are evaluated and compared. These are representative of acquisition designs a farmer could use for gathering information about the forest volumes and profits. Both of these designs have 49 observation sites. A random design and a center design are illustrated in Figure 4.5. The center design is based on acquiring measurements along the center lines (north–south and east–west) in the grid – i.e., 49 data locations in total – while the random measurement sites are simply generated by selecting 49 random data locations among the 25 × 25 grid cells.

Table 4.2 shows the maximum likelihood estimates for the regression and covariance parameters along with their asymptotic standard errors computed from the likelihood

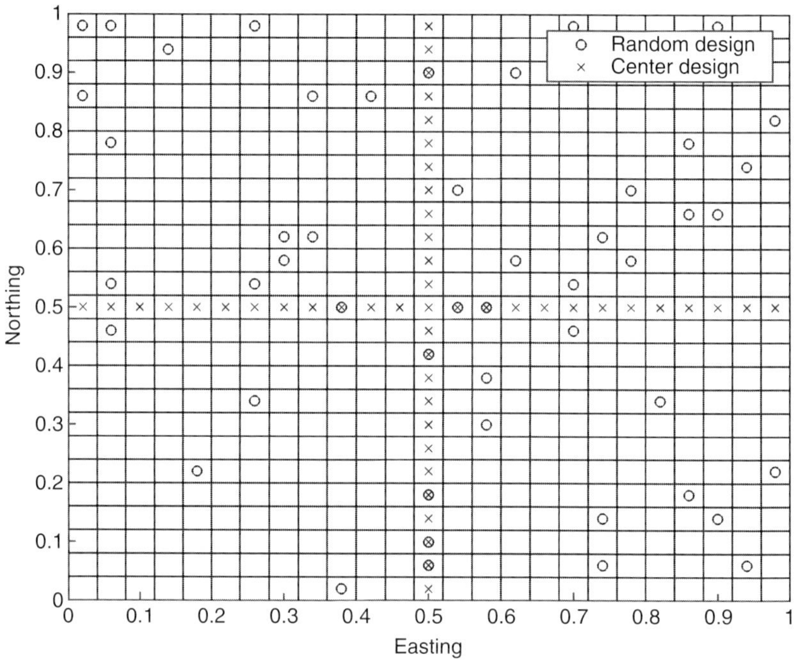

Figure 4.5 Random and center designs of spatial experiments. There are 49 data locations in both designs. The distinction of interest is represented on a regular grid of size 25 × 25.

Table 4.2. *Parameter estimates and standard errors (in parentheses) for the Gaussian regression model with center line and random acquisition design*

	β_1	β_2	β_3	σ^2	η	τ^2
Center	−2.1 (0.6)	3.4 (0.7)	0.4 (0.7)	0.3 (0.14)	7.2 (2.0)	0.002 (0.001)
Random	−2.0 (0.5)	3.4 (0.6)	0.8 (0.5)	0.3 (0.12)	7.9 (2.0)	0.005 (0.007)
Truth	−2	3	1	0.25	9	0.0025

expression. These are shown for the particular realization and for both center and random designs. We note that the regression parameters are estimated quite well from the data, and there is little difference in the variability of estimates. For traditional regression analysis, it is beneficial to acquire data at the edges of the domain, because that makes it easier to estimate the slope of the regression curve. The situation is a little different with spatial correlation because the data at common edges tend to be highly correlated.

In Figure 4.6, we show the pointwise prediction and prediction standard error based on the 49 measurements of the Gaussian process and using the parameter estimates in Table 4.2. The displays are represented on the 25 × 25 grid. They show the prediction surface (left) for

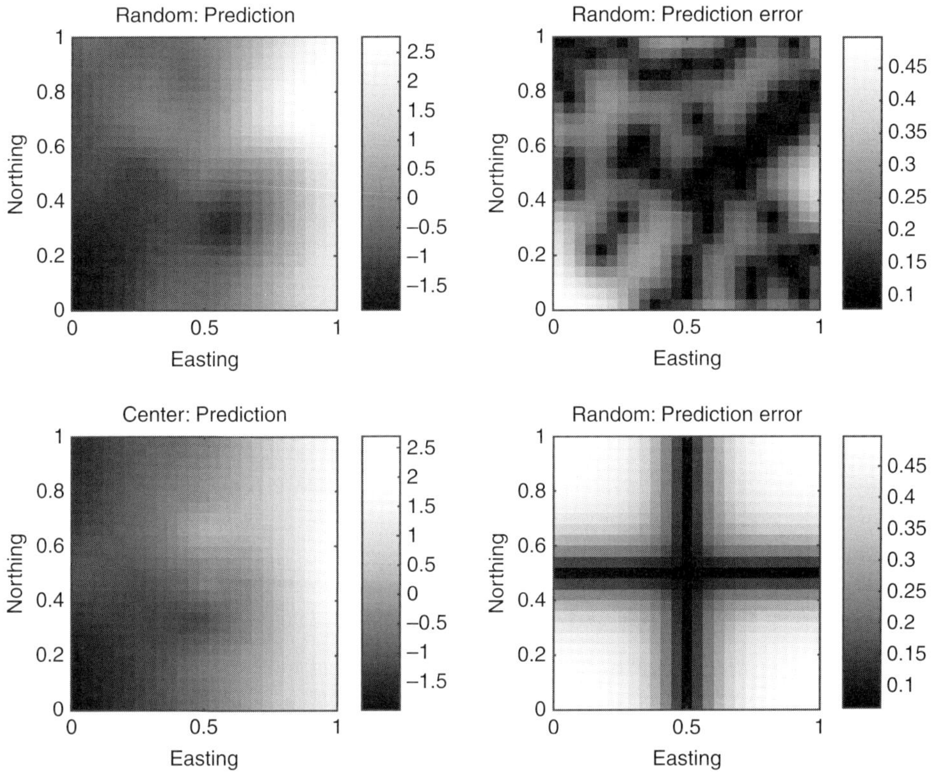

Figure 4.6 Prediction results for the random design (top) and the center design (bottom). The Kriging predictor is in the left display with the associated standard deviations in the right display.

both the random and center designs and the associated prediction standard errors (right). Note how the predictions are smoother than the realization in Figure 4.4. The Kriging predictions in Figure 4.6 are based on linear interpolation of the data $\boldsymbol{y}$, and in this case these predictions become very smooth with little data. The random design has a better spatial coverage of the domain, and the prediction results appear more realistic than for the center design. The prediction results for the center design reflect more details in the central parts but just return the effect of the east and north covariate trends in the corners of the grid. The prediction standard errors are very small near the observation sites. These increase to the level of the process noise at a distance indicated by the correlation range of the random field. For both designs, we see the clear imprint of the acquisition pattern along the acquisition lines or points.

Table 4.3 shows a summary of the prediction results for the center and random design data acquisition. This is summarized by the empirical MSPE and the 90% prediction interval in Equation (4.20) at two sites: $\boldsymbol{s}_0 = (0.5, 0.5)$ and $\boldsymbol{s}_0 = (0.1, 0.1)$. The empirical MSPE for this particular realization is defined by

$$\mathrm{MSPE} = \frac{1}{625} \sum_{i=1}^{625} \left(y(\boldsymbol{s}_{0,i}) - E(y(\boldsymbol{s}_{0,i}) \mid \boldsymbol{y}) \right)^2. \tag{4.23}$$

Table 4.3. *Summary of prediction results for the center data acquisition and the random data acquisition design for a Gaussian process*

	MSPE	Prediction Interval (90%) $s_0 = (0.5, 0.5)$	Truth $s_0 = (0.5, 0.5)$	Prediction Interval (90%) $s_0 = (0.1, 0.1)$	Truth $s_0 = (0.1, 0.1)$
Center	0.24	(−0.48, −0.27)	−0.24	(−2.47, −0.89)	−1.64
Random	0.03	(−0.52, −0.06)	−0.24	(−2.28, −0.99)	−1.64

The MSPE is much larger for the center design because of its poor data coverage away from the central domains. Naturally, the interval based on the center design is narrower at the central site. The prediction interval for the random design is narrower than the center design for site $s_0 = (0.1, 0.1)$. Note that these prediction intervals only cover the truth 9 out of 10 times by construction. For the central prediction site, using a center design, the truth is just outside the interval.

I love rock and ore: mining oxide grade example

Keywords: *oxide grade example, mining, X-ray data, Gaussian random field, spatial regression model, maximum likelihood estimation, spatial prediction, Kriging*

This data set consists of oxide grade measurements made along boreholes in a mine (Eidsvik and Ellefmo 2013). It was discussed briefly for parameter estimation in Chapter 2. The goal of the mining company is to use data from the current boreholes to predict the spatial distribution of the oxide grade. The data set consists of bivariate response variables as follows:

XMET data: Core samples measured with a handheld X-ray fluorescence meter (XMET). These XMET data are relatively fast to collect on the mining location. They are considered to be imperfect measurements of the true oxide grade of the samples. There are $m_1 = 1871$ locations with XMET data.

XRF data: Lab experiments of the cores using careful X-ray fluorescence (XRF) analysis. These XRF data are rather time consuming to analyze. They are considered to provide perfect information about the oxide grade at the location where they are acquired. There are $m_2 = 103$ locations with XRF data.

At the 103 locations where there are XRF data, there are also XMET data. Figure 4.7 shows a map view of the mine with the data locations indicated. All 1871 XMET data and 103 XRF data are used to predict the oxide along the vertical plane illustrated in the display.

A spatial regression model is used for the spatial oxide grade – i.e., $p(\boldsymbol{x}) = N(\boldsymbol{H\beta}, \Sigma)$, where the regression part is based on an intercept term and a categorical mineralization covariate. There are three mineralization classes (I, II, and III), where Class III is most likely to have high grades of oxide. Representing this situation into the mathematical framework outlined earlier, we have $K = 1$ variable of interest: the oxide grade at location s_0, denoted by $x(s_0)$. The observations are either XMET or XRF – i.e., $L = 2$. We have

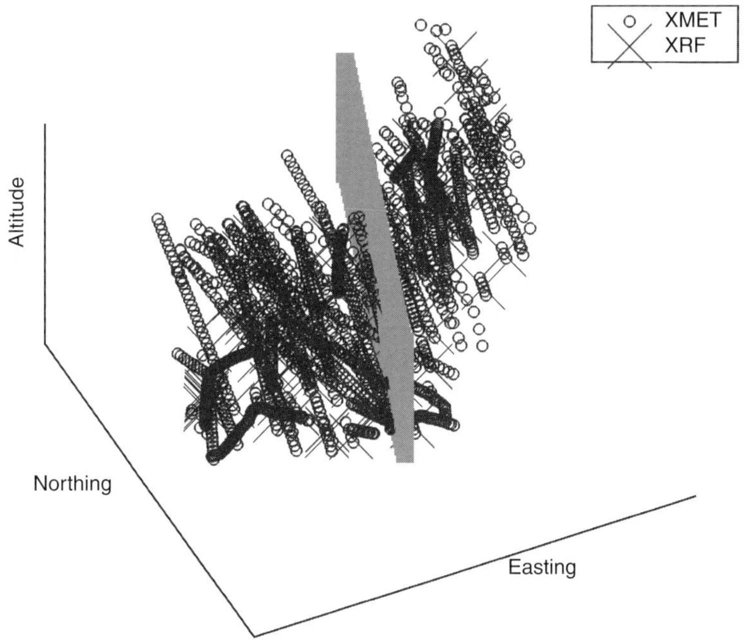

Figure 4.7 Map view of the oxide grade data in a mine. Two kinds of data are acquired from the boreholes: there are 103 XRF data (perfect information, illustrated by cross) and 1871 XMET data (imperfect information, illustrated by circles). The oxide grades are predicted on the grayscale vertical plane.

XRF data at $m_1 = 103$ locations: $\boldsymbol{y}_1 = \left(y\left(\boldsymbol{s}_{1,1}\right),\ldots,y\left(\boldsymbol{s}_{1,103}\right)\right)$ and XMET data at $m_2 = 1871$ locations: $\boldsymbol{y}_2 = \left(y\left(\boldsymbol{s}_{2,1}\right),\ldots,y\left(\boldsymbol{s}_{2,1871}\right)\right)$, where 103 of the locations in XRF and XMET are identical.

Even though the main goal is prediction, the mining company must first specify the statistical model parameters and study whether the Gaussian modeling assumptions seem appropriate. Similar to Eidsvik and Ellefmo (2013), the current description relies on using probability quantile plots to justify a Gaussian model for the oxide grade data and the measurements. The first step in this analysis is to fit a least-squares model to the XMET data and compare the residuals of the fit with the theoretical counterparts of the Gaussian pdf.

Figure 4.8 shows the histogram of the residuals (left) and a quantile–quantile plot of sorted empirical residuals against the theoretical percentiles. The fit is acceptable, possibly with some differences in the tail of the distribution.

Figure 4.9 shows the empirical variogram of the residuals after the least-squares estimation of the mean (regression) parameters. The empirical computation is done by binning the data into 5-m interval zones for the distance between measurement locations. Based on the variogram plot, it appears as if the nugget effect is about 0.3, the effective correlation range is between 50 and 100 m, and the structured variance term is about $\sigma^2 = 0.7 - 0.3 = 0.4$.

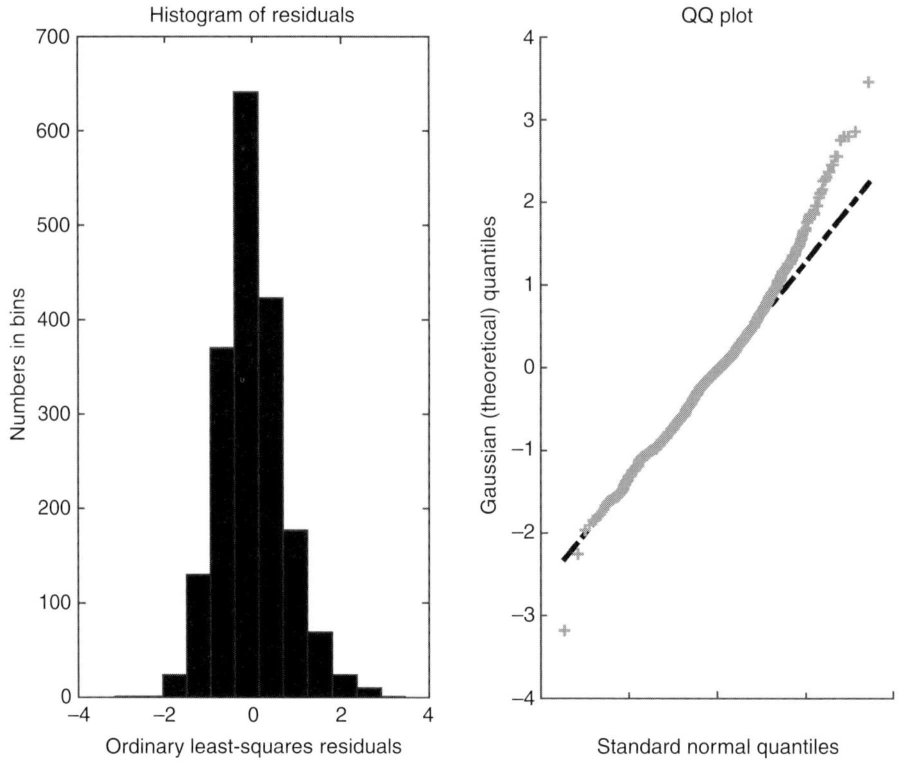

Figure 4.8 Histogram (left) and quantile–quantile plot of the residual of XMET data after a least-squares fit. The purpose of this exercise is to see whether the Gaussian regression model is appropriate for this data set. The fit looks reasonable, possibly with some missing skewness in the tails.

A Matern (3/2) spatial covariance model is used to describe the Gaussian random field for the oxide. The XMET data are assumed to be conditionally independent, given the grades at the measurement locations. We compute the maximum likelihood estimates of all the statistical model parameters using the Fisher scoring algorithm (Appendix A.1) with the least-squares regression estimates and variogram parameters as starting values. The regression parameters are $\hat{\beta}_0 = -0.18$ for the regression intercept and $\hat{\beta}_1 = 1.32$ for the slope with mineralization class. The covariance parameters are $\hat{\sigma}^2 = 0.62^2$ for the spatially varying oxide grade, with an effective range of about 50 m and a measurement noise variance of $\hat{\tau}^2 = 0.45^2$ for the XMET data.

One can predict the oxide grade at selected locations from the parameter estimates. As we discussed earlier, the basis for prediction is to form a joint Gaussian model for the variable of interest and all available data, which in this case involves the oxide grade at prediction sites as well as the XRF and the XMET data. In the prediction, we must incorporate how the mean value depends on the location through the mineralization covariate (Class I, II, or III). Moreover, we must account for the correlation in the oxide grade and the bivariate data (XRF and XMET).

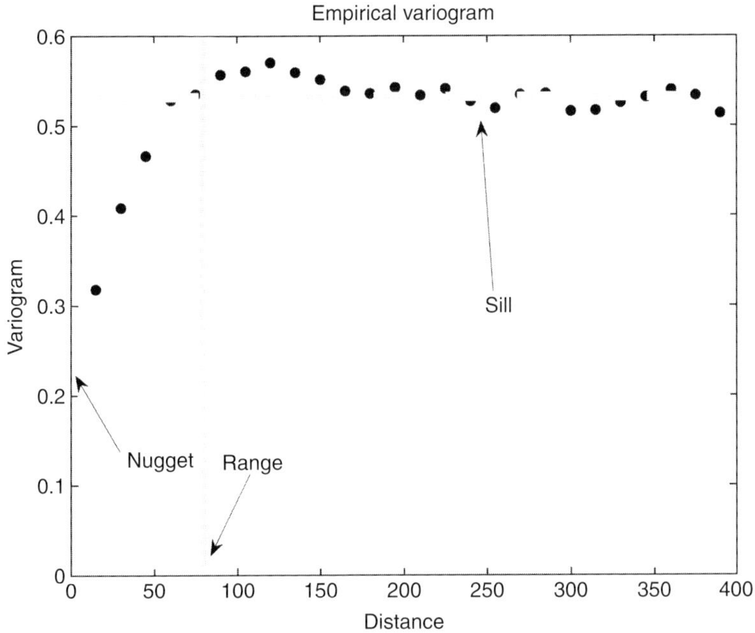

Figure 4.9 Empirical variogram estimate for the oxide grade data set. The variogram appears to meet the second axis above the origin (indicating a nugget effect). It increases to its sill at around a range of 50–100 m.

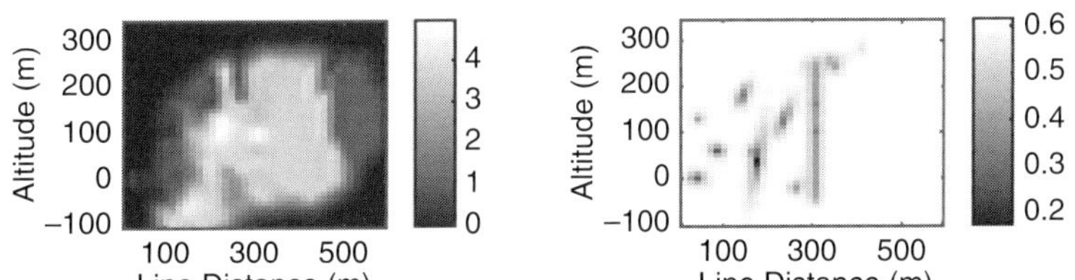

Figure 4.10 Predictions (left) and prediction standard error (right) of the oxide grade along the vertical grayscale profile in Figure 4.7.

In Figure 4.10, we show the prediction and the prediction standard error of oxide grade at the vertical cross section illustrated in Figure 4.7. The prediction clearly illustrates the ore body with higher grades but also shows spatial variability. In the standard error display, we recognize the nearby boreholes where there is low uncertainty. The prediction variance is much higher at prediction locations far away from the boreholes.

Chapter 6 describes VOI analysis for this case. The question there is whether the mining company should collect additional data (XRF or XMET) to improve the predictability of the oxide grade and to make better decisions about the development of the mine.

4.5 Non-Gaussian response models and hierarchical spatial models

The Gaussian model is possibly overused (and abused) in statistical applications, and it should be applied with care. Residual plots, cross-validation, and other techniques should be used to check whether the Gaussian modeling assumptions are appropriate. These approaches may also indicate which other models may be reasonable. Are there non-linearities in the data? Are the data skewed? Is there a clear increase in variance with increased response? Are the data discrete, and is a Gaussian assumption not suitable?

In some of these situations, one may benefit by working with models inspired by the Gaussian pdf or models using the Gaussian distribution as an important building block. In this section, we present some spatial models that incorporate skewness and count data. We focus on extending the analytical closed-form solution of the Gaussian distribution. Some other popular approaches for non-Gaussian response and hierarchical models are discussed in the bibliography.

4.5.1 Skew-normal models

Scientists use the Gaussian model extensively, largely due to the computational advantages gained from having closed-form solutions. For instance: the sum of two Gaussian variables is also Gaussian; when the joint pdf is Gaussian, the marginal pdfs are also Gaussian; when we have linear conditioning on data, the conditional pdf is Gaussian. We present and motivate a skew pdf that has a closed-form pdf under sums, marginalization, and conditioning.

The **skew-normal** distribution extends the Gaussian distribution by enforcing skewness in a selected direction. Define the pdf of the variable of interest $\boldsymbol{x} = (x_1, \ldots, x_n)$ by

$$p(\boldsymbol{x}) = 2N(\boldsymbol{\mu}, \Sigma)\Phi\left(\boldsymbol{\lambda}^t \Sigma^{-\frac{1}{2}}(\boldsymbol{x} - \boldsymbol{\mu})\right). \tag{4.24}$$

Here, $\Phi(\cdot)$ is the cumulative distribution function of the standard normal; i.e., $\Phi(x) = \int_{-\infty}^{x} \frac{1}{\sqrt{2\pi}} \exp\left(-\frac{z^2}{2}\right) dz.$

Thus, if the variable of interest $\boldsymbol{x}$ is in the dimension penalized by the skewness weights $\boldsymbol{\lambda} = (\lambda_1, .., \lambda_n)$, then the cumulative distribution is close to 0 and the Gaussian pdf $N(\boldsymbol{\mu}, \Sigma)$ is dampened. If the variable is in the dimension not penalized by the skewness weights, then the cumulative is close to 1. This leads to a skewed pdf that increases slower than the Gaussian pdf on one side and dies out like a Gaussian at the other side.

We motivate the use of skew-normal models with an example of saturation and porosity variables in a petroleum reservoir and seismic amplitude-versus-offset (AVO) response variables at the subsurface horizon representing the top of the reservoir. In the example, we consider porosity and saturation variables obtained from the Society of Petroleum Engineers (SPE) 10 data set, which has been used substantially as a test study in petroleum reservoir characterization (Christie and Blunt 2001). We use forward models for predicting the seismic responses from saturation and porosity variables. In our case, this entails rock

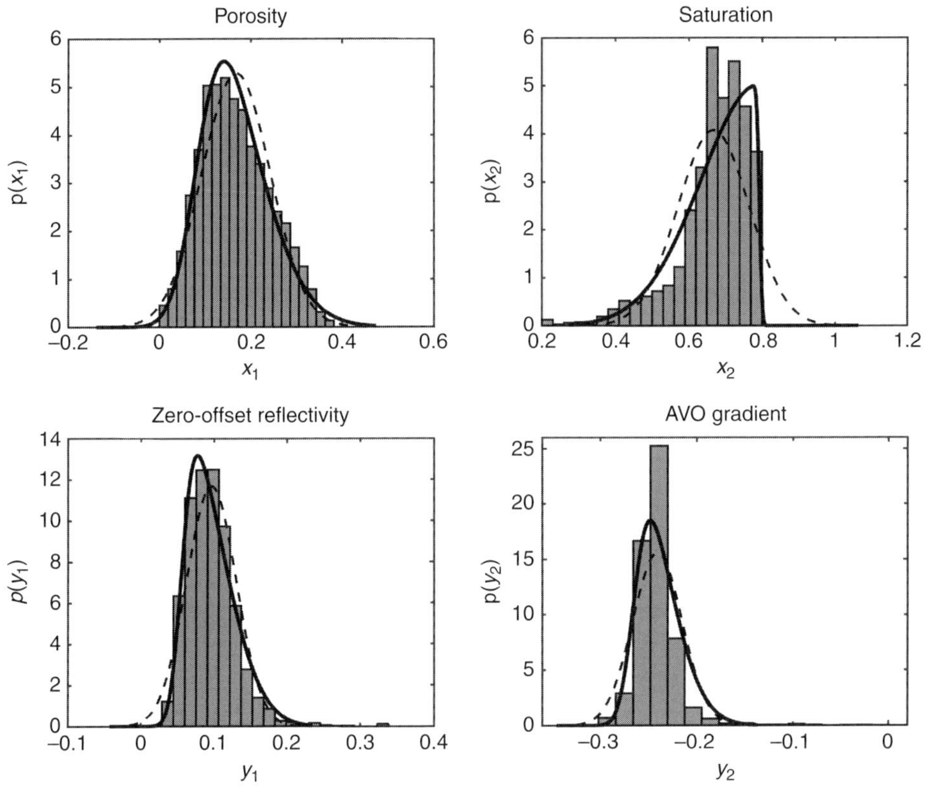

Figure 4.11 Gaussian and skew-normal fit to reservoir variable porosity (top left) and saturation (top right), as well as synthetic seismic amplitude data for zero offset (bottom left) and amplitude versus offset (AVO) gradient (bottom right). The modeling is based on the Society of Petroleum Engineers (SPE) 10 data set.

physics modeling to create elastic moduli from which synthetic seismic reflection data are computed at the top reservoir as a function of the incidence angle, resulting in two seismic amplitude attributes (the (AVO) attributes): the zero-offset reflectivity and the AVO gradient describing the amplitude variation with source-receiver offset.

The applicability of Gaussian or skew-normal pdfs may be seen by plotting the parametric modeling fit with the empirical distribution of saturation, porosity, and seismic reflection data. Figure 4.11 shows the Gaussian fit, the skew-normal fit, and the empirical histogram for this SPE 10 data set. The skew-normal pdfs give notably better fit than the Gaussian for the brine saturation and the seismic attributes. This indicates that the reservoir variables and data are skewed, possibly driven by lithologic changes in the reservoir. There is not much difference between the Gaussian and the skew-normal pdf for the porosity variable (Rezaie et al. 2014).

These skew models have been used in the spatial domain as straightforward extensions of the Gaussian pdfs – see, e.g., Kim and Mallick (2004) and Allard and Naveau (2007). In its simplest construction, the spatial skew-normal model borrows from the Gaussian regression model described in Section 4.4. First, the spatial variables $\boldsymbol{x} = (x(\boldsymbol{s}_1),\ldots,x(\boldsymbol{s}_n))$ have mean vector $\boldsymbol{\mu} = \boldsymbol{H}\boldsymbol{\beta}$, which includes regression parameters. Note, however, that this is no longer the mean, because the first moment or expectation would also depend on the other model parameters in this non-Gaussian situation. Second, a spatial covariance function is built into the matrix Σ. Note similarly that this is no longer the covariance matrix, because the second moment of the skew-normal distribution is defined via the skewness weights as well. Nevertheless, the matrix entries are indicative of the smoothness in the spatial variable. A particular low-dimensional form is often used for the skewness weights $\boldsymbol{\lambda}$. For instance, $\boldsymbol{\lambda} = \lambda_1 \mathbf{1}$ is only one additional model parameter, λ_1, on top of the Gaussian formulation, or $\boldsymbol{\lambda} = \lambda_1 \mathbf{1}\Sigma^{1/2}$ canceling the effect of the covariance in the cumulative function in the pdf for the skew normal. There are several ways of incorporating skewness.

A challenge of the usual skew-normal pdf is that skewness is only imposed in one direction: the direction defined by the length n vector $\boldsymbol{\lambda}^t \Sigma^{-\frac{1}{2}}(\boldsymbol{x}-\boldsymbol{\mu})$. To overcome this, the **closed skew-normal** formulation additionally allows multivariate (size q) skewness directions and a non-standardized cumulative distribution function. Let us denote the multivariate Gaussian pdf by $\phi_n(\boldsymbol{x};\boldsymbol{\mu},\Sigma) = N(\boldsymbol{\mu},\Sigma)$, and let the multivariate Gaussian cumulative distribution function be $\Phi_q(\boldsymbol{x}) = \iint_{z<x} \phi_q(z;\boldsymbol{\mu},\Sigma)\,dz$. Then, the closed skew-normal pdf is defined by

$$p(\boldsymbol{x}) = \frac{\phi_n(\boldsymbol{x};\boldsymbol{\mu},\Sigma)\Phi_q(\Lambda(\boldsymbol{x}-\boldsymbol{\mu});\boldsymbol{\zeta},\Delta)}{\Phi_q(0;\boldsymbol{\zeta},\Delta+\Lambda\Sigma\Lambda^t)}. \tag{4.25}$$

The skewness matrix $\boldsymbol{\Lambda}$ is of size $q \times n$, enabling much more flexibility than in the skew-normal pdf. Moreover, the size $q \times 1$ center point of the skewness dimension $\boldsymbol{\zeta}$ and the $q \times q$ scale matrix $\boldsymbol{\Delta}$ allow closed-form distributions under marginalization and conditioning. Suppose that we have a prior pdf $p(\boldsymbol{x})$ and a likelihood model $p(\boldsymbol{y} \mid \boldsymbol{x})$ that is also closed skew normal and with a linear relation for the center parameter of the likelihood model. Then, the posterior pdf $p(\boldsymbol{x} \mid \boldsymbol{y})$ will also be closed skew normal, as will the marginal likelihood $p(\boldsymbol{y})$ (see, e.g., Dominguez-Molina et al. (2003) and Rezaie et al. (2014) and a related exercise in Chapter 7).

It is useful to have a skew distribution with analytic, closed-form marginal and conditional pdfs. The closed skew normal is thus a pdf with added flexibility compared with the usual Gaussian assumptions, and it still allows for fast computation. However, it can be hard to identify the model parameters in the closed skew model from data – say, using MLE. Moreover, interpreting the results of the model may be demanding because of interactions between model parameters. For instance, the mean and variance are no longer defined directly by $\boldsymbol{\mu}$ and $\boldsymbol{\Sigma}$. Finally, it is not trivial to evaluate the multivariate cumulative distribution function in large dimensions, but useful approximations exist

(Genz and Bretz 2009). Thus, even though the model extends the Gaussian, the additional flexibility comes at a cost.

4.5.2 Spatial generalized linear models

Generalized linear models (GLMs) have become very popular in statistics to model non-Gaussian responses (see Appendix A.2.) When one has acquired or considers acquiring spatial count or absence/presence data, the spatial versions of the GLM become relevant. Examples include the number of joints in a mine (see example later), the counts or presence or absence of a plant at a possible biological conservation site, and so on.

If the data are non-Gaussian, the assumptions of the Gaussian regression model cannot be justified, but one can either transform the data or build a hierarchical model. The former approach transforms the data to responses that appear Gaussian. Commonly, a log or square root transform is used in applications. This approach is more difficult for response variables with a discrete sample space. The hierarchical models are then more attractive, using the Gaussian random field as a latent process incorporating spatial smoothness. For spatial GLMs, this approach was pioneered by Breslow and Clayton (1993) and Diggle et al. (1998).

The spatial GLM likelihood is based on the exponential family model, which includes the **binomial and Poisson distributions**. The parameters of the distributions are now spatially varying and typically incorporate spatial correlation according to a log or logistic Gaussian random field. The likelihood model is commonly assumed to be conditionally independent, given the latent Gaussian process. The usual statistical tasks of parameter estimation and prediction can then be done borrowing ideas from standard GLM analysis. In mathematical notation, $p(\boldsymbol{x}) = N(\boldsymbol{\mu}, \Sigma)$ for latent variable $\boldsymbol{x} = (x(\boldsymbol{s}_1),..., x(\boldsymbol{s}_n))$ represented at n locations, with mean $\boldsymbol{\mu} = \boldsymbol{H\beta}$ incorporating the effect of covariates and the spatial covariance matrix $\boldsymbol{\Sigma} = \boldsymbol{\Sigma}(\boldsymbol{\theta})$. The likelihood is denoted $p(y(\boldsymbol{s}_i) \mid x(\boldsymbol{s}_i))$ for $i = 1,..., m$, where we assume that m of the n sites of interest are observed. For the Poisson distribution, the spatially varying mean in the Poisson likelihood is $V_i \exp(x(\boldsymbol{s}_i))$ for a fixed time window or spatial volume V_i.

The main goals are estimation of statistical model parameters and predicting spatial variables. The specification of model parameters cannot be done from the empirical variogram because the data are counts and not intensity. MLE is also complicated because there is no closed-form expression for $p(\boldsymbol{y})$, $\boldsymbol{y} = (y(\boldsymbol{s}_1),..., y(\boldsymbol{s}_m))$. Similarly, the posterior pdf $p(\boldsymbol{x} \mid \boldsymbol{y}) \propto p(\boldsymbol{x}) p(\boldsymbol{y} \mid \boldsymbol{x})$ is not analytically available. However, approximate computation techniques are often useful for specifying model parameters and for spatial prediction in GLMs.

Consider first the prediction of the latent intensity $x(\boldsymbol{s})$ at location $\boldsymbol{s}$, given data $\boldsymbol{y}$, which can be summarized by the posterior pdf $p(\boldsymbol{x} \mid \boldsymbol{y})$. Since there is no closed-form expression, this pdf must be approximated, and one possibility is to fit a Gaussian approximation at the posterior mode. This approach requires solving an optimization problem to identify the mode of the posterior $\hat{\boldsymbol{\mu}}_{x|y} = \arg\max_x \{p(\boldsymbol{x})p(\boldsymbol{y} \mid \boldsymbol{x})\}$ and assessing the curvature at the mode of the posterior via the Hessian $\frac{d^2 \log(p(\boldsymbol{x})p(\boldsymbol{y} \mid \boldsymbol{x}))}{d^2\boldsymbol{x}}$, evaluated at the mode – i.e., $\boldsymbol{x} = \hat{\boldsymbol{\mu}}_{x|y}$. The mode can be detected by an iterative search that, at each step, expands the exponent of the GLM likelihood in a quadratic form. With a Gaussian prior, this results in a quadratic form for the posterior expression – i.e., a Gaussian approximation. Its mean is used to expand the GLM likelihood for the next iteration, and so on.

Consider next the estimation of parameters. An approximation to the marginal is provided by the Laplace approximation (see Chapter 2 and Appendix A.2):

$$\hat{p}(\boldsymbol{y};\boldsymbol{\beta},\boldsymbol{\theta}) = \frac{p(\hat{\boldsymbol{\mu}}_{x|y};\boldsymbol{\beta},\boldsymbol{\theta}) \prod_{i=1}^{m} p(y_i \mid \hat{\mu}_{x_i|y})}{\hat{p}(\hat{\boldsymbol{\mu}}_{x|y} \mid \boldsymbol{y};\boldsymbol{\beta},\boldsymbol{\theta})}, \tag{4.26}$$

where the numerator is defined by the model, and the denominator is the Gaussian approximation at the posterior mode. The evaluation of the marginal likelihood approximation in Equation (4.26) requires an optimization step for the latent variable $\boldsymbol{x}$ for the fixed parameter values $(\boldsymbol{\beta},\boldsymbol{\theta})$.

The Laplace approximation tends to perform rather well for spatial GLMs since likelihood models such as the Poisson are a part of the exponential family of distributions, which does not impose any severe non-linearities or non-Gaussianity. The approximation may not be done so easily for more complex likelihood models outside the GLM class. For instance, some likelihood models impose a multimodal posterior distribution for the variables of interest. In that case, it may be possible to use analytical solutions as proposal distributions in a Monte Carlo sampling scheme for parameter estimation or the prediction of spatial effects.

4.5.3 Example

We will rock you: rock hazard example

Keywords: *rock hazard example, mining, joint frequency data, spatial generalized linear model, Laplace approximation, maximum likelihood estimation, spatial prediction*

The data consist of joint counts acquired to study the stability of the rock mass in a Norwegian mine. The count data $\boldsymbol{y} = (y(\boldsymbol{s}_1),\ldots,y(\boldsymbol{s}_m))$ are collected at $m = 1615$ locations of the operating mine. Zones with large counts may require further analysis due to the danger of rock fall, and the mining company has interest in predicting the spatial joint intensity and the number of joints.

Figure 4.12 shows the locations of the count data in a three-dimensional (3-D) view and a histogram of the $m = 1615$ data. Most joint counts are between 10 and 40. The smallest count is 0, while the largest count is 93.

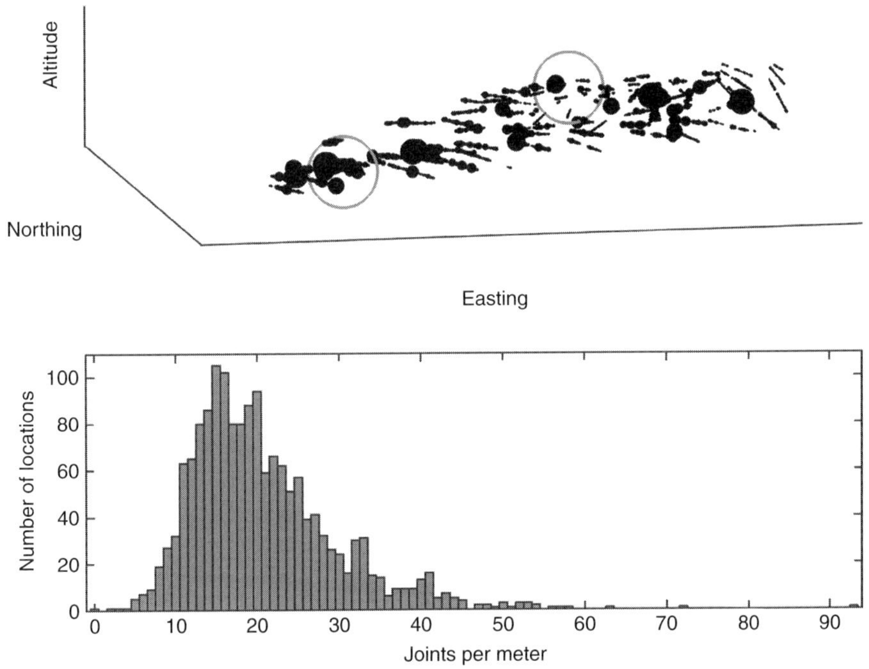

Figure 4.12 Top: map view of the joint count data acquired along boreholes in a Norwegian mine. The dot sizes are indicative of the number of joints at the location. The two open circles are locations for prediction. Bottom: histogram of the joint count data. There are 1615 measurements in total.

The data are modeled by a hierarchical model with a conditionally independent Poisson likelihood $p(y(\mathbf{s}_i) \mid x(\mathbf{s}_i)) = \text{Poisson}(4\exp(x(\mathbf{s}_i)))$, $i = 1,\ldots,m$, where the factor 4 is needed because the joint frequency data have been aggregated over 4-m intervals. We model the log joint intensities $\boldsymbol{x} = (x(\mathbf{s}_1),\ldots,x(\mathbf{s}_n))$ by a Gaussian random field – i.e., $p(\boldsymbol{x}) = N(\boldsymbol{\mu},\boldsymbol{\Sigma})$, where we use $n = 1617$ here. The first 1615 entries coincide with data locations, while the last two are the prediction sites marked as open circles in Figure 4.12. This allows efficient parameter estimation from the 1615 data as well as prediction at the two sites. The mean is constant, so $\boldsymbol{\mu} = \mu_0\mathbf{1}$. The spatial covariance is of a Matern type (see Table 4.1), with unknown variances and the range denoted by parameter $\boldsymbol{\theta}$.

The Laplace approximation of the marginal likelihood is

$$\hat{p}(\boldsymbol{y};\mu_0,\boldsymbol{\theta}) = \frac{p(\boldsymbol{x};\mu_0,\boldsymbol{\theta})\prod_{i=1}^{m} p(y_i \mid x_i)}{\hat{p}(\boldsymbol{x} \mid \boldsymbol{y};\mu_0,\boldsymbol{\theta})}, \tag{4.27}$$

which is evaluated at the posterior mode as in Equation (4.26). In Equation (4.27), the denominator is the Gaussian approximation for the joint intensity, while the numerator is defined by the prior for log intensity and the likelihood model for the joint counts. Numerical optimization is used to maximize the marginal likelihood in Equation (4.27).

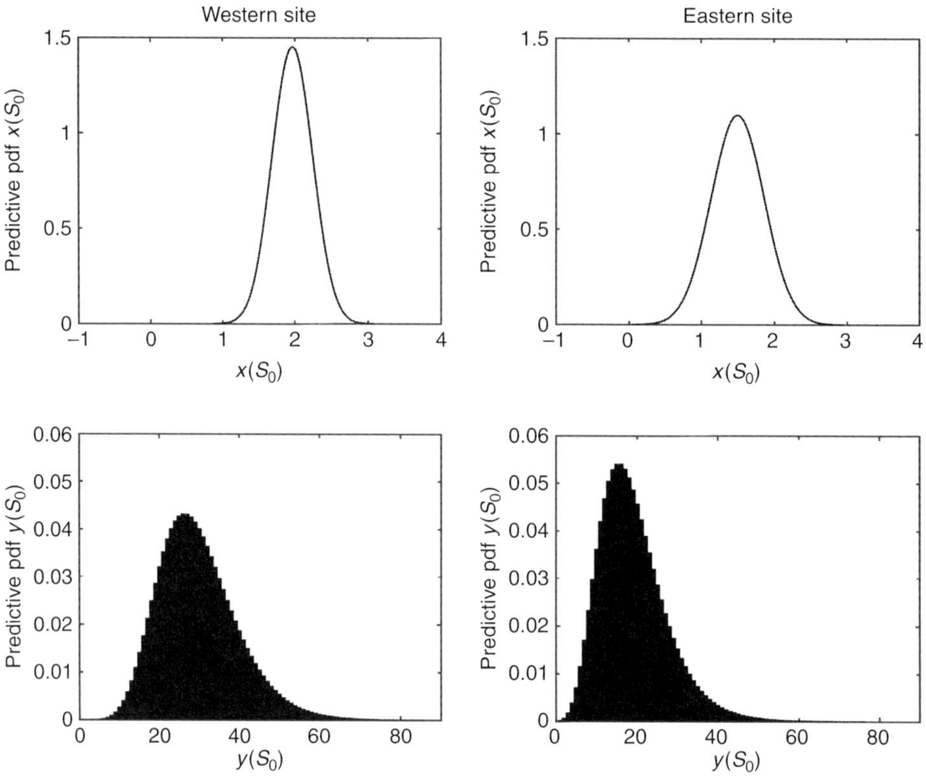

Figure 4.13 Prediction of joint frequency at the selected east and west locations in the mine. The top displays show the predicted intensity of the joint frequencies, while the bottom displays show the predicted joint count distribution.

The procedure results in a mean of 1.55 and covariance parameters in the Matern model as follows: a partial sill of 0.13, a nugget of 0.04, and an in-strike correlation range of 300 m.

The Gaussian approximation $\hat{p}\left(\boldsymbol{x} \mid \boldsymbol{y}; \hat{\mu}_0, \hat{\boldsymbol{\theta}}\right)$ is used to predict the joint frequency $y(\boldsymbol{s}_0)$ at unobserved locations $\boldsymbol{s}_0$. In the hierarchical modeling framework, this involves a marginalization over the latent intensity – i.e.,

$$\hat{p}\left(y(\boldsymbol{s}_0) \mid \boldsymbol{y}\right) = \int p\left(y(\boldsymbol{s}_0) \mid x(\boldsymbol{s}_0)\right) \hat{p}\left(x(\boldsymbol{s}_0) \mid \boldsymbol{y}; \hat{\mu}_0, \hat{\boldsymbol{\theta}}\right) dx(\boldsymbol{s}_0). \tag{4.28}$$

In this case, $\hat{p}\left(x(\boldsymbol{s}_0) \mid \boldsymbol{y}; \hat{\mu}_0, \hat{\boldsymbol{\theta}}\right) = N\left(\hat{\mu}_{x(s_0)|y}, \hat{\sigma}^2_{x(s_0)|y}\right)$ is the approximation obtained by picking entries 1616 or 1617 in the mean and covariance matrix of the Gaussian approximation $\hat{p}\left(\boldsymbol{x} \mid \boldsymbol{x}; \hat{\mu}_0, \hat{\boldsymbol{\theta}}\right)$ for the joint log intensities, given the data. The integral in Equation (4.28) can be solved numerically, and it must be evaluated for all discrete values of $y(\boldsymbol{s}_0) \in \{0, 1, 2, \ldots\}$.

Figure 4.13 shows the predictive distributions at the two selected locations marked with open circles in Figure 4.12. The western site is predicted to have much higher joint intensity (top), and the Gaussian approximation is narrower than at the eastern site. The predicted pdfs for the joint count distributions (bottom) show that large joint counts are more likely at the western site. Note that the pdf at the western site is now more dispersed than for the eastern site. This results from a characteristic of the Poisson distribution, where the variance increases with the expectation.

We will use this example in Chapter 6 for VOI analysis for the rock hazard situation in the mine. The question there is whether additional joint observations are worthwhile in helping the mining company make better decisions about adding rock support in the mining tunnels.

4.6 Categorical spatial models

If the distinction of interest is categorical, one can build a model by transforming or truncating Gaussian variables. It is also possible to explicitly model a categorical process over the spatial domain. The joint model for the process is then represented on a huge sample space d^n, where d is the number of categories (colors), and n is the number of locations (cells) used to represent the domain.

4.6.1 Indicator random variables

A natural way to model categorical random variables is through an indicator variable that can only assume values of 0 or 1 – e.g., indicating the absence or presence of a distinction such as lithofacies at every location. **Sequential indicator simulation** is the method applied to indicator variables. The spatial structure of the indicator random field is specified by the traditional two-point variogram or covariance function. The local conditional distribution at each simulation node is approximated by the indicator Kriging estimate at that node conditioned to previously simulated nodes and any hard data. This makes use of the property that the stationary mean of a binary indicator is its distribution function itself. Thus, the problem of evaluating the local conditional distribution is mapped to evaluating the conditional expectation, which is done by Kriging. Using this local conditional distribution preserves the mean and the covariance structure.

The sequential indicator algorithm allows for flexible handling of auxiliary data (e.g., facies probabilities derived from seismic interpretation) and inequality or interval constraints. As noted in the literature, there are theoretical problems with this method since the Kriging estimate may not lie in the interval [0, 1] and hence would not be a valid conditional distribution. Corrections have to be applied, which then does not preserve the covariance exactly. Moreover, the two-point covariance can be a very poor descriptor of the spatial properties of indicator random functions. The covariance function for an indicator and its complement (e.g., sand channels in mud background or pores and grains) is the same, but in terms of connectivity and flow response, an indicator set and its complement can have very different properties. Though a flexible and popular method available in many geostatistical softwares, sequential indicator simulations produce "blob"-like geometries unable to capture geological shapes and connectivity (Daly and Caers 2010).

4.6.2 Truncated Gaussian and pluri-Gaussian models

A class of models for categorical variables closely related to Gaussian random field models are the truncated Gaussian and pluri-Gaussian models. Here, the categorical random variable arises by applying one or more thresholds to a standard Gaussian random field. The underlying continuous variable could have physical meaning – e.g., ore grade or contamination level – and the associated categorical variable could be a certain cut-off grade or contamination warning threshold. In other cases, the underlying continuous variable might just be a mathematical construction of convenience for the model. The thresholds are chosen according to the proportions of each category. The correlogram of the standard Gaussian random field in Equation (4.5) is not the same as that of the indicator field obtained by truncating the Gaussian variable, but they are related to each other. If the standard Gaussian field with correlogram $\rho(|t|)$ is truncated at level λ, then the covariance of the indicator is given by

$$C_\lambda(|\mathbf{t}|) = \frac{1}{2\pi}\int_0^{\rho(|t|)} \exp\left(-\frac{\lambda^2}{1+u}\right)\frac{du}{\sqrt{1-u^2}}, \tag{4.29}$$

which for the median threshold of $\lambda = 0$ simplifies to $C_0(|\mathbf{t}|) = (1/2\pi)\arcsin\rho(|\mathbf{t}|)$. The derivations and other properties of truncated Gaussian models are discussed in Chiles and Delfiner (2012). Lantuejoul (2002) presents expressions relating the variograms of the two fields.

The truncated Gaussian model is generalized to the truncated pluri-Gaussian model by using thresholding rules on two Gaussian random fields, which can be independent or correlated. Complex spatial associations between different categories can be modeled by pluri-Gaussian simulations using appropriate thresholding rules. These methods have been applied to model petroleum reservoirs (turbidites and carbonates) as well as ore deposits. Armstrong et al. (2011) give a detailed presentation of the pluri-Gaussian method and its applications.

4.6.3 Categorical Markov random field models

Suppose that the spatial domain of interest has been split into n disjoint subdomains, units, or cells. A random field is represented on the resulting grid of these disjoint cells. In Section 2.3, we studied categorical Markov chains, where the Markov property was phrased in the sequence of time or some other ordering – i.e., the distribution today, given the past, depends only on yesterday. Here, for a spatial process, there is no natural ordering in time, and the **Markov property** is instead described in terms of the full conditional distributions breaking down to the conditional distribution given only the values in a **neighborhood**. The notion of a neighborhood makes sense in the spatial context, since variables

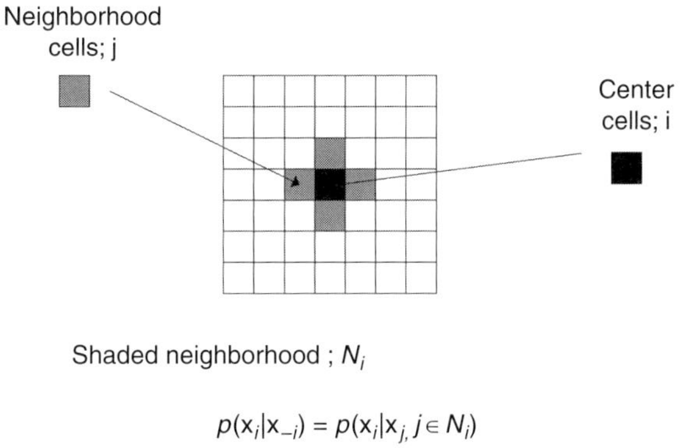

Figure 4.14 Schematic of a Markov random field model. The grayscale cells are the neighborhood of the center cell for a first-order Markov random field.

that are close in space tend to propagate information going from one geographical domain to another. Here we provide a brief background on Markov random fields (MRFs) (see Appendix A.4 for further details).

Let $\boldsymbol{x} = (x_1,\ldots,x_n)$ be the distinction of interest represented at the n cells covering the spatial domain of interest. We assume a set of possible categories: $x_i \in \{1,\ldots,d\}$, $i = 1,\ldots,n$. The number of classes is most commonly $d = 2$ – i.e., the random field can be regarded as black and white. In some applications, more categories are natural, and $d = 3,4,5$ add some grayscale to the field. Furthermore, let $\boldsymbol{x}_{-i} = \left(x_1,\ldots,x_{i-1},x_{i+1},\ldots,x_n\right)$ be all variables except the one at cell i. A conditional MRF definition is

$$p\left(x_i \mid \boldsymbol{x}_{-i}\right) = p\left(x_i \mid x_j; j \in N_i\right), \tag{4.30}$$

where N_i is the neighborhood of cell i. This equation tells us that once we know the outcomes within the neighborhood, there is no additional information in knowing the variables outside this neighborhood.

Consider the case shown in Figure 4.14, and suppose that one is interested in the full conditional distribution of cell i defined by $p\left(x_i \mid \boldsymbol{x}_{-i}\right)$. The display shows the situation with a first-order neighborhood, where the full conditional only depends on the outcome at the four nearest neighbors (north, east, south, and west). When the discrete field has two possible outcomes at each location, this model with a first-order neighborhood is called the Ising model.

An MRF model, defined in terms of neighborhoods, can be shown to have a one-to-one correspondence with a joint probability model represented over subsets of variables called cliques (Besag 1974). The joint probability for the Ising model is

$$p\left(\boldsymbol{x}\right) = \frac{\exp\left(\beta \sum_{i\sim j} I(x_i = x_j)\right)}{Z} \propto \exp\left(\beta \sum_{i\sim j} I(x_i = x_j)\right), \tag{4.31}$$

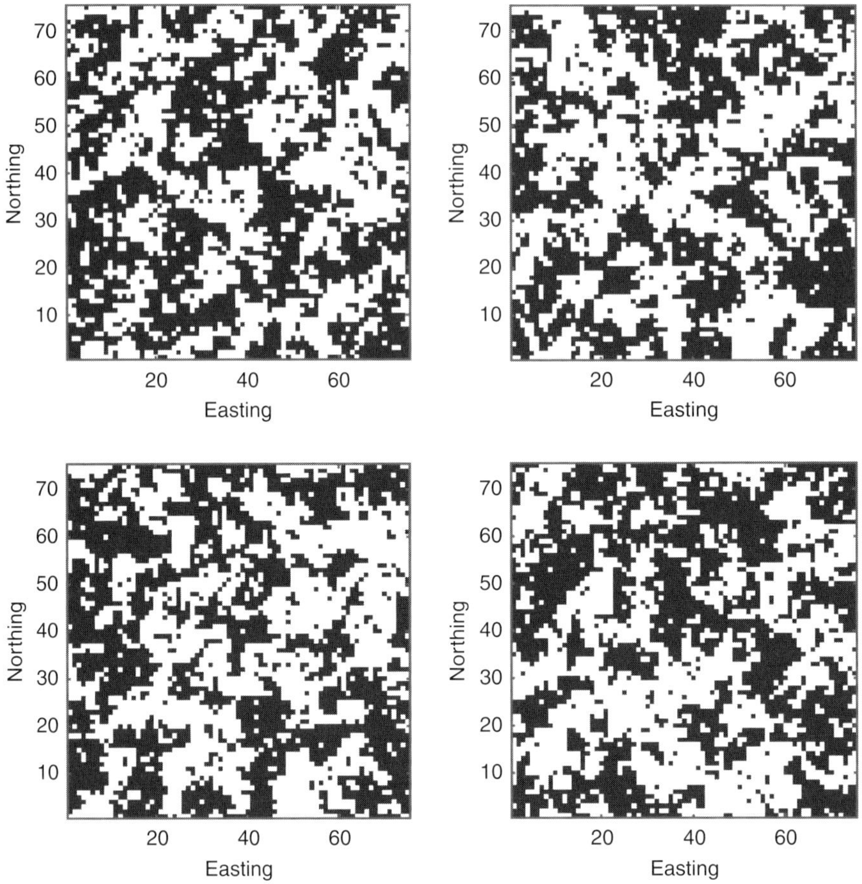

Figure 4.15 Four realizations of a Markov random field model with a first-order neighborhood and two colors (Ising model). The grid size is 75 × 75 and the spatial interaction parameter is $\beta = 0.75$.

where the sum $(i \sim j)$ is over all the north–south or east–west variable pairs – cliques – on the grid, and I is the indicator function. The clique potentials are β for equal colors in the neighbors, while they are 0 for different colors. A larger interaction parameter β implies more spatial continuity in the colors across the lattice. The normalization constant Z ensures that the expression is a valid pdf.

Figure 4.15 shows four realizations of an Ising model on a 75 × 75 lattice with interaction parameter $\beta = 0.75$. By using a larger neighborhood of the MRF, with added flexibility in the clique potentials, one can obtain more realistic realizations from the associated MRF models – for instance, channel structures, lobes, and so on. By using more than two colors, it is possible to represent additional geological classes – for instance, sand, shale, coal, chalk, etc.

Discussion: full conditionals for the Ising model

For the typical Ising model, the marginal probability of white (0) or black (1) at any cell is 0.5. The full conditional distributions depend on the number of equal colored variables in the neighborhood:

$$p(x_i = 1 \mid x_j; j \in N_i) = \frac{\exp\left(\beta \sum_{i \sim j} I(x_j = 1)\right)}{\exp\left(\beta \sum_{i \sim j} I(x_j = 1)\right) + \exp\left(\beta \sum_{i \sim j} I(x_j = 0)\right)}.$$

This can be split into five situations, depending on the number of equal neighbors:

If all neighbors with $x_j = 1$: $p(x_i = 1 \mid x_j; j \in N_i) = \frac{\exp(4\beta)}{\exp(4\beta) + 1}$

If three neighbors with $x_j = 1$: $p(x_i = 1 \mid x_j; j \in N_i) = \frac{\exp(3\beta)}{\exp(3\beta) + \exp(\beta)}$

If two neighbors with $x_j = 1$: $p(x_i = 1 \mid x_j; j \in N_i) = 0.5$.

If one neighbor with $x_j = 1$: $p(x_i = 1 \mid x_j; j \in N_i) = \frac{\exp(\beta)}{\exp(3\beta) + \exp(\beta)}$

If no neighbors with $x_j = 1$: $p(x_i = 1 \mid x_j; j \in N_i) = \frac{1}{\exp(4\beta) + 1}$

The MRF model can be extended from the basic formulation by introducing an external field, given as single site potentials. The joint pdf for the Ising model is then

$$p(\boldsymbol{x}) = \frac{\exp\left(\beta \sum_{i \sim j} I(x_i = x_j) + \sum_i \alpha_i(x_i)\right)}{Z}, \tag{4.32}$$

where $\alpha_i(l), i = 1, \ldots, n$, is a potential for each color $l = 1, \ldots, d$. This external field can either be specified a priori, or it can depend on data or covariates. The latter situation is common when there are conditionally independent data at the nodes.

Extending the idea of the hidden Markov model from the time domain leads to a hidden Markov random field model. Let us denote the likelihood by $p(y_i \mid x_i)$ for all cells $i = 1, \ldots, n$. The posterior pdf of interest, conditional on the data, becomes

$$p(\boldsymbol{x} \mid \boldsymbol{y}) \propto \exp\left(\beta \sum_{i \sim j} I(x_i = x_j) + \sum_i \alpha_i(x_i) + \sum_i \log p(y_i \mid x_i)\right). \tag{4.33}$$

We demonstrate this with an example later, where there are seismic reflection data y_i available at each cell.

MRF models of moderate size can be evaluated and sampled from with exact forward–backward algorithms for efficient marginalization – see, e.g., Reeves and Pettitt (2004) and Appendix A.4. For large grids, Markov chain Monte Carlo sampling may be useful (see Appendix B).

The interaction parameter β or other parameters in the MRF model must be inferred from the data. The data can be either perfect information $\boldsymbol{x}$ or imperfect information $\boldsymbol{y}$. MLE is one way to specify parameters, but this task is not straightforward for large grids (see Appendix A.4).

Even though categorical MRF models have been helpful in applications with two spatial dimensions, it has been cumbersome to apply them in full three-dimensional (north, east, depth) spatial models. One challenge is that Markov chain Monte Carlo sampling convergence can be very slow. Rather than following the strictest mathematical formulations of MRFs, there have been promising approaches trying to modify the MRF formulation.

One idea is to separate the depth dimension from the lateral directions. This can be reasonable in some settings, since special geological rules apply in the depth dimension. The Markov chain Monte Carlo sampler can now solve for the depth direction by a direct solver (like forward–backward recursions for Markov chains), while the Markov chain Monte Carlo scheme loops over the two-dimensional lateral space. The convergence of the Markov chain Monte Carlo sampler becomes much faster than doing a full loop in three dimensions. Recent applications for seismic reservoir prediction include Ulvmoen et al. (2010) and Rimstad et al. (2012).

Another idea for making the MRFs more applicable is to introduce ordering in the grid. Markov mesh models are a type of partially ordered Markov models, which consider the conditional distribution for a cell given the cells with a lower order and not the entire neighborhood. This view speeds up the computations, but care must be taken to parameterize the dependency realistically. Stien and Kolbjørnsen (2011) apply techniques from generalized linear models to specify parameters in Markov mesh models. See also Tjelmeland and Austad (2012), who follow ideas more similar to the exact forward–backward schemes but apply a truncation, where the higher-order interactions in the (forward) computation of the MRF are ignored.

4.6.4 Example

Black gold in a white plight: reservoir characterization example

Keywords: *reservoir characterization, Markov random fields, seismic data, petroleum, Monte Carlo sampling*

Seismic data are useful for reservoir characterization. By careful processing and interpretation of seismic amplitude data at the top of the reservoir, a petroleum company can predict (imperfectly) the spatial distribution of, for instance, sands or shales, or brine and oil saturation. The quality of the prediction, of course, depends on the prior knowledge of the reservoir and rock and fluid properties, as well as the quality of the seismic amplitude data. A similar example was discussed in Chapter 2 and will also be used for VOI analysis in Chapter 6.

In this example, the data and the rock-type (facies) variables are represented on a grid of size 75 × 75 at the top reservoir. Here, the petroleum company is working toward the

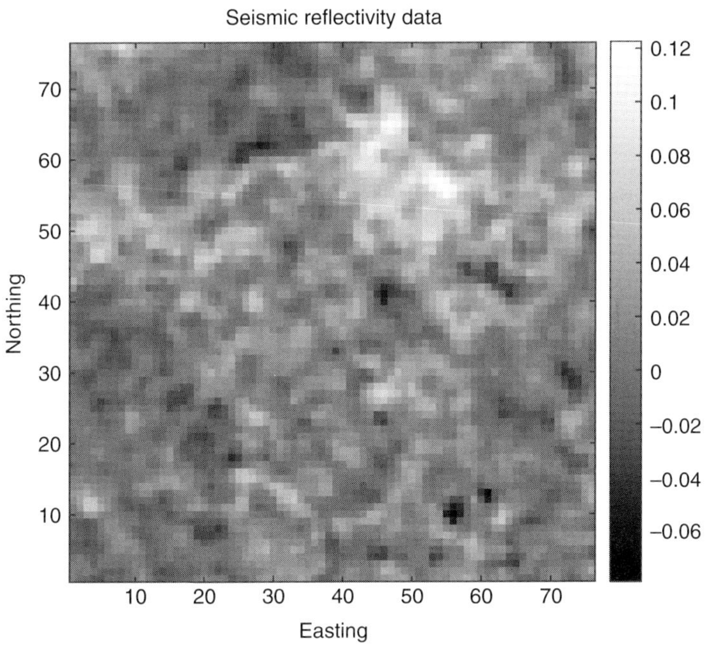

Figure 4.16 Seismic reflection data (zero-offset amplitudes) at the top reservoir horizon. The top reservoir is represented on a grid of size 75 × 75.

prediction of the discrete facies variable $\boldsymbol{x}$. The facies at a reservoir unit is either oil sand (black) or shales (white) – i.e., $x_i \in \{0,1\}$ and $i = 1,\ldots,5625$. We consider only the stacked zero-offset attribute y_i of the seismic amplitude data for all cells at the top reservoir. The goal is to demonstrate the applicability of an Ising model prior for the facies in this prediction exercise. This can be phrased as a Bayesian inference problem, where the likelihood builds on the established geophysical knowledge via a forward model, while the prior model incorporates smoothness or spatial interaction between the facies variables at different cells.

Figure 4.16 shows the seismic amplitude data at the top reservoir. The data are informative of the facies because sands and shales tend to give different seismic responses. Given the facies variables, the seismic data are modeled as conditionally independent. The likelihood $p(y_i \mid x_i)$ is a Gaussian pdf, where the mean value depends on the facies variable, while the variance is the same for both classes. Figure 4.17 shows the likelihood model for both classes.

Let us define an Ising model prior with the interaction parameter set to $\beta = 0.75$, similar to the model used in the realizations shown in Figure 4.15. For the posterior analysis, we run a Markov chain Monte Carlo algorithm for 500 000 updates. Specifically, we use a Metropolis–Hastings scheme, which updates the facies variable at one randomly selected cell at each iteration of the algorithm. The updating is easily done because of the neighborhood of the MRF model and the conditionally independent likelihood model. At each update, only the four nearest neighbors must be considered, as well as the seismic amplitudes at the selected cell (see Appendix B).

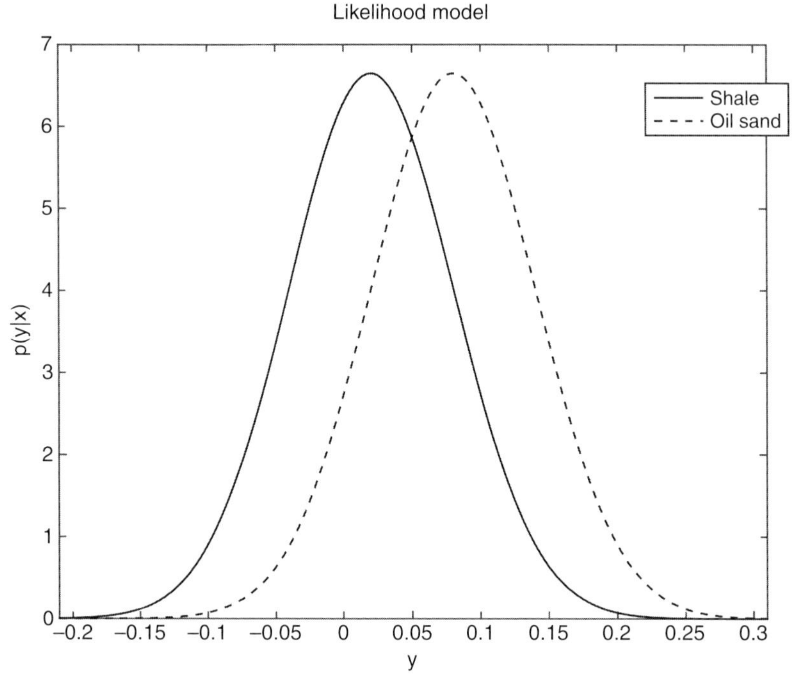

Figure 4.17 Likelihood models for the seismic zero-offset amplitudes. The likelihood model is conditional on the discrete variable.

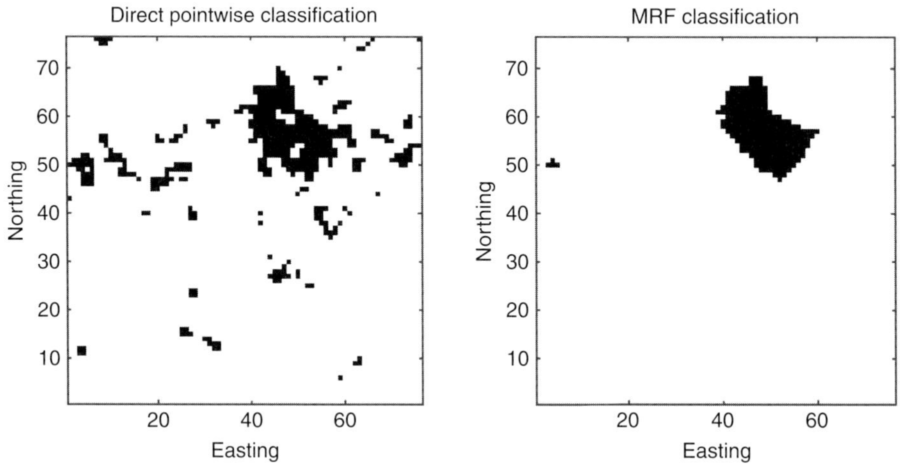

Figure 4.18 Classification of discrete variable (oil sand is black and shales is white) from the seismic data and the prior model. Left: pointwise prediction using no prior assumptions about spatial interaction. Right: Markov random field prediction with interaction $\beta = 0.75$.

Figure 4.18 (right) shows the maximum a posteriori solution of the facies variable using the Markov chain Monte Carlo algorithm. We compare the MRF solution with the results of a pointwise classifier based on the likelihood model alone (Figure 4.18, left). For the pointwise classifier, the algorithm simply picks the most likely class depending on the data at that cell – i.e., the highest likelihood value for the data is read directly out of Figure 4.17. If the data are larger than 0.05, a cell is classified as oil sands; otherwise, it is classified as shales. The pointwise result has many more black cells within regions dominated by white and vice versa. The MRF solution borrows spatial information from neighboring cells, and this entails that all data are involved in making a classification at a cell. The prediction becomes smoother than the pointwise classification.

4.7 Multiple-point geostatistics

Although Markov random field models are based on a solid statistical foundation and only require a few parameters, it is challenging to construct MRF models that produce realizations that visually appear like geological images. For instance, a large neighborhood is required to reproduce channel structures often seen in reservoirs. Careful computing is required to obtain reliable inference and prediction for MRFs with large neighborhoods. Markov chain Monte Carlo algorithms tend to be very slow here. An alternative approach to Markov random fields is that of multiple-point geostatistics.

4.7.1 Algorithms

The ideas of training images and multiple-point (MP) geostatistics were originally presented to overcome the limitations of two-point (Gaussian) models (Guardiano and Srivastava 1993). Although they were initially developed to construct discrete valued realizations of geology from training images, current MP algorithms can also construct realizations of continuous variables and handle non-stationarity. The original MP algorithms for categorical variables (e.g., single normal equation simulation (SNESIM) (Strebelle 2000, 2002)) extract the MP statistics – i.e., the probability of a state at a given position given the state of multiple neighbors – from a **training image**. The patterns seen in the training image are then reproduced (at random) according to the database of conditional distributions. This means that no new patterns can occur unless we assign a very small probability for non-existing patterns. The MP idea is very empirical, as it is constructed directly from the training image, different from Markov random fields, which are built from parametric assumptions of neighborhood configurations. MP simulations are certainly rich (given a rich training image) in the sense that realizations offer close resemblance to geology. However, when the patterns grow larger and the training image is not large enough, MP algorithms may be prone to overfitting since they basically reproduce what is available in the training image, and this may lack predictive power.

In many of the MP algorithms, the simulations using a training image follow the sequential simulation paradigm. Each node is visited sequentially, and the local conditional distribution (non-parametric) is obtained from the training image. In this sense, it can be classified as a sequential algorithm, as is common for Gaussian simulation of spatial processes. However, the difference is that it is conditioned to data events consisting of patterns of multiple data taken together. Some of the more recent MP algorithms are based more on computer graphics and do not rely much on probability theory. Instead of assessing parametric or non-parametric conditional distributions from the training image, these algorithms directly lift from the training image realizations of single nodes or multiple sets of nodes in a template (pattern simulation) (Mariethoz and Caers 2015). As an example, the direct sampling (DS) MP algorithm (Mariethoz et al. 2010) directly samples the training image conditioned to the MP data event surrounding the node to be simulated instead of building a conditional distribution from the training image. The algorithm randomly scans the training image, and as soon as a match (within a certain tolerance threshold) is found to the MP data event, the value from the training image is lifted and pasted at the simulation node. The match is defined in terms of some similarity distance – e.g., whether the facies categories at the nodes in the data event match the corresponding nodes from the training image scan. The threshold needs to be selected carefully, of course. Requiring a perfect match leads to an exact copy of the training image. A small fraction (<0.1) of mismatching nodes is taken to be acceptable. Pattern-based algorithms also compare similarities between the data event and the patterns from the training image, but instead of lifting one pixel at a time, they patch a finite-sized region of the training image into the simulation grid. Patch-based algorithms are generally faster than pixel-based algorithms. The DS algorithm can be used as a patch-based algorithm. In addition, some other patch- (or pattern-) based MP algorithms in the literature include image quilting (Efros and Freeman 2001), FILTERSIM (Zhang et al. 2006), SIMPAT (Arpat and Caers 2007), and CCSIM (also Tahmasebi et al. 2012). Mariethoz and Caers (2015) provide a detailed overview of all of these algorithms and many others.

As we mentioned earlier, MP statistics can simulate geologically realistic patterns. Note that this relies on an appropriate training image, which may not always be available. Coming up with a rich enough training image (or sets of training images) that characterize the prior uncertainty of the subsurface heterogeneity is challenging, often requiring close interaction with geologists familiar with the depositional setting of the area under study. Based on the geologists' inputs, the training image may be built using unconditional simulations such as object-based methods or process-based or process-mimicking methods. Though challenging, once an appropriate training image is available, modern MP simulation algorithms offer very effective ways of modeling realistic geologic heterogeneity that no traditional two-point geostatistical method can match.

We discuss more references to MP statistics in the bibliography that follows and finish this section with an example.

4.7.2 Example

Go with the flow: petroleum simulation example

Keywords: *petroleum simulation, petroleum development, petroleum production, reservoir characterization, multiple-point statistics*

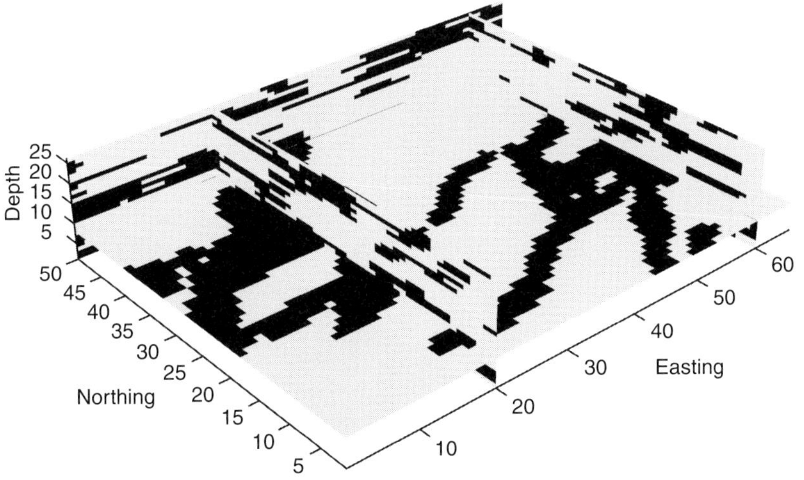

Figure 4.19 Training image for a subsurface meandering channel system with two classes (channel sands and floodplain). One can create multiple-point statistics (MPS) realizations of discrete spatial variables based on such training images.

Decisions regarding drilling and future production of reservoirs are based on several aspects of reservoir characterization. One important aspect is the ability to produce the reservoir from a relatively small number of wells, which is related to the connectivity and the permeability of the reservoir. These variables direct the flow of hydrocarbons to the producing wells. Petroleum engineers use fluid flow simulators to make production forecasts, to detect reservoir zones of bypassed oil, and to develop and manage the reservoir. As we discussed for Figure 4.2 at the beginning of this chapter, it has been shown to be beneficial to run flow simulations over multiple realizations of stochastic reservoir models. Along with the suite of other tools for spatial uncertainty modeling, MP geostatistics have played an important role in the modeling of geologic facies as inputs for reservoir simulation.

In Figure 4.19, we show a training image for a meandering river system. The display shows spatially dependent patterns of white and black in a three-dimensional (north, east, depth) view. It is common practice to use a training image like this one to realistically describe the spatial features and alternation styles of the rock types (facies) in the reservoir, which is a critical reservoir variable that determines much of the variations in the saturation (oil or gas content), porosity (tied to the pore volumes in the reservoir rock), permeability (tied to the flow properties in the reservoir), and elastic properties (tied to the seismic data).

There are two facies classes in this training image: channel sands and floodplain (shown as black and white, respectively, in the display). The training image is constructed from an object-based unconditional simulation. Multiple realizations are then simulated using the SNESIM MP algorithm by visiting each node in a sequential routine. When the algorithm visits a cell, it samples a color. The color is generated conditional on the colors at the previously visited cells. The events or patterns around the current node jointly influence the

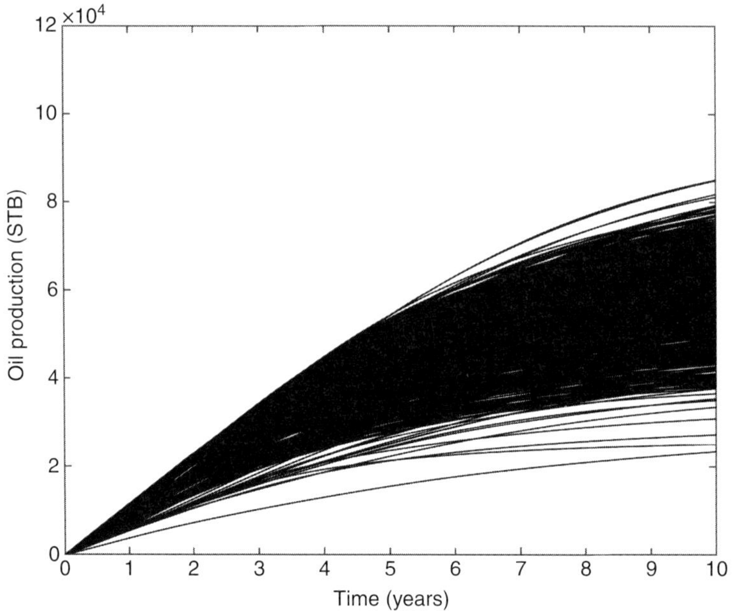

Figure 4.20 Production profiles calculated by multiphase flow simulation over many realizations of facies, porosity, and permeability fields.

conditional probability – i.e., there is conditioning on the outcome at multiple points. The computation is done in a straightforward manner: counting the number of times the configuration of the history occurred in the training image (database), with a white or black color at the selected configuration. This is Bayes' rule by counting.

We generate $B = 1000$ realizations of the discrete random process. At each cell, the stochastic facies variable is either channel or floodplain. The next step in petroleum reservoir simulation is to generate other important reservoir variables like porosity or permeability, conditional on the facies classes. We use sequential Gaussian simulation for the porosity and permeability in the reservoir, conditional on the facies.

If the goal of this exercise were to predict the recoverable petroleum resources, then the reservoir engineer would feed the deterministic numerical reservoir flow simulation model with the multiple stochastic porosity and permeability inputs. The output is production profiles in production wells over time. Of course, other geological and petrophysical inputs are also required here, such as well controls, boundary conditions, rock and pore fluid properties, relative permeability curves, capillary pressure curves, wettability, etc.

Figure 4.20 shows 1000 realizations of the total oil production from six producing wells. Each curve is representative of the production obtained by flow simulating one of the stochastic reservoir realizations for facies, porosity, and permeability. The variability in the production profiles is inherited from the stochastic elements of the reservoir variables.

Since the reservoir simulator is based on highly non-linear multiphase flow in porous media, one cannot rely on linearization and Gaussian assumptions for the output variables.

The realization-driven approach described earlier is necessary to quantify uncertainties; this method is popular in a number of Earth sciences applications. Rather than having a closed-form pdf for the output variable (petroleum production in this case), an approach like this would result in a discrete, empirical representation of the pdf, as determined by the samples.

4.8 Bibliographic notes

Traditional geostatistics books describing variogram methods, Kriging, and simulation

Books describing variogram-based geostatistics and Kriging in various forms are abundant. Excellent reference books include Journel and Huijbregts (1978), Isaaks and Srivastava (1989), Kelkar and Perez (2002), and Wackernagel (2003).

Deutsch and Journel (1992), in the *Geostatistical Software Library (GSLIB)*, provide a terse yet useful recall of the theory underlying the algorithms coded in the corresponding public domain software. The list covers most traditional two-point geostatistical algorithms commonly used in practice, from variograms to Kriging to simulations and their many flavors. GSLIB has been the source of many commercial geostatistics softwares. Goovaerts (1997) focuses only on traditional two-point methods and is remarkable for its careful and complete derivations and the continuity offered by the sequence of applications performed on the same multivariate soil data set. Through that data set, readers can readily appreciate the typical flow of a geostatistical study. Lantuejoul (2002) provides theory and algorithms for unconditional and conditional simulations of a wide variety of spatial models. Besides the classical Gaussian and pluri-Gaussian random functions, theory and algorithms for other spatial stochastic models such as point processes, Boolean models, tessellations, and substitution random function are described. Chiles and Delfiner (2012) is perhaps one of the best current graduate-level books on the theory and applications of geostatistical methods based on covariances and random function theory. The authors cover a breadth of topics, including not only the traditional Kriging-based technniques but also spatiotemporal models, multivariate methods, change of support, and pluri-Gaussian models, along with many examples from petroleum geosciences, mining, and hydrogeology. There are concise but useful sections on geostatistical inversion and multiple-point simulations from training images. Though the mathematics is at an advanced level, it is presented in a readable manner. Pyrcz and Deutsch (2014) review geostatistical modeling tools, paying attention to careful quantification of geological concepts. They also describe process-mimicking models and event-based models used in geostatistical modeling.

Sequential simulation techniques and related topics

Borgault (1997) introduced the direct sequential simulation algorithm, where there is no need to perform a normal score transform and back-transform of the data. The advantage of this approach is that it can handle conditioning to linear averages of the original data. The disadvantage is that there is no guarantee that the simulated realizations reproduce the data histogram. Indicator Kriging is a technique that could be used to estimate the conditionals at a site. A common reference paper for this technique is Journel (1983). In this case, no assumption is made about the shape of the distribution, which is assessed by directly estimating the probability of being below a series of thresholds (for continuous variables) or by estimating the probability of being within a set of discrete categories. When this method is used for sequential simulation, the algorithm is usually called sequential indicator simulation. In the probability field simulation (P-field) algorithm, conditionals are estimated based only on the hard data and not on indirect noisy observations. A common reference paper is Froidevaux (1992).

Spatial statistics

There are several excellent books treating spatial data analysis from a more statistical perspective. For instance, Cressie (1993) covers a wide range of spatial statistics topics not only limited to Gaussian- and Kriging-type techniques. Over the years, it has served as an excellent reference book for spatial statistics topics. Stein (1999) provides a solid background on the more theoretical aspects underlying parameter estimation and prediction (Kriging) in spatial Gaussian processes. Banerjee et al. (2004) provide an overview of more recent methodologies for analyzing spatial processes. For instance, they include a description of areal-based models, a number of novel tools for non-stationary processes, and the use of Markov chain Monte Carlo simulation for inference. The focus of the book is on hierarchical modeling in a fully Bayesian perspective. Le and Zidek (2006) take an information theory-based approach, paying attention to the design of spatial experiments to improve predictions with a focus on the entropy measure. The main applications studied involve environmental data about air quality, ozone, and temperature. Schabenberger and Gotway (2009) and Gaetan and Guyon (2010) are other great books covering diverse topics in spatial statistics.

There has lately been a large focus in the statistical community on methods for large data sets. Topics such as predictive processes (Banerjee et al. 2008), fixed-rank Kriging (Cressie and Johannesson 2008), and localized models (Stein et al. 2004) are promising modeling approaches.

Non-Gaussian response models

In our presentation, we focused on the skew-normal model and the generalized linear models, as we will later use them in models for VOI analysis. There are of course several non-Gaussian response models that we did not cover here. Perhaps the most common technique is to transform response variables to achieve Gaussian distributions for the empirical marginal. This has been referred to as anamorphosis (Cressie 1993). This approach seemed to die out with the advent of hierarchical models in statistics but has lately garnered some interest again in another similar form known as copulas (Joe 2014).

The Gaussian copula has been used in geostatistics – see, e.g., Bardossy and Li (2008) and Kazianka and Pilz (2011).

Non-parametric models such as spatial Dirichlet processes – e.g., Gelfand et al. (2005) – or stick-breaking mixture formulations – e.g., Reich and Fuentes (2007) – have also gained popularity recently.

There is a large body of literature on point-process models in spatial statistics. These are somewhat different from other spatial models because they focus on modeling the random location of spatial events rather than the random variable at a fixed location. A popular modeling approach is the log-Gaussian Cox process, which can be regarded as a hierarchical Gaussian model – see, e.g., Illian et al. (2008).

Markov random fields

The most influential publications in Markov random fields are Besag (1974) and Besag (1986). These papers sparked a lot of interest in the statistical community – not limited to spatial statistics but also in graphical modeling and image processing. Markov chain Monte Carlo (MCMC) has been a popular approach for sampling Markov random fields, as we have described, and the original paper on Gibbs sampling studied the Ising model (Geman and Geman 1984). See also the discussion of multiple-point geostatistics mentioned earlier, which in a similar vein provides very useful discrete valued spatial models.

Gaussian Markov random fields are very useful for fast computations of Gaussian processes. An overview is provided in the book by Rue and Held (2005). Connections between Gaussian Markov random fields on tessellations and continuously described Gaussian random fields are drawn in Lindgren et al. (2011).

Multiple-point geostatistics

Guardiano and Srivastava (1993) initially proposed multiple-point geostatistics. Strebelle (2000) and Strebelle and Journel (2001) were among the first to describe and use multiple-point geostatistics formally. Caers (2005) focuses on reservoir characterization techniques. The book includes applications of multiple-point geostatistics to stochastic reservoir modeling and uncertainty assessment. Caers' (2005) book can be considered the first book of the multiple-point generation. Remy et al. (2008) were the first to deliver both theory and software related to multiple-point geostatistics. Although this book is a users' guide to the Stanford Geostatistical Modeling Software (SGeMS), its presentation focuses not on the modeling of covariance and Kriging but on the representation of a general random function. That representation takes the form of equiprobably sampled simulated realizations mimicking the actual spatial distribution of the variable(s) under study. The spatial law (MP statistics) of that random function is retrieved from a training image that can carry prior knowledge about the physics controlling the phenomenon under study.

Mariethoz and Caers (2015) is a recent comprehensive book on spatial modeling using training images and multiple-point statistics. The authors provide a detailed overview of the various algorithms for spatial modeling based on training images. Inverse modeling using training images is also covered. Example applications include reservoir modeling, mineral resources, and climate modeling. This is an excellent modern reference for stochastic spatial modeling in the physical sciences.

Spatiotemporal models

The recent book by Cressie and Wikle (2011) provides a great overview of recent approaches for spatiotemporal statistical modeling. The focus is on low-dimensional basis representations in combination with differential equations modeling. There is currently significant interest in the dynamic updating of spatial models. For linear Gaussian models, the celebrated Kalman filter yields the optimal solution, but this updating becomes very complicated for non-linear time-varying models, and there is no single solution. Ensemble Kalman filtering (Evensen 2009) has become a popular tool for approximating the distribution over many time steps with data. A more generally applicable technique for data assimilation over time, with its own computational drawbacks, is that of particle filtering or sequential Monte Carlo updating (Doucet et al. 2001).

5

Value of information in spatial decision situations

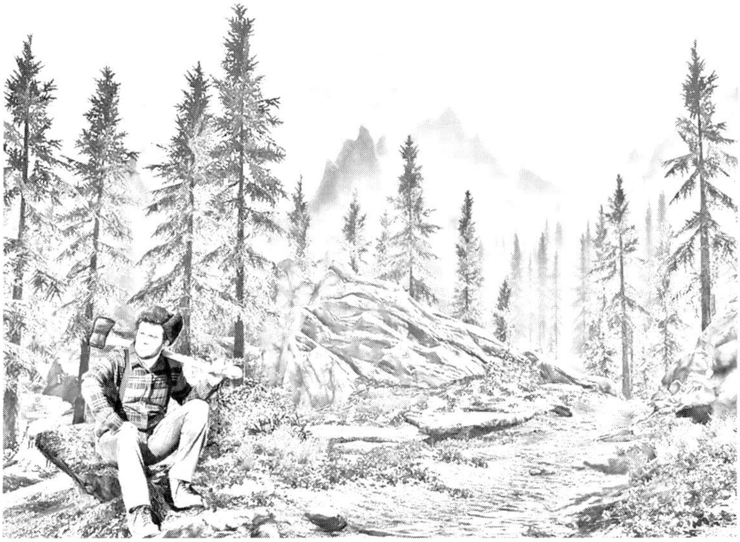

Figure 5.1 harvesting these woods
means felling our leafy friends
so let's choose wisely

In this chapter, we integrate concepts from decision analysis (Chapter 3) and spatial statistics (Chapter 4) to model decision situations that are typical of applications in the Earth sciences. These models are then developed to support decisions pertaining to information gathering in the Earth sciences.

We start by introducing the concepts of spatial decision situations and information-gathering schemes and frame the various models introduced in the chapter in Section 5.1. In Section 5.2, we present a fairly general formulation for the value of information (VOI) for a class of problems that is studied most comprehensively in the book.

We elaborate on a number of special cases in Sections 5.3–6 by categorizing decision situations based on the number of available alternatives as well as the value function model. Sections 5.7–8 describe more complex decision situations: more general risk preferences, explicit constraints in the decision situation, and sequential decisions or sequential information gathering. Section 5.9 discusses some other information measures that are popular for evaluating information-gathering schemes in Earth sciences applications. We demonstrate the concepts using several examples throughout the chapter. Our focus here is on using the examples to illustrate the assumptions of the various models; later in Chapter 6, we describe case studies where the emphasis lies more on the applications themselves.

5.1 Introduction

Decision situations in the Earth sciences involve the exploration, utilization, management, retrieval, restoration, and sustenance of the Earth's resources. Decisions are often complex due to uncertainties related to the Earth's subsurface. Furthermore, decisions may have a lot at stake, and therefore it may be worthwhile to obtain more information before the decision is actually made – i.e., before an irrevocable allocation of resources. A crucial question to answer is: how much information should one purchase? Gathering the right kind and the right amount of information is crucial for the decision maker so as to obtain as much value from the Earth's resources as possible. In this section, we first present a formulation of decision situations that are typical of applications in the Earth sciences, referring to them as spatial decision situations. Then, we show how VOI analysis can support various information-gathering questions that are relevant for such applications.

5.1.1 Spatial decision situations

Applications in the Earth sciences often inherently involve **spatial** variables – i.e., variables that are defined over several locations in space, as well as alternatives that are associated with these spatial locations. Consider a spatial random variable $\boldsymbol{x} = \{x_i : i = 1,2,...,n\}$ and an alternative (or, synonymously, action) $\boldsymbol{a} = \{a_j : j = 1,2,...,N\}$, where indices i and j are associated with spatial locations – i.e., $x_i = x(\mathbf{s}_i)$ and $a_j = a(\mathbf{s}_j)$ using the notation of Chapter 4. Alternatives are chosen from an available set $\boldsymbol{a} \in \boldsymbol{A}$. Let $v(\boldsymbol{x}, \boldsymbol{a})$ be the value derived by the decision maker from the decision situation; note that this is a function of both the spatial variable $\boldsymbol{x}$ as well as the chosen alternative $\boldsymbol{a}$. As in the previous chapters, we refer to the spatial variable as the **distinction of interest** when it is a variable that is fundamental to the decision maker's value.

Consider the illustrative example shown in Figure 5.2, where a farmer's forest has been segmented into several irregular units. The farmer must decide which units to utilize – which entails cutting the timber and then transporting, producing, and selling the wood – and which units to leave as is for the forthcoming season to maintain a sustainable resource. In this example, x_i is the spatial variable that denotes the volume of timber at unit i, and a_j is chosen from a binary alternative set at unit j. The two alternatives are: should the

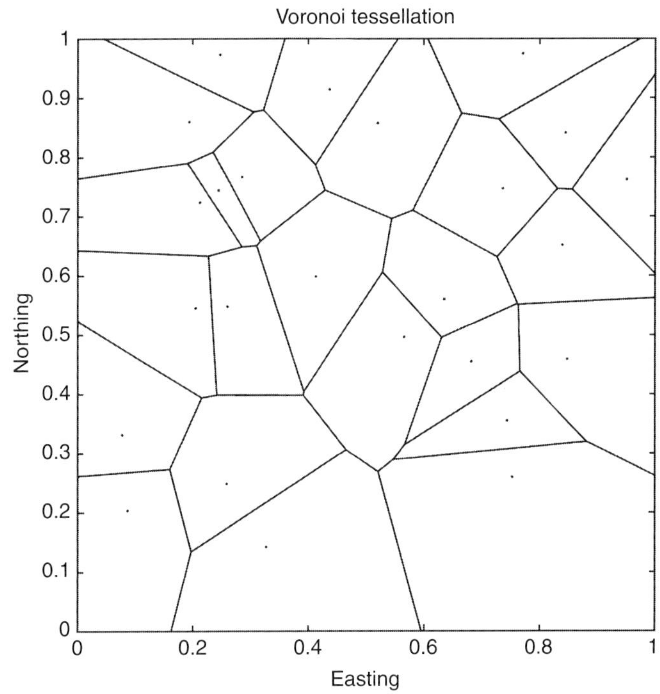

Figure 5.2 Illustration of a spatial domain: a farmland that is segmented into units. The decision maker must decide which units to harvest.

farmer harvest the wood at this unit or not? There is uncertainty around the actual volume of timber available at each unit. The farmer's value for the season $v(\boldsymbol{x}, \boldsymbol{a})$ is a function of the volume of timber as well as the choice of utilized units.

Recall that both the spatial variable $\boldsymbol{x} = \{x_i : i = 1, 2, ..., n\}$ as well as the set of alternatives $\boldsymbol{a} = \{a_j : j = 1, 2, ..., N\}$ are associated with spatial locations. The number of units n and the number of available alternatives N need not necessarily be the same; they depend on the spatial resolution and the assumptions of the model. For binary alternatives (harvest/do not harvest) at each unit $i = 1, 2, ..., N$, the set $\boldsymbol{A}$ consists of 2^N possible alternatives. Decision situations may, of course, have more than two alternatives at each unit – e.g., harvest, do not harvest, or sell the unit.

As another example, consider Figure 5.3, which depicts a heterogeneous oil reservoir. The decision maker must choose where to place injection and production wells to recover oil. Here, x_i is the spatial variable that denotes oil reserves at unit i of a three-dimensional (3-D) reservoir model representing the subsurface. The recoverable oil reserves are typically complex functions of spatially heterogeneous variables such as rock types, porosity, and permeability.

In practice, the set of possible alternatives could potentially be complex; it could involve tuning production well rates and injection rates of water or gas, as well as the placement and

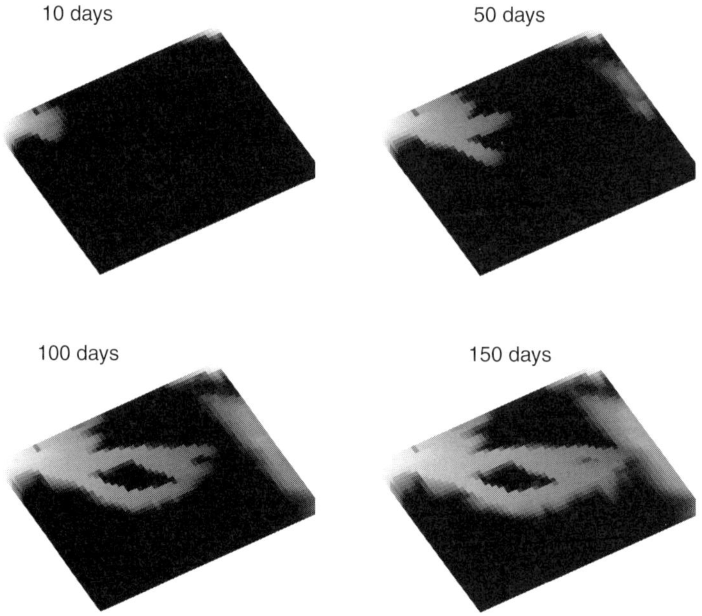

Figure 5.3 Illustration of a spatial domain: a heterogeneous oil reservoir. The decision maker must decide where to place injection and production wells so as to recover oil. Here, an injection well is positioned in one corner, and a production well is positioned diagonal from the injection corner. The water replaces the oil, and the displays show the fluid flow at four stages during the first 150 days of production.

spatial configuration of these wells. The alternative set may be discrete (well type: injector or producer) or continuous (well rates) or a combination of both. The figure depicts one possible reservoir model for one particular alternative $\boldsymbol{a}$, where an injection well and a single production well are placed at corners diagonal from each other. The decision maker's value from the reservoir $v(\boldsymbol{x},\boldsymbol{a})$ depends on the oil that is recovered, which is a function of the oil reserves $\boldsymbol{x}$ as well as the chosen production strategy $\boldsymbol{a}$. The value can be computed from a subsurface flow scenario that depends on the underlying heterogeneous porosity and permeability.

Both examples highlight common themes around the spatial aspects of decisions in the Earth sciences. We refer to such decision situations as **spatial decision situations**, identifying three primary characteristics that are common:

1. Spatial variables typically exhibit spatial dependence.
2. There is typically high decision flexibility since alternatives are associated with spatial locations.
3. The decision maker's value is typically highly coupled and can be a complex function of properties of spatial locations.

Property 1 is inherently present in most Earth sciences applications due to physical processes that explain the formation and changes of geographic features. Spatial variables are often best described using models that incorporate **spatial dependence**, because a location that is closer to another may have more similar properties than a location that is farther away. In Chapter 4, we reviewed a host of models that may be appropriate for capturing such effects, such as Markov random fields, multivariate Gaussian models, multiple-point statistics, and training image-based models.

Property 2 is concerned with the alternatives that are available to the decision maker. Typically, Earth sciences applications involve selecting locations and subsequently utilizing and managing resources at these location(s). In the forestry example from Figure 5.2, the farmer must choose the set of units to cultivate – this leads to a combinatorial (2^N) number of alternatives. We refer to this property as that of **high decision flexibility**, because the decision maker has many alternatives to choose from.

Property 3 involves the decision maker's value from the decision situation. Due to physical processes, the value in Earth sciences applications is often best described as a complex function of the properties and the chosen alternative. In the oil reservoir example from Figure 5.3, the decision maker's profit is a function of the oil that is recovered at the production well, which might be best described by complex reservoir simulations computed by numerically solving multiphase fluid flow differential equations in a heterogeneous Earth model. We refer to this property as that of **coupled value**, because the value can be a non-trivial function of properties at various locations.

From the perspective of the field of operations research, spatial decision situations can be viewed as a type of dependent portfolio problem where dependence is of a spatial nature (Bhattacharjya et al. 2013). We stress the generality of these decisions in the Earth sciences and suggest models for their analysis.

Discussion: cogency versus verisimilitude in models

Howard (2007) notes two major desiderata of models for decision situations (and in general): **cogency** – the extent to which the model satisfies its functions – and **verisimilitude** – the extent to which the model reflects reality. It is important to formulate models such that they capture the most important features of the application being served without being overly complex. This is a common trade-off that modelers face, and it is no different in Earth sciences applications. In this book, we attempt to frame problems in a realistic fashion using models at granular levels of spatial resolution, but we recognize the need for reasonable, problem-dependent simplifications. Property 1 of spatial decision situations is retained for most of the book, as we believe that this is a crucial aspect of Earth sciences applications. However, incorporating both Properties 2 and 3 in addition can result in models with significant computational complexity – which can affect the cogency of these models. In Section 5.2, we discuss various formulations resulting from relaxing Properties 2 or 3 or both.

5.1.2 *Information gathering in spatial decision situations*

Decision situations in general, including those that we refer to as spatial decision situations, are associated with auxiliary decisions pertaining to information gathering. For instance, in the forestry example from Figure 5.2, the farmer can perform certain tests to study the volume of timber before deciding which units to utilize. Similarly, in the oil reservoir example from Figure 5.3, the decision maker can conduct seismic or electromagnetic geophysical surveys to further study the geology of the reservoir before choosing a production strategy to recover oil. When there is a lot at stake in a decision situation, it may be worthwhile to obtain more information to better understand the critical uncertainties that have a bearing on the decision.

Since spatial variables play an important role in decisions in the Earth sciences, there is a spatial aspect to information gathering as well, resulting in numerous practical complexities and opportunities. Just like the underlying decision situation itself, information gathering may include a combinatorial number of alternatives. Consider, for instance, the alternatives that might be available to the farmer with the forest shown in Figure 5.2 with regard to inferring the volume of timber at the various units by carrying out a few sampling tests. To make matters more interesting, the decision maker often must choose from among various types of data sources, sometimes at varying spatial resolutions and varying reliability. Moreover, the information is available only at some expense and effort: consider the attempts at surveying by trekking through thick, dense forests or the extensive infrastructure and computation required for acquiring, processing, and interpreting seismic information to study the geology of reservoirs. When does the information then become worth its cost?

The complexities surrounding information-gathering decisions create opportunities for decision makers in the Earth sciences. Since there are so many possibilities, the decision maker might be able to tune the spatial coverage of the test or the accuracy of the measurements. In addition, due to spatial dependence that is typical in spatial decision situations (Property 1), information about only a select few locations may be sufficient for learning enough about the entire spatial domain. The decision maker can therefore be creative in designing high-value information-gathering experiments.

Our main focus in this book is on describing techniques for evaluating and comparing different information-gathering schemes, providing the decision maker with powerful tools for analysis. These techniques are formal and are based on foundations of probability and decision theory, as well as practical considerations, in that they try to address actual concerns of decision makers in real-world situations. Integrating decision theoretic notions with those from spatial statistics ensures that the connection between the information itself and how it will be useful is always maintained.

5.1.3 Overview of models

VOI analysis is performed for information-gathering schemes that provide the decision maker the opportunity to observe uncertainties in their underlying decision situation before making their decision(s). The remaining sections of this chapter, up to and including Section 5.8, are devoted to describing models for VOI analysis in spatial decision situations. Here, we provide a framework for these models by making a high-level distinction between types of decision situations and information-gathering schemes.

First, we distinguish between static and sequential (or dynamic) decision situations. In **sequential decision situations**, the decision maker makes several decisions over time; it is typical for the decision maker to choose an alternative for a particular decision, observe some uncertainties, and make subsequent decisions by possibly adapting to whatever outcomes have been observed thus far. The notion of a strategy, which specifies a plan for the decision maker for various decisions given the information that is available at the time of the decisions, is central to sequential decision situations. The reader is referred to Section 3.3, where we discuss strategies in the context of influence diagrams. **Static decision situations**, on the other hand, are decision situations where the decision maker makes a one-shot decision. For instance, in the static model of the decision situation for the oil reservoir example from Figure 5.3, the decision maker is assumed to choose the production plan, including well locations, control rates, etc., all in one go. In the sequential version, the decision maker would have the capability to adapt the plan based on the information from production data.

In a similar fashion, we also distinguish between static and sequential information-gathering schemes. In a **sequential information-gathering scheme**, the decision maker can acquire data in a sequential fashion – i.e., they can observe an uncertainty and then decide whether to pursue further information based on whatever has been observed. In a **static information-gathering scheme**, the decision maker is assumed to obtain all the information together. In the oil reservoir example, a static information-gathering model would assume that the decision maker selects all exploration wells up front.

In this book, we predominantly focus on static models for both the decision situation and information gathering. Although sequential models are more realistic in general, such models can involve serious modeling and computational challenges, and often a static model is a reasonable and sufficient model for the application under consideration. Since our focus is on performing analysis for practical problems, the efficiency of the proposed approach is an important consideration, and as we will see, there are plenty of interesting insights to be gleaned from VOI analysis using static models.

Table 5.1 categorizes the models as per the static versus sequential framework and also specifies various sections of the chapter that deal with various models. The literature on

Table 5.1. *Categorizing models: static versus sequential decision situations and information-gathering schemes*

	Static Decision Situation	Sequential Decision Situation
Static Information Gathering	Section 5.2 provides a fairly general formulation. Sections 5.3–6 are special cases.	Section 5.7.3 discusses these models briefly.
Sequential Information Gathering	Section 5.8 discusses these models briefly.	This case is not discussed in the book.

sequential models for VOI analysis for Earth sciences applications is currently fairly limited; we refer to a few relevant articles in the appropriate sections. We suspect that the literature in these areas will continue to grow.

Even for the static models that are explored extensively in the book, we make some simplifying assumptions about the decision situation and the information-gathering schemes. Specifically:

- The decision maker is assumed to be risk neutral, therefore the optimal alternative is computed by maximizing the expected value. Implications of relaxing this assumption are examined in Section 5.7.1.
- It is assumed that there are no additional constraints in the decision situation (beyond those that have already been explicitly modeled through the available set $\boldsymbol{A}$) – i.e., the decision maker can select any alternative from the available set of alternatives. We explore the effect of additional constraints for a special case in Section 5.7.2.

5.2 Value of information: a formulation for static models

We assume that the reader is familiar with the VOI-related concepts that were explained in Chapter 3. Recall that the VOI is computed by comparing the situations with and without information; we therefore first study the prior value of the decision situation, proceeding subsequently to a posterior value that incorporates additional information.

5.2.1 *Prior value*

A priori, the decision maker is faced with the following spatial decision. He or she must choose an alternative from the available set, $\boldsymbol{a} \in \boldsymbol{A}$, and his or her beliefs about the uncertain spatial variable $\boldsymbol{x}$ are captured by the prior probability distribution $p(\boldsymbol{x})$. The value derived from the decision situation is denoted $v(\boldsymbol{x},\boldsymbol{a})$.

Assuming that the decision maker is risk neutral, it is optimal to choose the alternative that maximizes the expected value. Therefore, the prior value of the decision situation, before any additional information has been observed, is:

$$PV = \max_{\boldsymbol{a}\in A}\left\{E\left(v(\boldsymbol{x},\boldsymbol{a})\right)\right\} = \max_{\boldsymbol{a}\in A}\left\{\int_{x} v(\boldsymbol{x},\boldsymbol{a})p\left(\boldsymbol{x}\right)d\boldsymbol{x}\right\}. \tag{5.1}$$

Computing the expected value in this fashion may be computationally challenging for problems in the Earth sciences. The high decision flexibility in the decision situation (Property 2) may result in a combinatorial set of alternatives $\boldsymbol{A}$ to maximize over. Moreover, the coupled value arising from a complex value function (Property 3) may require time-consuming numerical simulations for assessing $v(\boldsymbol{x},\boldsymbol{a})$. Equation (5.1) provides a general formulation; more specific simplifications are provided in subsequent sections.

5.2.2 Posterior value

How does the value of the decision situation change when the decision maker can observe some information related to the uncertainties before making the decision? We make a distinction between **perfect** and **imperfect** information based on whether the actual distinction of interest, or merely a noisy measurement related to the distinction of interest, is observed. In the case of the oil reservoir example in Figure 5.3, the perfect information case for one location occurs when the oil accumulation, porosity, and permeability in that location are observed before the drilling strategy is determined. This sort of information might be provided by an exploration well. Imperfect sources of information include seismic and electromagnetic data and interpretations, since they are noisy measurements that provide indirect, remotely sensed information about the pore fluid saturations and rock types. Note that perfect information can be viewed as a special case of imperfect information – it is the case when there is no measurement inaccuracy.

We also make a distinction between **total** and **partial** information, depending on the spatial coverage of the information. The case of partial information occurs when information is provided for a strict subset of the spatial locations, as opposed to total information, where information for all spatial locations is observed before the decision is made. In the oil reservoir example, exploration wells are typically partial sources of information, as they are only drilled at certain locations in the reservoir. Three-dimensional reflection seismic surveys, on the other hand, typically cover the entire spatial domain of the reservoir and are therefore sources of total information.

Most information sources in the Earth sciences are arguably imperfect or partial sources of information. Common data-gathering schemes include surveys by humans that involve visual observations (such as counting species of plants or animals) or recovering samples (such as cores of minerals) that are brought back to the laboratory for testing. Such tests are often noisy measurements of the distinction of interest due to human errors in measurement, instrument inaccuracy, challenges associated with the spatial scale of the domain, etc. Surveys could also be performed remotely, where instruments automatically measure features that are relevant to the distinction of interest. Examples include seismic, electromagnetic, radar, and satellite data. Table 5.2 summarizes the various cases in the form of a matrix, providing a few examples. Although the case of total perfect information is rare, it can often be an efficient theoretical construct because it provides an upper bound that can be useful for evaluating information sources. Other, more general information-gathering schemes are also possible. For instance, the decision maker could perform experiments that provide information about averages or sums of properties over many spatial units. Replicated sampling at locations may also be possible, perhaps with varying data qualities. We explore each case in Table 5.2 individually.

Perfect information

When there is **total perfect information**, the decision maker has the opportunity to observe the distinction of interest $\boldsymbol{x}$ before making the decision. When the instantiation $\boldsymbol{x}$ is

Table 5.2. *The matrix of information-gathering cases: perfect versus imperfect and total versus partial information*

	Perfect	**Imperfect**
Total	Exact observations are gathered for all locations. This is rare, occurring when there is extensive coverage and highly accurate data gathering.	Noisy observations are gathered for all locations. This is common in situations with remote sensors with extensive coverage – e.g., seismic, radar, or satellite data.
Partial	Exact observations are gathered at some locations. This might occur, for instance, when there is careful analysis of rock samples along boreholes in a reservoir or a mine.	Noisy observations are gathered at some locations. Examples include handheld (noisy) meters to observe grades in mine boreholes, electromagnetic testing along a line, biological surveys of species, etc.

observed, the decision maker should choose the alternative to maximize $v(\boldsymbol{x},\boldsymbol{a})$. However, there is uncertainty around what will eventually be observed before the information is acquired, and the expected value is computed by integrating over the various outcomes of the test. The posterior value, for a risk-neutral decision maker, is therefore:

$$PoV(\boldsymbol{x}) = \int_{x} \max_{\boldsymbol{a}\in A}\{v(\boldsymbol{x},\boldsymbol{a})\}\, p(\boldsymbol{x})\, d\boldsymbol{x}. \tag{5.2}$$

Since the decision maker is risk neutral, the VOI is computed as the difference between the posterior and prior values. The value of the total perfect information is:

$$VOI(\boldsymbol{x}) = PoV(\boldsymbol{x}) - PV. \tag{5.3}$$

The argument in the parentheses for posterior value and VOI indicates the uncertainty for which VOI is being computed. (The same notation was used in Chapter 3.)

Consider next the case of **partial perfect information**, where the distinction of interest at a subset of the locations is observed. As was done in Chapter 2, suppose that the locations at which the observations are made is denoted by $\mathbb{K}$, where $\mathbb{K} \subset \{1,2,...,n\}$, and that the decision maker observes $\boldsymbol{x}_{\mathbb{K}}$ before making the decision. Denote the complementary set $\mathbb{K}^C$ by $\mathbb{L}$. When the instantiation $\boldsymbol{x}_{\mathbb{K}}$ is observed, the decision maker should choose the alternative to maximize the conditional expected value $E(v(\boldsymbol{x},\boldsymbol{a}) \mid \boldsymbol{x}_{\mathbb{K}})$. With total perfect information in Equations (5.2) and (5.3), this conditional expectation was the value itself, $E(v(\boldsymbol{x},\boldsymbol{a}) \mid \boldsymbol{x}) = v(\boldsymbol{x},\boldsymbol{a})$. Again, there is uncertainty around what will eventually be observed, and the expected value is computed by integrating over the various outcomes of the test. The posterior value is:

$$PoV(\boldsymbol{x}_{\mathbb{K}}) = \int \max_{\boldsymbol{a}\in A}\{E(v(\boldsymbol{x},\boldsymbol{a}) \mid \boldsymbol{x}_{\mathbb{K}})\}\, p(\boldsymbol{x}_{\mathbb{K}})\, d\boldsymbol{x}_{\mathbb{K}}. \tag{5.4}$$

The value of the partial perfect information is:

$$VOI\left(\boldsymbol{x}_{\mathbb{K}}\right) = PoV\left(\boldsymbol{x}_{\mathbb{K}}\right) - PV. \tag{5.5}$$

Note that this case requires computing the distribution $p\left(\boldsymbol{x}_{\mathbb{K}}\right)$ by marginalizing out the distinction of interest at locations where the information is not observed. Additionally, the computation for the conditional expected value $E\left(v(\boldsymbol{x},\boldsymbol{a}) \mid \boldsymbol{x}_{\mathbb{K}}\right)$ is not necessarily straightforward.

Imperfect information

Suppose that there is another uncertain variable $\boldsymbol{y} = \{y_i : i = 1,2,...,n\}$ that is relevant to the distinction of interest $\boldsymbol{x}$. The probabilistic dependence between these variables is represented through the likelihood $p(\boldsymbol{y} \mid \boldsymbol{x})$. When there is **total imperfect information**, the decision maker can potentially observe $\boldsymbol{y}$ at all locations before making the decision. The posterior value can be obtained by integrating over the outcomes of $\boldsymbol{y}$:

$$PoV\left(\boldsymbol{y}\right) = \int \max_{\boldsymbol{a} \in A} \left\{E\left(v\left(\boldsymbol{x},\boldsymbol{a}\right) \mid \boldsymbol{y}\right)\right\} p\left(\boldsymbol{y}\right) d\boldsymbol{y}. \tag{5.6}$$

The computation for the posterior value in this case requires finding the conditional expected value $E\left(v\left(\boldsymbol{x},\boldsymbol{a}\right) \mid \boldsymbol{y}\right)$ as well as the pre-posterior marginal probability density function (pdf) $p\left(\boldsymbol{y}\right)$, which can be obtained from the prior $p\left(\boldsymbol{x}\right)$ and the likelihood $p\left(\boldsymbol{y} \mid \boldsymbol{x}\right)$ using Bayes' rule. This likelihood that is required for formulating the case of imperfect information introduces complexities that we will return to when we discuss the special cases.

The value of the total imperfect information is:

$$VOI\left(\boldsymbol{y}\right) = PoV\left(\boldsymbol{y}\right) - PV. \tag{5.7}$$

The posterior value and the VOI for the case of **partial imperfect information** can be computed in a fashion similar to that of partial perfect information in Equations (5.4) and (5.5) after replacing $\boldsymbol{y}_{\mathbb{K}}$ for $\boldsymbol{x}_{\mathbb{K}}$. For completeness, we write the equations below:

$$PoV\left(\boldsymbol{y}_{\mathbb{K}}\right) = \int \max_{\boldsymbol{a} \in A} \left\{E\left(v(\boldsymbol{x},\boldsymbol{a}) \mid \boldsymbol{y}_{\mathbb{K}}\right)\right\} p\left(\boldsymbol{y}_{\mathbb{K}}\right) d\boldsymbol{y}_{\mathbb{K}}, \tag{5.8}$$

$$VOI\left(\boldsymbol{y}_{\mathbb{K}}\right) = PoV\left(\boldsymbol{y}_{\mathbb{K}}\right) - PV. \tag{5.9}$$

5.2.3 Special cases: an overview

The formulation described in the previous section applies generally to all situations. However, simplifications may be required for computational tractability, thereby enabling tools for efficient VOI analysis. In Section 5.1, we described three distinctive properties of spatial decision situations. In the following four sections, we present special cases that are obtained by relaxing and enforcing these properties. Property 1 is never relaxed (spatial dependence); relaxing and enforcing the other two properties provides the various combinations presented in Table 5.3.

Table 5.3. *Categorization of different spatial decision situations*

Section	Special Case	Assumption: Decision Flexibility	Assumption: Value Function	Main Example
5.3	Low decision flexibility; decoupled value	Alternatives are easily enumerated	Total value is a sum of value at every unit	Norwegian wood – forestry example
5.4	High decision flexibility; decoupled value	None	Total value is a sum of value at every unit	The tree amigos – conservation biology example
5.5	Low decision flexibility; coupled value	Alternatives are easily enumerated	None	Go with the flow – reservoir simulation example
5.6	High decision flexibility; coupled value	None	None	Frozen – hydropower example

5.3 Special case: low decision flexibility and decoupled value

A potential way to enable computational tractability for VOI analysis is to relax both Properties 2 and 3.

Relaxing Property 2 leads to a decision situation where there are a small set of alternatives – i.e., set A has a low cardinality. As a result, computations are easier because maximizations are performed over small and manageable sets. Such an assumption may be reasonable in a host of real-world applications for a variety of reasons. For instance, there may be factors that have not been explicitly modeled, enforcing constraints that only make a small set of alternatives feasible, or it may be that previous analysis has filtered out most of the alternatives. Or it may be the case that only a few alternatives are considered for the sake of simplicity of analysis and computational tractability.

Relaxing Property 3 leads to a decision situation where the value function can be easily "decoupled," in the sense that the total value can be decomposed using a separable function of the value from each individual spatial unit. In the simplest case, the total value is the sum of the value from all units in the spatial domain. This assumption allows computations to be performed at the local level of individual units and then combined in a convenient matter. Again, this assumption may either be natural for a particular application, or it may be a simplification that is not necessarily entirely realistic but still a reasonable approximation for computational reasons.

Applications where these assumptions have been made include domains such as agriculture (Wiles 2004), fisheries (Forsberg and Guttormsen 2006), mining (Eidsvik and Ellefmo 2013), and forestry (Kangas 2010). In this book, we focus on the explicitly modeled spatial uncertainties.

Our example at the end of this section is motivated by what has been the running example in this chapter from the domain of forestry. As a reminder to the reader, the situation is as follows: a farmer owns several units of a forest. In this special case, the farmer only has two available alternatives: to harvest the forest at all units or to leave the forest as is. The value obtained from the wood, if the farmer decides to harvest, would depend on the volume of timber, as well as the selling price of wood and paper and the operational and processing costs. Before making the decision, the farmer considers purchasing information pertaining to the volume of timber. We use VOI analysis to study various information-gathering schemes.

5.3.1 *Prior value*

We choose the alternative $a \in A$ that optimizes the overall value. Since the value function decouples for the n spatial locations,

$$v(\boldsymbol{x}, a) = \sum_{i=1}^{n} v_i(x_i, a). \tag{5.10}$$

In the forestry example, the value is assumed to be proportional to the volume of timber, and the total volume is a sum of volumes from the n units – hence the sum in Equation (5.10). Note that the value function $v_i(x_i, a)$ can vary from unit to unit because there may be important differences from site to site. For instance, the spatial units of a forest could be at varying altitudes, slopes, distances from infrastructure, etc., and may therefore require different local costs to develop.

With the decoupling of value in Equation (5.10), the prior value for this special case becomes

$$\begin{aligned} PV &= \max_{a \in A}\left\{E\left(v(a, \boldsymbol{x})\right)\right\} = \\ &\max_{a \in A}\left\{\sum_{i=1}^{n} E\left(v_i\left(x_i, a\right)\right)\right\} = \max_{a \in A}\left\{\sum_{i=1}^{n} \int v_i\left(x_i, a\right) p(x_i) dx_i\right\}, \end{aligned} \tag{5.11}$$

where the final equation holds if the distinction of interest is a continuous random variable with marginal pdf at site i denoted $p(x_i)$. Note that the spatial dependence between variables is not relevant for the prior value in Equation (5.11) because the decoupling of the value function leads to n univariate integrals.

5.3.2 *Posterior value*

Suppose that we can purchase further information about the distinction of interest. The information may take any of the forms discussed previously: total versus partial, perfect versus imperfect. Note that it is possible for the information to be at a less granular level than the spatial units. For instance, in the forestry example, a surveyor could collect information about the average tree height or timber volume along a chosen path in the forest, and not necessarily about the height or timber volume at particular locations along this path. Such average properties may be valuable, and they might be less expensive to acquire.

The posterior value is now an integral over the data, assuming continuous sample space. For the case of total perfect information:

$$PoV(\boldsymbol{x}) = \int \max_{a \in A} \left\{ \sum_{i=1}^{n} v_i(x_i, a) \right\} p(\boldsymbol{x}) d\boldsymbol{x}. \tag{5.12}$$

The associated VOI is the difference between the posterior and the prior value, as expressed in Equation (5.3), which represents the upper bound on the value of any information-gathering scheme in this situation. In the forestry example, perfect and total information occurs when the decision maker observes the timber volume at every unit before making the decision. In practice, forest surveys would likely be partial sources of information because only selected units are observed, and they would be imperfect sources due to inaccuracies in measurements.

Instead of repeating the expression for the posterior value for the various cases, here we only present the case with partial and imperfect information – i.e., when the data is $\boldsymbol{y}_{\mathbb{K}}$ with marginal likelihood $p(\boldsymbol{y}_{\mathbb{K}})$, as it can be viewed as a generalization. The posterior value is

$$PoV(\boldsymbol{y}_{\mathbb{K}}) = \int \max_{a \in A} \left\{ \sum_{i=1}^{n} E(v_i(x_i, a) \mid \boldsymbol{y}_{\mathbb{K}}) \right\} p(\boldsymbol{y}_{\mathbb{K}}) d\boldsymbol{y}_{\mathbb{K}}, \tag{5.13}$$

and the VOI is the difference between the posterior and prior value, as in Equation (5.9).

5.3.3 Computational notes

Computations for the prior and posterior value depend on the local value functions $v_i(x_i, a)$, $i = 1, \ldots, n$, and the statistical model. This special case is typically simpler than the more complex cases described in Sections 5.4–6.

For the forestry example that follows, there are analytical solutions for the VOI. This is possible because the example assumes a Gaussian distribution for the wood volumes as well as an additive value function. With a more complex statistical model formulation or value function, one may need to approximate the value in one way or another. A general approach to computing the prior and posterior values is Monte Carlo sampling. We will discuss Monte Carlo methods further for the more complex situations later in this chapter.

Pitfalls: expected value of a non-linear function

The expected value of a non-linear function is not necessarily the same as the non-linear function of the expectation. We cannot interchange the function and the expectation in general – this only holds for linear functions. The value calculations are based on the expectation of the value function: $E(v(\boldsymbol{x}, a) \mid \boldsymbol{y}) = \int v(\boldsymbol{x}, a) p(\boldsymbol{x} \mid \boldsymbol{y}) d\boldsymbol{x} \neq v(E(\boldsymbol{x} \mid \boldsymbol{y}), a)$.

5.3.4 Example

We present a synthetic example to gain insight about the VOI for this special case. The statistical model is identical to an example from Section 4.4. In Chapter 7, we present a hands-on project based on the same example.

Norwegian wood: forestry example

Keywords: *forestry example, Gaussian random field, total information, partial information, low decision flexibility, decoupled value function*

Let us study the forestry example in greater detail, with numbers.

Framing the decision situation

Suppose that a farmer owns a forest, which is modeled as a set of spatial units, shown as a spatial grid of size 25 × 25 in Figure 5.4. There are thus a total of $n = 625$ units in this example. The uncertain distinctions of interest are the profits x_i, $i = 1,\dots,n$ from the spatial forest units. The farmer must make a decision about whether to harvest all the trees or to

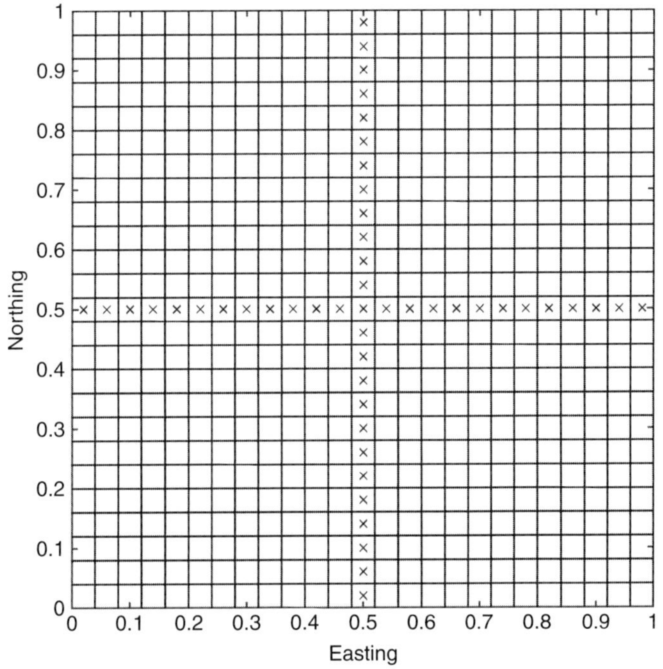

Figure 5.4 Forestry example: illustration of a size 25 × 25 grid of forest units. The crossed vertical and horizontal lines indicate locations for partial testing.

leave the forest as is. This means that the farmer has only two alternatives $a \in \{0,1\}$ defined as follows: to utilize all units ($a = 1$) or to do nothing ($a = 0$). The value function decouples to the sum over the spatial units. The total value if all units are utilized – i.e., $a = 1$ – is $v(\boldsymbol{x}, a = 1) = \sum_{i=1}^{n} x_i$, whereas the value $v(\boldsymbol{x}, a = 0) = 0$ if the farmer leaves the forest as is.

Information gathering

Before making the decision, the farmer can pay for tests that are informative of the timber volume and therefore the profits. Potential tests one could imagine in this case entail sending a surveyor to selected units and carefully measuring trees and the timber volume in selected spatial units. The surveyor would likely walk along a chosen path through the field with a device, observing the light or canopy variables that are relevant to the volume of timber. There are also other possibilities for imperfect measurements about the timber volume at all units, such as processed satellite data or other inventory data (Bergseng et al. 2015).

We consider three testing options outlined in detail in the following text. They are all considered imperfect information experiments and include a total test, a partial test with data at spatial units along two paths in the domain, and a partial test with average timber volumes or profits along the paths.

Modeling

We assume that the farmer has a reasonably good idea about the selling price of wood and the operational costs, but there is uncertainty about the timber volume. This volume determines the uncertain profit, which can be positive or negative. These profits at the n spatial units are the distinction of interest $\boldsymbol{x} = (x_1, \ldots, x_n)$. As discussed earlier, the total value decouples as a sum over the units. The total profit the farmer receives is simply the sum of the profits from all units.

We assume that the profits are Gaussian distributed a priori – i.e., $p(\boldsymbol{x}) = N(\boldsymbol{\mu}, \boldsymbol{\Sigma})$. The prior mean is set to be the same for all units, $\mu_i = \mu, i = 1, \ldots, n$, and the covariance is defined via an exponential-type correlation function, $\Sigma_{ij} = \sigma^2 \exp\left(-3\frac{t_{ij}}{\xi}\right)$, where t_{ij} is the Euclidean distance between two units i and j, the prior marginal variance is σ^2 at every cell, and the spatial correlation range is ξ. Recall that we use $\eta = \frac{3}{\xi}$ for the spatial correlation decay parameter; it may be easier to interpret the effect of the correlation range ξ directly since the correlation decreases to only $\exp(-3) = 0.05$ for distance $t_{ij} = \xi$. We perform sensitivity analysis for VOI with respect to all these model parameters.

The tests are imperfect observations of timber volume providing imperfect information about the profits, and the results of the observations are distributed according to a Gaussian pdf. The type of test (denoted 1–3 later) and the number or size of measurements (denoted m) vary as follows:

1. Total test of forest timber volume and hence profits at every unit. This is denoted using the likelihood $y_j = x_j + N(0,\tau^2)$ for $j = 1,\ldots,m = n$.
2. Partial test of the forest along the north–south and east–west center transects of the forest domain. We denote these by C_v and C_h, respectively. This is denoted using the likelihood $y_j = x_j + N(0,\tau^2)$, $j \in C_v \cup C_h$, and $m = 49$.
3. Partial test of aggregated forest timber volume (and hence aggregated profits) along the north–south and east–west center lines of the forest domain. This is denoted using the likelihood $y_1 = \frac{1}{25}\sum_{j\in C_v} x_j + N(0,\tau^2)$ for the north–south line and $y_2 = \frac{1}{25}\sum_{j\in C_h} x_j + N(0,\tau^2)$ for the east–west line, and $m = 2$.

The partial testing options in Tests 2 and 3 are illustrated in Figure 5.4. These tests would be representative of a surveyor walking along the north–south and east–west lines in the forest carrying a surveying device to measure the accumulated timber volume via, for instance, the light coming from the canopy. For Test 3, the instrument only gauges the average over a certain interval, and the surveyor does not want to reset this all the time. Such situations are also common in tomography where only the aggregated response along lines or ray paths is measured. Note that Test 2 is similar to the one considered in Section 4.4. An experiment such as the other random design studied in that section would likely be more expensive to acquire in practice.

VOI analysis

The prior value of this two-action utilization decision for the forest is

$$PV = \max\left\{E\left(v(\boldsymbol{x},a=0)\right), E\left(v(\boldsymbol{x},a=1)\right)\right\} = \max\left\{0, \sum_{i=1}^{n} \mu_i\right\}. \tag{5.14}$$

The posterior value depends on the testing scheme. Using generic notation, we define a likelihood model by $\boldsymbol{y} = \boldsymbol{F}\boldsymbol{x} + N(0,\boldsymbol{T})$, where the matrix $\boldsymbol{F}$ of size $m \times n$ defines the acquisition design of the test, with measurement noise covariance matrix $\boldsymbol{T} = \tau^2\boldsymbol{I}$, assuming conditionally independent tests defined by Tests 1–3 mentioned earlier. For the total test, this design matrix $\boldsymbol{F}$ is the identity matrix. For the partial test, it picks the entries which identify the center cells of the grid. For Test 3, the entries are $1/25$ at the selected units. The marginal likelihood model is $p(\boldsymbol{y}) = N(\boldsymbol{F}\boldsymbol{\mu}, \boldsymbol{F}\boldsymbol{\Sigma}\boldsymbol{F}^t + \boldsymbol{T})$.

The conditional mean profits are

$$\boldsymbol{w} = \boldsymbol{w}(\boldsymbol{y}) = E(\boldsymbol{x} \mid \boldsymbol{y}) = \boldsymbol{\mu} + \boldsymbol{\Sigma}\boldsymbol{F}^t\left(\boldsymbol{F}\boldsymbol{\Sigma}\boldsymbol{F}^t + \boldsymbol{T}\right)^{-1}(\boldsymbol{y} - \boldsymbol{F}\boldsymbol{\mu}). \tag{5.15}$$

For the posterior value in Equation (5.13), we use the sum of terms from Equation (5.15) in the decision. We then obtain the following posterior value:

$$PoV(\boldsymbol{y}) = \int \max\left\{0, E\left(\sum_{i=1}^{n} x_i \mid \boldsymbol{y}\right)\right\} p(\boldsymbol{y})d\boldsymbol{y} = \int \max\left\{0, \sum_{i=1}^{n} w_i(\boldsymbol{y})\right\} p(\boldsymbol{y})d\boldsymbol{y}. \tag{5.16}$$

Note that the profit variables are Gaussian distributed because of the Gaussian linear modeling functions and the decoupling – i.e., for variable $w = \sum_{i=1}^{n} w_i(\boldsymbol{y})$, we have

$$p(w) = N\left(\mu_w, r_w^2\right), \ \mu_w = \sum_{i=1}^{n} \mu_i, \quad r_w^2 = \sum_{i=1}^{n}\sum_{j=1}^{n} r_{ij}, \tag{5.17}$$

where r_{ij} is element (i, j) in the matrix $\boldsymbol{R}$ given by

$$\boldsymbol{R} = \Sigma \boldsymbol{F}^t \left(\boldsymbol{F}\Sigma\boldsymbol{F}^t + \boldsymbol{T}\right)^{-1} \boldsymbol{F}\Sigma. \tag{5.18}$$

Based on a result for Gaussian variables (see Appendix A.1), the posterior value in Equation (5.16) becomes

$$PoV(\boldsymbol{y}) = \int \max\left\{0, E\left(\sum_{i=1}^{n} x_i \mid \boldsymbol{y}\right)\right\} p(\boldsymbol{y})d\boldsymbol{y} = \mu_w \Phi(\mu_w/r_w) + r_w \phi(\mu_w/r_w). \tag{5.19}$$

where the mean parameter μ_w and the standard deviation parameter s_w are defined in Equation (5.17). Note that the mean μ_w only depends on the prior mean values, while the standard deviation s_w depends on the sampling design of the data, as well as on the prior and likelihood covariance matrices. Here the symbols Φ and ϕ denote the cumulative distribution function and the pdf of a standard Gaussian variable, respectively.

Let us compare the VOI for the three forest measurement designs 1–3 mentioned earlier. The comparison is done over a range of prior and likelihood parameters. For the baseline model, we use a constant prior mean $\mu = 0$ for all sites, a constant prior marginal variance $\sigma^2 = 1$, a spatial correlation range $\xi = 0.3$, and a measurement noise variance of $\tau^2 = 0.5^2$.

Figure 5.5 shows the VOI for Tests 1–3, as we change the different parameters of the statistical model. The main trends are the same for all test schemes: the VOI is largest for the zero prior mean value. At this prior mean, we are indifferent about the decision and information is likely to change our decision. The VOI increases as a function of the prior standard deviation parameter σ since data become more valuable when there is more prior uncertainty. The VOI decreases with larger measurement uncertainty as the data become less informative about the profits, and the VOI increases with more spatial dependence as the data become more informative about other nearby sites. Note that the spatial correlation in the timber volumes has a clear marked influence on the VOI. Ignoring the spatial correlation would lead to an erroneous assessment of the VOI.

Among the different test schemes, the aggregated test (partial, 3) has the smallest VOI, but this is reasonable given that the aggregated test consists of just two measurements compared with 49 (partial, 2) or 625 (total, 1). Presumably, the aggregated test would be much cheaper compared to the other two information-gathering schemes, and the decision maker could then make a decision based on comparing the VOI versus the cost for each scheme. With increasing spatial correlation, the VOI for even the aggregated scheme becomes relatively large because this partial information of sums along the center lines is informative of the sum from the entire domain due to spatial dependence. However, when there is more measurement noise (upper right), this scheme loses its value faster than the other schemes.

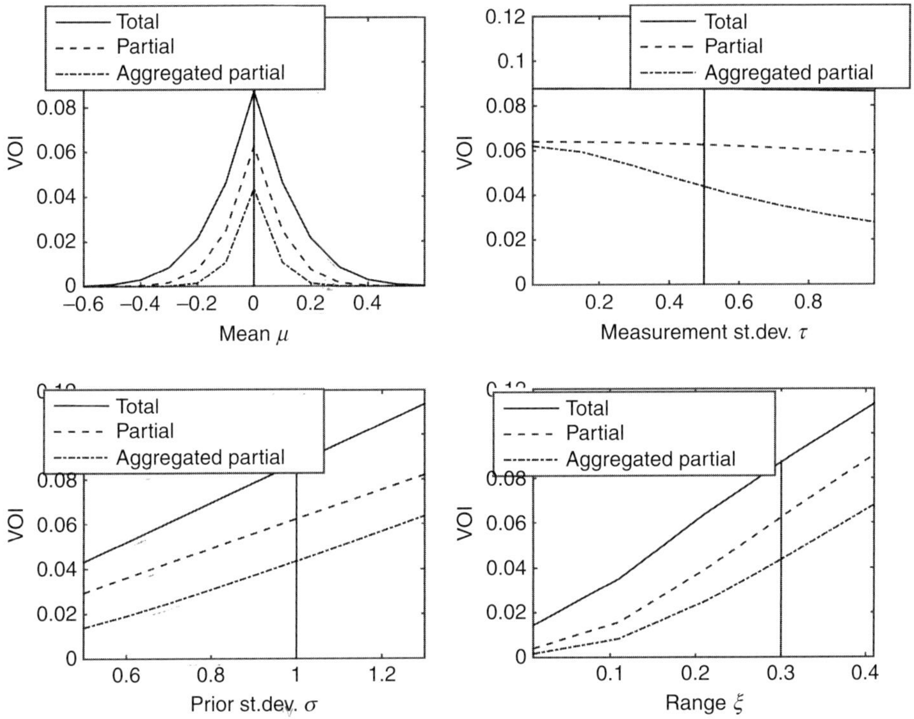

Figure 5.5 Forestry example: the VOI as a function of model parameters. Upper left: VOI as a function of the prior mean. Bottom left: VOI as a function of the prior standard deviation. Upper right: VOI as a function of the measurement noise standard deviation. Bottom right: VOI as a function of the spatial correlation range. The baseline case is shown by vertical lines.

5.4 Special case: high decision flexibility and decoupled value

We now turn to a spatial decision situation where we only relax Property 3. The total value is therefore a separable function of value from the individual spatial units. There may be initial costs related to equipment or man-hours that are common for some or all units, but these may be considered to be negligible. We sometimes refer to this decision situation as a "free selection of sites" because as per the assumptions of the model, the decision maker can choose to utilize as many sites as is profitable.

Examples of this case include a conservation biology situation where a park is modeled as a set of units and the decision is: which units should the decision maker maintain? See Polasky and Solow (2001) and Bhattacharjya et al. (2010) for applications. It is also relevant for the selection of large-scale prospects in petroleum development and exploration (Martinelli et al. 2011). One may also encounter this situation in the forestry example discussed in Section 5.3 if the farmer has the opportunity to harvest trees in selected units and this action does not influence the value at other forest units. We present a few illustrative examples in detail at the end of this section.

5.4.1 Prior value

Without loss of generality, we assume that the decision maker can make a decision at any site where the distinction of interest $\boldsymbol{x} = (x_1,...,x_n)$ is represented. This implies that we can denote alternatives at spatial locations by $a_i \in A_i$ for $i = 1,..,N = n$. More generally, we could have alternatives at blocks of spatial variables. These blocks may be defined based on the geography and infrastructure – we discuss such examples at the end of this section.

The value from each site is denoted by $v_i(x_i, a_i)$, or simply $v(x_i, a)$. Because the value decouples, the decision maker can optimize the decision locally, and the prior value equals

$$PV = \sum_{i=1}^{n} \max_{a_i \in A_i} \{E(v_i(x_i, a_i))\}. \tag{5.20}$$

We assume two alternatives in the following presentation – i.e., $|A_i| = 2$, but the analysis can be easily extended to incorporate several alternatives. Like we discussed, two alternatives at every site results in 2^n alternatives in total. If we further assume that x_i is the profit at site $i = 1,..,n$ and that the value function is $v_i(x_i) = x_i$ when $a_i = 1$ and zero otherwise, the prior value becomes

$$PV = \sum_{i=1}^{n} \max\{0, E(x_i)\}. \tag{5.21}$$

5.4.2 Posterior value

Total perfect information at all sites means that we can make optimal decisions before observing the outcome of the uncertain variable. This allows us to switch the integral and the max term in the prior value, and therefore the posterior value is

$$PoV(\boldsymbol{x}) = \sum_{i=1}^{n} E\left(\max_{a_i \in A_i} \{v(x_i, a_i)\}\right). \tag{5.22}$$

The VOI is computed as in Equation (5.3). As discussed earlier, this situation with total perfect information represents an upper bound for the value of any experiment. In the conservation biology example, this situation arises if we know the presence or absence of all trees at every unit.

Consider next the case of partial perfect information, with information only at a subset of the sites. When we calculate the posterior value for this case, we can split the expression into two parts: one associated with the information set $\mathbb{K}$ and one with the unobserved units $\mathbb{L}$. The posterior value of partial perfect information is then

$$\begin{aligned} PoV(\boldsymbol{x}_{\mathbb{K}}) = & \sum_{i \in \mathbb{K}} \int \max_{a_i \in A_i} \{v(x_i, a_i)\} p(x_i) dx_i \\ & + \sum_{i \in \mathbb{L}} \int \max_{a_i \in A_i} \{E(v(x_i, a_i) \mid \boldsymbol{x}_{\mathbb{K}})\} p(\boldsymbol{x}_{\mathbb{K}}) d\boldsymbol{x}_{\mathbb{K}}, \end{aligned} \tag{5.23}$$

assuming continuous sample space.

We make a few comments about the posterior value:

1. Note that just like for total perfect information in Equation (5.22), the posterior value in Equation (5.23) depends only on the prior pdf. We use the joint prior pdf to determine marginal pdfs $p(x_i)$ and $p(\boldsymbol{x}_{\mathbb{K}})$ and to get the expected conditional value $E(v(x_i, a_i) \mid \boldsymbol{x}_{\mathbb{K}})$.
2. The latter part of the posterior value in Equation (5.23), the sum over the sites in the $\mathbb{L}$ set, contains a multivariate integral over the information set $\boldsymbol{x}_{\mathbb{K}}$. Together with the conditional expectation inside the decision rule, this integral over $p(\boldsymbol{x}_{\mathbb{K}})$ tells us how the observations $\boldsymbol{x}_{\mathbb{K}}$ can have a bearing on our decisions and hence our posterior value at unobserved sites, arising from the possible spatial dependence between observed and unobserved sites.
3. If a variable x_i, $i \in \mathbb{L}$, is independent of $\boldsymbol{x}_{\mathbb{K}}$, the observations have no bearing on the variable, and we gain no value from the prior information. This can be seen mathematically as follows:

    a. The conditional expectation is not a function of the observation set when we have independence – i.e., $\mathrm{E}(v(x_i, a_i) \mid \boldsymbol{x}_{\mathbb{K}}) = \mathrm{E}(v(x_i, a_i))$, and the max term can go outside the integral.
    b. The integral over the pdf is $\int p(\boldsymbol{x}_{\mathbb{K}}) d\boldsymbol{x}_{\mathbb{K}} = 1$. We are only left with the $\max_{a_i \in A_i} \{E(v(x_i, a_i))\}$ term, which equals the part from site i in the prior value in Equation (5.20).

4. If we have a more realistic spatial statistical model, a variable x_i is dependent on the observation set $\boldsymbol{x}_{\mathbb{K}}$, and we would on average make better decisions and increase the value compared to the prior value without the observation set.

In models with complex spatial dependencies, it is common to study what-if scenarios, which entails conditioning on various outcomes of selected variables and studying what happens to the model outputs. What if we have a particular outcome of $\boldsymbol{x}_{\mathbb{K}}$? In the current setting, with perfect information at a subset of the variables, the VOI calculation can be regarded as revealing the what-if results for optimal decisions under the outcome. This is done for all possible outcomes and averaged over the marginal probabilities of the experimental results.

Consider next imperfect information about the distinction of interest. We refer to imperfect data (partial or total) as $\boldsymbol{y}$. For the conservation biology example, this might entail a biologist surveying the sites and checking for the presence or absence of a plant or animal. The observer may see nothing and this indicates absence, but some plants or animals may go unnoticed, and this would lead to a false negative test. On the other hand, the observer would notice presence, but this may be the result of misclassification, and the test may lead to a false positive. Imperfect information accounts for such errors in the likelihood model $p(\boldsymbol{y} \mid \boldsymbol{x})$. When the imperfect information depends only on the local value of the distinction of interest, conditional independence may be realistic – i.e., $p(\boldsymbol{y} \mid \boldsymbol{x}) = \prod_i p(y_i \mid x_i)$.

In other words, the observation at location i depends only on the value of $\boldsymbol{x}$ at location i and not at other locations.

The mathematical expressions for the posterior value of imperfect partial and total information are similar. We have

$$PoV(\boldsymbol{y}) = \sum_{i=1}^{n} \int \max_{a_i \in A_i} \left\{ E\left(v\left(x_i, a_i\right) \mid \boldsymbol{y}\right)\right\} p(\boldsymbol{y}) d\boldsymbol{y}. \tag{5.24}$$

The complexity of computing this posterior value with total or partial imperfect information depends on the likelihood model $p(\boldsymbol{y} \mid \boldsymbol{x})$ and the prior pdf $p(\boldsymbol{x})$. Both are needed to find the marginal distribution of the data $p(\boldsymbol{y})$ and for computing the conditional expectation of profits at all sites $E\left(v\left(x_i, a_i\right) \mid \boldsymbol{y}\right)$, $i = 1,..,n$.

Unlike the situation with total and partial perfect information, there is no point in splitting the sum in Equation (5.24) into observed and unobserved sites as was done in Equation (5.23). With imperfect information, the conditional expectation is required inside the decision rule regardless of whether it is an observation site or a non-observed site.

5.4.3 Computational notes

The posterior value and VOI calculations are more involved than in the situation described in Section 5.3. For some prior and likelihood models, the posterior value expressions are still analytically tractable, but in many situations one must use Monte Carlo solutions or a combination of analytical, numerical, and Monte Carlo approximations.

The prior value computation is usually not a significant challenge in the free selection case because only the marginal probabilities are used for computations. In some situations, these marginal probabilities may be hard to extract from a joint probability model, but it is more common that the marginal expectations are directly available from the definition of the pdf $p(\boldsymbol{x})$. Even when the value function $v(x_i, a_i)$ has a complex form, the required integrals (or sums) for the prior value are univariate,

$$PV = \sum_{i=1}^{n} \max_{a_i \in A_i} \left\{ \int_{x_i} v\left(x_i, a_i\right) p\left(x_i\right) dx_i \right\}, \tag{5.25}$$

and numerical solutions are possible.

Similarly, for free selection, the posterior value with total perfect information is based on a series of univariate integrals:

$$PoV(\boldsymbol{x}) = \sum_{i=1}^{n} \int \max_{a_i \in A_i} \left\{ v\left(x_i, a_i\right)\right\} p\left(x_i\right) dx_i. \tag{5.26}$$

These integrals are usually more challenging to compute than the prior value because the maximum is now inside the integral. Nevertheless, it is still a univariate integral, which means that numerical approaches can be efficient.

The posterior value for partial perfect information or for imperfect information can be much harder to calculate, but in general the complexity depends on the computational efficiency of marginalization and conditioning for the particular statistical model. Note that the conditional expected value is the key here; with imperfect information $\boldsymbol{y}$ (which could be partial or total), the decision is based on $w_i(\boldsymbol{y}) = E(v(x_i, a_i) \mid \boldsymbol{y}), i = 1,\ldots,n$. These conditional expectations have a univariate function output with a multivariate input as an argument. If one can compute the pdf of $w_i = w_i(\boldsymbol{y})$, the multivariate integral reduces to a univariate integral. Analytical or numerical solutions are likely available for this univariate integral. However, finding the pdf of w_i may be difficult, except for very special prior pdfs and likelihoods and simple value functions.

When the prior pdf or likelihood models are more complicated, Monte Carlo methods are useful for approximating the posterior value.

Algorithm: Monte Carlo approximation of the posterior value under free selection

1. Draw realizations of the data $\boldsymbol{y}^1,\ldots,\boldsymbol{y}^B$ from $p(\boldsymbol{y})$. This can be done by first sampling $\boldsymbol{x}^b$ from $p(\boldsymbol{x})$, then $\boldsymbol{y}^b$ from $p(\boldsymbol{y} \mid \boldsymbol{x}^b), b = 1,\ldots,B$.
2. Compute the conditional expectation $E(v(x_i, a_i) \mid \boldsymbol{y}^b)$ for $b = 1,\ldots,B$.
3. Approximate the posterior value by the average

$$PoV = \frac{1}{B}\sum_{b=1}^{B} \max_{a_i \in A_i} \left\{E(v(x_i, a_i) \mid \boldsymbol{y}^b)\right\}.$$

This Monte Carlo solution is generally applicable, and it could potentially run in parallel computations. In Step 1, which solves the outer integral over the data, efficient sampling is needed. In Step 2, the posterior expectation $E(v(x_i, a_i) \mid \boldsymbol{y}^b)$ is calculated. Step 3 is a simple average over the Monte Carlo samples. Step 2 is the most difficult step; it may be possible to perform this analytically for some models, or one could explore analytical approximations. If the conditional expectation in Step 2 requires Monte Carlo sampling, a Monte Carlo loop is inside another, and this can become demanding from a computational point of view. One could instead try to re-use prior samples in the approximation of $E(v(x_i, a_i) \mid \boldsymbol{y}^b)$. We will discuss this in more detail in later sections.

5.4.4 *Examples*

We present a number of illustrative examples to gain insights about VOI analysis for this special case. In Chapter 7, we present hands-on projects as extensions of these examples.

Never break the chain: Markov chain example

Keywords: *Markov chain, dependent binary variables, perfect information, partial information, high decision flexibility, free selection of sites, decoupled value function, design of experiments*

In this example, we explore a situation where the decision maker owns multiple dependent oil prospects that are modeled as a Markov chain.

Framing the decision situation

Consider a situation where the decision maker can choose to develop prospects and as a result encounters binary (uncertain) outcomes: success or failure. We consider the case with $N = n = 100$ hydrocarbon prospects. The decision to utilize a prospect is based on whether the expected profits are positive or not. There are known costs associated with failure and revenues associated with success. The oil and gas company will choose a prospect if its expected a priori profit is positive. Let outcome $x_i = 0$ represent a failure at location i, while $x_i = 1$ indicates success.

The decision maker is free to select as many prospects as are profitable, so the alternatives are given by a_i, $i = 1,\ldots,n$. Here, $a_i = 0$ indicates the alternative associated with not developing a prospect, while $a_i = 1$ indicates developing a prospect. There are 2^n possible alternatives the decision maker can choose from, making this a situation with high decision flexibility. The decision maker has the flexibility to select any k of the n sites with $k = 0,\ldots,n$.

The value function decouples between prospects. Under the development alternative, the uncertain value of a prospect is $v(x_i, a_i = 1)$. The expected profits at prospects $i = 1,\ldots,n$ are

$$(\text{Rev} - \text{Cost}) \cdot p(x_i = 1) - \text{Cost} \cdot p(x_i = 0) = \text{Rev} \cdot p(x_i = 1) - \text{Cost}, \qquad (5.27)$$

for a fixed revenue (Rev) and cost of development (Cost).

Information gathering

Before making decisions at prospects, should the company purchase information about the prospects? If so, which site should it select for testing? In the petroleum industry, such information may be acquired through an exploration well at a selected prospect. The exploration well informs the company whether hydrocarbon is present at the location where the well is drilled. Here, we will assume that the company selects one location for drilling and that this provides perfect information at that site. Since there is dependence among hydrocarbon prospects, this partial information may also be valuable in informing the company about other sites. VOI analysis is useful for designing such experiments – i.e., for prioritizing one testing site over others.

Modeling

The hydrocarbon prospect outcomes are dependent and modeled as a Markov chain. We used Markov chains and hidden Markov models to illustrate evidence propagation in Section 2.3 and for parameter estimation in Section 2.4. (See Appendix A.3 for more background on Markov chains and hidden Markov models.) In this example, we study a Markov chain with non-stationarity incorporated via an external field. We consider a situation where some currently available data govern this external field. Let us denote this data by $\tilde{y}_1,...,\tilde{y}_n$. We use the data from the hidden Markov model example in Section 2.4. The probability model is proportional to

$$p(\mathbf{x}) \propto p\left[\prod_{i=2}^{n} p(x_i \mid x_{i-1})\right]\prod_{i=1}^{n} f_i(x_i) \tag{5.28}$$

and the local function $f_i(0) = \exp\left(-\frac{(\tilde{y}_i+1)^2}{2\sigma^2}\right)$ and $f_i(1) = \exp\left(-\frac{(\tilde{y}_i-1)^2}{2\sigma^2}\right)$. If the current data $\tilde{y}_i$ at some location i are very large, the prior probability of success at this location is also large because the external field favors Class 1 in this situation.

Figure 5.6 shows the marginal prior probabilities of success $p(x_i = 1)$ at the prospects $i = 1,...,n$ and 10 realizations of the Markov chain from the prior pdf.

VOI analysis

The prior value for the free site selection situation at individual prospects is the expected profit of development in Equation (5.27). We have

$$PV = \sum_{i=1}^{n} \max\left\{0, \text{Rev}\cdot p(x_i = 1) - \text{Cost}\right\}. \tag{5.29}$$

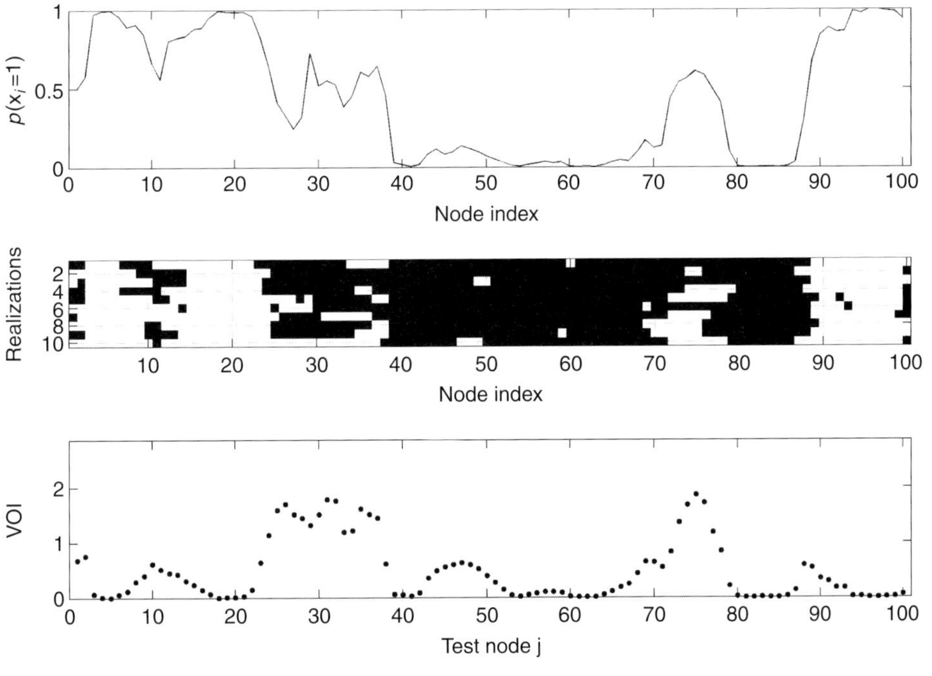

Figure 5.6 Markovian prospects example – top: the plot shows the prior probability of success at each prospect. Center: the plot shows 10 realizations of the Markov chain, each row being one realization. White indicates success. Bottom: VOI as a function of the test sites along the prospect chain.

For this example, we use Rev = 3 and Cost = 1.

We consider collecting perfect information (an exploration well) at one of the prospects, indexed j. The posterior value of a test at Prospect j is

$$PoV(x_j) = \sum_{i=1}^{n} \sum_{y=0}^{1} \max\left\{0, \text{Rev} \cdot p(x_i = 1 \mid x_j = y) - \text{Cost}\right\} p(x_j = y). \tag{5.30}$$

The VOI is equal to the difference in the posterior and prior values as defined in Equation (5.23).

Let us compare the VOI of observing different prospects $j = 1,...,n$. For a Markov chain, the conditional probabilities can be computed efficiently (Appendix A.3), and all evaluations are exact.

The VOI results of testing at a single prospect are shown in Figure 5.6 (bottom). We note a tendency of higher VOI when we collect perfect information at prospects where we are a priori most indifferent – i.e., $p(x_i = 1)$ close to Cost/Rev = 0.33 – while the VOI is very low when a priori we are either fairly sure of a success or a failure at a prospect. Since there is dependence between the prospects, the VOI is highest when we are unsure about several prospects near one another, like the prospects around indices 30 and 75.

The tree amigos: conservation biology example

Keywords: *conservation biology, Markov random field, total information, imperfect information, free selection of sites, high decision flexibility, decoupled value function*

This example is inspired from situations in conservation biology that were discussed briefly previously. The distinction of interest is the presence or absence of a species such as a type of tree. The decision maker has the flexibility of selecting individual sites for conservation, if it is indeed beneficial to do so.

Framing the decision situation

We study a small spatial domain split into a 3 × 3 grid – i.e., $n = N = 9$ spatial sites or cells that can be selected for conservation. At each cell, the distinction of interest (presence or absence of species) is a binary variable. Figure 5.7 illustrates the model, showing a graph of the grid. Let $\boldsymbol{x} = (x_1,\ldots,x_9)$ denote the uncertain presence or absence of the species at all cells in the grid. The binary outcome x_i is 1 if the species is present at cell i and 0 if the species is absent.

If the decision maker selects a cell, he or she must pay a cost for the construction of the natural reserve. The decision maker will not know for sure whether the species is present or not until the cell has been selected. If the species is present at a cell that is chosen, the decision maker obtains revenue. Both costs and revenues are assumed to be the same across cells. Here, revenue represents the value or benefits to society due to conservation of the species. Since the sites can be freely selected for conservation, the alternatives

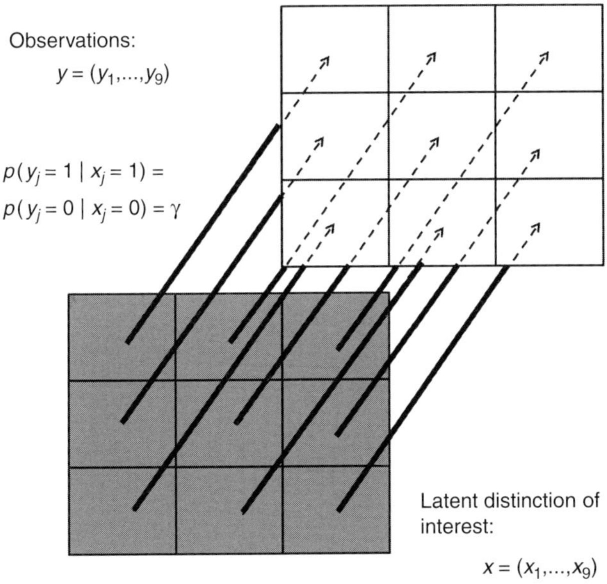

Figure 5.7 Conservation biology example: illustration of a spatial model for the distinction of interest on a 3 × 3 grid and a likelihood model with conditional independence for a total imperfect test.

are defined a_i, $i = 1,\ldots,n$. Here, $a_i = 0$ indicates leaving a site as is, while $a_i = 1$ indicates conservation.

The value function is defined in a fashion similar to the previous example, assuming known costs and revenues. Under the alternative of conserving a unit, the value is $v(x_i, a_i = 1) = \text{Rev}\cdot I(x_i = 1) - \text{Cost}$, where the indicator function $I(x_i = 1) = 1$ in the event of presence and zero otherwise. There is neither cost nor revenue associated with the other alternative. The costs of conservation are set to Cost = 1, while we compare VOI for different revenues.

Information gathering

Biologists may perform surveys at some units (partial testing) or all units (total testing) to check whether the species of interest is present or absent. We will consider a survey where a team explores every cell and indicates whether they believe that the species is present or not for each cell. This results in a scheme with total information $\boldsymbol{y} = (y_1,\ldots,y_9)$. At each cell, this survey result may be falsely positive or falsely negative (imperfect testing).

Modeling

The binary spatial variables are modeled using a Markov random field (Section 4.6 and Appendix A.4). We have a prior pdf defined by

$$p(\boldsymbol{x}) = \frac{1}{Z}\exp\left(\beta \sum_{i \sim j} I(x_i = x_j)\right), \tag{5.31}$$

where the denominator Z ensures that this is a valid pdf summing to 1. Note that this defines an uninformative prior – i.e., there is no prior pointwise information on the field and thus there is an equal chance of species presence or absence in every cell. There is some spatial correlation for $\boldsymbol{x}$ determined by the interaction parameter β. When this interaction parameter is zero, the cell variables are independent. When it gets larger, there is more dependence in the spatial random field. We will study the sensitivity of this spatial interaction parameter on the VOI result.

The experiments are assumed to have binary outcomes. The survey result $y_i = 1$ implies that the team believes that the species is present at cell i, and $y_i = 0$ suggests otherwise. The experimental results need not be accurate. The cellwise likelihood $p(y_j = k \mid x_j = k)$, $k = 0,1$, for $j = 1,\ldots,9$, might be different conditioned on whether the latent variable is 1 or 0. We set the specificity to $p(y_j = 0 \mid x_j = 0) = \gamma$ and the sensitivity to $p(y_j = 1 \mid x_j = 1) = \gamma$ – i.e., the test has the same proportion of false positives as false negatives. We further assume that this parameter is constant for all spatial locations. A value of γ close to 1 indicates a test with high accuracy. We will study the sensitivity of VOI to this likelihood parameter.

VOI analysis

The formula for the prior value is identical to that of the previous example in Equation (5.29). The posterior value with total imperfect information is defined by

$$PoV(\boldsymbol{y}) = \sum_{i=1}^{n} \sum_{\boldsymbol{y}} \max\{0, \text{Rev} \cdot p(x_i = 1 \mid \boldsymbol{y}) - \text{Cost}\} p(\boldsymbol{y}) d\boldsymbol{y}. \tag{5.32}$$

We compute this posterior value directly by first calculating the marginal probabilities $p(\boldsymbol{y})$ for all $2^9 = 512$ possible data outcomes and evaluating the posterior probabilities inside the max term for each data set. The marginal probability $p(\boldsymbol{y})$ and the conditional probabilities $p(x_i = 1 \mid \boldsymbol{y})$ are computed from the forward–backward recursions useful for Markov random fields (see Appendix A.4). Note that this approach would, however, not scale well. For larger grids, one would need to use Monte Carlo approaches, first for the outer loop over data $\boldsymbol{y}$. See Chapter 7 for a hands-on example.

We vary the spatial interaction and test accuracy for sensitivity analysis. Figure 5.8 shows VOI as a function of these parameters for revenue Rev = 2 and Rev = 5 monetary units and for different accuracies of the test. As expected, the test accuracy is a critical parameter. The curve for $\gamma = 0.9$ shows higher VOI than $\gamma = 0.7$. The VOI with perfect information is a horizontal line and has the same value (4.5 units in this case) for all β and for both revenue levels. This occurs because we have probability $p(x_i = 1) = 0.5$ for all $i = 1,\ldots,9$, and with $\text{Rev} \geq 2 > \text{Cost} = 1$, the prior value is $\max\{0, \text{Rev} \cdot p(x_i = 1) - \text{Cost}\} = \text{Rev} \cdot p(x_i = 1) - \text{Cost}$. For computing $VOI(\boldsymbol{x})$, the revenue terms cancel each other, and only the difference in the cost terms remains:

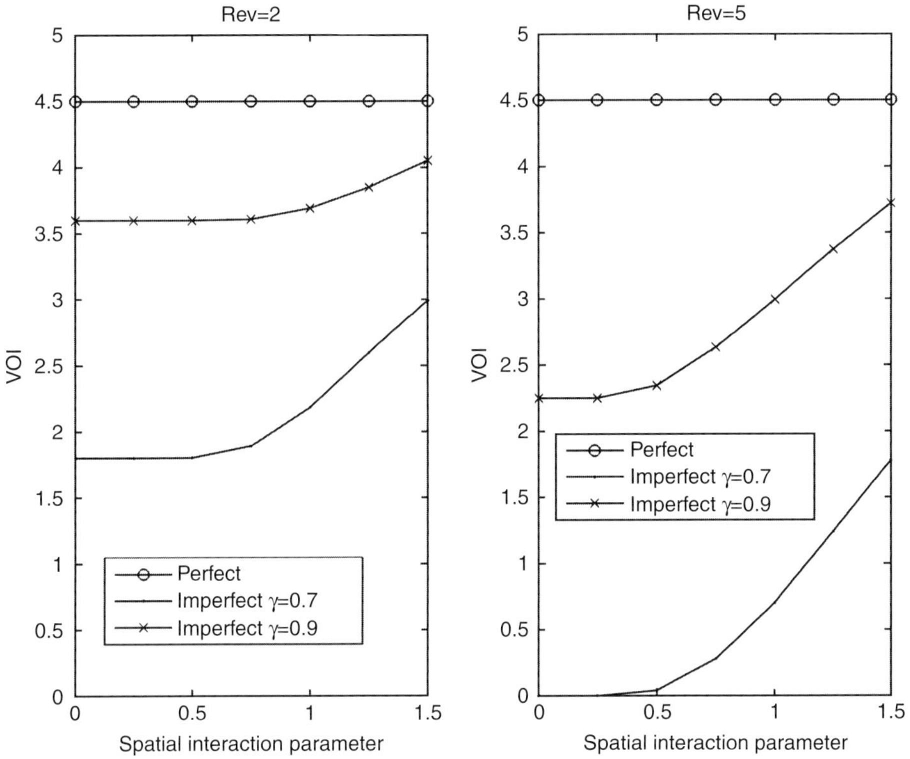

Figure 5.8 Conservation biology example: the VOI for total information on the 3 × 3 grid, plotted as a function of the spatial coupling in the distinction of interest (first axes). The three tests that are compared vary in the accuracy of the measurements.

$$\begin{aligned} VOI(\boldsymbol{x}) &= \sum_{i=1}^{9} p(x_i = 1)(\text{Rev} - \text{Cost}) - \sum_{i=1}^{9}\left(\text{Rev} \cdot p(x_i = 1) - \text{Cost}\right) \\ &= -\sum_{i=1}^{9} \text{Cost} \cdot p(x_i = 1) + 9 \cdot \text{Cost} = \frac{9}{2}\text{Cost}. \end{aligned} \tag{5.33}$$

The VOI increases as the spatial interaction β increases. VOI is not very sensitive to β for lower values, which is useful to know if one is unsure about the interaction parameter or cannot spare much effort estimating it. However, for larger interaction levels, the VOI increases rapidly. This is because the chance that the entire grid will either contain the species at all cells or in no cells becomes higher as β increases. The experiment becomes more valuable as β increases because it can tell you about a possible jackpot (all cells favorable) or prevent a huge loss (all cells unfavorable). As there are no constraints on the number of cells that can be selected, the decision maker is free to choose all the cells or none; for large values of β, this all-or-nothing policy is optimal. The experiment can really make a difference in the decision for large β.

Note that when Rev = 2, $p(x_i = 1) = 0.5$, and Cost = 1, the decision maker is indifferent about the choice a priori. The experimental result could change the decision, and in this way the experiment is more valuable than for Rev = 5, as it is able to have an impact on the decision. When the accuracy of the measurement is small $\gamma = 0.7$, and the spatial interaction is small, $VOI(\boldsymbol{y}) = 0$ for the case with Rev = 5 (right display). In comparison, the VOI has its smallest value around 1.8 money units for the independent case when Rev = 2.

Norwegian wood: forestry example

Keywords: *forestry example, Gaussian random field, total information, partial information, high decision flexibility, decoupled value function*

We return to the forestry example with a spatial domain split into 25 × 25 units (see Section 5.3). This example will now be studied from the perspective of free site selection – hence, high decision flexibility. We will compare these results with the previous results that assumed low decision flexibility.

Framing the decision situation

There are $n = N = 625$ units, with uncertain profits x_i, $i = 1,\ldots,625$. Unlike the situation in Section 5.3, the decision maker is now free to harvest individual forest units. The alternatives at spatial forest units are denoted a_i, $i = 1,\ldots,n$, and at each unit i the alternative a_i is either 0 (not harvest) or 1 (harvest).

Information gathering

The data-gathering options are the same as those in Section 5.3, where we defined three testing options denoted 1–3.

Modeling

We use exactly the same statistical modeling assumptions as in Section 5.3, with a multivariate Gaussian model for the profits of timber volumes for forest units.

VOI analysis

The prior value is

$$PV = \sum_{i=1}^{n} \max\{0, E(x_i)\}. \tag{5.34}$$

The information-gathering schemes are as 1–3 in Section 5.3. Again, using generic notation with data $\boldsymbol{y}$, the posterior value is

$$PoV(\boldsymbol{y}) = \sum_{i=1}^{n} \int \max\{0, E(x_i \mid \boldsymbol{y})\} p(\boldsymbol{y}) d\boldsymbol{y} = \sum_{i=1}^{n} \int \max\{0, w_i\} p(w_i) dw_i, \tag{5.35}$$

where the variable $w_i = w_i(\boldsymbol{y}) = E(x_i \mid \boldsymbol{y})$. Recall from the example in Section 5.3 that the posterior value calculation could be calculated analytically since the value function is linear and the Gaussian modeling assumptions result in a Gaussian distribution for w_i, $i = 1,\ldots,n$. Using the analytical results (Appendix A.1), the posterior value is:

$$PoV(\boldsymbol{y}) = \sum_{i=1}^{n}\left(\mu_i\Phi\left(\mu_i/\sqrt{r_{ii}}\right) + \sqrt{r_{ii}}\phi\left(\mu_i/\sqrt{r_{ii}}\right)\right), \tag{5.36}$$

where r_{ii}, $i = 1,\ldots,n$ are the diagonal elements of the matrix $\boldsymbol{R}$ in Equation (5.18), and the cumulative distribution function Φ and the pdf ϕ of the standard Gaussian distribution are as in Equation (5.19). The difference is that we now have a free selection of forest units to harvest or not, and Equation (5.36) contains a sum over all sites.

Figure 5.9 shows the VOI as a function of the input parameters. Clearly the VOI is much larger for the free selection problem than for the decision situation described in Section 5.3 with less spatial decision flexibility. With free selection, the decision maker has the flexibility to make decisions at each unit. The information may show that some units are profitable while others are not, and the decision maker can now select only the profitable ones. With less spatial decision flexibility in Section 5.3, the decision maker would have to consider all

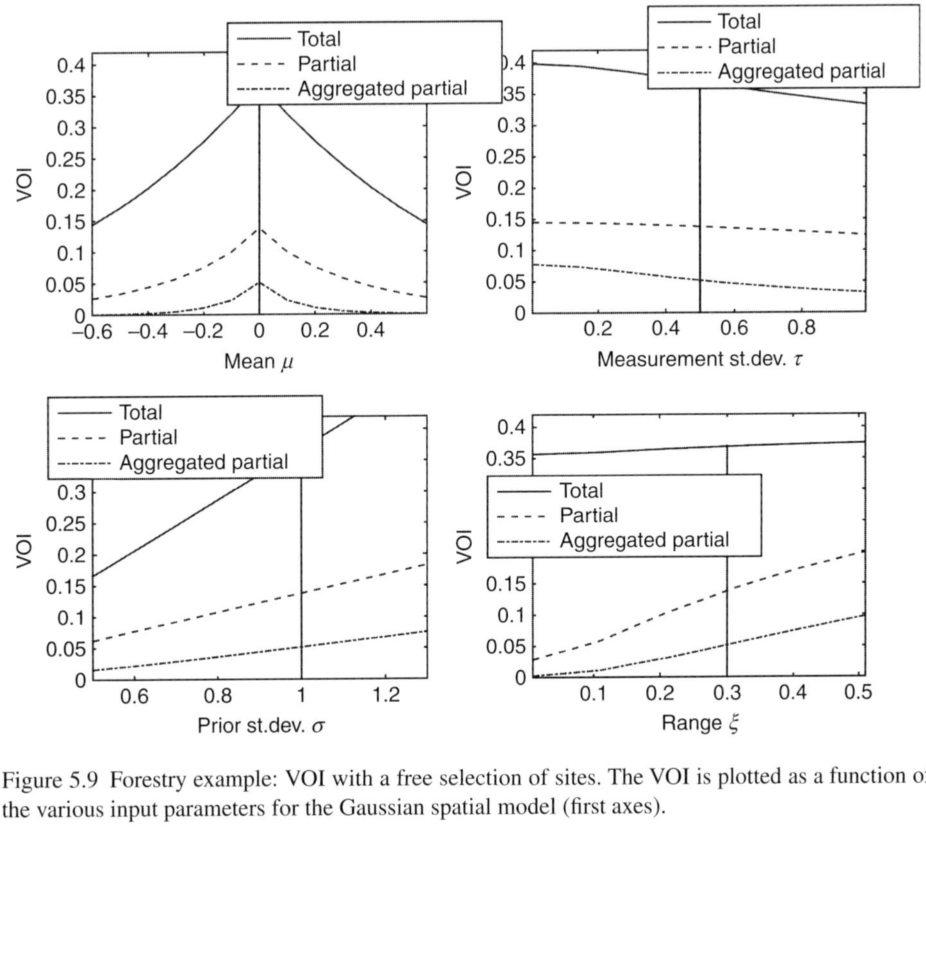

Figure 5.9 Forestry example: VOI with a free selection of sites. The VOI is plotted as a function of the various input parameters for the Gaussian spatial model (first axes).

units as one entity and would be compelled to choose the entire domain when the expected sum over all units is positive.

The spatial units to model the distinction of interest are determined by the finest granularity required for realistically modeling the heterogeneity of the spatial process. The spatial granularity required for the spatial alternatives is determined by demographic units or environmental conditions or constraints imposed by the infrastructure. There is always added value in having the flexibility to select smaller units if there is not a proportionally higher cost per area of the unit. However, this additional value induced by free selection of sites will also depend on the spatial nature of the phenomena being studied.

We perform a sensitivity analysis to bridge the differences between decision situations with less and high spatial flexibility. Suppose that the decision maker is now only free to harvest blocks of size 5 × 5, so it is no longer possible to make decisions at every spatial unit. The alternatives are then $a_j \in \{0,1\}$, $j = 1,...,N$, where $N = 25$, and $a_j = 1$ signifies harvesting block b_j. The prior value in this situation with block decisions becomes

$$PV = \sum_{j=1}^{N} \max\left\{0, \sum_{i \in b_j} E(x_i)\right\}, \tag{5.37}$$

where the notation $i \in b_j$ means the forest units i within block b_j.

Note the combination of a sum both inside the maximum as for the situation with less spatial flexibility in Equation (5.14) as well as outside the maximum for selecting blocks [as for units in Equation (5.34)]. The posterior value with information $\mathbf{y}$ acquired by one of the 1–3 designs is

$$PoV(\mathbf{y}) = \sum_{j=1}^{N} \int \max\left\{0, \sum_{i \in b_j} E(x_i \mid \mathbf{y})\right\} p(\mathbf{y}) d\mathbf{y}. \tag{5.38}$$

The VOI is plotted in Figure 5.10 as a function of the various statistical model parameters.

As anticipated, the VOI is between that achieved by the decision situation with low spatial flexibility (Figure 5.5) and free selection of forest units (Figure 5.9). For a very large spatial correlation (bottom right displays), the VOI is close to the VOI with flexibility to harvest individual units because there is little to lose by making the decision on the block level instead of the unit levels. The profits for units within the block are similar, and the same alternative tends to be optimal for all these units.

5.5 Special case: low decision flexibility and coupled value

In this section, we describe the special case with low decision flexibility in spatial decision situations. This resembles Section 5.3, but here the value function no longer decouples. A more complex value function results in more computationally demanding expressions for the prior and posterior values.

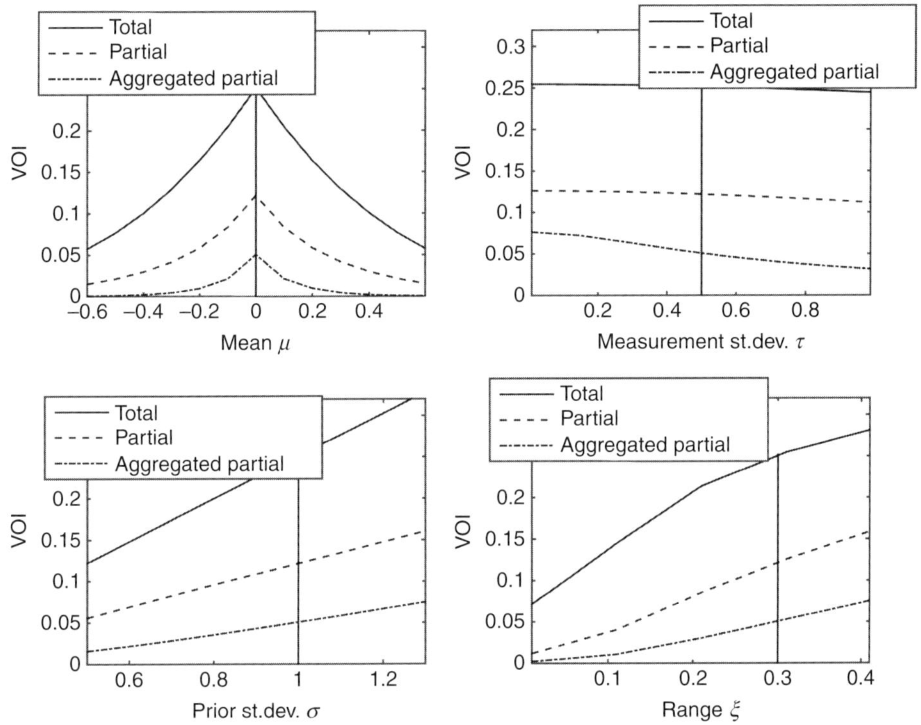

Figure 5.10 Forestry example: VOI with a free selection of blocks of size 5 × 5. The VOI is plotted as a function of the various input parameters for the Gaussian spatial model (first axes).

This case is applicable when there are only a few alternatives, but the value from choosing any of the different alternatives involves a complex coupling of the spatial variables. This setting is, for instance, appropriate in petroleum reservoir development decisions (Bratvold et al. 2009) and for decisions concerning the recharge and management of groundwater resources (Trainor-Guitton et al. 2011). In such applications, the value is often related to the result of fluid flow in the porous subsurface. The fluid flow depends on complex interactions of subsurface variables such as the heterogeneous rock porosity and permeability, flow connectivity between different lithologic units, and presence or absence of faults that can act as either barriers or conduits to flow. The flow response at one site depends on the unobserved uncertain variables at many (potentially all) other sites. Subsurface flow modeling requires the numerical solution of partial differential equations in a heterogeneous porous media. Modeling the flow (and hence value) at one site cannot be decoupled from other sites. There is low decision flexibility when there are limited alternatives, possibly due to infrastructural, logistical, legal, or other practical issues.

The petroleum reservoir simulation example will be used for illustration in this section. A petroleum company must evaluate whether to develop the reservoir. There is uncertainty around the reservoir variables (spatially dependent facies, porosity, and permeability), and hence the profits from production are uncertain. In our example, we assume that the main uncertainty is the geological scenario that dictates the nature of spatial distribution of the high-permeability channel facies. A priori, the geologists believe that the spatial characteristics of the reservoir are governed by two possible depositional scenarios: a river delta geological system or a meandering river channel geological system. We consider the simplest situation with only two alternatives: (i) no development and (ii) development with a particular drilling strategy.

Note that in most of these complex coupled value function settings, the decision maker would clearly benefit from more decision flexibility (Section 5.6), but this is not always possible, perhaps because of limitations in the way the existing infrastructure must be used. It could also be conceptually easier for management to reflect upon and discuss only a select few alternatives. Finally, the added computational requirements would, of course, be greater when there is more spatial decision flexibility.

5.5.1 Prior value

The decision maker chooses the alternative $a \in A$ that maximizes expected value. In the petroleum reservoir example, the decision maker could either develop the reservoir or avoid development. The prior value over set A is

$$PV = \max_{a \in A} \left\{ \int v(\boldsymbol{x}, a) p(\boldsymbol{x}) d\boldsymbol{x} \right\}. \tag{5.39}$$

The value function $v(\boldsymbol{x}, a)$ of the reservoir simulation example signifies the revenues obtained from the oil production, minus the cost of development and production. Since $v(\boldsymbol{x}, a)$ involves the solution of partial differential equations governing the flow in porous media, it cannot be decoupled. Hence, it is not possible to simplify the expression for the prior value in Equation (5.39) any further. We must evaluate the expected value within the parentheses for all alternatives $|A|$. We return to computational methods later.

5.5.2 Posterior value

Suppose that the decision maker can gather information. The posterior value of perfect total information is computed by switching the integral and the maximization from the prior value in Equation (5.39). We have

$$PoV(\boldsymbol{x}) = \int \max_{a \in A} \{ v(\boldsymbol{x}, a) \} p(\boldsymbol{x}) d\boldsymbol{x}, \tag{5.39}$$

because the decision maker can choose the best alternative upon observing every possible outcome of the spatial variables $\boldsymbol{x}$. For the petroleum reservoir simulation example, this entails knowing the spatial reservoir variables at all locations. This kind of information scheme is unrealistic for the petroleum reservoir application, but the resulting VOI, which like before is the difference between the posterior value in Equation (5.40) and the prior value in Equation (5.39), would be an upper bound on the value of any kind of information-gathering scheme. Thus, if we get to know the VOI of perfect total information, and we know that some experiment is more expensive than this upper bound, it would surely not be worthwhile collecting this data. This upper bound obtained from the VOI of perfect total information is easier to compute than the VOI of imperfect or partial information.

The posterior value expressions of partial perfect information $\boldsymbol{x}_{\mathbb{k}}$, partial imperfect information $\boldsymbol{y}_{\mathbb{K}}$, or total imperfect information $\boldsymbol{y}$ are similar to each other for this case. We will simply denote data by $\boldsymbol{y}$. Like before, the posterior value involves the conditional expected value function for every data outcome:

$$PoV(\boldsymbol{y}) = \int \max_{a \in A} \{E(v(\boldsymbol{x},a) \mid \boldsymbol{y})\} p(\boldsymbol{y}) d\boldsymbol{y}. \tag{5.41}$$

For the petroleum reservoir simulation example, relevant data include reservoir monitoring in wells, geologic information, seismic data acquisition and processing, electromagnetic surveys, etc. Such data would be informative of the reservoir variables and hence the value obtained from development. The VOI is the difference between the posterior and prior values.

One challenge in Equation (5.41) is finding the conditional expectation of the values. Unlike the case of total perfect information, this expression now involves an inversion step to get the conditional pdf $p(\boldsymbol{x} \mid \boldsymbol{y})$, and one must compute the expectation of the complex value function with respect to this pdf. We discuss computational approaches in the next section.

5.5.3 Computational notes

The coupled value function means that one must use some sort of approximation to assess the VOI for this case. There are many ways to approximate this; we will discuss a couple of approaches (see also the discussion in Section 5.6).

Monte Carlo sampling can be used to approximate the prior and posterior values. As the number of Monte Carlo samples goes toward infinity, the approximation converges to the true theoretical integral. In practice, trial and error is used to tune the number of samples required for accurate approximation.

The prior value in Equation (5.39) is approximated as follows:

Algorithm: Monte Carlo approximation of the prior value

1. Draw samples $\boldsymbol{x}^1,\ldots,\boldsymbol{x}^B$ from the prior model $p(\boldsymbol{x})$.
2. Compute the value $v(\boldsymbol{x}^b,a)$ for each sample and every action $a \in A$.
3. Approximate the expected value for each alternative:

$$\widehat{E(v(\boldsymbol{x},a))} = \frac{1}{B}\sum_{b=1}^{B} v(\boldsymbol{x}^b,a)$$

4. Approximate the prior value $PV \approx \max_{a \in A}\left\{\widehat{E(v(\boldsymbol{x},a))}\right\}$.

The posterior value in Equation (5.40) with total perfect information can be approximated using a similar algorithm:

Algorithm: Monte Carlo approximation of the posterior value with total perfect information

1. Draw samples $\boldsymbol{x}^1,\ldots,\boldsymbol{x}^B$ from the prior model $p(\boldsymbol{x})$.
2. Compute the value $v(\boldsymbol{x}^b, a)$ for each sample and every action $a \in A$.
3. Approximate the posterior value $PoV(\boldsymbol{x}) \approx \frac{1}{B}\sum_{b=1}^{B} \max_{a\in A}\{v(\boldsymbol{x}^b, a)\}$.

The samples and value calculations could be re-used from the prior value approximation. The only difference with the algorithm for the prior value is in Step 3, where we switch the sum and the max. If resources permit, the value computation $v(\boldsymbol{x}^b, a)$ could be done in parallel, thereby making these computations much faster.

It is more difficult to approximate the posterior value of total or partial imperfect information in Equation (5.41). This calculation requires realizations of data and the posterior expectation of the value function given the data. The core algorithm is as follows:

Algorithm: Monte Carlo sampling for approximating the posterior value with imperfect or partial information

1. Draw samples $\boldsymbol{y}^1,\ldots,\boldsymbol{y}^B$ from $p(\boldsymbol{y})$.
2. Approximate the posterior value $PoV(\boldsymbol{y}) = \frac{1}{B}\sum_{b=1}^{B} \max_{a\in A}\{E(v(\boldsymbol{x},a) \mid \boldsymbol{y}^b)\}$.

In Step 1, it is common to first sample $\boldsymbol{x}$ from $p(\boldsymbol{x})$, then $\boldsymbol{y}$ from the likelihood $p(\boldsymbol{y} \mid \boldsymbol{x})$, which provides a sample from the marginal model for the data. The main challenge lies in assessing the posterior conditional mean in Step 2, which may not be easy for a complex value function. If this were done only once, it would amount to inferring the value for the single data set $\boldsymbol{y}^b$. This could involve, for example, inverting for realizations of reservoir variables given the data (stochastic inversion for facies) and then running a flow simulator on the realizations. This is computationally demanding because it must be done for all data outcomes $\boldsymbol{y}^b$, $b = 1,\ldots,B$. The direct approach would then be to perform another round of Monte Carlo sampling to approximate the conditional expectation. This results in a double Monte Carlo loop, which is of computational complexity $B^2|A|$, assuming that we use B samples for the inner loop as well. We argue that a double Monte Carlo loop may be too slow to approximate the posterior value here. On the other hand, the number of alternatives is not very large (since there is low decision flexibility), so it may be possible to use the direct Monte Carlo approximation method, especially if parallel computing is possible. (Another Monte Carlo approach will be discussed in Section 5.6 for the more flexible decision situation.)

A promising approach for approximating the VOI is to embed the spatial variables in a low-rank spatial representation. We will use such an approach in the reservoir simulation example in the following text. Instead of working with the full spatial description of the distinction of interest, we focus on a few discrete underlying geological scenarios. The approach could be practical for decision situations with few spatial alternatives, especially when the geological scenario has a large influence on the spatial variables, the coupled value function, and the downstream decision. Trainor-Guitton et al. (2011) use channel directions as the geologic scenario in a groundwater application, reasoning that this geologic attribute is an important variable for decisions regarding recharge. Petroleum reservoir simulation studies have considered scenarios with open or closed faults (Bratvold et al. 2009).

The idea is to define a set of discrete geographic or **geologic scenarios** $x \in \Omega$, where Ω denotes the sample space of the scenario. Based on the scenario or geological features, one can generate multiple realizations of a random spatial process $\tilde{\boldsymbol{x}}(x)$, inheriting properties defined by the scenario, but at a full spatial resolution. This would typically involve one of the spatial modeling methods described in Chapter 4. Note that we use x to denote the low-rank discrete scenario and $\tilde{\boldsymbol{x}}$ to denote the full spatial process under a specific scenario. We approximate the expected value function for all actions a under every scenario via Monte Carlo sampling of the many spatial realizations.

This approach requires $B|A||\Omega|$ computations of the value function, where $|\Omega|$ denotes the number of geologic scenarios.

Algorithm: Monte Carlo procedure for prior value for scenario-based methods

1. Draw Monte Carlo realizations of spatial process $\tilde{\boldsymbol{x}}^1(x),\ldots,\tilde{\boldsymbol{x}}^B(x),\ x \in \Omega$.
2. Approximate values $\hat{v}(x,a) = \frac{1}{B}\sum_{b=1}^{B} v(\tilde{\boldsymbol{x}}^b(x),a),\ a \in A\ ,\ x \in \Omega$.
3. The prior value is $\widehat{PV} \approx \max_{a\in A}\left\{\sum_{x\in\Omega} \hat{v}(x,a)p(x)\right\}$.

The posterior value with total and perfect information is again approximated by re-using the values computed from the sampling procedure for the prior value. For the posterior value, the decision maker chooses after knowing the scenario, given the total and perfect information.

$$\widehat{PoV}(x) \approx \sum_{x\in\Omega} \max_{a\in A}\left\{\hat{v}(x,a)\right\}p(x). \tag{5.42}$$

Since there are only a few geological scenarios and alternatives, the expected value and prior and posterior values can be reasonably approximated even with moderate evaluations of the value function. As before, the VOI is the difference between the posterior and prior values.

Trainor-Guitton et al. (2011) suggested a method for the approximate inversion of features, which makes it applicable to partial information or total and imperfect information. It is based on the construction of a **reliability measure**.

Algorithm: approximation of the reliability measure

1. Simulate spatial processes $\tilde{\mathbf{x}}^1(x),\ldots,\tilde{\mathbf{x}}^B(x)$ for each class of geologic scenario $x \in \Omega$.
2. Simulate data sets $\mathbf{y}^b$, $b = 1,\ldots,B$ conditional on the realizations from Step 1.
3. Construct a low-rank inversion or interpretation of every data set, possibly based on some summary statistics. The output of this interpretation is a classification, $y^{*b} \in \Omega$ for all data sets, into one of the defined classes of geologic scenarios.
4. Based on the simulation scenario and the classification, form a reliability measure:

$$p(y^* = l \mid x = k) \approx \frac{\sum_{b=1}^{B} I(x = k, y^{*b} = l)}{B}$$

The reliability measure represents a classification confusion (or transition) matrix, $p(y^* = l \mid x = k) = P_{kl}$, where large entries on the diagonal imply that the data are correctly classified to belong to the scenario from which they were derived. Off-diagonal terms provide the misclassification errors. When we study the petroleum reservoir simulation example in more detail in the next section, we will use a reliability measure that is a function of the accuracy of interpreted seismic attributes y^*, given geological environments x.

The discrete sample spaces of both the input geologic scenario and the interpreted inversion results ensure that the posterior value and VOI with imperfect information are straightforward to calculate using a sum over the sample space for the interpretation. Let $\mathbf{y}$ denote data acquired with a generic information-gathering scheme. The associated interpretation result is denoted y^*. The posterior value is

$$PoV(y^*) = \sum_{y^* \in \Omega} \max_{a \in A} \{E(v(x,a) \mid y^*)\} p(y^*). \tag{5.43}$$

The marginal distribution for the interpretation is computed by marginalization – i.e.,

$$p(y^*) = \sum_{x \in \Omega} p(x) p(y^* \mid x). \tag{5.44}$$

The expected values given the data are computed using Bayes' rule for the conditional probabilities and the approximated values for each scenario. We have

$$\hat{E}\left(v(x,a) \mid y^*\right) = \sum_{x \in \Omega} \hat{v}(x,a) p(x \mid y^*), \quad p(x \mid y^*) = \frac{p(x)p(y^* \mid x)}{p(y^*)}. \tag{5.45}$$

In summary, the posterior value in Equation (5.43) is computed using the same samples and value evaluations as those for the computation of the prior value and the posterior value with total perfect information in Equation (5.42). The computation now weights the values with the posterior probabilities given the interpreted data.

Recall that there is a large reduction of spatial dimension here. The interpretation is a scalar variable, as is the geologic scenario. However, full-scale spatial modeling is used as an intermediate step to model the value function and to model the full-scale seismic data $\boldsymbol{y}$. If we want to compute the VOI of total testing, the data variable is obtained at all sites, and we perform the interpretation based on this full spatial coverage test. For a partial test, the data are only represented at a subset of the sites, which would provide a less accurate interpretation – i.e., smaller diagonal entries in the reliability measure transition matrix.

5.5.4 Example

Go with the flow: petroleum simulation example

Keywords: *petroleum simulation, petroleum development, petroleum production, reservoir characterization, multiple-point statistics, geologic scenario, low decision flexibility, coupled value function*

Let us consider an example about the development and production of a petroleum reservoir. The spatial setting is similar to the case described in Section 4.7. The decision has a lot at stake, and it may be worthwhile to collect geophysical data before making the decision. We will perform VOI analysis for seismic data.

Framing the decision situation

A company considers developing a hydrocarbon reservoir. The development decision entails large investments in infrastructure: platforms, wells, pipelines, etc. If the company is fortunate, extensive production of oil or gas results in profits. On the other hand, the petroleum company could end up with a lower production rate, and this may lead to a large loss. There is significant uncertainty around the spatial reservoir variables. We assume that most of the uncertainty is driven by the geological scenario, which is classified as either $x = 0$: a delta system, or $x = 1$: a meandering river system.

There are a number of decisions regarding the development of a petroleum reservoir. We will consider a simplified decision situation where the decision maker can either (i) develop the reservoir by drilling a fixed configuration of wells or (ii) avoid development. Thus, there is low flexibility, with only two alternatives. Development is represented by alternative $a = 1$, while the alternative to avoid development is denoted $a = 0$. The only spatial aspect of the decision situation is the known well configuration under the development decision. If the decision about drilling production wells is made, a fixed design of the production wells is used. The decision situation is not nearly as flexible as that of optimizing the well configuration or tuning the production and injection rates by well controls. It describes a much simpler decision situation. However, a petroleum company could have limited decision flexibility for various reasons: there could be existing infrastructure, making a few well configuration alternatives much less expensive than others, or there could be limited availability in well equipment and the ability to perform complicated well controls.

The petroleum reservoir we consider here is modeled with the geologic scenario as the main uncertain variable. Given the geological scenario, we generate the three-dimensional spatial distribution of facies classes, porosity, and permeability. The value function is highly coupled, and described in more detail after the spatial modeling.

Information gathering

The decision maker may obtain valuable information about the reservoir variables by collecting seismic data. For instance, seismic interpretation could be useful for identifying large-scale geological structures such as key reflectors representative of major geological events, subsurface horizons, and faults. At a finer granularity, quantitative rock physics interpretations of seismic amplitude data are also useful for characterizing the spatial distribution of facies and pore fluid in the reservoir. In this example, we will look at the information content in seismic amplitude data using a reliability measure. Thus, the interpreted seismic attributes will be considered as the data, and we are interested in studying how informative the interpreted seismic data might be about certain geologic scenarios.

Modeling

The two uncertain geologic scenarios are considered equally likely a priori – i.e., $p(x=0)=p(x=1)=1/2$. The geologic scenario governs a lot of the heterogeneous variability in the reservoir. Given the geological scenario, the decision maker knows the statistics of the spatial facies distribution, the porosity, and the permeability. Along with the well design, these variables describe the flow and production of petroleum. The value function is the total production until a stopping time, minus the cost of development. We will study sensitivity to the costs, but the main part of the task is to compute the production profiles for various reservoir variables for both scenarios $x \in \{0,1\}$.

The spatially varying distinctions of interest $\tilde{\boldsymbol{x}}(x)$ are facies, porosity, and permeability. First, we use multiple-point statistics (see Section 4.7) to generate $B=1000$ realizations of facies for each geologic scenario. Next, for each realization, we populate the geo-model with porosity and permeability values, given the facies realizations, using appropriate rock physics relations for the channel sands and the shale facies. The permeability is related to the porosity via a standard Kozeny–Carman-type relation – see, e.g., Mavko et al. (2009).

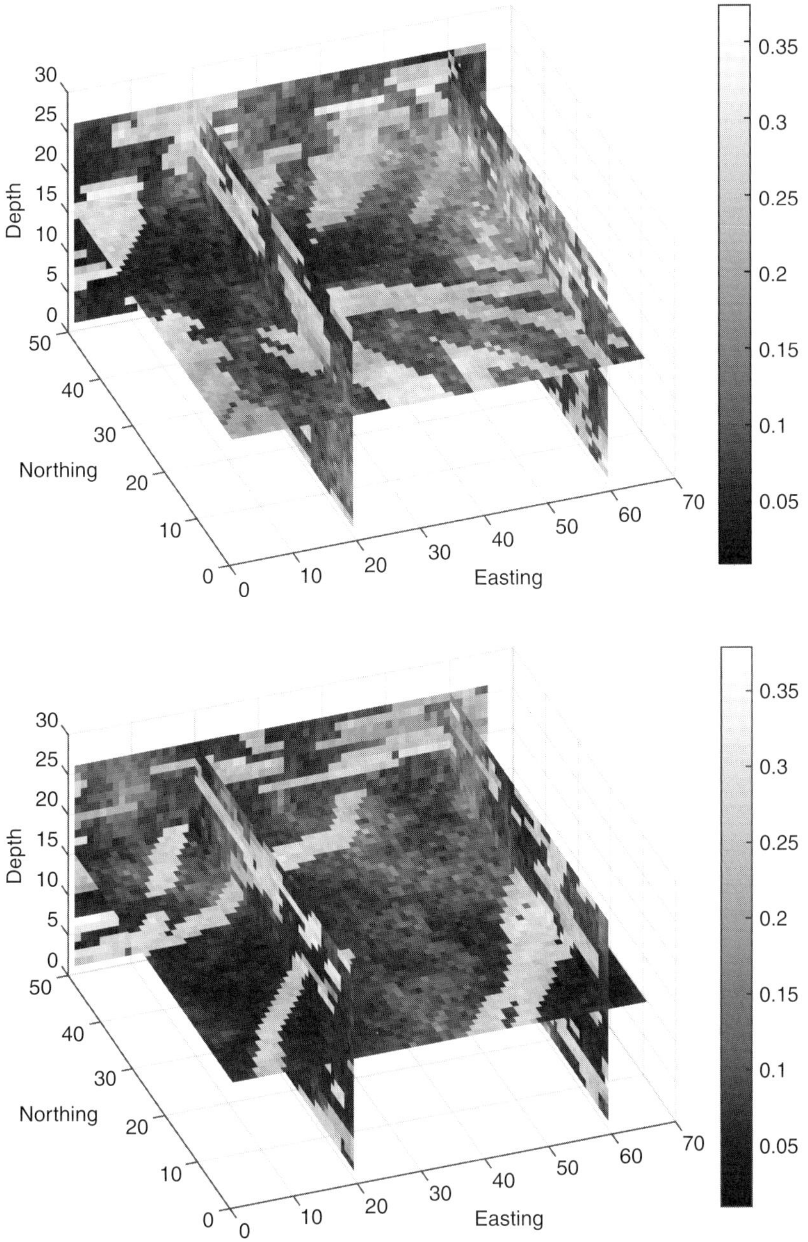

Figure 5.11 Petroleum simulation example: the base case porosity maps derived from a training image of the reservoir facies. Top: display represents a delta system. Bottom: display represents a meandering river system.

Figure 5.11 shows an example of porosity realizations for the two geological scenarios. The delta system (top) clearly has less of a channel structure than the meandering river system (bottom). Also, the delta system has more trends in the facies in the eastern parts. The subsurface fluid flow and production profiles would be different under the two geologic scenarios.

The value of developing the reservoir $v(x, a = 1)$ is based on the uncertain production of oil under each geologic scenario, with the costs subtracted. The amount of oil being produced is obtained by established software for fluid flow simulation in porous media. Among other variables, they take the porosity and permeability fields as inputs. For each geological scenario, the value is approximated from the 1000 realizations of spatial facies, porosity, and permeability. We used the well-known Eclipse software to compute the production profiles of each realization. In the hands-on exercises in Chapter 7, the reader can either use Eclipse or the MATLAB reservoir simulation toolbox to try out similar situations.

Seismic data $\boldsymbol{y}$ are informative of the reservoir parameters. It may be worthwhile to purchase seismic data to learn more about the reservoir properties. Rock physics relations and seismic wave propagation equations can be used for modeling the expected response of seismic amplitudes. We may write $\boldsymbol{y} = \boldsymbol{f}(\tilde{\boldsymbol{x}})$, where the function $\boldsymbol{f}$ represents the forward model – rock physics and wave propagation. The seismic data must be inverted or interpreted before they can be used for the prediction of reservoir variables. This interpretation is commonly done by looking at smooth attributes valid at the seismic resolution scale. The interpretation is imperfect for many reasons, including challenges of seismic imaging, non-uniqueness in rock physics relations between seismic attributes and reservoir properties, and the presence of reservoir heterogeneities below the scale of seismic resolution. Here, we will consider a variety of simple reliability measures for the interpreted seismic data, denoted by y^*. We set the reliability measure to be equal for the two scenarios – i.e., $p(y^* = 0 \mid x = 0) = p(y^* = 1 \mid x = 1)$. Combined with the equal marginal probabilities of the two geological scenarios, this implies that the marginal for the interpreted data is $p(y^* = 0) = p(y^* = 1) = 1/2$.

VOI analysis

Figure 5.12 shows the total oil production over time for the delta system (top) and the meandering river system (bottom) computed for the 1000 realizations for each scenario.

We see that the delta system produces more oil than the meandering system, with some uncertainty, largely driven by the facies realizations. The meandering system has a larger variation in the production, whereas for the delta system, the production curves cluster into three groups. After 10 years of simulated production, the delta system has an average production of 113,000 standard barrels, whereas it is only 60,000 for the meandering river system. Note that the actual value may depend on other factors besides oil production. For instance, if the wells get early water breakthrough, the costs of processing may become so large that it is not profitable to continue production for long.

If the cost (in units of a thousand barrels) of development and production is less than 60 (assuming that the revenue per barrel is one unit), then the decision maker would develop the reservoir in any case. If the cost is larger than 113, then the decision maker would avoid development. For intermediate costs, the deal is profitable only if it turns out that the

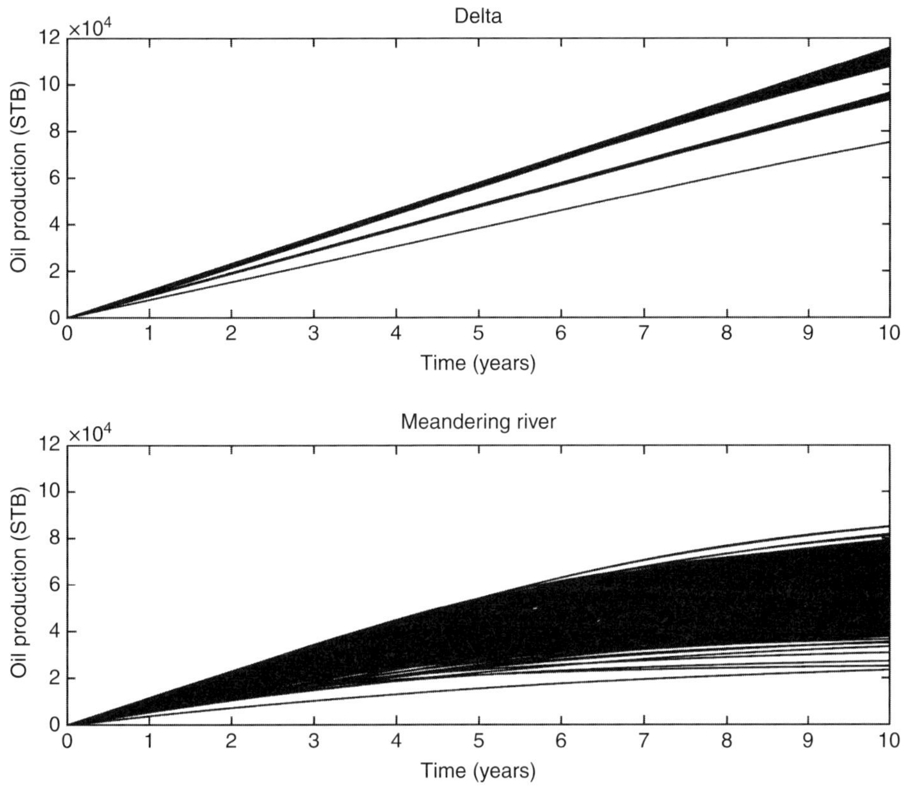

Figure 5.12 Petroleum simulation example: realizations of the production profiles for the delta system (top) and the meandering river system (bottom).

underlying geologic system is a delta system. Of course, the decision maker does not know the geologic scenario a priori.

The prior value is the average profit taken over the geologic scenarios. It is

$$PV = \max\left\{0, \frac{1}{2}(113 - \text{Cost}) + \frac{1}{2}(60 - \text{Cost})\right\} = \max\{0, 86.5 - \text{Cost}\}, \tag{5.46}$$

where we have subtracted the cost of development and production. We will assume three development cost levels, high: Cost = 100, medium: Cost = 85, and low: Cost = 70. The prior value is slightly positive for the medium development and production level and is clearly positive for the low level. For the high level, the decision is to avoid development.

The posterior value with perfect information about the geologic scenario is defined by Equation (5.42), and for all the specified values of the cost of development we have

$$PoV(x) = \frac{1}{2}\max\{0, 113 - \text{Cost}\} + \frac{1}{2}\max\{0, 60 - \text{Cost}\} = \frac{1}{2}\max\{0, 113 - \text{Cost}\}. \tag{5.47}$$

Table 5.4. *VOI results for perfect information and imperfect information in the reservoir simulation example with different reliability measures for the seismic data and for different costs of production and development*

	Perfect Information	Reliability of 0.9	Reliability of 0.7
VOI – Low Cost	5	2.4	0
VOI – Medium Cost	12.5	9.9	4.6
VOI – High Cost	6.5	3.9	0

The posterior value is positive for all levels of development and production cost. The VOI is the difference between the posterior and prior values. We compare the VOI with perfect information versus imperfect information in the following text.

For the posterior value with imperfect information in Equation (5.43), the reliability measure is needed in addition to the prior probabilities. We compute the posterior value of imperfect information and the associated VOI for different reliabilities.

The VOI results for perfect information and imperfect information are summarized in Table 5.4. As expected, we see that the VOI is lower for less reliability in the seismic interpretation. If the accuracy of the interpretation is only 70% – i.e.,$p(y^* = 0 \mid x = 0) = p(y^* = 1 \mid x = 1) = 0.7$, the development cost must be near medium for VOI to be positive. At this medium cost, the expected profits are near 0, and the data are valuable since they could influence the decision maker either toward or away from developing.

5.6 Special case: high decision flexibility and coupled value

This case entails a much larger set of alternatives than in Section 5.5. The decision maker can choose among alternatives at various locations. However, unlike the case in Section 5.4, the value can no longer be decoupled into a sum over sites. The value function is complex and jointly involves the alternatives chosen at different sites. This situation occurs, for instance, in hydropower production from several dams that are connected in a river system. Decisions must be made about the water level in all dams. If water flows over one dam, it will continue downstream to the next dam and influence the value from the alternatives chosen there. See, e.g., Alemu et al. (2011) for a more general discussion of this hydropower production problem. This case is also applicable for the petroleum simulation example in Section 5.5 when there is more flexibility in the spatial decision situation. Such flexibility could include optimization of well locations or tuning the well production rates at different locations. See, e.g., Barros et al. (2014) for a promising VOI approach for this application.

A hydropower production example is used throughout this section, where we analyze the VOI of snow measurements before the melting period. The underlying decision situation is how should the decision maker set the levels of several reservoirs (dams) before the snow melts? The goal is to fill up all dams during the melting season and then sell the energy later at a higher selling price. However, if water runs over dams, the hydropower company loses income. By collecting snow measurements at selected locations, the decision maker obtains more information about the water that could flow to the dams during the melting period, and this information could potentially be worthwhile.

5.6.1 Prior value

We denote alternatives by $a_j \in A_j$, for $j = 1,.., N$, and the total set of alternatives by $\boldsymbol{a} \in \boldsymbol{A}$. As in Section 5.4, alternatives are associated with spatial locations and may or may not be of the same dimension as the distinction of interest $\boldsymbol{x} = (x_1,...,x_n)$. In the hydropower example, $n > N$. Unlike the decoupled value situation in Section 5.4, the value calculation now depends on the alternatives at many locations.

The prior value is the same as in the most general formulation:

$$PV = \max_{\boldsymbol{a}\in A}\left\{E\left(v(\boldsymbol{x},\boldsymbol{a})\right)\right\} = \max_{\boldsymbol{a}\in A}\left\{\int_{x} v(\boldsymbol{x},\boldsymbol{a})p(\boldsymbol{x})d\boldsymbol{x}\right\}. \tag{5.48}$$

Only in special cases can the expression be simplified further – the structure of the optimization problem depends very much on the situation. For instance, in the hydropower example, the value function decouples a little based on the design of dams in the river basin, where water running over one dam flows downstream to the next.

5.6.2 Posterior value

The posterior value is also the same as in the most general formulation:

$$PoV(\boldsymbol{x}) = \int_{x} \max_{\boldsymbol{a}\in A}\left\{v(\boldsymbol{x},\boldsymbol{a})\right\} p(\boldsymbol{x})d\boldsymbol{x} \tag{5.49}$$

for total and perfect information. Again, this provides an upper bound for any sort of information-gathering scheme.

For partial perfect or imperfect information (denoted $\boldsymbol{y}$),

$$PoV(\boldsymbol{y}) = \int \max_{\boldsymbol{a}\in A}\left\{E\left(v(\boldsymbol{x},\boldsymbol{a}) \mid \boldsymbol{y}\right)\right\} p(\boldsymbol{y})d\boldsymbol{y}. \tag{5.50}$$

The associated VOI is the difference between the posterior and prior values as in Section 5.2.

In the context of hydropower production, weather models and precipitation measurements acquired during the winter season provide a priori knowledge about the snow levels and therefore about how much water will run into the dams. However, there is still some uncertainty about the spatial distribution of snow. It may be useful to measure the snow depth at selected locations (partial testing).

5.6.3 Computational notes

The case of coupled value function and high flexibility in the spatial alternatives is extremely computationally demanding in general. Methods similar to those in Section 5.5 could be useful for approximating the VOI, but since there is high decision flexibility, it may not be that appropriate to study large-scale discrete spatial characteristics as was done with the geologic scenarios. It seems more useful to consider approximations that fully incorporate the spatial distinction of interest.

Suppose that we have a sample-based representation of the prior. This is represented as $\boldsymbol{x}^1,\ldots,\boldsymbol{x}^B$ realizations of spatial variables sampled from $p(\boldsymbol{x})$. The prior value can be approximated over the samples:

$$PV \approx \max_{\boldsymbol{a}\in A}\left\{\frac{1}{B}\sum_{b=1}^{B} v\left(\boldsymbol{x}^b,\boldsymbol{a}\right)\right\}. \tag{5.51}$$

This prior value approximation requires $B|A|$ evaluations of the value function $v\left(\boldsymbol{x}^b,\boldsymbol{a}\right)$. The number of alternatives can be large in this setting with high decision flexibility, and it may be necessary to limit the number of Monte Carlo runs required in some way. For instance, one may realize that some combinations of spatial alternatives will surely result in less value than some another combination, in which case it is not necessary to perform the value calculation for alternatives with lower value.

As usual, the posterior value of perfect information is computed by switching the integral and the max term from the prior value:

$$PoV\left(\boldsymbol{x}\right)=\int \max_{\boldsymbol{a}\in A}\left\{v\left(\boldsymbol{x},\boldsymbol{a}\right)\right\}p\left(\boldsymbol{x}\right)d\boldsymbol{x} \approx \frac{1}{B}\sum_{b=1}^{B}\max_{\boldsymbol{a}\in A}\left\{v\left(\boldsymbol{x}^b,\boldsymbol{a}\right)\right\}. \tag{5.52}$$

The calculation requires no additional value calculations if we have computed it for all samples and all alternatives for the prior value approximation in Equation (5.51).

For other kinds of information $\boldsymbol{y}$, the posterior value calculation is more challenging. As we discussed in Section 5.5, one must first find the conditional pdf $p\left(\boldsymbol{x}\mid\boldsymbol{y}\right)$ given data outcomes $\boldsymbol{y}$ and then compute the expectation of the value function under this conditional pdf for all alternatives. Finally, this is averaged over all data outcomes. Instead of running a double Monte Carlo scheme for this posterior value approximation, we outline a computational method that aims to re-use the Monte Carlo samples generated from the prior value approximation. The approach we use here is inspired by approximate Bayesian computing (ABC) – see Chapter 2 and Appendix B.

Suppose that the data $\boldsymbol{y}$ are related to the spatial variables of interest by $\boldsymbol{y}=\boldsymbol{f}\left(\boldsymbol{x}\right)$. The function $\boldsymbol{f}$ could involve (non-linear) physical relations like in a total imperfect test (say, seismic data), or it could pick a subset of variables (partial testing) like we assume in the setting with snow measurements at some selected sites. The relation $\boldsymbol{f}$ between the data and spatial distinctions of interest could, of course, include random noise elements associated with the experimentation.

For each realization from the prior, one can obtain simulated data $\boldsymbol{y}^1,\ldots,\boldsymbol{y}^B$, where $\boldsymbol{y}^b = \boldsymbol{f}(\boldsymbol{x}^b)$. This sample is representative of the marginal distribution of the data $p(\boldsymbol{y})$. The posterior value of imperfect information is therefore

$$PoV(\boldsymbol{y}) = \int \max_{\boldsymbol{a}\in A}\left\{E\left(v(\boldsymbol{x},\boldsymbol{a})\mid \boldsymbol{y}\right)\right\}p(\boldsymbol{y})d\boldsymbol{y} \approx \frac{1}{B}\sum_{b=1}^{B}\max_{\boldsymbol{a}\in A}\left\{E\left(v(\boldsymbol{x},\boldsymbol{a})\mid \boldsymbol{y}^b\right)\right\}. \tag{5.53}$$

We now present an approach for the inner expectation that re-uses samples from the prior. This approximation is based on the nearest neighbors to each data realization according to some loss function or metric.

We define a mismatch, score, or loss $l(\boldsymbol{y}^b,\boldsymbol{y}^c)$ to represent the "distance" between a data set $\boldsymbol{y}^b$ computed from prior sample $\boldsymbol{x}^b$ and data set $\boldsymbol{y}^c$ based on the prior sample $\boldsymbol{x}^c$, b, $c = 1,\ldots,B$. This defines a $B\times B$ loss or dissimilarity matrix. We generate posterior samples by picking the prior samples that satisfy a small loss $l(\boldsymbol{y}^b,\boldsymbol{y}^c) < \varepsilon$. We can tune the threshold ε so that we have at least a fraction – say, $B/10$, of the original prior samples.

The diagonal is ignored in the analysis. The focus is on the remaining samples that are independent of the realization for the data. We use these losses to approximate the conditional expectation of value. Samples within the specified threshold are thus uniformly weighted, not unlike in rejection sampling (see Section 2.5).

The associated VOI for imperfect information is the difference between the posterior value and the prior value in Equation (5.48):

$$\begin{aligned} VOI(\boldsymbol{y}) &= PoV(\boldsymbol{y}) - PV \\ &\approx \frac{1}{B}\sum_{b=1}^{B}\max_{\boldsymbol{a}\in A}\left\{\hat{E}\left(v(\boldsymbol{x},\boldsymbol{a})\mid \boldsymbol{y}^b\right)\right\} - \max_{\boldsymbol{a}\in A}\left\{\frac{1}{B}\sum_{b=1}^{B}v(\boldsymbol{x}^b,\boldsymbol{a})\right\}. \end{aligned} \tag{5.54}$$

Algorithm: Monte Carlo approximation of posterior value using approximate Bayesian computing for the conditioning

1. Sample $\boldsymbol{x}^1,\ldots,\boldsymbol{x}^B$ from prior $p(\boldsymbol{x})$.
2. Compute the value $v(\boldsymbol{x}^b,\boldsymbol{a})$ for each sample $b = 1,\ldots,B$ and all alternatives $\boldsymbol{a}\in \boldsymbol{A}$.
3. Generate synthetic data variables $\boldsymbol{y}^b = \boldsymbol{f}(\boldsymbol{x}^b)$, $b = 1,\ldots,B$.
4. Compute the loss, $l(\boldsymbol{y}^b,\boldsymbol{y}^c)$, $b,c = 1,\ldots,B$, between all data pairs.
5. Approximate the conditional expectation as follows:

$$\hat{E}\left(v(\boldsymbol{x},\boldsymbol{a})\mid \boldsymbol{y}^b\right) = \frac{1}{B_{b,\varepsilon}}\sum_{c:l(\boldsymbol{y}^b,\boldsymbol{y}^c)<\varepsilon} v(\boldsymbol{x}^c,\boldsymbol{a}),\quad B_{b,\varepsilon} = \sum_{c\neq b} I\left(l(\boldsymbol{y}^b,\boldsymbol{y}^c)<\varepsilon\right).$$

6. Approximate the posterior value as follows:

$$PoV(\boldsymbol{y}) \approx \frac{1}{B}\sum_{b=1}^{B}\max_{\boldsymbol{a}\in A}\left\{\hat{E}\left(v(\boldsymbol{x},\boldsymbol{a})\mid \boldsymbol{y}^b\right)\right\}.$$

This procedure is computationally efficient because we re-use samples from the prior value computation. Steps 1 and 2 are identical to those for the prior value in Equation (5.51). The number of value computations is $B|A|$. In some situations, the forward modeling in Step 3 can be quite demanding as well, depending on the complexity of the data-gathering scheme. For instance, seismic data could require complex modeling in the relation $\mathbf{y} = f(\mathbf{x})$. The snow measurements, on the other hand, simply provide the snow level at the locations they are made, with noise.

Even though this approximate conditioning scheme based on loss or dissimilarity saves plenty of computer time, one must be careful because the statistical properties of such an approach are not fully understood. Moreover, these properties are difficult to check because the approximate Bayesian computing procedure enforces no statistical modeling assumptions on the likelihood. Instead, we simply use samples that are close to the data in some sense. There are several ways to select the loss function (see Chapter 2).

Barros et al. (2014) use a related computational approach, trying to learn from one set of realizations in creative ways, for VOI analysis in reservoir management. The approach evaluates the effect of data assimilation (information) for closed-loop reservoir optimization, which includes the locations and tuning or recovery factors of wells. They use an ensemble of models and a workflow that loops over the realizations or ensemble members as the truth while finding the optimal strategy on all the other ensemble members.

5.6.4 Example

Frozen: hydropower example

Keywords: *hydropower, snow measurements, transformed Gaussian variables, Monte Carlo sampling, approximate Bayesian computing, partial information, high decision flexibility, coupled value function*

We study the VOI of snow measurements for decisions about water levels in a system of dams for hydropower production. For more background on the subject of snow measurements for hydrology scheduling, see, e.g., Lundberg et al. (2010) and Bruland et al. (2015).

Framing the decision situation

Hydropower production benefits from advanced planning of water levels in dams. In our example, as well as in many real hydropower systems, dams are connected such that when water from one dam is used or lost due to flooding, it flows into the next dam downstream. The connection of dams makes the optimal scheduling (control) of the system challenging. The situation is further complicated by uncertainties around snow storage, groundwater levels, future precipitation, and energy prices. Note also that value may be a function of factors other than hydropower production – for instance, flood protection and water storage for agricultural purposes.

We will consider a simplified problem of short-term decisions about the water level in dams right before the snow melting period. At the time of snow melting, selling prices of energy are very low, while they get much higher later in the year. The goal for the hydropower company is thus to fill up all the dams and sell later. If the water levels before melting season are high, then the dams will likely exceed their maximum water level, and water will be lost for hydropower production. Hence, the hydropower company could actually have made more money previous to the melting period by selling, even though this would have been at a lower price. If the water levels before the melting season are too low, the dams are unlikely to fill, and future revenue is lost.

The example studied here consists of nine dams in a river basin (Figure 5.13). The alternatives pertain to the water levels of the dams. We assume that the decision maker is free to lower the water level at each dam before the snow melting starts (Figure 5.13, right). The high and low levels of the dams are denoted alternatives $a_j \in \{0,1\}$, $j = 1,\ldots,9$. A low level implies that the dam can handle an additional volume vol(1) before it runs over, while a high level means that the dam can handle a smaller additional volume vol(0). The units are in million cubic meters. We assume that the current dam level corresponds to the high alternative. By choosing the alternative of reducing the dam level to the lower level, the hydropower company receives income $\text{Rev}_1\left(\text{vol}(1)-\text{vol}(0)\right)$, where Rev_1 is the selling price per volume unit during this time of year. However, going through with this low alternative implies that there would be a smaller chance of filling up the dam later. The future selling price per unit volume is Rev_2. We set $\text{Rev}_2 \geq \text{Rev}_1$ because there is plenty of energy production and relatively little demand during the snow melting season. We assume that the production of energy during the rest of the year will bring the water level down to the high level again $(\text{vol}(0))$ at a constant selling price Rev_2. Note that in practice, the prices would be set by the market. Moreover, the energy per unit could also depend on the water level in the dam and the production profiles of dams, etc.

The map in Figure 5.13 (left) displays the river basin – i.e., the water outside this region will not run off to the dams. The catchments define the water that accumulates to the dam. In addition, the run-off from the dams above will flow along the river to the next dam. The profit made from hydropower production depends on the price of energy and the water volumes. The uncertain distinction of interest is the water volume equivalents of snow at spatial cells in the river basin. We split the spatial domain defining the river basin into a grid composed of 50 × 50 regular cells. Each cell is of size 1 × 1 km². The optimal decisions for dam levels are based on the uncertain water volume equivalents denoted x_i, $i = 1,\ldots,2500$.

The value function is spatially coupled through the connection between the various dams. We define it in further detail as we present expressions for the prior value in the following text.

Information gathering

It is common to acquire measurements of snow depth and density before the snow melts. A team would travel with snow scooters to measure the snow depth and properties of the snow at select locations.

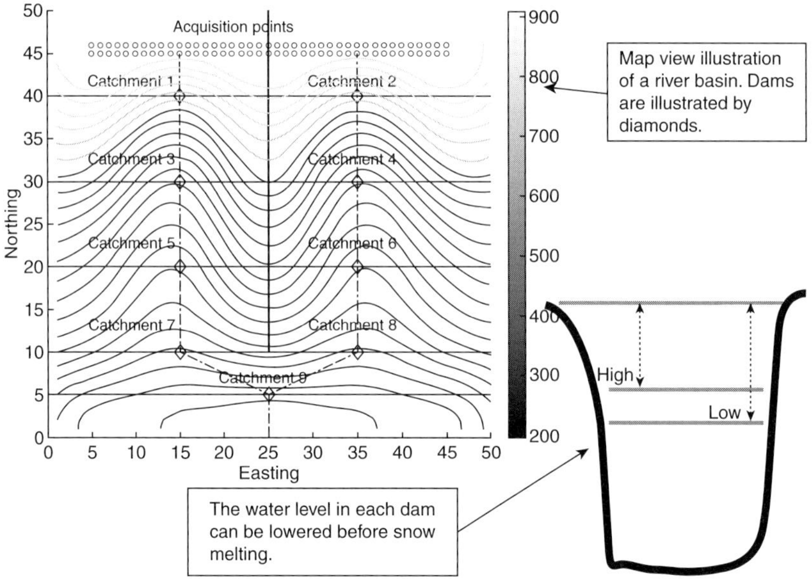

Figure 5.13 Hydropower example: a map view of nine dams with catchments. The acquisition line (north, center) indicates where the decision maker can collect snow measurements to better predict the water that will run into the catchments. The decision maker chooses the water level of the dams before the snow melts.

Figure 5.13 (left) illustrates the acquisition sites (circles) in this example. They are at the northernmost part of the basin at high altitude. The snow measurements acquired at these locations could influence the decisions about water levels at the dams. If it has a large effect on the decisions, the snow data acquisition may perhaps be worth its price.

Note that the snow measurements include the depth of the snow column as well as certain properties of the snow such as the snow water equivalent. For simplicity, we assume that the snow water equivalent is 0.5 and focus on the snow depth measurements that are related to the water volume equivalents at the spatial cells where the measurements are made.

Modeling

The water volume equivalent x_i of snow at cells $i = 1,\dots,2500$ is estimated by multiplying the snow column depth at that cell, the area of the cell, and the snow water equivalent (set to 0.5). In practice, the prior knowledge of the uncertain snow depth relies on numerical weather modeling, as well as precipitation and temperature observations at a few monitoring sites. We model the snow depth as a random field with a spatial correlation and a spatial trend. Spatial correlation is incorporated by an exponential covariance function for a Gaussian random field. These Gaussian variables at all sites are transformed to truncated power law variables (see Section 2.2) that represent the snow depth. The power law distributions have location-specific parameters, making large snow depths more likely at high altitudes and in the eastern part of the domain.

Figure 5.14 shows two realizations of the snow depth distribution (in meters). As modeled, there is a tendency to find deeper snow at higher altitudes and in the eastern parts. The cells with deep snow are seen to be spatially correlated. This correlation means that the aggregated snow in a catchment could vary significantly from one realization to another.

We denote the spatial cells for different catchments (basins of flow attraction) by $C_1,\dots,C_9$ and the accumulated water volumes at the dams by $x_{C_1},\dots,x_{C_9}$. For Dams 1 and 2, which are at the highest altitude, the accumulated snow melt contribution from the catchments is

$$x_{C_1} = \sum_{i \in C_1} x_i, \; x_{C_2} = \sum_{i \in C_2} x_i. \tag{5.55}$$

For the dams farther down the river system, the accumulation from snow melting is more complicated. For catchments $j = 3,\dots,8$,

$$x_{C_j} = \sum_{i \in C_j} x_i + \max\left\{0, x_{C_{pa(j)}} - \mathrm{vol}\left(a_{pa(j)}\right)\right\}, \tag{5.56}$$

where $pa(j)$ denotes the nearest upstream (or "parent") dam. Note that the melting contribution here depends on the alternatives chosen at the upstream dams. If the water volume equivalents are larger than the volume the parent dam can handle, the run-off flows to the next dam.

For the dam at lowest altitude, the sum over its catchment and the contribution from water flowing over dams 7 and 8 is

$$x_{C_9} = \sum_{i \in C_9} x_i + \max\left\{0, x_{C_7} - \mathrm{vol}\left(a_7\right)\right\} + \max\left\{0, x_{C_8} - \mathrm{vol}\left(a_8\right)\right\}. \tag{5.57}$$

The decision maker's monetary value depends on these uncertain water volumes in the dams, as described by Equation (5.55)–(5.57) after snow melting, as well as the alternatives chosen at the dam. It is also a function of the selling price Rev_1 before melting and Rev_2 in the future.

Data are acquired at 82 cells in the northern parts of the river basin (Figure 5.13). For these snow measurements, we assume that the depth measured at a cell is representative

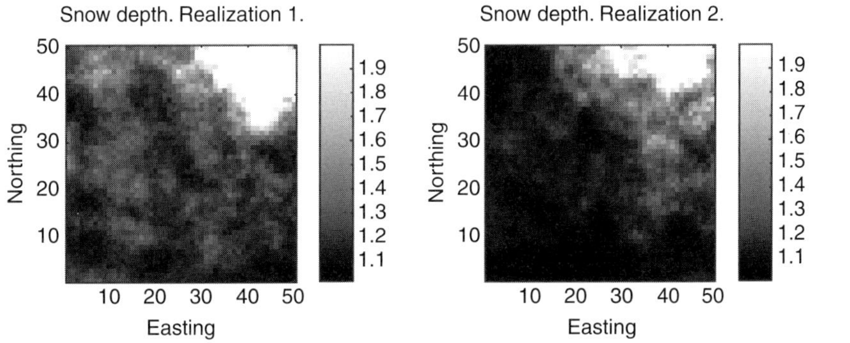

Figure 5.14 Hydropower example: two realizations sampled from the model for snow thickness.

of that spatial 1×1 km^2 cell. We also assume that the measurements provide perfect information at the cells where they are made – i.e., the data consist of $\boldsymbol{x}_{\mathbb{K}}$, where $\mathbb{K}$ denotes the subset of partial testing locations among the total of 2500 cells.

VOI analysis

The value is a function of the sum of water volumes sold prior to snow melting (if alternatives include lowering the water level) and the selling price of volume capacity after snow melting. This can be summarized by

$$v(\boldsymbol{x},\boldsymbol{a}) = \sum_{j=1}^{9} v_j(\boldsymbol{x},\boldsymbol{a}), \tag{5.58}$$

where the contributions at dams $j = 1,\ldots,9$ are defined by

$$\begin{aligned} v_j(\boldsymbol{x},\boldsymbol{a}) = {} & \mathrm{Rev}_1\left(\mathrm{vol}(1) - \mathrm{vol}(0)\right) I\left(a_j = 1\right) + \\ & \mathrm{Rev}_2 \min\left\{x_{C_j} - I\left(a_j = 1\right)\left(\mathrm{vol}(1) - \mathrm{vol}(0)\right), \mathrm{vol}(0)\right\}. \end{aligned} \tag{5.59}$$

Note that the value from a dam in Equation (5.59) depends on the uncertain snow amount in its catchment and the chosen dam level, as well as the uncertain snow contribution at higher altitudes and the chosen levels at these upstream dams.

In this example with high flexibility in the spatial alternatives as well as a coupled value function, the calculation of the prior value is rather demanding. For calculating the **prior value approximation,** we use Equation (5.51) with $B = 5000$ samples, where each value calculation $v(\boldsymbol{x}^b, \boldsymbol{a})$ is based on Equation (5.58). Thus, we generate B realizations of snow depths using the random field model with power law marginal distributions for the depth. For each of these realizations, we run the dams for all possible combinations of alternatives – i.e., $|A_1|\ldots|A_9|$. With the two alternatives at each dam, there are $2^9 = 512$ alternatives. The optimal alternative is the configuration of alternatives that provides the largest value, on average, over the realizations.

Assuming an average snow depth of 1.2 m, there is a volume of $1000^2 \cdot 1.2 \cdot 0.5 \cdot 250 = 150$ million m^3 from the 250 cells in a catchment. Obviously, if the higher level corresponds to smaller volumes than this, it is likely best to lower the dams. On the other hand, when the higher level corresponds to larger volumes than this, the decision maker is better off by not lowering the dam levels. Aside from these extreme situations, it is not straightforward to tell the effects of the various alternatives because of the complicated interactions, and it also depends on the price levels before melting and in the future.

For calculating the **posterior value**, we use the approximate Bayesian computing-type algorithm presented earlier. We condition on the data at the 82 acquisition points by choosing a subset of realizations that have snow depths close to what is observed in the data realization. The similarity measure between realizations is calculated from the average depth at the acquisition cells. The specified threshold of the approximate Bayesian computing algorithm picks about $B/10$ realizations for posterior assessment.

Table 5.5. *VOI for the snow measurements for hydropower planning. The results are over various prices per water unit of current water and that of future water*

Prior Value	$\mathrm{vol}(0)=140$, $\mathrm{vol}(1)=160$	$\mathrm{vol}(0)=120$, $\mathrm{vol}(1)=180$	$\mathrm{vol}(0)=150$, $\mathrm{vol}(1)=180$	$\mathrm{vol}(0)=120$, $\mathrm{vol}(1)=150$	$\mathrm{vol}(0)=100$, $\mathrm{vol}(1)=200$
$\mathrm{Rev}_1=1, \mathrm{Rev}_2=1$	1320	1340	1340	1300	1350
$\mathrm{Rev}_1=1, \mathrm{Rev}_2=2$	2520	2350	2600	2340	2160
$\mathrm{Rev}_1=2, \mathrm{Rev}_2=5$	6290	5750	6480	5740	5210
VOI	$\mathrm{vol}(0)=140$ $\mathrm{vol}(1)=160$	$\mathrm{vol}(0)=120$, $\mathrm{vol}(1)=180$	$\mathrm{vol}(0)=150$, $\mathrm{vol}(1)=180$	$\mathrm{vol}(0)=120$, $\mathrm{vol}(1)=150$	$\mathrm{vol}(0)=100$, $\mathrm{vol}(1)=200$
$\mathrm{Rev}_1=1, \mathrm{Rev}_2=1$	0	0	0	0	0
$\mathrm{Rev}_1=1, \mathrm{Rev}_2=2$	20	28	18	14	27
$\mathrm{Rev}_1=2, \mathrm{Rev}_2=5$	50	64	36	35	53

We compute the prior and posterior values and the VOI for a variety of parameters, as indicated in Table 5.5. The prior values clearly increase with larger revenues, and the same trend is seen for the VOI. The largest VOI is achieved when the energy prices per unit are $\mathrm{Rev}_1=2$ and $\mathrm{Rev}_2=5$ and for high and low water levels around the typical water volume in a catchment of 150. When high and low volumes are $\mathrm{vol}(0)=120$ and $\mathrm{vol}(1)=180$, the best a priori combination of alternatives is to lower the level in the east and south parts of the basin, in which case the prior value is 5750. If all dams are lowered, the average value is only 5100, because the decision maker will likely not be able to fill up the dams, thereby losing high-revenue water. The large VOI noted here indicates that the snow depth information is clearly useful for decisions pertaining to the dam level. When the prices are the same, $\mathrm{Rev}_1=\mathrm{Rev}_2=1$, it is optimal to lower the dams in most situations. When water levels are too high, $\mathrm{vol}(0)=120$ and $\mathrm{vol}(1)=150$, lowering the dams is prudential.

Figure 5.15 shows the prior and posterior values and the VOI as a function of the difference in the low and high water levels – i.e., $\mathrm{vol}(1)-\mathrm{vol}(0)$, symmetric around the typical volume of 150. The two curves represent different sets of revenues. The displays show that the prior and posterior values decrease as the high and low levels become farther apart. Having extreme alternatives prevents the decision maker from tuning the levels. On the other hand, when the high and low water levels are almost the same, there is little to gain by collecting the snow depth information because it does not have enough coverage to affect the decisions. The highest value of the snow measurements appears to occur for

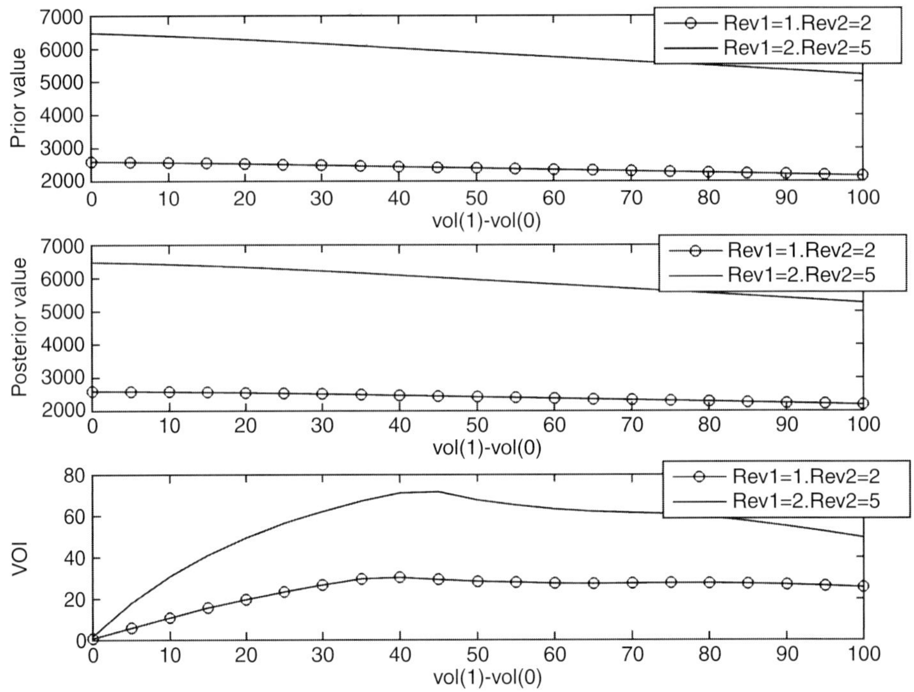

Figure 5.15 Hydropower example: prior value, posterior value, and VOI as a function of the difference in high and low values around the typical volumes in catchments. The plots are shown for two different sets of current revenue Rev_1 and future revenue Rev_2.

intermediate high and low differences when the decision maker gains sufficient flexibility of lowering dams at the right level after observing the snow depth information.

5.7 More complex decision situations

In Section 5.1.3, we described a high-level framework for a suite of models, and in Section 5.2, we presented a fairly general formulation for static models. We then proceeded to discuss various special cases in Sections 5.3–6. Here, we relax some of the previous assumptions: specifically, we consider more general risk preferences (beyond risk neutrality), constraints in the decision situation, and sequential decision situations. The reader should note that our discussion in this section and the next is fairly cursory, since research on models for VOI analysis in the Earth sciences for more complex decision situations is at a nascent stage.

More complex decision situations are very much relevant for real applications. However, they should be used wisely, because the price of generality in the formulation is often a significant increase in the computations required. We focus on the essential ideas to provide a flavor of some of the modeling and computational complexities using examples that are essentially special cases of more general formulations.

5.7.1 Generalized risk preferences

We previously assumed a risk-neutral decision maker – i.e., someone who makes decisions by maximizing expected value. For prior value calculations, we computed the expectation of the value function, and for posterior value calculations, we computed the conditional expectation of the value function. The VOI was obtained by subtracting the prior value from the posterior value. When the decision maker has a more general utility function, the expected value is no longer the only statistic that is pertinent. Moreover, the VOI computation is not necessarily the difference between the prior and posterior values. Here, we return to aspects from Chapter 3 regarding general utility functions and discuss these in the context of spatial decision situations.

In Chapter 3, we presented a general approach for computing VOI for a general utility function. Let us return to first principles and apply these techniques to equations from Section 5.2. Suppose that the decision maker has wealth w and utility function $u(.)$ and faces a spatial decision situation where the distinction of interest is $\boldsymbol{x}$ and any alternative from the available set, $\boldsymbol{a} \in \boldsymbol{A}$, can be chosen.

Prior to receiving any information, the decision maker chooses the alternative that maximizes the expected utility. Incorporating the initial wealth and of course the utility function, the maximum expected utility is

$$MEU = \max_{\boldsymbol{a} \in A} \left\{ E\left(u\left(v(\boldsymbol{x},\boldsymbol{a}) + w\right)\right) \right\} = \max_{\boldsymbol{a} \in A} \left\{ \int_{x} u\left(v(\boldsymbol{x},\boldsymbol{a}) + w\right) p(\boldsymbol{x}) d\boldsymbol{x} \right\}. \quad (5.60)$$

Suppose that the decision maker is offered total perfect information about the distinction of interest at a price P. The decision maker will lose this price for all prospects, since this price will be paid for certain. Again, as usual, the integral and the maximization are switched, and the maximum expected utility of this new situation is

$$MEU' = \int_{x} \max_{\boldsymbol{a} \in A} \left\{ u\left(v(\boldsymbol{x},\boldsymbol{a}) + w - P\right) \right\} p(\boldsymbol{x}) d\boldsymbol{x}. \quad (5.61)$$

The value of perfect information is the price at which the maximum expected utilities for the situations from the preceding two equations are equal. A similar approach can be taken for computing the value of partial perfect or imperfect information.

As we described in Chapter 3, when the decision maker has an exponential utility function, then the VOI can indeed be computed as a difference between the prior and posterior values (see Appendix A.1). Only the linear and exponential utility functions have this convenient property.

5.7.2 Additional constraints

Recall that the decision maker can choose alternatives from the available set $\boldsymbol{a} \in \boldsymbol{A}$. This notation captures the effect of constraints to a degree – if some alternatives are no longer feasible for whatever reason, they will no longer be in set $\boldsymbol{A}$, and the analysis can be replicated without making further changes to the model. However, what if there are additional constraints that enforce further restrictions on the alternatives that are available to the decision maker by necessitating underlying changes to the model? We will explore this aspect by extending an example from a previous subsection.

Consider the special case from Section 5.4 – a decision situation where there is high decision flexibility and a decoupled value function. We also referred to this case as that of the "free selection of sites," because the decision maker was free to select as many sites as profitable. Now suppose that there is a budget that puts limitations on the number of selections that can be made. Suppose that the decision maker is only allowed to select $k < n$ sites. The possible way to select subsets of size k out of n sites is $\binom{n}{k}$. We denote the set of such subsets by $W_{\binom{n}{k}}$. An element of this set, $\boldsymbol{\varpi} = (\varpi_1, \ldots, \varpi_k) \in W_{\binom{n}{k}}$, indicates which of the k cells are selected. Here, $\varpi_i \in \{1, \ldots, n\}$ and $\varpi_i \neq \varpi_j$, i, j = 1,...,k. For the chosen set, the decision maker can choose among alternatives $a_{\varpi_i} \in A_{\varpi_i}$.

If the value function can be decoupled, like in Section 5.4, the prior value is the maximum over all subsets of size k among the n sites – i.e.,

$$PV_{\binom{n}{k}} = \max_{\boldsymbol{\varpi} \in W_{\binom{n}{k}}} \left\{ \sum_{i=1}^{k} \max_{a_{\varpi_i} \in A_{\varpi_i}} \left\{ E\left(v\left(x_{\varpi_i}, a_{\varpi_i}\right)\right) \right\} \right\}. \tag{5.64}$$

Due to the decoupling of value, the computation of Equation (5.64) is simply done by ranking the n sites and picking the best k. If the value function is coupled, the calculation would be much more difficult.

The posterior value with imperfect information $\mathbf{y}$ is

$$PoV_{\binom{n}{k}}(\mathbf{y}) = \int \max_{\boldsymbol{\varpi} \in W_{\binom{n}{k}}} \left\{ \sum_{i=1}^{k} \max_{a_{\varpi_i} \in A_{\varpi_i}} \left\{ E\left(v\left(x_{\varpi_i}, a_{\varpi_i}\right) \mid \mathbf{y}\right) \right\} \right\} p(\mathbf{y}) d\mathbf{y}. \tag{5.65}$$

The critical difference with the prior value computation in Equation (5.64) is that the selection of sites now depends on the data – the decision maker can choose different sites depending on the data. The VOI is the difference between the posterior and prior values, as the decision maker is risk neutral.

The tree amigos: conservation biology example

Keywords: *conservation biology, Markov random field, constraints, total information, imperfect information*

We revisit the example inspired by a situation in conservation biology in Section 5.4. The spatial domain is represented as a 3 × 3 grid, and the presence $(x_i = 1)$ or absence $(x_i = 0)$ of a species of tree at the cells is modeled as a binary Markov random field model.

Now we allow for constraints on the number of reserve sites that can be selected. Let k denote the maximum number of cells that can be selected (based on a budget). The prior value computation is Equation (5.64), while the posterior value is computed in Equation (5.65). The value function for the two alternatives is as in Section 5.4 – i.e.,

$v(x_i, a_i = 1) = \text{Rev} \cdot I(x_i = 1) - \text{Cost}$ and $v(x_i, a_i = 0) = 0$, $i = 1,...,9$. From the prior or posterior probability at all cells, the profits from the field are maximized by choosing the best k prospects.

Suppose that the decision maker considers purchasing a total test. Furthermore, suppose that the outcome of this experiment is a discrete variable at each cell, $y_i \in \{0,1\}$, $i = 1,...,9$, which is indicative of the true absence or presence of the species of tree. As in Section 5.4, the accuracy of the experiment is defined by the likelihood model. Here, we set this accuracy to $\gamma = 0.9$.

The VOI for the different parameter settings is calculated using forward–backward recursions for the marginal likelihood $p(\mathbf{y})$ evaluation and the marginal probability $p(x_i = 1 \mid \mathbf{y})$, $i = 1,...,9$.

Figure 5.16 shows the VOI as a function of the spatial interaction parameter β. The display has three curves, valid for different values of k (the maximum number of cells that can be selected): $k = 1$, $k = 5$, and $k = 9$. The revenues are set to Rev = 2 (left) and Rev = 5 (right), while Cost =1, as in Section 5.4. The case with $k = 9$ corresponds to the unconstrained case in Figure 5.8, since there are $n = 9$ cells in the field.

Previously, we noted that the VOI tends to increase with spatial interaction in the unconstrained situation. However, for $k = 5$ and $k = 1$, the VOI seems to decrease as β increases. This is most prominent in the graph for large revenues on the right. We explain this tendency as follows: the experiment conducted over the entire field has a certain facet that can be relatively more valuable when there is both low spatial dependence and a limit on the number of cells that can be selected. It tells you which cells are likely to be favorable, thereby guiding the decision maker about the choice of site location. There is a little more leeway for the decision maker regarding the selection of an appropriate location when there is high spatial dependence. Thus, VOI can be relatively smaller when β is high.

VOI is not always highest when $k = 9$ and lowest when $k = 1$. The prior value is larger without constraints, but as we see in Figure 5.16 (right), the VOI need not follow such a trend. It is indeed possible for the experiment to be more valuable in the situation when there are constraints, as opposed to when there are none. In our experience, it is often complicated to interpret the VOI in situations with constraints, even in this simple example with a decoupled value function and a grid of relatively small size.

5.7.3 Sequential decision situations

In Section 5.1.3, we laid down a high-level framework for the various models by making a distinction between static and sequential decision situations as well as static and sequential information-gathering schemes. In this section, we briefly consider models of sequential decision situations. Our intention is not to be general, but only to provide the reader a sense of the complexities involved. We focus largely on the special case of high decision flexibility and decoupled value from Section 5.4.

In static spatial decision situations (for a risk-neutral decision maker with binary alternatives), the selection of sites is a one-time decision where the decision maker chooses

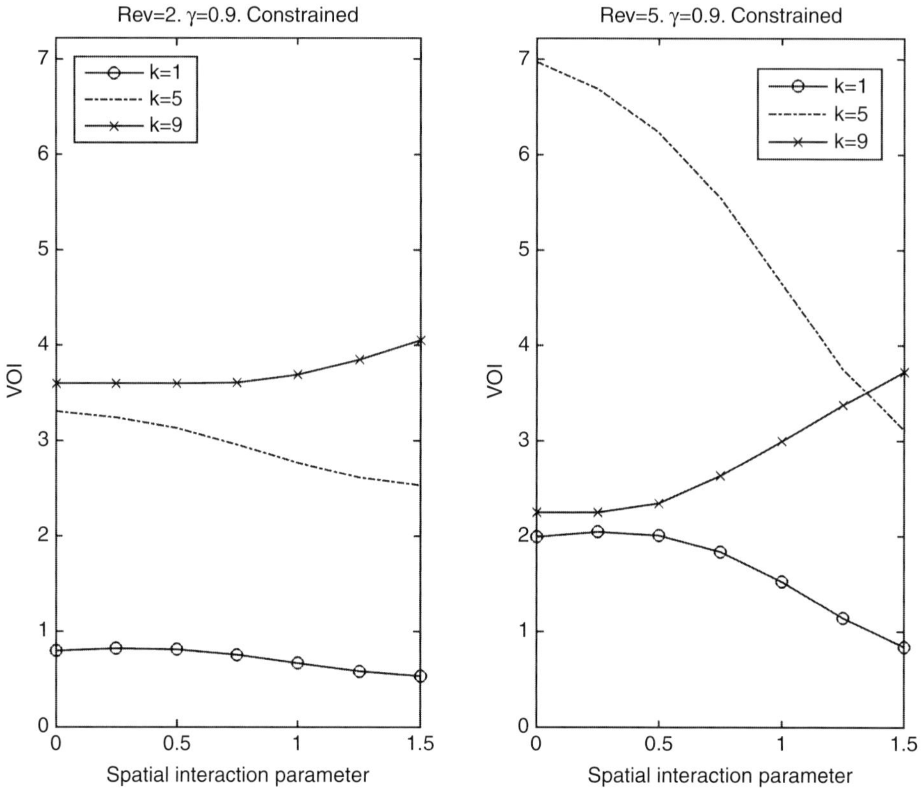

Figure 5.16 Conservation biology example: constrained selection of conservation biology units. VOI for total information on the 3 × 3 grid as a function of the spatial coupling in the distinction of interest (first axes). The three curves represent the number of cells that can be selected in the constrained decision situation. Revenues are 2 (left) and 5 (right) while cost is 1.

alternatives from $a_i \in \{0,1\}$, indicating no selection and selection, to maximize the total expected profits. Instead, suppose that the decision maker is now allowed to make sequential decisions; in this way, they can adapt to outcomes as they are observed. Sequential decisions therefore provide additional flexibility in making the decision. However, the corresponding model dramatically increases computational tasks. This increase is partly due to the need for evaluating many (sequential) conditional probabilities, given the results, but it is also caused by a large increase in the number of alternatives in each strategy of the sequential decision situation.

Figure 5.17 illustrates a decision tree for this sequential decision situation when there are only two uncertain variables x_1 and x_2. At the first stage, the decision maker can select one of these projects or none. If Project 1 or 2 is selected, they get to observe the result of the chosen one. Given the observation, the decision maker can now either select the other project as well or stop. The optimal selection strategy depends on the profits associated with the projects as well as the conditional probabilities of random variables x_1 and x_2. If

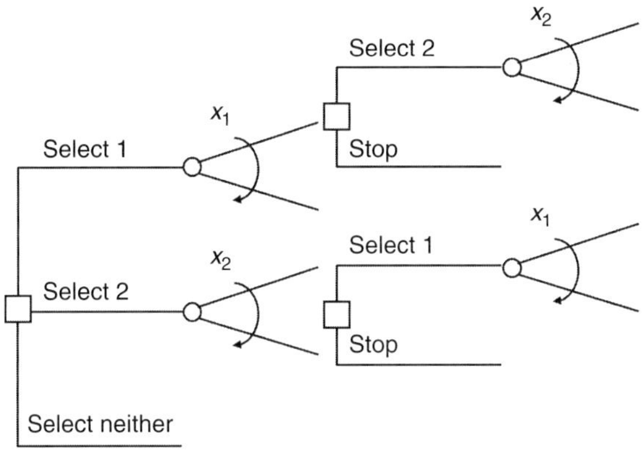

Figure 5.17 Illustration of sequential decisions for a two-variable situation: at the first branch, the decision maker selects 1, 2, or neither. After seeing the result of the first decision, the decision maker can go on selecting or stop.

observing the variable x_1 has an impact on the decision about Project 2, the decision maker would benefit from making the decision sequentially.

Assume a situation with n sites and two alternatives, select or not, for all projects. Selecting project number i is denoted $a_i = 1$. The prior value is

$$PV_{seq} = \max_{i_1}\left\{E\left(v\left(x_{i_1}, a_{i_1} = 1\right)\right) + \int_{x_{i_1}} \max_{i_2 \neq i_1}\left\{0, \text{ContVal}_{i_2}\left(x_{i_1}\right)\right\} p\left(x_{i_1}\right) dx_{i_1}, 0\right\}, \quad (5.66)$$

where the **continuation value** is defined by

$$\text{ContVal}_{i_2}\left(x_{i_1}\right) = E\left(v\left(x_{i_2}, a_{i_2} = 1\right) \mid x_{i_1}\right) + \int_{x_{i_2}} \max_{i_3 \neq i_1, i_2}\left\{0, \text{ContVal}_{i_3}\left(x_{i_1}, x_{i_2}\right)\right\} p\left(x_{i_2} \mid x_{i_1}\right) dx_{i_2}. \quad (5.67)$$

We have assumed a continuous sample space for these equations. For a discrete sample space, the integrals would become sums.

This prior value is computed by a dynamic program, which can be solved using Bellman's equations. These equations are nested, and their solution requires going through all possible sequences of selections. The optimal sequence depends on the outcome of selected sites at earlier stages. At the final stage, the continuation value is

$$\text{ContVal}_{i_n}\left(x_{i_1}, \ldots, x_{i_{n-1}}\right) = \max\left\{0, E\left(v\left(x_{i_n}, a_{i_n} = 1\right) \mid x_{i_1}, \ldots, x_{i_{n-1}}\right)\right\}. \quad (5.68)$$

The prior value in Equation (5.66) is always at least as large as the prior value for the associated static decision situation. When there is dependence, it is always helpful to see the outcome of another variable before making a selection. Bickel and Smith (2006) describe an example where the static decision situation has $PV = 0$, while the sequential decision situation has a value that is clearly positive.

The computations depend on the probabilistic model in general. The integrals are often not solvable using analytical methods. For a discrete sample space or for a discretized numerical approach, it is possible to sum out the entire multivariate sample space when there are relatively few sites – see, e.g., Bickel and Smith (2006) for an example with prior value calculation on a size $n = 6$ problem inspired from the petroleum industry. (A similar example is provided as an exercise in Chapter 7.) When there are many spatial sites, solving the integrals or sums can be very computationally demanding, and one may need to turn to approximate solutions. The method of choice depends very much on the situation and the probabilistic model. There is a large body of literature on dynamic programs and approximate dynamic programs – see, e.g., Powell (2011). A myopic (or greedy) solution is often used in practice, where the algorithm looks one step ahead, then updates, then looks one step further, updates, and so on. An alternative is to consider multiple steps into the future. This is often better than the myopic approach but is more computationally demanding – see, e.g., Martinelli et al. (2013a).

For imperfect information $\boldsymbol{y}$, one can compute the posterior value using a similar recursion, but this time conditional on the information. Data could be partial, total, imperfect, or perfect. The posterior value becomes

$$PoV_{seq}(\boldsymbol{y}) = \int_{\boldsymbol{y}} \max_{i_1} \left\{ \begin{array}{l} 0, E\left(v\left(x_{i_1}, a_{i_1} = 1\right) \mid \boldsymbol{y}\right) + \\ \int_{x_{i_1}} \max_{i_2 \neq i_1} \left\{0, \mathrm{ContVal}_{i_2}\left(x_{i_1}, \boldsymbol{y}\right)\right\} p\left(x_{i_1} \mid \boldsymbol{y}\right) dx_{i_1} \end{array} \right\} p(\boldsymbol{y}) d\boldsymbol{y}, \qquad (5.69)$$

where the continuation value must again be computed in a nested manner, and now includes the information provided by the data $\boldsymbol{y}$. The VOI for a risk-neutral decision maker is the difference between the posterior and prior values. Computations for obtaining the posterior value and the VOI are much harder than those for the prior value, and simplifications inspired by approximate dynamic programming will likely be required.

Gotta get myself connected: Bayesian network example

Keywords: *Bayesian network, exploration wells, binary outcomes, Bayes' rule, sequential decisions*

This example revisits the two-node network example in Section 2.3. Two prospects are assumed to be dependent via a common parent node in a Bayesian network. The marginal success probability at both nodes is $p(x_i = 1) = 0.1$. The conditional probability, given a success at the other, is $p(x_{i_2} = 1 \mid x_{i_1} = 1) = 0.5$, while $p(x_{i_2} = 1 \mid x_{i_1} = 0) = 0.055$. The revenues and costs associated with success and failure are specified to be the same for both prospects. The revenues are set to Rev = 3, while we vary the costs to study sensitivity. For static selection, the prior value is $PV = 2\max(3 \cdot 0.1 - \mathrm{Cost}, 0)$. This is plotted in Figure 5.18 (dashed line). It decreases as a straight line with a slope of 2 until it reaches the first axis at Cost = 0.3.

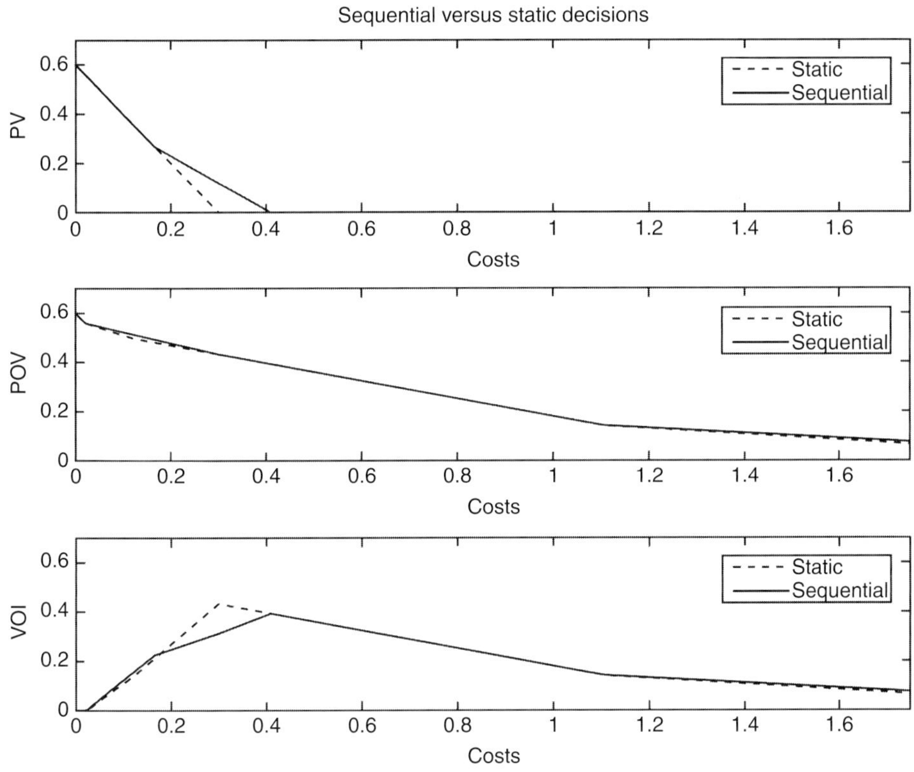

Figure 5.18 Prior and posterior values and VOI for the situation with two dependent prospects for a sequential decision situation. The dashed lines are for the situation with static decisions; the solid lines are for the situation with sequential decisions.

When the decision maker has the ability to make sequential decisions, there is a greater prior value, as given by Equation (5.66). When we plug in the probabilities for this case, we obtain:

$$PV_{seq} = \max\left\{0, 3\cdot 0.1 - \text{Cost} + \begin{pmatrix} \max\{0, 3\cdot 0.055 - \text{Cost}\}\cdot 0.9 + \\ \max\{0, 3\cdot 0.5 - \text{Cost}\}\cdot 0.1 \end{pmatrix}\right\}. \tag{5.70}$$

The prior value with sequential decisions in Equation (5.70) is piecewise linear, and it goes slower toward 0 than for the static case. At Cost $= 3\cdot 0.055 = 0.17$, the slope changes because one enters a range of costs where the second project will not be selected if the first was a failure. Figure 5.18 (top display) shows how the prior value of the sequential decisions is clearly larger in the cost range where the decision maker is most indifferent.

Suppose that the decision maker considers gathering imperfect information at both projects. The test has a discrete result, but it may yield a false positive or false negative. The accuracy is assumed to be the same for the two and is denoted by $\gamma = 0.9$, as we have done in previous examples. Following Equation (5.69), the posterior value of the total imperfect information is

$$PoV_{seq}(\boldsymbol{y}) = \int_{y} \max_{i \in 1,2} \left\{ \begin{matrix} 0, \text{Rev} \cdot p(x_i \mid \boldsymbol{y}) - \text{Cost} + \\ \sum_{k=0}^{1} \max\left\{0, \text{ContVal}_{j \neq i}(k, \boldsymbol{y})\right\} p(x_i = k \mid \boldsymbol{y}) \end{matrix} \right\} p(\boldsymbol{y}) d\boldsymbol{y}, \quad (5.71)$$

where $\text{ContVal}_j(k, \boldsymbol{y}) = \text{Rev} \cdot p(x_j = 1 \mid x_i = k, \boldsymbol{y}) - \text{Cost}$. Note that one must now account for the order in which the decisions are made. For some observations, the decision maker may select one prospect before the other and vice versa. When the prior value was computed, this order was not considered due to the symmetry in this specific situation. Conditional on the data $\boldsymbol{y}$, this symmetry is broken.

The posterior values and the associated VOIs are plotted in Figure 5.18. The posterior value of the sequential strategy is only slightly more than that for the static decision situation, and it is larger for different cost ranges than for the prior value (top display). Therefore, the VOI could either be lower or higher for sequential decisions as compared to the VOI for static decisions.

5.8 Sequential information gathering

We briefly discuss models related to sequential information gathering – i.e., when information need not be collected in one go, and therefore the decision maker can use the information obtained from one test to evaluate whether subsequent tests should be performed. Sequential information schemes provide additional value due to the additional flexibility around planning information gathering. Miller (1975) is an early reference on this subject.

We consider the case of a static spatial decision situation – i.e., where the decision maker must make a one-shot selection of sites – but with sequential information gathering, so the decision maker can perform experiments sequentially. Furthermore, we consider the special case of high decision flexibility and decoupled value from Section 5.4. The prior value is the same as in Section 5.4 – i.e.,

$$PV = \sum_{i=1}^{n} \max_{a_i \in A_i} \left\{ E\left(v(x_i, a_i)\right) \right\}. \quad (5.72)$$

The posterior value is evaluated for all possible strategies of sequential testing. Let the possible tests be $\boldsymbol{y}_j$, $j = 1, \ldots, m$. This is generic notation that represents all sorts of tests – for instance, the tests could be univariate, aggregated, or made site by site. They could be total, partial, perfect, imperfect, etc. In any event, we assume a price P_j, $j = 1, \ldots, m$ of tests. For calculating the VOI of sequential tests, one must know the prices of these experiments up front, since they are needed to evaluate whether future tests are worthwhile. Miller (1975) clearly states some assumptions about these prices P_j of experiments, which we will also use. First, we assume that the prices are fixed – i.e., they are not random variables. Second, we assume that they are additive, implying that the price of shared elements in the acquisition, processing, or some initial common price is negligible compared with the prices of each experiment.

For now, consider two tests $\boldsymbol{y}_1$ and $\boldsymbol{y}_2$. Suppose that the decision maker acquires the first test, with the ability to follow up with the other. For some outcomes of the first test, he or she may not acquire the second. For other outcomes, he or she may indeed acquire the second test. Thus, the value of a sequential information-gathering scheme cannot be less than doing both tests at once – i.e., a static testing scheme. A risk-neutral decision maker can determine the best **test sequence** by maximizing the expected value. With two tests, the decision maker has the following four alternatives: Test 1 and stop, Test 1 and continue with Test 2, Test 2 and stop, or Test 2 and continue with Test 1.

Assume that Test $\boldsymbol{y}_1$ has been performed. The posterior value of this static test is

$$PoV(\boldsymbol{y}_1) = \int \sum_{i=1}^{n} \max_{a_i \in A_i} \{E(v(x_i, a_i) \mid \boldsymbol{y}_1)\} p(\boldsymbol{y}_1) d\boldsymbol{y}_1. \quad (5.73)$$

With sequential testing, there is a more complicated integrand for the solution because one must consider all possible sequences of information gathering. The decision maker will only acquire the test if the additional value minus the price of the second experiment P_2 is larger than the value of the first test alone – i.e., we require

$$\int_{\boldsymbol{y}_2} \sum_{i=1}^{n} \max_{a_i \in A_i} \{E(v(x_i, a_i) \mid \boldsymbol{y}_1, \boldsymbol{y}_2)\} p(\boldsymbol{y}_2 \mid \boldsymbol{y}_1) d\boldsymbol{y}_2 - P_2 > \sum_{i=1}^{n} \max_{a_i \in A_i} \{E(v(x_i, a_i) \mid \boldsymbol{y}_1)\}. \quad (5.74)$$

We refer to the left side of Equation (5.74) as continued testing, while the right side is associated with the decision maker stopping testing. The posterior value of the first test having the option of continuing with Test 2 becomes

$$PoV_{seqtest}(\boldsymbol{y}_2 \mid \boldsymbol{y}_1) = \int \max \left\{ \begin{array}{l} \int_{\boldsymbol{y}_2} \sum_{i=1}^{n} \max_{a_i} \{E(v(x_i, a_i) \mid \boldsymbol{y}_1, \boldsymbol{y}_2)\} p(\boldsymbol{y}_2 \mid \boldsymbol{y}_1) d\boldsymbol{y}_2 - P_2, \\ \sum_{i=1}^{n} \max_{a_i} \{E(v(x_i, a_i) \mid \boldsymbol{y}_1)\} \end{array} \right\} p(\boldsymbol{y}_1) d\boldsymbol{y}_1, \quad (5.75)$$

and, since the decision maker is risk neutral, the VOI is the difference between the posterior and prior values. This VOI should now be compared to the price P_1 of Test $\boldsymbol{y}_1$. If the VOI is larger than this price of the experiment, it is worthwhile doing a sequential test starting with $\boldsymbol{y}_1$. The optimal sequential testing scheme may be to perform Test 2 first and then Test 1. The posterior values of all strategies must be evaluated and compared after incorporating the prices of the experiments.

More generally, for sequential tests $\boldsymbol{y}_j$, $j = 1, \ldots, m$, the VOI can be written as a sequence of interchanging integrals and maximizations, not dissimilar to the expression for the prior value for sequential decisions in Equation (5.66). The integrals (or sums for the discrete sample space) are assumed to be over the outcomes of the tests. The maximizations are assumed to be over the alternatives of continued testing or stopping testing. The situation is illustrated as a partially drawn decision tree in Figure 5.19.

The posterior value of sequential information starting with Test $\boldsymbol{y}_{j_1}$ is

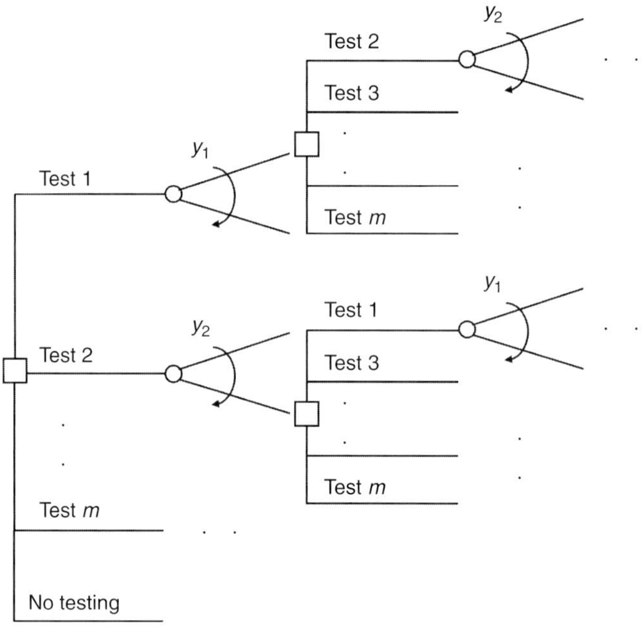

Figure 5.19 Illustration of sequential testing: at each branch, the decision maker can continue testing or stop. When he or she stops, the decision is made at all sites.

$$PoV_{seqtest}\left(\bullet | \boldsymbol{y}_{j_1}\right)=\int \max \begin{Bmatrix} \max_{j_2 \neq j_1}\left\{\text{ContVal}_{j_2}\left(\boldsymbol{y}_{j_1}\right)\right\}, \\ \sum_{i=1}^{n} \max_{a_i}\left\{E\left(v\left(x_i,a_i\right) | \boldsymbol{y}_{j_1}\right)\right\} \end{Bmatrix} p(\boldsymbol{y}_{j_1})d\boldsymbol{y}_{j_1}, \tag{5.76}$$

where the continued value when doing more sequential testing is as follows:

$$\text{ContVal}_{j_2}\left(\boldsymbol{y}_{j_1}\right)=\int_{\boldsymbol{y}_{j2}} \max \begin{Bmatrix} \max_{j_3 \neq j_1, j_2}\left\{\text{ContVal}_{j_3}\left(\boldsymbol{y}_{j_1},\boldsymbol{y}_{j_2}\right)\right\}, \\ \sum_{i=1}^{n} \max_{a_i}\left\{E\left(v\left(x_i,a_i\right) | \boldsymbol{y}_{j_1},\boldsymbol{y}_{j_2}\right)\right\} \end{Bmatrix} p\left(\boldsymbol{y}_{j_2} | \boldsymbol{y}_{j_1}\right)d\boldsymbol{y}_{j_2} - P_{j_2}. \tag{5.77}$$

Due to the nesting with continued values, the computational solution to the posterior value and the VOI is extremely demanding in general. The solution to the general setup is provided by dynamic programming. For small models and toy problems, this can be solved exactly, but for larger models, heuristic strategies are required in practice. There are numerous approximate solutions that can be leveraged for different cases – see, e.g., Powell (2011).

For whom the bell tolls: Gaussian projects example

Keywords: *Gaussian projects example, bivariate Gaussian distribution, correlation, linear Gaussian model, perfect information, partial information, total information, sequential information gathering*

We return to the bivariate Gaussian pdf, which was also used in Sections 2.3 and 3.4. Two projects have Gaussian profits with mean $\mu_i = 0$, variance $\sigma_i^2 = 1$, and correlation ρ. We study imperfect information-gathering schemes with a conditionally independent likelihood model $p(y_i \mid x_i) = N(x_i, \tau^2)$, $\tau^2 = 0.5^2$. We consider a static decision situation where we can freely select a project if it is profitable.

This example has been constructed specifically for comparing the cases of static and sequential information gathering. We plot the VOI minus the price of the experiments. When the decision maker can pursue sequential information gathering, the price of the second test is subtracted wherever testing is continued. Figure 5.20 shows the results as a function of the correlation parameter ρ. We compare the VOI of the sequential testing model with that of static testing at both prospects as well as static testing at one prospect.

The figure indicates that sequential testing is always of greater value than the results of static testing at one or both sites. For small correlations, the results of sequential testing seem to follow that of static testing in the sense that for most data outcomes of the first test, the sequential scheme will continue testing. For very high correlations, the sequential testing curve follows the static test at one of the sites. In this case, there is so much dependence that performing one test is sufficient – i.e., the sequential scheme will stop testing for most

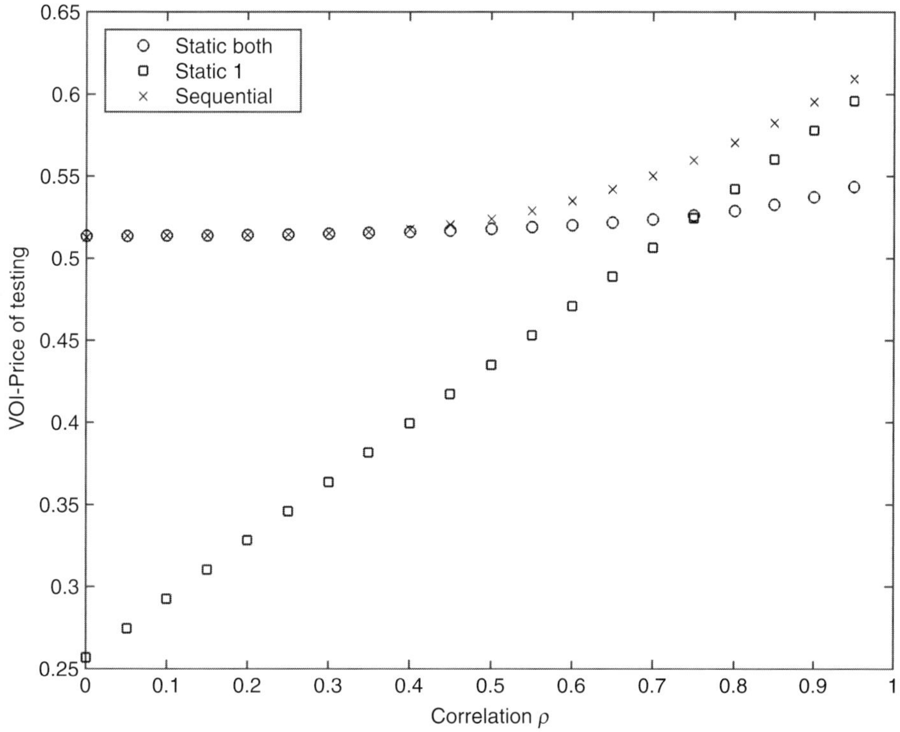

Figure 5.20 Gaussian projects example: VOI minus the price of the experiment for different testing options plotted versus the correlation between the projects (first axis). The plot compares the results of static and sequential information gathering.

outcomes of the first test. Intuitively, the added value obtained from sequential testing is largest for intermediate correlations, where it is worthwhile for the decision maker to have the additional flexibility of determining whether to continue testing.

5.9 Other information measures

We have described VOI analysis for the comparison and evaluation of various information-gathering schemes. In this section, we take a diversion and discuss some alternative approaches for measuring information content. We advocate the use of VOI analysis when faced with a clear decision situation involving uncertain distinctions of interest that can be modeled reasonably. However, when there is ambiguity around the decision situation or situations that the information might be useful for, alternative information measures such as the ones discussed in this section may be useful.

Our discussion focuses on three information measures that are commonly used for selecting and designing spatial experiments: entropy, prediction variance, and prediction error. Other measures are also used in practice, and we mention these briefly. Dobbie et al. (2008) review the different types of spatial monitoring designs. Geometric designs are routinely used in practice (Royle and Nychka 1998). They are also denoted as the space-filling design of experiments since they do not rely on any aspects of the spatial statistical model or spatial correlation. Other designs attempt to have a degree of space filling while maintaining some design locations close together (Diggle and Lophaven 2006). This is useful when the goal is to estimate spatial correlation parameters and at the same time maintain good spatial coverage for prediction purposes. We only treat the situation with fixed statistical model parameters here, assuming that auxiliary data exist that allow us to specify unknown nuisance parameters. Most of the literature on spatial experimental design relies on a Gaussian model for the distinction of interest and the data. Under this assumption, both the entropy and variance criteria are easily computed in a prospective sense – i.e., before the data are actually collected. For other probability models and for high-dimensional spatial variables, the computations required to evaluate a suggested design may be very demanding. Approximate solutions are available for some models, like spatial generalized linear models (GLMs) (Evangelou and Zhu 2012), but in general the computations involve an outer loop drawing realizations of data and an inner loop evaluating the information measure for each data realization. The search for the best design in a large set of spatial candidate sites becomes a combinatorial problem, and various heuristics have been applied. One of these is the exchange algorithm (Royle 2002), which iteratively picks candidate sites and tries to replace the selected site with another.

There is a gulf of difference between the philosophies of the information measures presented in this section and VOI. For instance, the information-theoretic notion of entropy provides a sense of how much uncertainty can be reduced by performing an experiment, but it cannot directly indicate how much the decision maker should pay for it. The decision-analytic notion of VOI is tied inexorably with decisions and the preferences of the decision maker, and can inform the decision maker about how much he or she should pay. VOI is a more complete measure for valuing information and therefore is also more difficult to obtain. Recall from Chapters 1 and 3 that from a decision analysis perspective, an experiment should be conducted when it is: (i) relevant to the distinction of interest, (ii) material to the decision that brings value, and (iii) economic for the decision maker. A material experiment is one that can affect the decision – i.e., the action chosen by the decision maker is not identical for different outcomes of the experiment. By the decision-analytic philosophy, information from experimentation may reduce uncertainty but is not valuable until it can change the decision. The information measures we present in this section only address aspects of the relevancy of the experiment without addressing the other two requirements. Entropy measures may be used as a guide in designing the most relevant experiment (e.g., Mukerji et al. 2001), but it cannot be used to say anything about the experiment being material or economic.

5.9.1 Entropy

Information entropy was originally introduced by Shannon (1948) as a measure of uncertainty in problems related to signal communication across noisy channels. Later, the concepts of entropy and mutual information were applied in other fields, including statistics, physics, biology, engineering, and computation (Cover and Thomas 2006). The notion of entropy, introduced from information theory (e.g., Ash 1965) is useful for measuring the reduction in uncertainty of $\boldsymbol{x} = (x_1, \ldots, x_n)$ on observing the outcome of an experiment. It is a commonly used information measure to gauge the relevancy of experiments and has been used for constructing useful designs of spatial experiments – see, e.g., Shewry and Wynn (1987), Zimmerman (2006), or Fuentes et al. (2007). For a fixed-size design, one selects the spatial experiment sites which, when combined, result in the smallest overall entropy for the distinction of interest.

Recall the definition of entropy, which was briefly discussed in Section 2. For a continuous sample space, $Ent(\boldsymbol{x}) = -\int p(\boldsymbol{x}) \log p(\boldsymbol{x}) d\boldsymbol{x}$, where $p(\boldsymbol{x})$ is the pdf of $\boldsymbol{x}$. Consider experimental observations $\boldsymbol{y} = (y_1, \ldots, y_m)$, which are informative of the distinction of interest $\boldsymbol{x}$. The data could represent perfect or imperfect information and partial or total testing. Without loss of generality, we assume a likelihood model $p(\boldsymbol{y} \mid \boldsymbol{x})$, with the marginal distribution $p(\boldsymbol{y}) = \int p(\boldsymbol{x}) p(\boldsymbol{y} \mid \boldsymbol{x}) d\boldsymbol{x}$. The conditional entropy, given the experiment $\boldsymbol{y}$, is given by $Ent(\boldsymbol{x} \mid \boldsymbol{y}) = -\int p(\boldsymbol{x} \mid \boldsymbol{y}) \log p(\boldsymbol{x} \mid \boldsymbol{y}) d\boldsymbol{x}$.

The expected reduction in entropy is obtained by subtracting the expected conditional entropy from the original (marginal) entropy. We have

$$\mathrm{EMI}(\boldsymbol{y}) = Ent(\boldsymbol{x}) - \int Ent(\boldsymbol{x} \mid \boldsymbol{y}) p(\boldsymbol{y}) d\boldsymbol{y}, \tag{5.78}$$

where EMI denotes the expected mutual information between $\boldsymbol{x}$ and $\boldsymbol{y}$.

Entropy can be used to compare various information-gathering schemes. For instance, EMI can be used to study if one test reduces entropy more than another test of similar size or complexity. One can also compute the sequential reduction in entropy – for instance, by augmenting the current design with one experiment or site at a time. In the beginning, one may choose to pick the sites that are expected to reduce entropy the most, but as more sites are sequentially selected, the entropy reduction becomes smaller. One could potentially stop selecting sites after a fixed number m is reached. Note, however, that a forward selection of m spatial sites chosen in this fashion may not be identical to the optimal size m set of sites.

The entropy information criterion has been popular in spatial statistics – see, e.g., Le and Zidek (2006). Entropy has close ties with likelihood theory, and this makes it a natural measure for a statistician when there is no extrastatistical knowledge. The criterion relies solely on the joint probability models. Entropy has been particularly popular for Gaussian geostatistical models, where a common assumption is that of stationarity. When the spatial model is non-stationary, there may be other criteria that are more intuitive. For instance, Lilleborge et al. (2015) suggest using sums of marginal entropy rather than the EMI in Equation (5.78) for a petroleum exploration example. Lindberg and Lee (2015) argue for an asymmetric entropy measure to guide the selection of experiments. In their applications, the experiments are very computationally demanding, and the traditional entropy measure appears to select a lot of wasteful experiments.

The goal of Lilleborge et al. (2015) is to select a subset of petroleum fields for exploration drilling. A Bayesian network (BN) model is used to model the outcome of n petroleum fields. An exploration well provides perfect information about the presence or absence of hydrocarbons at the location where it is drilled. Suppose that the decision maker can perform one experiment here – i.e., drill one exploration well to learn as much as possible about the petroleum resources. We will denote the perfect information at the testing node by $x_{\mathbb{K}}$, while the uncertain outcomes at other nodes are $\boldsymbol{x}_{\mathbb{L}}$. By conditional probability, $p(\boldsymbol{x}) = p(x_{\mathbb{K}})p(\boldsymbol{x}_{\mathbb{L}} \mid x_{\mathbb{K}})$; therefore, $\log p(\boldsymbol{x}) = \log p(x_{\mathbb{K}}) + \log p(\boldsymbol{x}_{\mathbb{L}} \mid x_{\mathbb{K}})$. Using this formula in Equation (5.78)

$$\begin{aligned} EMI(x_{\mathbb{K}}) &= Ent(\boldsymbol{x}) - \int Ent(\boldsymbol{x}_{\mathbb{L}} \mid x_{\mathbb{K}}) p(x_{\mathbb{K}}) dx_{\mathbb{K}}, \\ &= Ent(x_{\mathbb{K}}) + \int Ent(\boldsymbol{x}_{\mathbb{L}} \mid x_{\mathbb{K}}) p(x_{\mathbb{K}}) dx_{\mathbb{K}} - \int Ent(\boldsymbol{x}_{\mathbb{L}} \mid x_{\mathbb{K}}) p(x_{\mathbb{K}}) dx_{\mathbb{K}} = Ent(x_{\mathbb{K}}). \end{aligned} \tag{5.79}$$

The largest reduction in entropy is achieved by selecting the node with the highest marginal entropy. Since the marginal entropy is largest at success probability of 0.5, the EMI criterion focuses experimentation on the prospects where the decision maker is most indifferent. In some situations, this is natural, but it does not necessarily value learning about the outcomes at other nodes. Consider a network consisting of $n-1$ very dependent prospects, where the n-th node is independent of the rest. If this last node has marginal probability closer to 0.5 than any of the other nodes, the algorithm based solely on EMI would select this node. This seems counter-intuitive since the information acquired at this single node is completely uninformative about the rest of the nodes. In comparison, an observation at another node could be very informative about the entire network except the one independent node.

5.9.2 Prediction variance

Another commonly used measure for comparing experimental designs is prediction variance. The goal is to construct spatial designs that provide the smallest possible overall (integrated) variance for the distinction of interest. It is most commonly used for the Gaussian model or hierarchical models with GLM likelihood – see, e.g., Diggle and Lophaven (2006) and Evangelou and Zhu (2012).

As before, let us represent the distinction of interest at n sites as $\boldsymbol{x} = (x_1,\ldots,x_n)$. We denote the covariance matrix of $\boldsymbol{x}$ by $\boldsymbol{\Sigma}$. The sum of marginal prior variances at the sites is:

$$\mathrm{SV} = \sum_{i=1}^{n} Var(x_i) = \sum_{i=1}^{n} \Sigma_{ii} = \mathrm{trace}(\boldsymbol{\Sigma}). \tag{5.80}$$

An experiment yields data $\boldsymbol{y}$, and we can compute the posterior variances. The expected reduction in variance is used to compare various experimental designs. To construct a criterion, we must therefore compute the expected variance reduction (EVR):

$$\mathrm{EVR}(\boldsymbol{y}) = \sum_{i=1}^{n} Var(x_i) - E\left(\sum_{i=1}^{n} Var(x_i \mid \boldsymbol{y})\right) = \sum_{i=1}^{n} Var(x_i) - \int \sum_{i=1}^{n} Var(x_i \mid \boldsymbol{y}) p(\boldsymbol{y}) d\boldsymbol{y}. \tag{5.81}$$

When we compare designs of equal size, the one with the largest $\mathrm{EVR}(\boldsymbol{y})$ would be preferred. Again, one could use this criterion in a sequential manner, as discussed for the entropy criterion.

Pitfalls: reduction of uncertainty by data

It is a misconception that data always reduce uncertainty. Consider, for instance, the binary univariate model, $x \in \{0,1\}$, with success probability $p = p(x=1)$. We can easily construct likelihood models and data realizations y such that $p(x=1 \mid y)$ is closer to 0.5 than the prior probability. For instance, in the pirate example in Section 2.3, $p = 0.01$, but when we get a positive test, $p(x=1 \mid y=1) = 0.16$. This means that the variance $Var(x \mid y) > Var(x) = p(1-p)$. However, notably, this will not hold when we average over all possible data realizations y. On average, of course, data will reduce the uncertainty. Using the formula of conditional variance:

$$Var(x) = E(Var(x \mid y)) + Var(E(x \mid y)) \geq E(Var(x \mid y)).$$

Similar arguments hold for the other information criteria.

Norwegian wood: forestry example

Keywords: *forestry example, Gaussian random field, prediction variance, entropy, design of experiments, spatial design*

Let us compare the prediction variance for different spatial designs of experiment for the forestry example discussed in Sections 5.3 and 5.4. We maintain the same discretization of the spatial domain: the sites are the $25^2 = 625$ cells covering the unit square.

The a priori Gaussian distribution of the forest profits is denoted $p(\boldsymbol{x}) = N(\boldsymbol{\mu},\boldsymbol{\Sigma})$. We assume a homogeneous structure with no trend (mean 0), but as we will see, this is not relevant for the prediction variance since it is a function of the variance–covariance alone and not the mean.

When the decision maker collects data $\boldsymbol{y}$, which are acquired according to one of the three test designs described in Section 5.3, the conditional covariance matrix matters. Using generic notation, data are obtained via $\boldsymbol{y} = \boldsymbol{F}\boldsymbol{x} + N(\boldsymbol{0}, \boldsymbol{T})$ for some design matrix $\boldsymbol{F}$ defined by the spatial experiment.

The posterior covariance matrix, conditional on the data, is given by

$$\boldsymbol{\Sigma}_{x|y} = \boldsymbol{\Sigma} - \boldsymbol{\Sigma}\boldsymbol{F}^t\left(\boldsymbol{F}\boldsymbol{\Sigma}\boldsymbol{F}^t + \boldsymbol{T}\right)^{-1}\boldsymbol{F}\boldsymbol{\Sigma}. \tag{5.82}$$

Note that for the Gaussian distribution, the conditional variance does not depend on the data, nor does it depend on the prior mean. It only depends on the prior variance, the accuracy of the experiment, and the design of data acquisition via the matrix $\boldsymbol{F}$. The variance reduction criterion includes the expected trace of the matrix in Equation (5.82), where the expectation is taken over by the data $\boldsymbol{y}$. However, since the data are not directly involved, the EVR is simply given by the trace of the prior covariance minus the trace of the posterior covariance. We have

$$\text{EVR}(\boldsymbol{y}) = \text{trace}(\boldsymbol{\Sigma}) - \text{trace}(\boldsymbol{\Sigma}_{x|y}) = \text{trace}\left(\boldsymbol{\Sigma}\boldsymbol{F}^t\left(\boldsymbol{F}\boldsymbol{\Sigma}\boldsymbol{F}^t + \boldsymbol{T}\right)^{-1}\boldsymbol{F}\boldsymbol{\Sigma}\right). \tag{5.83}$$

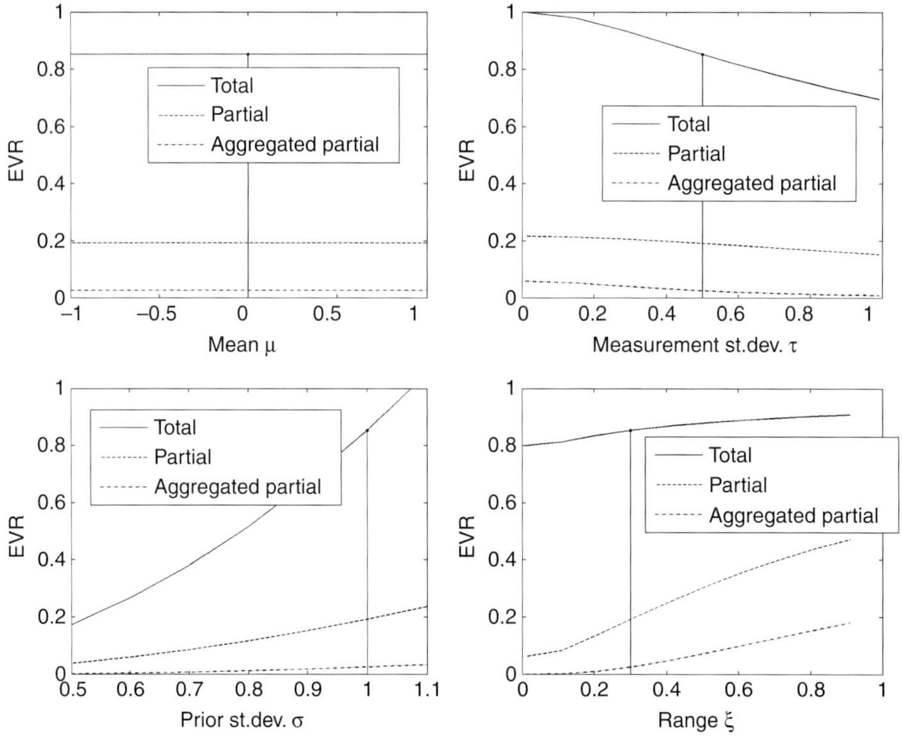

Figure 5.21 Forestry example: the expected reduction in variance for three different spatial designs of experiment. The results are plotted as a function of the various input parameters for the Gaussian spatial model (first axes).

In Figure 5.21, we plot the variance reduction for the three acquisition schemes as a function of the model parameters. As expected, we see that the variance reduction does not depend on the mean (top left), while it decreases as a function of the measurement noise (top right) since more accurate measurements would tend to pin down the distinctions of interest more than an inaccurate test. We also see that the EVR increases as a function of the prior noise level (bottom left) and the spatial correlation range parameter (bottom right). When there is more prior noise, the data are more informative, and it is relatively easier to make a large reduction. For a univariate model, the reduction in variance is $\sigma^4 / \left(\sigma^2 + \tau^2\right)$, where the increasing trend with σ^2 is apparent. The large spatial correlation means that the variables are more connected, and the data would have more ability to reduce overall uncertainty. The increase of EVR with the spatial range parameter is very pronounced for the partial tests compared with the decrease or increase as a function of the mean, measurement noise, or prior noise parameters. A partial test becomes relatively more informative when there is a stronger correlation. There is not much point in testing cells adjacent to each other with high correlation, as the two cells have very similar properties.

5.9.3 Prediction error

The prediction error is commonly studied for discrete models – see, e.g., Peyrard et al. (2013) and Lilleborge et al. (2015). The intent is that data collection on average should pull predictive probabilities closer to 1 or 0.

Consider again a distinction of interest $\boldsymbol{x} = \left(x_1, \ldots, x_n\right)$ represented at n sites. The variables are represented on a finite discrete sample space. The prediction error is defined via classification based on the most likely outcome – i.e., for each $i = 1,\ldots,n$, a class $k^* = \arg\max_k \left\{p\left(x_i = k\right)\right\}$ is selected. If the complement occurs, there is an error. A priori, before collecting any data, the prediction error is:

$$\mathrm{PE} = \sum_{i=1}^{n} \left(1 - \max_k \left\{p\left(x_i = k\right)\right\}\right). \tag{5.84}$$

Conditional on observing data $\boldsymbol{y}$, the expected prediction error is:

$$\mathrm{PE}\left(\boldsymbol{y}\right) = \mathrm{E}\left(\sum_{i=1}^{n} \left(1 - \max_k \left\{p\left(x_i = k \mid \boldsymbol{y}\right)\right\}\right)\right) = n - \mathrm{E}\left(\sum_{i=1}^{n} \max_k \left\{p\left(x_i = k \mid \boldsymbol{y}\right)\right\}\right), \tag{5.85}$$

where the expectation is over the data pdf $p\left(\boldsymbol{y}\right)$. Within the expectation, for each data outcome, the largest posterior probability is computed at all cells. For some data outcomes, the largest probability may be Class 1 at a particular cell; for other data outcomes, the largest may be for Class 2 at the same cell, and so on. In this way, the data would allow a change in the classification at cells, and this would be more likely if the data were collected at that cell or at a cell in its vicinity. Since the probability at a cell is constant in the prior but can change depending on data in the posterior, $\mathrm{PE}\left(\boldsymbol{y}\right) < \mathrm{PE}$. The improvement made by data collection can be compared over various experimental designs, and according to this criterion, one should select the design with the smallest expected prediction error.

The tree amigos: conservation biology example

Keywords: *conservation biology, Markov random field, prediction error*

This example revisits the conservation biology case with the selection of reserve sites for trees; see Sections 5.4 and 5.7. A Markov random field model is used for the binary distinctions of interest x_i on a 3×3 grid – i.e., $i = 1,\dots,9$.

We evaluate the expected prediction error for various interaction parameters β in the binary Markov random model for two different likelihood accuracies: $\gamma = 0.7$ and $\gamma = 0.9$. This is a case of total imperfect information – the decision maker collects binary data at all nine cells in the grid, but there is a chance that the data at a cell will give the wrong interpretation of the latent distinction of interest (in this case, whether a particular species of tree is present or not).

The prior probability is assumed to be $p(x_i = 1) = 0.5$ at each cell $i = 1,\dots,9$, so the prediction error without data is $\mathrm{PE} = \sum_{i=1}^{9}(1-0.5) = 4.5$.

We compute the expected posterior prediction error in Equation (5.85) by the forward–backward algorithm (Appendix A.4) for the marginal likelihood $p(\boldsymbol{y})$ and the conditional probabilities $p(x_i = k \mid \boldsymbol{y})$, $k = 0,1$, and $i = 1,\dots,9$.

Figure 5.22 shows the expected prediction error as a function of the spatial interaction on the first axis. Both tests reduce the prior prediction error of 4.5 by a substantial amount. The prediction error is smaller for the more accurate experiment. The difference between the more accurate tests ($\gamma = 0.9$) and the less accurate tests ($\gamma = 0.7$) is largest for small spatial interaction. Both curves decrease when the spatial interaction gets larger. When there is larger dependence,

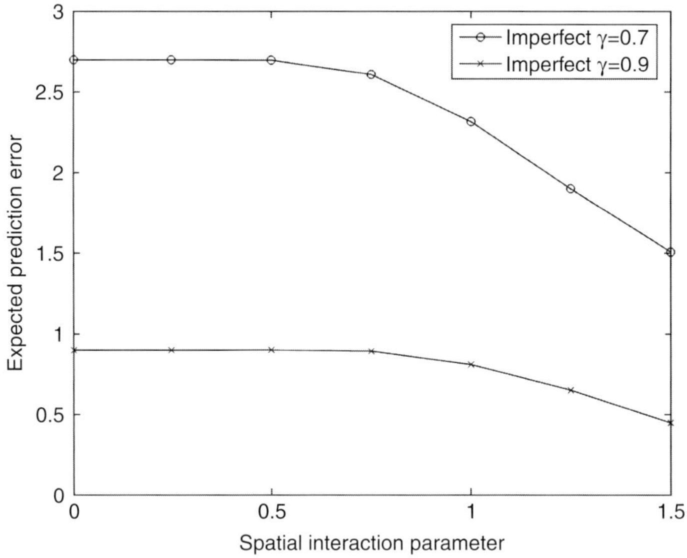

Figure 5.22 Conservation biology example: prediction error plotted as a function of the spatial interaction parameter in the Markov random field.

the latent variables are more coupled, and a test has greater impact. The decrease is more pronounced for the less accurate test, since this test gains relatively more from the dependence in the latent field. Similar tendencies were seen for the VOI analysis of this example in Section 5.4. The challenge with the prediction error criterion, just like other criteria that only explore relevance, is that it does not relate directly to the ultimate impact of the experiment on the decision.

5.10 Bibliographic notes

We present real-world applications of VOI analysis in the next chapter, so in this section, we will focus on a bibliography related to conceptual and methodological attempts at quantifying VOI in spatial decision situations.

Low decision flexibility and decoupled value

VOI analysis for spatial applications is commonly done for the simplified decision situation, which we have here referred to as low decision flexibility – i.e., few alternatives. Most of these describe a situation of "all" or "nothing" or "utilize now" or "wait." Analytical results for the Gaussian distribution were shown by Schlaiffer (1959) and Bickel (2008). Eidsvik and Ellefmo (2013) place such models in the spatial context.

Low decision flexibility and coupled value

Trainor-Guitton et al. (2011) and Trainor-Guitton et al. (2013) describe a spatial decision situation with few spatial alternatives for groundwater recharge. They carefully incorporate spatial heterogeneity to model a complex value function. Their approach requires the inversion of transient electromagnetic data, and Trainor-Guitton et al. (2011) introduce the reliability measure for interpreted geologic scenarios to simplify the inversion step.

High decision flexibility and decoupled value

For the situation with high decision flexibility, Polasky and Solow (2001) suggest a measure similar to VOI for the selection of reserve sites. The approach assumes a decoupled value giving a sum over the sites considered, but the criterion pertains more to a coverage measure rather than decisions. Eidsvik et al. (2008) and Bhattacharjya et al. (2010) describe the free selection problem as a spatial decision situation; their value computation includes no coupling. Bhattacharjya et al. (2013) outline closed-form expressions for this high decision flexibility situation under Gaussian modeling assumptions. The context is not dissimilar to that of portfolio management, where VOI has been studied, for instance, by Zan and Bickel (2013).

High decision flexibility and coupled value

When there is high spatial decision flexibility and coupling in the value function, new approaches and reliable approximations are required. In Section 5.6, we presented an approach that re-used Monte Carlo samples. Barros et al. (2014) apply a similar strategy of re-using Monte Carlo samples in a closed-loop reservoir management setting. They compute the VOI for production data using samples and treat one of these samples as the truth, while the others were used to find an optimal strategy. We believe that there is likely to be a large growth of tools and methods for high decision flexibility and coupled value decision situations in the future. We envisage that clustering techniques or sophisticated designs of experiments could be useful. For instance, multidimensional scaling – see, e.g., Scheidt and Caers (2009) for a geostatistical application – is useful for the subset selection of realistic models or realizations or clustering the sample space of high-dimensional situations. Approximate computer models – i.e., proxy models or surrogate models (Ginsbourger et al. 2013) – and multiscale models (Ferreira and Lee 2007) may also be important. New ideas around analyzing or emulating computer experiments – see, e.g., Santner et al. (2003) and Gramacy and Apley (2015) – could be applied for approximating the prior and posterior values. VOI computations lend themselves well to parallel computing, and we suspect that there will be future research along that direction.

More complex decision situations

Bickel (2008) outlines solutions to VOI analysis for a two-action situation for a decision maker with an exponential utility function. Bickel et al. (2008a) present VOI analysis for seismic data when there are multiple (spatial) drilling targets and constraints on the number of wells that can be drilled. There is a large body of literature on dynamic programing for sequential decision problems such as Markov decision processes and the policy iteration algorithm – see, e.g., Puterman (2005) and Bertsekas (2012) for extensive coverage of these topics. Even though these are very powerful tools, there has been limited use in spatial applications. Bickel and Smith (2006), Bickel et al. (2008b), and Martinelli et al. (2013a, 2014) present cases for optimal exploration among dependent prospects. Their cases were limited to rather small sizes. For problems of larger dimensions, approximate dynamic programming methods are available – see, e.g., Powell (2011) and Bertsekas (2012). At this larger scale, creative approaches for special cases are required. For instance, Brown and Smith (2013) use bounds for efficient sequential strategies for a Bayesian network model, and Martinelli and Eidsvik (2014) apply this approach for networks and Markov random field models. Frazier and Powell (2010) study VOI based on ranking of projects and sequential strategies. Krause and Guestrin (2007) present a non-myopic strategy based on entropy gain for Gaussian models, with an application to river monitoring. Alkhatib and King (2014) use a technique called least-squares Monte Carlo to approximate and solve a dynamic program around decision flexibility in reservoir production. Most of these approaches attempt to find the optimal strategy to optimize value rather than studying the VOI of experiments. Another new direction of important VOI research might be spatiotemporal models. This would require new methodologies that would likely be associated with sequential models.

Other information measures

There are several good books describing the challenges related to spatial design of experiments – see, e.g., Muller (2007) and Le and Zidek (2006). Dobbie et al. (2008) is an excellent tutorial paper on the subject. The book by Cressie and Wikle (2011) has some exposition around the use of information measures other than VOI for decision making and management in spatiotemporal models.

6
Earth sciences applications

Figure 6.1 before mining ore
it would help to test its grade
how much should we spend?

The examples presented in this chapter are based on real cases from the Earth sciences. In each case, we describe the spatial decision situation, discuss possible data-gathering schemes, present the spatial statistical modeling, and highlight the applicability of value of information (VOI) analysis. Some aspects of the examples have been introduced earlier in the book to motivate spatial decision making and for illustrating multivariate and spatial statistical modeling. Now we consider these examples from the perspective of performing a comprehensive VOI analysis. We consider these real-world data sets from across domains and apply the models and concepts of previous chapters to analyze their value to a decision maker. Projects based on these examples and corresponding data sets are presented in Chapter 7. Before presenting the cases, we start by describing a common workflow template.

6.1 Workflow

According to decision-theoretic principles, information is valuable to the decision maker only if it can influence the decision(s). Therefore, in any VOI analysis, the first step is to frame the underlying decision situation with clarity, because it is important to understand what the decision maker can do with the potential information. This is followed by assessing the different types of potential information that might be relevant. Spatial decision situations will typically involve some aspects of spatial modeling incorporating spatial dependencies due to geology and geography. Finally, all of these come together for VOI calculations and analysis. The key ingredients of this workflow are:

1. **Framing the decision situation:** For each case, one must define the underlying decision situation. It is crucial to understand the domain of application and how potential data might help in making better decisions. This process typically entails identifying the critical distinctions – i.e., the most relevant uncertainties involved and the decisions. The next stage is to understand which distinctions affect the decision maker's value. Typical questions to consider in spatial decision situations are related to the complexity of the decision situation, as we discussed in Chapter 5: what is the decision flexibility? Is the value function coupled, or can it be decoupled from one spatial site to another? Are there shared costs? Are there constraints? What are the decision maker's risk preferences, as represented by a utility function? Are sequential decisions possible? There are often several input parameters that are either known to be constant or are assumed to be fixed for convenience, as revealed in the initial discussions, such as costs and revenues. It is good practice to make a note of these and later check the sensitivity of these various input parameters on the VOI calculations.
2. **Studying the potential information-gathering schemes:** One should consider the various types of data that could be acquired and that may be relevant to the uncertain distinctions of interest. There are usually already some data or prior knowledge about the distinctions of interest. The question is which additional data to gather, if any. As was outlined in Chapter 5, the spatial data could be partial or total and could entail perfect or imperfect information. It may also be possible to perform sequential testing. One should conduct the VOI analysis for the different feasible data-gathering options.
3. **Building the spatial model:** One should specify the sample space of the uncertain distinctions of interest and the prior probability distribution. Spatial modeling typically involves statistical dependence, as discussed in Chapter 4. One must also specify the sample space of the potential data variables and a likelihood model linking the uncertain distinctions of interest to the data. When forming these models, we suggest using existing data from similar settings as well as expert knowledge. Techniques from Chapters 2 and 4 can often be applied when tuning the model parameters.

4. **Conducting VOI analysis:** At this stage, one has all the required inputs for performing VOI computations. The computations for prior and posterior values can sometimes be done exactly, or one may have to rely on approximations, as discussed in Chapter 5. The posterior values are obtained for the different data-gathering schemes, while the prior value is the same for a given decision situation, as it does not incorporate the additional data being evaluated. The VOI for the case of perfect total information is often useful as a limiting upper bound. This case is often easier to compute than the more realistic case of imperfect partial information. Sometimes the easier upper-bound calculation can short-circuit the workflow, bypassing the more difficult VOI analysis with imperfect information, by showing that even perfect complete information does not increase the posterior value sufficiently to make it worthwhile.

   The VOI should be compared with the potential price of the experiment in the final analysis. A test is worthwhile if the VOI is greater than the price of the experiment. With various alternatives for data-gathering schemes, one can compare them and identify schemes that are more valuable to purchase. This may be done by plotting the VOI under various information-gathering schemes or by displaying decision regions for various price ranges of the experiments. To better explore the possible solutions, VOI analysis should include sensitivity to a variety of input parameters or models. Rather than viewing this VOI analysis as a final conclusion, we believe that this should be used as a basis for discussion to guide decisions about information gathering. In fact, the main benefit of VOI analysis is that the decision maker can start considering various information-gathering schemes before the prices are known with any clarity to proactively consider how to improve the value from the situation.

The applications presented in this chapter are from basin-scale exploration of petroleum prospects (Section 6.2), geophysical reservoir characterization (Section 6.3), mine planning and safety (Section 6.4), and groundwater management (Section 6.5). We use a variety of spatial models in the applications. The aim is not to stress the particular models but rather to illustrate a range of possibilities. The chosen models and applications are in no way exhaustive. Using these case studies, together with the tutorial examples in Chapter 5 and the hands-on material in Chapter 7, we aim to provide motivating insights about VOI analysis for spatial situations, enabling practitioners to integrate more decision analysis and spatial modeling in their own applications.

6.2 Exploration of petroleum prospects

The examples in this section pertain to basin-scale petroleum exploration. A petroleum company is interested in selecting promising petroleum fields, which can be explored further and eventually developed to produce oil and gas. The goal is to drill successful wells at the petroleum prospects and to develop and produce the recoverable oil and gas at minimum cost. The company has established quite some knowledge about the regional geology, but there is still uncertainty about the geological properties. Since decisions about drilling wells and building infrastructure for petroleum production can be extremely expensive, it may be worthwhile to find creative ways to gather more information, potentially leading to better decisions.

At this stage of hydrocarbon exploration, the company considers gathering information through exploration drilling campaigns. The company must choose where to target their exploration wells. An offshore exploration well costs \$10–\$100 million, along with large planning costs. Then again, future revenues could be huge if an exploration well discovers hydrocarbons.

The main goal of an exploration well is, of course, to discover hydrocarbons at the drilling location. Another goal is to get information about the geological variables at the selected drilling target, and to use the results from an exploration well at one prospect to learn about the other petroleum prospects geologically related to the drilling location. This learning is possible because of common large-scale geological mechanisms that induce statistical dependence between petroleum prospects and segments. Therefore, the decision about one prospect should depend on whatever is known at other prospects. For instance, if we observe oil in an exploration well associated with Prospect A, this can boost the chances of finding oil at nearby Prospect B, because they have similar geological settings and are perhaps charged from the same source rock. The observation at A does not, however, completely remove all the uncertainty at B. Here, we will use VOI analysis to assist in the selection of exploration drilling locations. In all examples, we assume a risk-neutral decision maker and no constraints for the decision situation, and we consider the static situation instead of sequential decision making.

Next we discuss some common background aspects of the spatial modeling in the forthcoming examples. Petroleum geologists study the geological history of regions to understand the mechanisms forming the basin and petroleum systems. An important step is to use advanced computer models for numerically simulating the geological events over time. Several dynamic processes are modeled together, including sediment deposition, burial, compaction, erosion, faulting, kerogen maturation kinetics, heat flow, and multiphase fluid flow. The comprehensive text by Hantschel and Kauerauf (2009) describes the theories and workflows used in basin and petroleum system modeling. The models allow us to vary geological input parameters such as heat flow, total organic carbon content, lithology, porosity, and others and then study the associated output variables such as the generated hydrocarbons, hydrocarbon accumulations, and their spatial distributions. The motivating illustrations in Chapter 2 examined basin modeling using a simple flowchart.

An important goal of basin and petroleum system modeling is to study whether the required geological elements for a petroleum prospect are present and whether the timing of geologic processes is favorable for trapping and accumulating hydrocarbons. These elements are often summarized by a few categories. Some companies operate with three geological attributes: source, trap, and reservoir. In addition, sufficient overburden is necessary to facilitate the burial and preservation of the other elements. The source variable is related to the presence of organic-rich source rocks, their burial, and heating, causing the organic matter to be chemically transformed, ultimately generating hydrocarbons. Once generated, the oil and gas start to flow and migrate through the porous subsurface formations. It is essential to have a trap or seal to prevent the hydrocarbons from leaking out. The reservoir variable represents the reservoir facies, with good porosity and permeability to facilitate accumulation and ultimate production. All attributes – source, trap, and reservoir (along with overburden) – must be present to have a petroleum prospect "play." If an element or process is missing or a process occurs out of the required timing, the prospect is no longer viable. Since there is uncertainty, spatial and geological modeling is used to assign probabilities for the uncertainties of interest. In addition to the presence or absence of geological attributes, this modeling also includes the volume of hydrocarbons in prospects or segments of prospects under different configurations. Decision makers must use these geologic models and combine them with probabilistic models for the planning of future investments. We present various models and demonstrate VOI analysis for each example.

In Sections 6.2.1 and 6.2.2, we discuss Bayesian network (BN) models for the hydrocarbon at multiple prospects. The first case (6.2.1) is an application from the North Sea, with uncertain charge (source) attributes at several prospects. The charge variables are discrete random variables, and the data obtained from exploration wells are assumed to provide perfect information about this source attribute. We compare the VOI for different partial tests using single or multiple exploration wells. The second example (6.2.2) is based on a case used for teaching at Norwegian University of Science and Technology (NTNU), Norway. There are four geological petroleum prospects. The uncertain variables of interest are the source, reservoir, and trap elements. We use basin and petroleum system modeling runs to build and train (or learn) the BN model for all three attributes at the four prospects and to approximate the volumes associated with different outcomes. We compare the VOI of one or two exploration wells and use VOI for sequential data gathering. In Section 6.2.3, we present scenario-based models for prospect evaluation. The data set consists of 524 equally likely scenarios with revenues at 27 prospects in a region of the North Sea. Here, we take an empirical approach and work directly with the volumes or revenues derived from Monte Carlo runs (risking) of basin modeling scenarios. We compare the VOI results of this empirical model with the VOI obtained using a Gaussian approximation for the probability distribution of prospect revenues.

6.2.1 Gotta get myself connected: Bayesian network example

Keywords: *Bayesian networks, exploration wells, petroleum prospects, high decision flexibility, perfect information, partial information, decoupled value function*

Let us consider a part of the North Sea where a petroleum company has identified a number of candidate prospects. In Section 2.3, we showed a graphical representation for this situation. Here, we describe the case in detail and perform VOI analysis for exploration wells. This case has been studied for many purposes – see, e.g., Martinelli et al. (2011)

and Lilleborge et al. (2015). The idea of using BNs for the modeling of petroleum prospect dependencies is presented in van Wees et al. (2008). We assume that the risk-neutral decision maker is free to select as many prospects as are profitable – i.e., there are no constraints. Moreover, we consider the static decision situation, where the selection of prospects is done once up front without any sequential learning. The situation is similar to that described in Martinelli et al. (2011). Finding an optimal dynamic strategy is a sequential optimization problem, studied for this example in Brown and Smith (2013), Martinelli et al. (2013a), and Martinelli and Eidsvik (2014).

We view the highlights of this example in the context of the framework outlined in Chapter 5:

- The **exploration wells** are assumed to provide **perfect information** about the hydrocarbons at the location where they are drilled. Only one or two exploration wells are drilled – the company will not drill at every location of interest. This means that the exploration information gathering gives **partial perfect information**. We focus on static information gathering.
- Regarding the three properties of spatial decision situations:

  1. The **spatial modeling** is done using a Bayesian network model built from expert knowledge about the geology of the basin. The focus is on the uncertain geological charge variable.
  2. We assume that the decision maker is free to select profitable prospects for development, so there is **high decision flexibility**. The prospect consists of several subunits called segments, which cannot be selected independently, and this means that there are some restrictions on the flexibility in development decisions.
  3. Decisions are made at the prospect level, while exploration is done for segments associated with prospects. The **value function** is thus a sum of segment results for the prospect.

Framing the decision situation

The geologists have prior qualitative and quantitative knowledge about this region in the North Sea. There are wells in the vicinity of this region, the subsurface geometry of formations is known from seismic surveys, and the petroleum company has a reasonable understanding of the geological mechanisms involved in the petroleum system. Based on all this background information, the petroleum company has identified **13 geological prospects** that are development candidates. The petroleum company would like to develop the prospects that are profitable, but there is uncertainty around the outcome (oil, gas, or dry) at the prospects.

The background knowledge has led to the geological prospect regions being split into subunits called segments. Each prospect has between one and three associated segments, and in total there are **25 segments**. These segments are the actual oil reservoirs at known geographic coordinates, and the segments corresponding to the same prospect are close to each other. If a prospect is selected for development, the company will build infrastructure to produce from all segments for the prospect. The development decision must consider the uncertainty in the outcome at each segment.

Figure 6.2 shows a graph (BN) where the 13 prospect and 25 segment variables are represented as nodes. This graph summarizes the qualitative geological knowledge relevant for this decision situation, which is encoded in a network of nodes.

There is uncertainty around the geological outcomes at each node in the graph. In this region, the geologists are sure about the existence of a trap. They are also convinced that there are sufficiently good reservoir properties. This information about the trap and reservoir attributes is based on the regional knowledge about the geology obtained from wells and seismic data. The key remaining uncertainty is the **charge variable** – i.e., whether migration of generated hydrocarbons from the source rock kitchens have been effective enough to reach and fill the reservoir segments. The states are binary: are the segments dry (brine filled) or filled with oil or gas? This uncertain variable is the distinction of interest, denoted $\boldsymbol{x} = (x_1, \ldots, x_n)$, where $n = 25$ is the number of unknown segment variables.

Decisions are made at the prospect level – i.e., at nodes denoted P1,...,P13. Since development would occur at the prospect regions, the alternatives are defined $a_j \in \{0,1\}$, $j = 1,\ldots,13$, where the two alternatives per prospect j are whether to develop ($a_j = 1$) or not ($a_j = 0$). If a prospect is developed, there is a large known initial cost, and uncertain revenues result from production at the selected segments attached to each prospect. There is a fixed segment-specific cost associated with this production from a segment. The value function decouples to depend only on the segment variables associated with the selected prospects.

We use a value function that depends on the known revenues associated with finding oil and gas at the prospects and the costs of drilling and production of petroleum. We will discuss the value function in greater detail later.

Information gathering

In order to make better decisions about the possible development at prospects, an oil and gas company drills exploration wells at selected segments. Although this is not nearly as expensive as a development plan, an exploration well has a high price: about \$300 million in this area when the planning and segment or prospect evaluation is included. It is therefore important for a company to find a valuable information-gathering scheme – i.e., decide where to place the exploration wells. The evaluation includes several aspects, but here we will focus on a generic exploration well giving perfect information about the uncertainty at the segment where it is drilled. The well is assumed to perfectly discover whether that particular segment is brine, oil, or gas filled.

The company considers drilling one or two exploration wells, which corresponds to a partial testing scheme that selects only a subset of the segments. Using the subset notation from previous chapters, the information is denoted by $x_{\mathbb{K}}$ when we have one exploration well at segment $\mathbb{K}$, while we have information $\boldsymbol{x}_{\mathbb{K}} = (x_{\mathbb{K}1}, x_{\mathbb{K}2})$ when we drill two exploration wells.

We will assume static information gathering – i.e., when two exploration wells are drilled, the decision maker is not in a position to stop testing once the outcome of the first exploration well has been observed. The information-gathering alternatives are to drill one exploration well, two exploration wells, or none. VOI is used to compare the sets of two exploration wells. Note that one could also consider other data-gathering schemes in practice. In some exploration situations, it could be useful to purchase detailed seismic processing at a prospect. In other situations, it could be useful to pay to learn the outcome of some key underlying geological variables (such as source, reservoir, or trap).

Modeling

The probabilistic model is based on the BN shown in Figure 6.2. This display shows a number of edges between the nodes in the BN that describe the dependence structure in the model. The **network** edges probabilistically mimic the geological mechanisms associated with the migration of the hydrocarbons from the source to charge the reservoirs. We describe this in some detail using the notion of kitchens, prospects, and segments for the nodes. Each node is assumed to have a discrete sample space with three possible states (dry, gas, or oil), representing the actual state of the charge in that node. Our network includes three possible kinds of nodes:

1. **Kitchen nodes:** We define the kitchens as areas where source rock has reached appropriate conditions of pressure and temperature to generate hydrocarbons; in our model, the kitchens are nodes that with probability 1 assume the state "gas." Kitchens are denoted by K in Figure 6.2.
2. **Prospect nodes:** We define prospects as larger-scale geological structures that may contain hydrocarbons that have been fully evaluated. The prospects are the key nodes of the network, because they define the spatial relationships and the dependence setting, which is easier to elicit from experts. Prospects are denoted by P in Figure 6.2.
3. **Segment nodes:** We define the bottom or leaf nodes as segments representing the potential exploration drilling sites associated with a certain prospect. For each prospect, there can be one or more segments that share part of the infrastructure needed for development drilling and production. In Figure 6.2, segments inherit the number of their prospect with an alphabetical listing.

The edges define a dependence structure for the nodes in a BN formulation. The edge structure and conditional probabilities are defined by expert geologists together with statisticians. Experience in the complex basin modeling of this area is the foundation for the edge structure. For instance, in Figure 6.2, the prospect $\{P4\}$ has just one directed edge from the kitchen $\{K2\}$, because there is a strong prior belief indicative of a unique kitchen for this prospect. On the other hand, prospect $\{P1\}$ has directed edges from two kitchens ($\{K1\}$ and $\{K3\}$). The experts in the local geology assume that there is the possibility of migration pathways from any of these kitchens to provide hydrocarbons in prospect $\{P1\}$. A BN model makes it relatively easy to ask conditional "what if?" questions, which helps in the elicitation of conditional probabilities, along with the expert knowledge about the geology. We make a few assumptions about the geological mechanisms. For instance, dry nodes are assumed to propagate dry outcomes to the children nodes. This occurs because the kitchens generated the hydrocarbons, which then migrate to prospect areas and finally

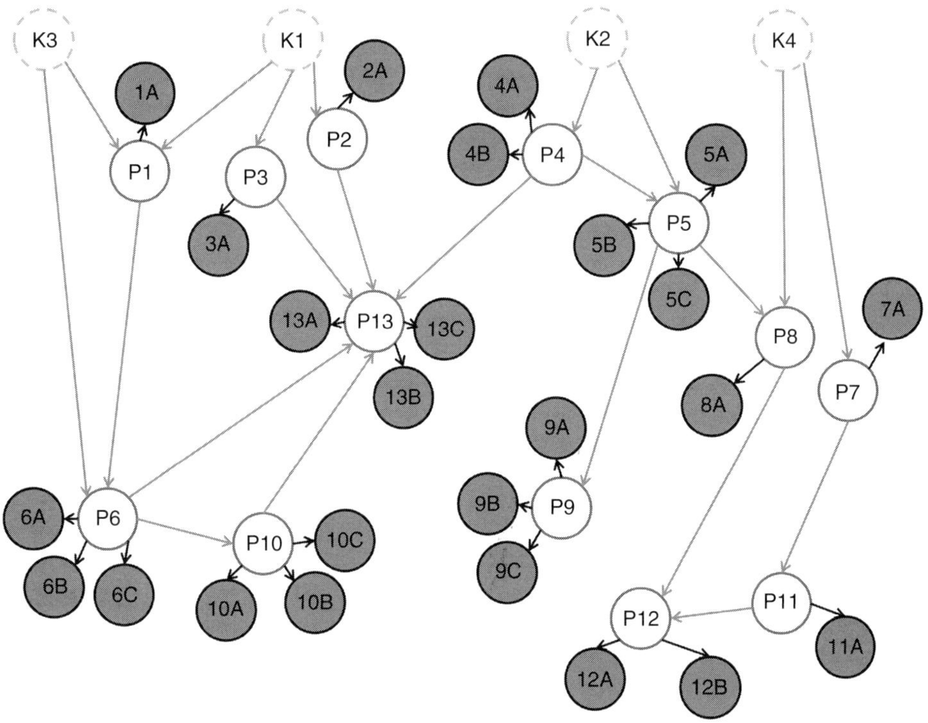

Figure 6.2 Bayesian network example: Bayesian network model for 4 kitchens (marked K1–K4), 13 prospects (marked P1–P13), and 25 segments (marked 1A–13C) in the North Sea.

to segments. If there are no traces of oil in the parents, the children nodes will for sure be dry. In the opposite situation, when a parent node contains oil or gas, there is still a chance that the child node may be dry because the migration failed.

The distinction of interest is $\boldsymbol{x} = (x_1, \ldots, x_{25})$, which is the uncertain charge variables at the segment nodes. These are the variables at nodes numbered 1A,...,13C in Figure 6.2. The nodes can have three discrete states, $x_j \in \{1,2,3\}$, $j = 1, \ldots, 25$, where the three discrete states are dry ($x_j = 1$), gas, ($x_j = 2$), or oil ($x_j = 3$) at segment j. The prospect nodes are also uncertain, but they are only included to realistically describe the probability density function (pdf) of the real distinctions of interest, which are the segment nodes. The kitchen nodes are assumed to be fixed in the subsequent analysis.

We show the marginal probabilities of all segments in Figure 6.3. Note that there is a large variability in the probabilities for the states "oil," "gas," and "dry" for the different segments. This holds true even for the segments tied to the same prospects because of the local failure rates between prospects and segments. Nevertheless, there is some similarity in the marginal probabilities at the prospect level – for instance, at Segments 5A–C tied to Prospect P5 and Segments 12A–B tied to Prospect P12.

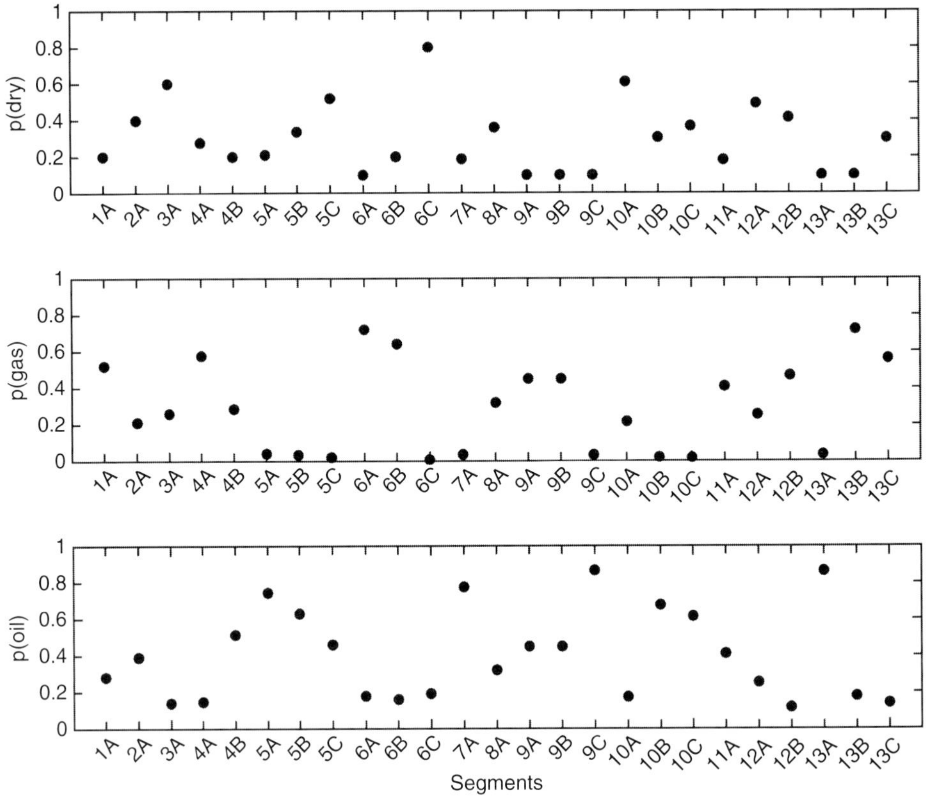

Figure 6.3 Bayesian network example: marginal probabilities of states at the 25 segments.

In Figure 6.4, we plot the conditional probabilities, having observed different outcomes at Segment 13B. In this display, the identical markers sum to 1 for each segment. For instance, the cross markers represent probabilities when Segment 13B is dry. For Segment 10C in the top display, the cross marker is at 0.57, while the cross marker for Segment 10C in the middle display is 0.01 and the cross marker for Segment 10C in the bottom display equals 0.42. These probabilities sum to 1, and they indicate that dry (probability of 0.57) is the most likely outcome for the state of 10C when Segment 13B is dry. The marginal probability of Segment 10C being dry, when one does not know the outcome at Segment 13B, is only 0.37. This means that the observation at Segment 13B has an effect on the probabilities at 10C. This effect of conditioning at 13B is most visible at segments tied to P10, P6, P4, P3, P2, and P1. This comes as no surprise since Prospect P13 is rather connected to these prospects in the network edge structure. For some other prospects, the effect on conditioning is very small. When we observe 13B to be dry, the chance of having a dry P13 node is increased, and this results in increased probabilities of 13A and 13C being dry. When we observe oil or gas at 13B, we know that P13 for sure contains hydrocarbons, and

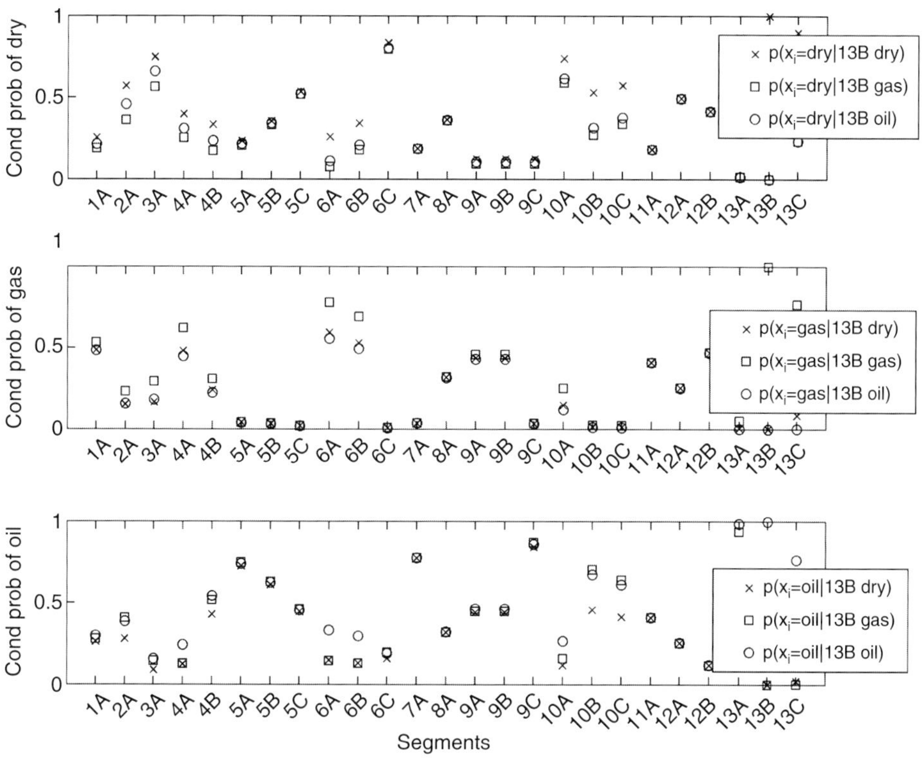

Figure 6.4 Bayesian network example: conditional probabilities of dry (top), gas (middle), and oil (bottom) at the 25 segments, conditional on different information at Node 13B.

this increases the gas and oil probability. The distinction between gas and oil at 13A and 13C appears to be largely influenced by the local failure probability. The computations of marginal and conditional probabilities are done through the junction tree algorithm (see Appendix A.5).

VOI analysis

The decision maker's value depends, of course, on the development decision at the prospects. We refer to the prospect-related cost of establishing infrastructure and production as the development fixed cost (DFC), which is assumed to be the same for all prospects. The prior value is

$$PV = \sum_{r=1}^{13} \max\left\{0, \sum_{i \in \Pr} IV(x_i) - DFC\right\}, \tag{6.1}$$

where the sum inside the maximum is over all segments belonging to prospect Pr. The intrinsic values are defined mathematically as

$$IV(x_i) = \sum_{k=1}^{3} \left(\text{Rev}_{i,k}\, p(x_i = k) - \text{Cost}_{i,k}\, p(x_i = k) \right) - \text{Cost}_{i,0}, \tag{6.2}$$

where the fixed expected revenues of oil $\text{Rev}_{i,3}$ and gas $\text{Rev}_{i,2}$ are associated with the volume, recovery rates, and market price of oil and gas, while the revenue of the dry case $\text{Rev}_{i,1} = 0$. Here, $\text{Cost}_{i,2}$ and $\text{Cost}_{i,3}$ represent proportionality factors of the produced oil and gas, while again $\text{Cost}_{i,1} = 0$. The fixed segment cost denoted $\text{Cost}_{i,0}$ is that of drilling and production at the segment level.

Figure 6.5 shows the intrinsic values in Equation (6.2) for each segment. These are the expected profits at the segment level, ignoring the shared costs at the prospect level.

From the display, we see a large variability in the intrinsic value. The largest intrinsic value is at Segment 10B. Segments for Prospects P2, P3, P4, P8, and P9 have negative or very small intrinsic values. They are not likely to be development targets, unless there is a clear indication that a neighboring prospect or segment contains hydrocarbons.

The posterior value for single-segment exploration wells is shown next. Assume that the company considers drilling an exploration well at any one of the 25 segments. The posterior value of perfect information at a single segment $\mathbb{K} = 1,\ldots,25$ is

$$PoV(x_{\mathbb{K}}) = \sum_{l=1}^{3} \sum_{r=1}^{13} \max \left\{ 0, \sum_{i \in \text{Pr}} IV(x_i \mid x_{\mathbb{K}} = l) - DFC \right\} p(x_{\mathbb{K}} = l), \tag{6.3}$$

where the conditional intrinsic values are defined as in Equation (6.2), but now using the conditional probabilities $p(x_i = k \mid x_{\mathbb{K}} = l)$ rather than the marginal $p(x_i = k)$.

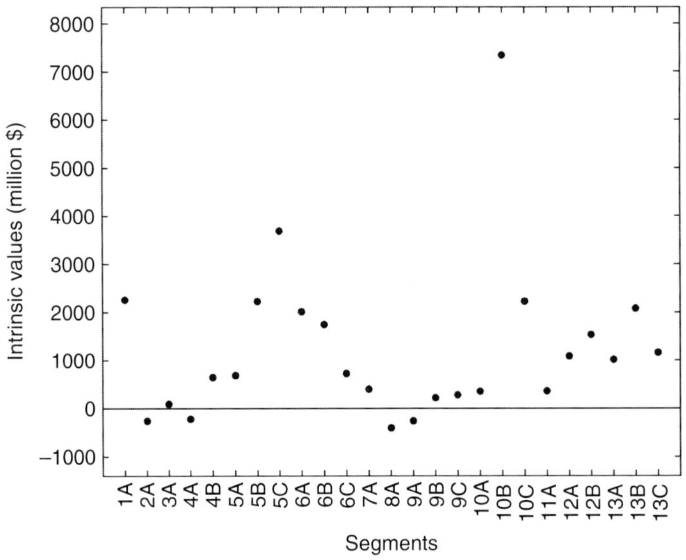

Figure 6.5 Bayesian network example: intrinsic values are the marginal expected values for the 25 segments.

For the case when two exploration wells can be drilled, the posterior value is

$$PoV(\boldsymbol{x}_{\mathbb{K}}) = \sum_{\boldsymbol{x}_{\mathbb{K}}} \sum_{r=1}^{13} \max\left\{0, \sum_{i \in \Pr} IV(x_i \mid \boldsymbol{x}_{\mathbb{K}}) - DFC\right\} p(\boldsymbol{x}_{\mathbb{K}}), \tag{6.4}$$

where the intrinsic values inside the maximization are now based on the conditional probabilities given the outcome of two exploration wells – i.e., $p(x_i = k \mid \boldsymbol{x}_{\mathbb{K}})$.

The VOI is the difference between the posterior values in Equations (6.3) or (6.4) and the prior value in Equation (6.1). Let us compare the VOI for perfect information at the different segment nodes for a range of costs.

In Figure 6.6, we show the VOI of perfect information at a single segment when the DFC is fixed at $1000 million. The VOI is plotted as a function of the 25 segments.

The most valuable node is 12B, followed by 12A and 10B. Therefore, the petroleum company should drill the exploration well at 12B if it is going to drill one well during a campaign. The VOI is rather large (about $700 million), which is clearly higher than the cost of an exploration well, so this information is worth the price. We see little association between the intrinsic values and the VOI results. The largest intrinsic value was for 10B, which also has a large VOI, but not as high as for the segments associated with P12. The tendency is that information is most valuable at prospects of great prior uncertainty, and the data can influence the decision toward or away from development. For the P12 segments, we see that the intrinsic values are near $1000 million, which is the DFC here. Similarly, for P1, P6, and

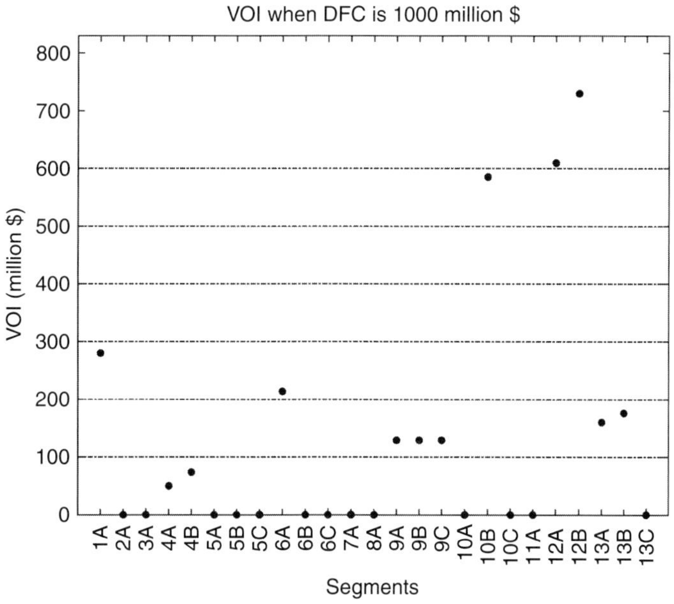

Figure 6.6 Bayesian network example: VOI of a single exploration well at a segment. The segments with perfect information gathering are plotted on the first axis.

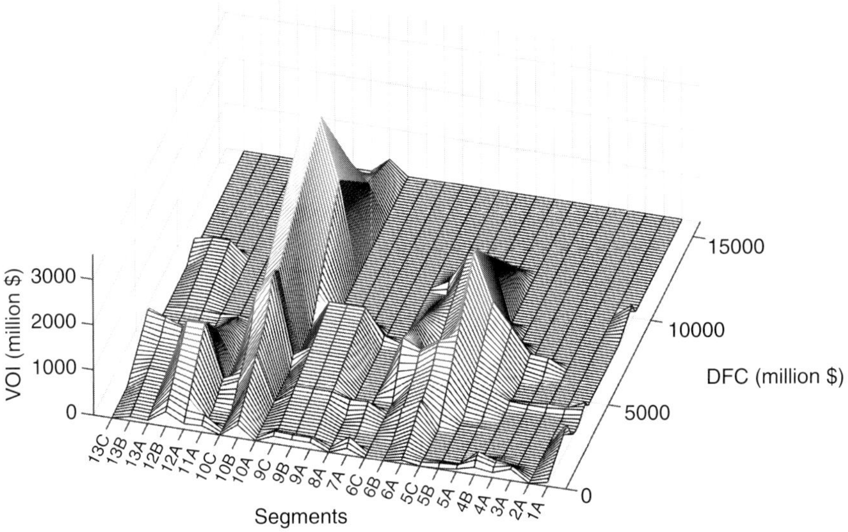

Figure 6.7 Bayesian network example: VOI of a single exploration well at a segment. The VOI is displayed on a grid of DFC costs and for the 25 segments.

P13, we observe positive VOI because the segment (and prospect) values are close to the DFC level, and a local exploration well will have impact. The VOI is not always easy to interpret, because there is dependence in the BN model and there are shared costs at the prospects.

Figure 6.7 shows the VOI of perfect information at single segments as a function of the segment number and as a function of increasing DFC. The VOI clearly depends on the DFC. A very small DFC means that most prospects are lucrative, and they can be developed without exploring any further. As a result, the VOI is small. Similarly, a very large DFC means that most prospects are too expensive and that development should be avoided. Data are unlikely to affect this decision, so the VOI is again small. For intermediate DFC levels, the VOI can be large because data have a large impact. The VOI plots further display similarities in segments belonging to the same prospect. For instance, the segments for P9 are very aligned in terms of VOI, and so are the ones tied to P13. When there is information at one segment, this binds up the probabilities for that prospect. Moreover, there are the shared prospect costs. We further note that the VOI is multimodal – it has local maxima when the information at that segment is likely to influence the decision not only at its own prospect but also at other prospects because of the dependence in the BN model. This effect kicks in at different DFCs – hence the multiple modes. For Segment 1A, there are three clear modes: one for DFC near \$200 million, another for \$4000 million, and one for \$10 000 million. The largest VOI by far shown in this display is for Segment 10B, which was also seen to have the largest intrinsic value. When the DFC is at a level where the decision maker is indifferent about P10, it will be very valuable to acquire the single-segment information at Segment 10B.

Let us now consider VOI for two exploration wells drilled at different segments. This is done by computing the value conditional on the perfect information at two wells and then marginalizing over this information in an outer loop. Figure 6.8 shows the VOI for perfect information at two segments for fixed DFC equal to \$1000 million. We assume that Segment 12B is always explored, since it was the most valuable in the single well assessment, and check which of the other wells to include with Segment 12B.

The largest VOI result in Figure 6.8 is for Prospects 12B and 10B. Therefore, Segment 12A, which had the second-largest VOI in Figure 6.6, should not be explored together with 12B. This is not surprising, as they are both for Prospect P12, and hence 12A provides very limited additional information over 12B.

The solid horizontal line in Figure 6.8 represents the VOI of a single exploration well at 12B. The VOI of an additional exploration well (together with 12B) must be larger than or equal to this level. In this example, the VOI results of perfect information at two wells are almost the same as the sum of the VOI at two single wells. For instance, the VOI of Segment 6A was around \$200 million in Figure 6.6, and when we compute the VOI of 6A and 12B together, we see that it is about \$200 million larger than that of 12B alone (two dashed horizontal lines up from the solid horizontal line in Figure 6.8). This additional effect could be useful to study in practice. Here, it holds since 12B is isolated from most of the other segments. The segments attached to the neighboring Prospects P7, P8, and P11 are too small to be developed for this level of DFC with or without information at 12B.

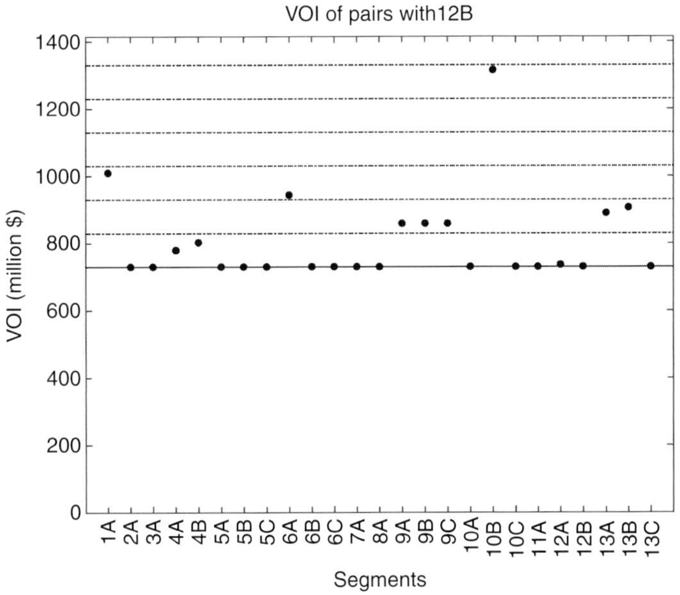

Figure 6.8 Bayesian network example: VOI of two exploration wells where one of the wells is at Segment 12B.

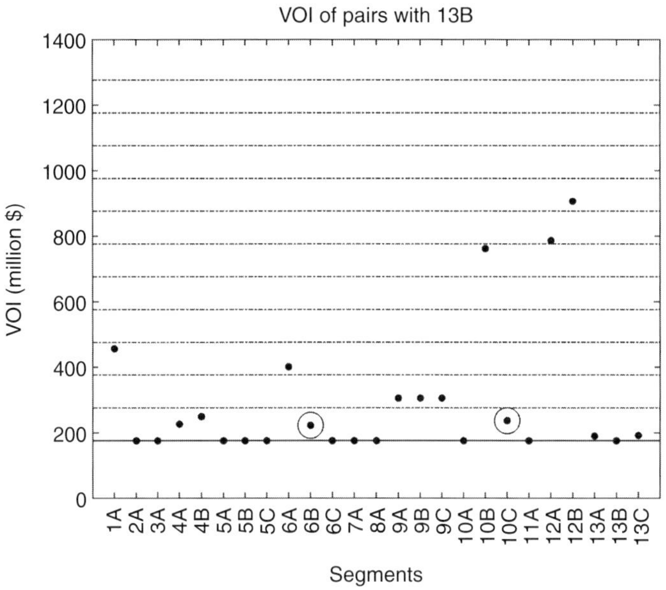

Figure 6.9 Bayesian network example: VOI of two exploration wells where one of the wells is at Segment 13B.

In Figure 6.9, we show the VOI of perfect information at two exploration wells when 13B is always included in the pair. We have included this segment just to relate the VOI results to the conditional probability illustrations in Figure 6.4 earlier.

This figure highlights the fact that the VOI results do not necessarily add up from the single exploration results in Figure 6.6. For instance, Segments 6B and 10C had zero VOI for single well exploration in Figure 6.6, but when explored together with 13B, they result in a VOI larger than 13B alone. Figure 6.4 shows how the conditional probabilities at Segments 6B and 10C are influenced by the information at Segment 13B. Considering the network structure in Figure 6.2, note that P13 is directly connected to P10 and P6. Figure 6.9 shows how this dependence between prospects and segments propagates to VOI.

6.2.2 *Basin street blues: basin modeling example*

Keywords: *Bayesian networks, basin and petroleum systems modeling, exploration wells, petroleum prospects, perfect information, partial information, sequential information*

This case study also pertains to prospect evaluation and valuing the information that can be obtained from exploration wells. The example consists of fewer prospects than in the previous one, but the modeling involves complex geological model assessment, including that of trap and reservoir properties. The baseline geological model is used for teaching purposes at NTNU, Norway. More background is provided in Tviberg (2011) and Martinelli et al. (2013b). Here, we will focus on the applicability of BNs to capture the key geological elements and on the VOI analysis.

We view the highlights of this example in the context of the framework outlined in Chapter 5:

- The **exploration wells** will be assumed to provide **perfect information** about the hydrocarbon properties at the location where they are drilled. Only one or two exploration wells are drilled; thus, there is **partial perfect information**. We mainly study static information gathering but also briefly mention sequential information gathering in this example.
- Regarding the three properties of spatial decision situations:

1. The **spatial modeling** uses outputs from **basin modeling** that are subsequently encoded as a probabilistic model by a Bayesian network for the key geological attributes **source, trap, and reservoir** at the prospects.
2. We assume that the decision maker is **free to select profitable prospects** for development, so there is **high decision flexibility**.
3. The total value is the sum of the individual values from the prospects. The value at each prospect only depends on the geological attributes, volumes, revenues, and costs at that prospect alone. Thus, we have a situation where the **value function decouples**.

Framing the decision situation

The case is a controlled basin environment where there are four identified petroleum prospects. Two of the prospects are at the eastern parts of the basin, while the other two are at the western parts. The eastern prospects are called **anticlinal prospects**. One of the prospects (called Top East (TE)) is at a shallower depth than the other (Bottom East (BE)). The western prospects are called **fault prospects**. Again, one of the prospects (called Top West (TW)) is at a shallower depth than the other (Bottom West (BW)).

The distinction of interest $\boldsymbol{x}$ is represented in an augmented space, including modeling elements for all three geological attributes – **source, trap, and reservoir –** at all four prospects. We will use terms such as success, partial success, partial failure, and failure for the prospect variables. The basin modeling simulations used to learn the inputs for a probabilistic representation of these attributes are discussed in the spatial modeling subsection.

Figure 6.10 illustrates the geographical locations of the prospect in a three-dimensional (3-D) numerical basin model. The names of the geological layers (Eek, Mlf, Mmd, and Ou) are consistent with the ones used in a course at NTNU, Norway (Tviberg 2011).

As in Section 6.2.1, the question is whether the prospects are valuable enough to justify expensive investments in planning and development. We consider the situation with a free selection of prospects. The petroleum company has alternatives $a_i \in \{0,1\}$, $i = 1,\ldots,4$, where the two alternatives are to develop ($a_i = 1$) or not ($a_i = 0$) at a prospect i. The

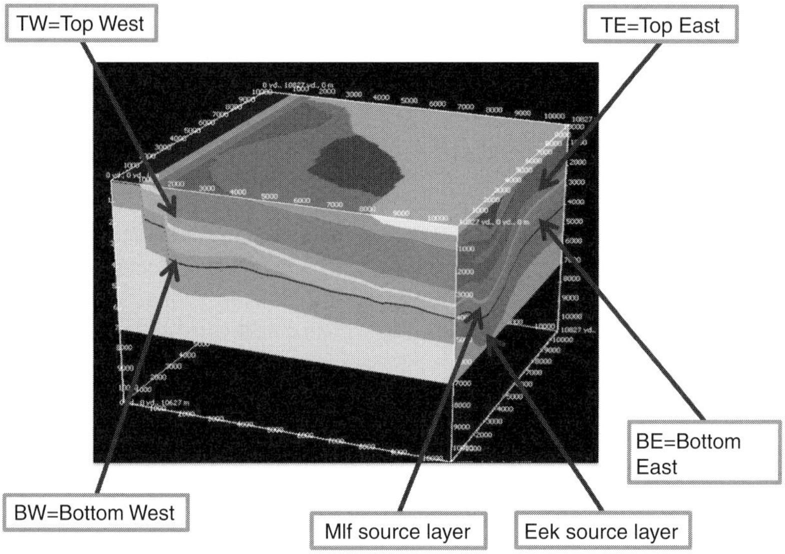

Figure 6.10 Basin modeling example: illustration of the basin under consideration and the four prospects (Top East (TE), Bottom East (BE), Top West (TW), and Bottom West (BW)).

decision maker will select individual prospects if they are profitable, and this selection is done once and for all, without opportunities for sequential decisions. We assume that the value function depends on fixed revenues or cost that vary with the uncertain discrete distinction of interest. The value function decouples to depend only on the outcome at every prospect $v(x_i, a_i)$.

Information gathering

Again, the information the petroleum company considers acquiring is that obtained from drilling exploration wells. This will provide partial perfect information about the presence of oil ("success") at the prospect where the well is drilled. In the established notation, the result of one exploration well at selected prospect $\mathbb{K} \in \{1,..,4\}$ is denoted $x_{\mathbb{K}}$, while two exploration wells that result in perfect information are denoted $\boldsymbol{x}_{\mathbb{K}} = (x_{\mathbb{K},1}, x_{\mathbb{K},2})$.

We will consider both **static and sequential information gathering**. Sequential information gathering means that the decision maker can first drill an exploration well at Prospect TE, with the capability of drilling a new exploration well at Prospect BE later. Since there is a depth ordering here, this is relevant – the exploration team could drill a second exploration well nearby without too many logistic costs. Through sequential testing, the company can choose to drill or not drill the second exploration well based on what is seen at the first well. Note that even though information gathering is sequential here, the underlying decision situation is assumed to be static – i.e., the decision maker must still make a one-shot decision regarding developing the prospects.

Modeling

We build a BN consisting of source, reservoir, and trap variables with conditional probabilities learned from multiple scenarios of basin modeling runs. Our focus will be on ways to encode this geo-modeling in a framework useful for decision making. One of the critical points in the network described in Section 6.2.1 was the substantial use of expert opinion when designing the BN. By running **basin and petroleum system modeling** (BPSM) – numerical simulations over a set of initial conditions and forcing conditions – we can integrate quantitative geology into the network construction. We train the probabilistic structure of the BN from multiple BPSM runs, and the resulting BN model incorporates the geological processes and their responses. The compression of a complex basin modeling setup into a BN allows fast computation of what-if scenarios and VOI analysis.

The resulting qualitative network is shown in Figure 6.11. Notice the three arms: one for source, one for reservoir, and one for trap. The attributes of trap, source, and reservoir that are used in the BN are not identical to the factors involved in the experimental setup for the basin modeling runs, even though there are usually connections. A variable like the heat flow interacts at both the source and trap levels. The outer (root) nodes in the BN are needed to build the probabilistic model, but we do not consider gathering information at these nodes; they are only considered to be latent variables. The BPSM simulations are used to understand the geological mechanisms underlying the structure of the BN.

BPSM is useful in exploration risk assessment. As mentioned earlier, BPSM numerically simulates the geological, physical, and chemical processes in a basin through geological time and helps to identify the critical aspects of the hydrocarbon generation, migration, and accumulation. The critical factor for sufficient hydrocarbon accumulation is the geological timing of events. BPSM allows one to quantitatively study the impact of different conditions on the accumulations of petroleum. We use established software (Petromod) to construct the BPSM simulations.

In this basin, there are two main source rocks producing hydrocarbons, the deepest being the coal bed layer denominated "Eek" and the shallowest being a shale rich in organic content denoted "Mlf." The layers above these are the reservoirs: "Ou" (bottom) and "Mmd" (top), which may contain sufficient hydrocarbons when the geological conditions are suitable. Since we have identified the anticlinal zone and the fault zone, we have the four prospects: BE, TE, BW, and TW.

The basin modeling runs are chosen based on various scenarios:

- We use three levels for the heat flow: cool, normal, or hot. It is expected that a cool basin will stay mainly in the oil window, consequently generating mostly oil, while a warm basin will reach the gas window at an earlier stage and will therefore generate more gas.
- We use two levels for the porosity of the reservoir rock: high or low.
- We use two levels for the total organic carbon (TOC) content of both source rocks, with TOC ranging from 8% (high) to 4% (low) for the Mlf black shale and from 20% (high) to 10% (low) for the Eek coal.
- We use two levels, open or close, for the presence of a main fault.

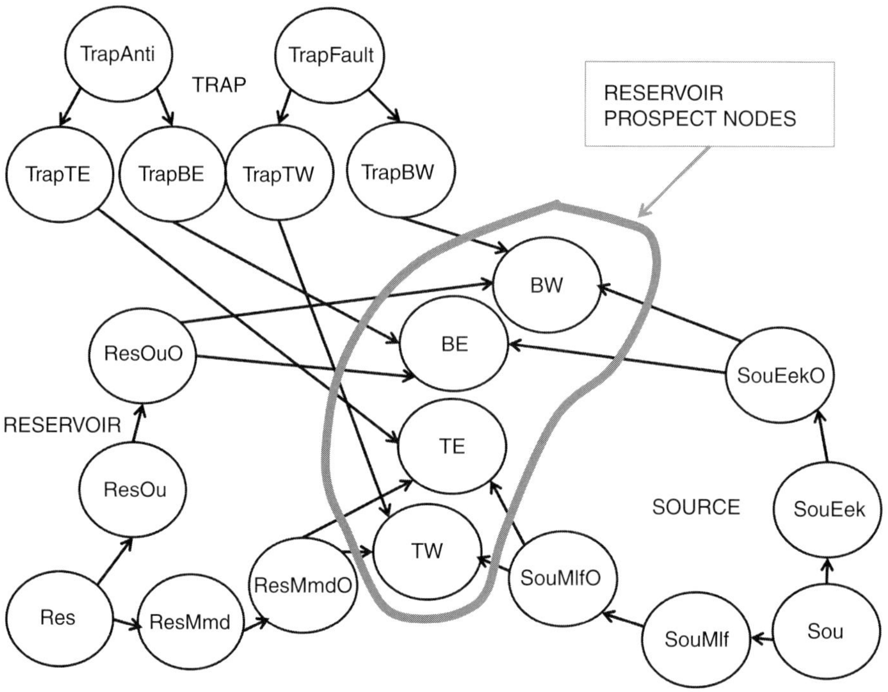

Figure 6.11 Basin modeling example: network of trap, reservoir, and source attributes. The distinctions of interest are the bottom nodes in the network. Their joint probability distribution is represented by the Bayesian network model, including more source, trap, and reservoir variables. The Bayesian network model is learned from basin and petroleum systems modeling.

Altogether, this yields 24 BPSM simulations. Figures 6.12 and 6.13 summarize the results for oil generation and accumulation.

In addition to these results, the simulations keep track of the contributions of each source rock to the hydrocarbon accumulations and the amounts of hydrocarbon that have leaked. Based on the totality of these outputs, we fit the edge structure and the conditional probability tables for the trap, source, and reservoir network (Figure 6.11).

The BN has one sub-network for each attribute (trap, source, reservoir), and we associate them to the bottom level nodes, which will be the prospect oil presence. The trap nodes are binary (on/off), and the structure is divided in two: the anticlinal zone and the fault zone. The reservoir nodes are binary (on/off), and the structure of the reservoir sub-network is tied to the Ou and Mmd reservoirs. For the source network (high/medium/low outcomes), the sub-network is based on the Eek and Mlf source rock layers, which are common for the bottom and top prospects, respectively.

The quantitative learning of the BN is partly based on the categorization of the output data and counting the number of successes/failures to estimate the conditional probability

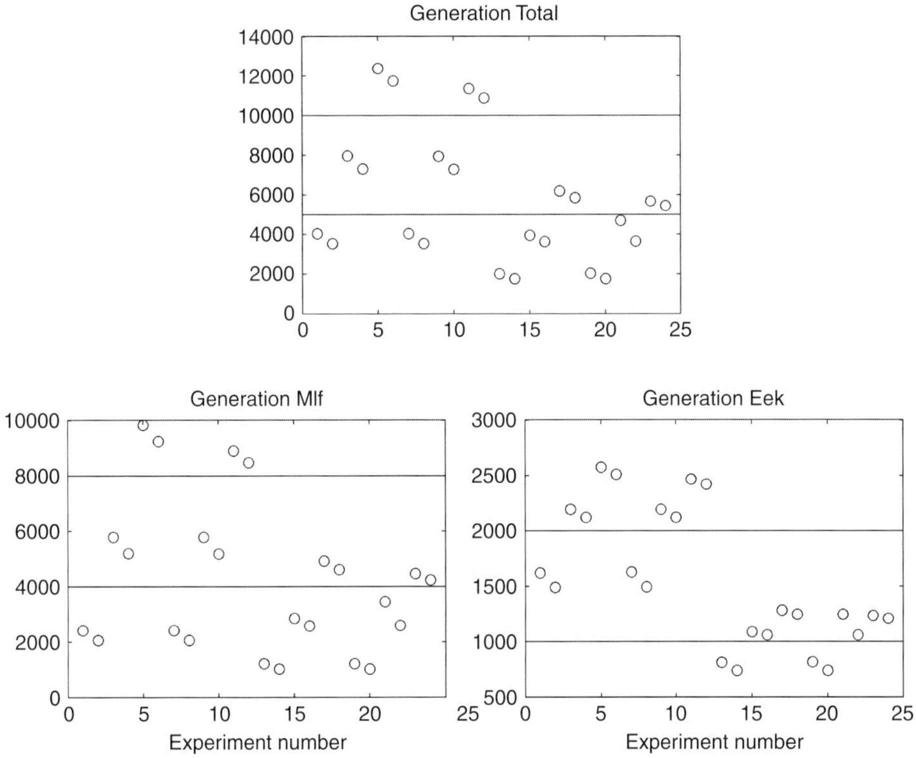

Figure 6.12 Basin modeling example: generation of hydrocarbons over many basin modeling runs. The horizontal lines indicate cluster boundaries used when building the Bayesian network model for the source attribute.

structure. We learn the probabilities at each edge in the network using various outputs from the BPSM runs. Since the nodes have discrete states, we also cluster the outputs into discrete groups. The categorization of the data clusters the 24 outputs for different variables. The horizontal lines shown in Figures 6.12 and 6.13 indicate splitting points of clusters. We use a k-means clustering algorithm with $k = 2$ (accumulation low or high) or $k = 3$ (generation low/medium/high) for categorizing the data. The reservoir network is learned from the categorized accumulation outputs. For example, we assign the probability that the reservoir root node Res is "on" = 9/24 since we see that 9 out of the 24 accumulation outputs fell in the "high accumulation" category (Figure 6.13). The conditional probabilities of success are based on the categorization of the partial accumulations in the Mmd and Ou units, as well as the ones at the total level. Similarly, the source network is learned from the generation data. For example, the marginal probability that the source Mlf is in the state "low" is assigned to be 12/24, while the source Eek is in state "low" with a probability of 4/24 (Figure 6.12). For the trap network, the anticlinal trap is always on, while the marginal probabilities are {0.5, 0.5} for the fault trap. The conditional probability

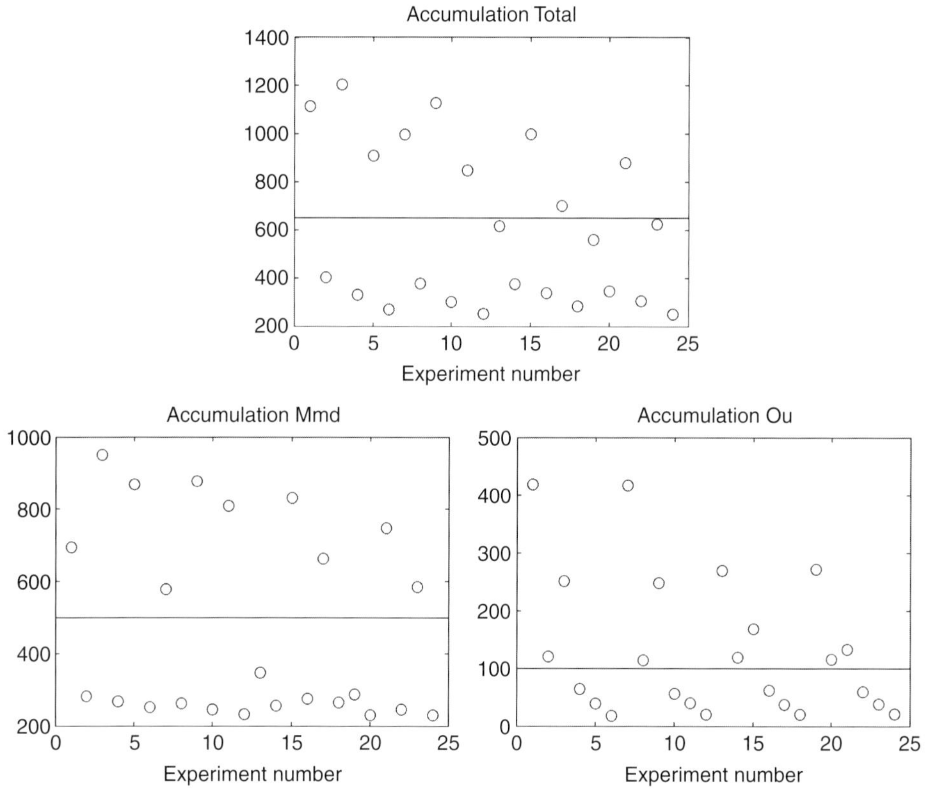

Figure 6.13 Basin modeling example: accumulation of hydrocarbons over many basin modeling runs. The horizontal lines indicate cluster boundaries used when building the Bayesian network model for the reservoir attribute.

for the children nodes includes the possibility of a local failure, quantified in the success probability of 0.9. A failure at the parent trap node will for sure result in a failure at the child node.

Finally, for the bottom nodes that are of primary interest, we summarize the possibilities for hydrocarbon accumulations broadly by assigning **four categories: success, partial success, partial failure, and failure**. These categories are defined depending on the different configurations of the discrete states for the reservoir, source, and trap nodes. We define a failure when any two of the variables fail. A partial failure occurs when the source is medium and either the reservoir or trap is on. A partial success occurs when the source is high and either the reservoir or trap is off or when the source is medium and both the reservoir and trap are on. Here, we have approximately modeled the situation that when the source is very prolific – even if there is some seal leakage or poor reservoir rock – there might still be some small accumulation available. A full success occurs when the source is high and the reservoir and trap are both on. The working assumption here is that

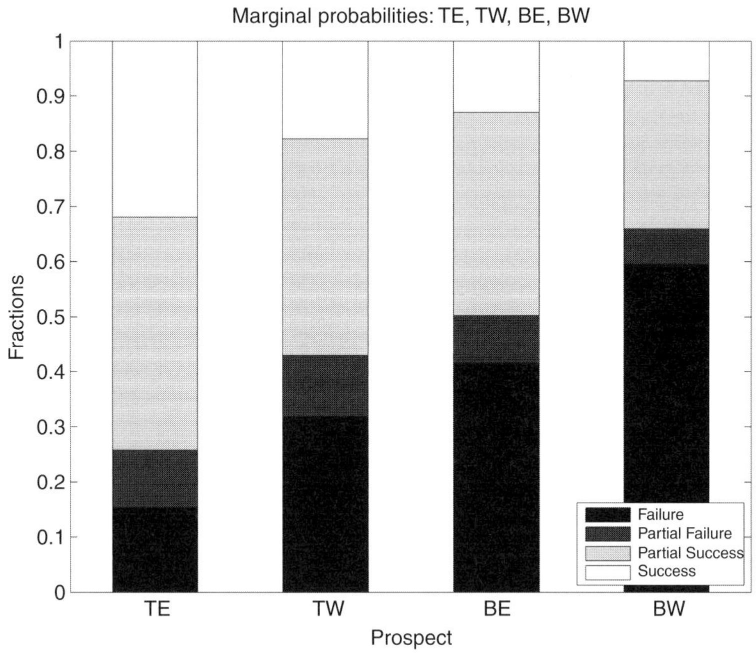

Figure 6.14 Basin modeling example: marginal probabilities of failure, partial failure, partial success, and success at the four prospects.

partial success and partial failure allow more flexibility than the simplest on/off configurations. A partial success is rarely lucrative at the exploration site but could carry very useful information about the source or trap of other prospects. These bottom-level nodes are the distinctions of interest. We will denote the outcomes at Prospects TE, TW, BE, and BW by $\boldsymbol{x} = (x_1, x_2, x_3, x_4)$, where each $x_i \in \{1,2,3,4\}$, $i = 1,...,4$. Martinelli et al. (2013b) used a more sophisticated approach with a Gaussian distribution for these bottom-level leaf nodes.

We describe the model by showing the marginal and conditional distributions computed using the junction tree algorithm (Appendix A.5). The marginal distribution for the four configurations at the bottom level is shown in Figure 6.14.

Next, we condition on Prospect BE being a failure or a success. In Figure 6.15, we see that there is a large dependence between the prospects in the event of failures. Of course, the probability is now 1 that BE is a failure, but the probability of failure is also much larger at the other three prospects. The information about the trap for the bottom unit and the migration from the Eek source rock are critical, particularly for Prospect BW.

Figure 6.16 shows the conditional probability when BE is a success. In this case, the chance of success is boosted in the other three prospects. There is now a chance of 0.7 for a full success at TE.

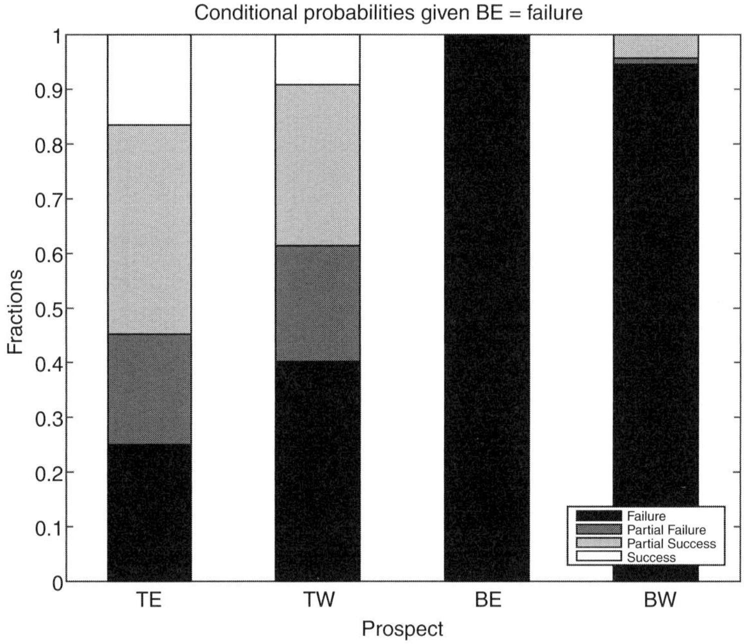

Figure 6.15 Basin modeling example: conditional probabilities of failure, partial failure, partial success, and success at the four prospects given a failure at Prospect BE.

VOI analysis

The value function under the selection of a prospect depends on the total accumulation values under the partial failure, partial success, and success configurations. We further take into account the recovery factor, which is estimated to be 0.45 for oil accumulations. The resulting revenues are summarized in Table 6.1 along with the marginal probabilities.

The VOI of exploration wells is computed by using the established BN model that has been trained with the BPSM results. We first compute the VOI for the static information-gathering scheme with one exploration well. VOI analysis can be used to guide the location of that exploration well.

From the BN model and given the revenue inputs, the prior value is

$$PV = \sum_{j \in \{\text{TE,TW,BE,BW}\}} \max\left\{0, \sum_{x=1}^{4} \text{Rev}_{j,x}\, p\left(x_j = x\right) - \text{Cost}\right\}, \tag{6.5}$$

where $\text{Rev}_{j,x}$ are the revenues associated with outcome x at prospect j, shown in Table 6.1, and Cost is that of the development and production at a prospect.

The posterior value of perfect partial information at Prospect $\mathbb{K} \in \{\text{TE,TW,BE,BW}\}$ is

$$PoV\left(x_{\mathbb{K}}\right) = \sum_{j \in \{\text{TE,TW,BE,BW}\}} \sum_{e=1}^{4} \max\left\{0, \sum_{x=1}^{4} \text{Rev}_{j,x}\, p(x_j = x \mid x_{\mathbb{K}} = e) - \text{Cost}\right\} p(x_{\mathbb{K}} = e), \tag{6.6}$$

Table 6.1. *Revenues for the four Prospects TE, TW, BE, and BW (in million barrels of oil equivalent [MMBOE]). The prospects have four possible outcomes. The recoverable volumes and the revenues grow with the level of success*

	Volume of TE	Probability of TE	Volume of TW	Probability of TW	Volume of BE	Probability of BE	Volume of BW	Probability of BW
Failure	0	0.15	0	0.32	0	0.41	0	0.59
Partial Failure	63.0	0.11	4.5	0.11	22.5	0.09	0.9	0.07
Partial Success	126.0	0.42	9.0	0.39	45.0	0.37	1.8	0.27
Success	315.0	0.32	22.5	0.18	112.5	0.13	4.5	0.07
Mean Volume	160.6		8.0		33.3		0.9	

where the decision maker may now change his or her decision depending on the evidence e at node $\mathbb{K}$. Note that when $\mathbb{K} = j$, the innermost probability becomes 0 or 1, because there is perfect information at this prospect. For the other prospects, $\mathbb{K} \neq j$, information is only obtained indirectly. Partial testing allows learning because of the statistical dependence (arising from the underlying geological dependence) between the prospects.

The conditional probability displays in Figures 6.15 and 6.16 show that information has a large effect for this network and that the petroleum company may hope to gain value by exploration – i.e., achieve large VOIs. In addition to changes in probability, the monetary values also play a role. If the cost is much smaller than the revenues, the prior value is large, and the data are unlikely to change the development decision. This is also the case when the cost is very large compared with the revenues, resulting in 0 prior value. The four prospects differ a lot in the revenues, and information at each prospect will be valuable for different cost ranges.

In Figure 6.17, we show the VOI for an exploration well acquired at each prospect for cost ranging from 0 to 200 MMBOE. The VOI varies a lot depending on the cost. Recall from Table 6.1 that the expected values of the two biggest prospects are about 33 and 160. We see that the VOI plots peak near these cost levels. For some of the curves, we also observe a small spike around Cost $\approx$ 10, which is close to the expected value for Prospect TW. The peaks occur here because at these costs, both alternatives (develop or

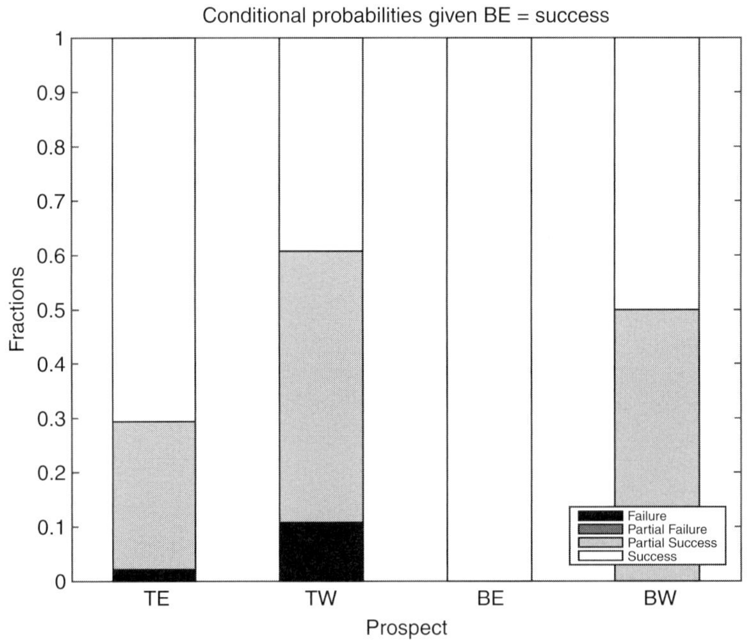

Figure 6.16 Basin modeling example: conditional probabilities of failure, partial failure, partial success, and success at the four prospects given a success at Prospect BE.

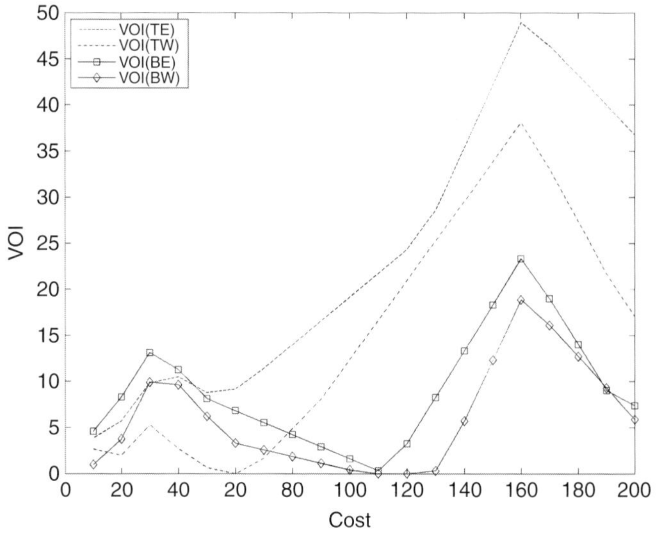

Figure 6.17 Basin modeling example: VOI for single exploration wells at the prospects.

not) have similar values; therefore, perfect information from an exploration well would be more likely to help the petroleum company make better decisions. The largest VOI is obtained when information is acquired at Prospect TE and the cost is 160. At a cost around 30, the VOI when the information is acquired at Prospect BE is smaller because the possible revenues associated with Prospect BE are smaller than for Prospect TE.

It is interesting to study the conditioning effect via Figure 6.17. When the cost is around 160, the VOI is large for observing TE, but it is also large for TW because there is a strong dependence between the two Mmd prospects. In particular, TW carries information about the source rock in Mlf and the reservoir properties in Mmd. There is less of a conditioning effect when BE or BW is observed, but an effect is still noticeable because the observation at BE would be informative about the anticlinal trap at the eastern prospects, and this is valuable for the TE prospect. Similarly, when the cost is around 30, the VOI is largest for BE but is also very large for the other Ou prospect (BW). An interpretation of the VOI results of this sort can be useful in situations with strong dependence imposed by geological processes. The processes are captured through the basin modeling simulations, while the interaction of information at different locations is captured via the BN model. Inferring the interaction effects of different observations directly from the basin modeling results would be hard. BNs are very useful for information propagation, while the BPSM simulations are very useful to learn the conditional probabilities needed by the BN model.

The VOI for any test should be compared with its price. Suppose that we express the price of an exploration well in MMBOE units. If the exploration well has a cost of 10 MMBOE and the development cost is equal to 30 MMBOE, it is optimal (more informative) to focus on Prospect BE. In this case, TE is not as informative since its high accumulation and expected revenue make it profitable anyway. If the development cost is equal to 150, it is most informative to explore TE. If the development cost is 50 MMBOE, the VOI is smaller than the price of the exploration well, and the decision maker should stick with the original choices without acquiring any additional information. Note, however, that we do not view the VOI results as a final output. Rather, plots of the VOI as a function of cost, or as a function of other input parameters, should be used as a basis for discussion. It can guide the decision maker toward looking for new and creative alternatives.

Let us now study the VOI of two exploration wells and only consider Prospects TE and BE. When gathering information at TE and BE, the petroleum company knows the bivariate outcome at these nodes. The posterior value calculation sums over all possible bivariate outcomes of the two exploration wells, denoted $\boldsymbol{x}_{\mathbb{K}} = (x_1, x_3)$. The VOI will again vary as a function of the development cost.

Figure 6.18 shows the VOI of bivariate (two-well) information gathering displayed as a solid line along with curves for the VOI of univariate (single-well) information gathering at TE or BE alone. Bivariate information seems to have no added value over univariate information gathering at TE when Cost is large. For very high development costs, Prospect TE is the only one that may be profitable, while the other prospects are too expensive. Indirect information about TE, via BE in this case, does not add anything to the information already obtained by knowing TE. This causes the overlap of VOI for TE

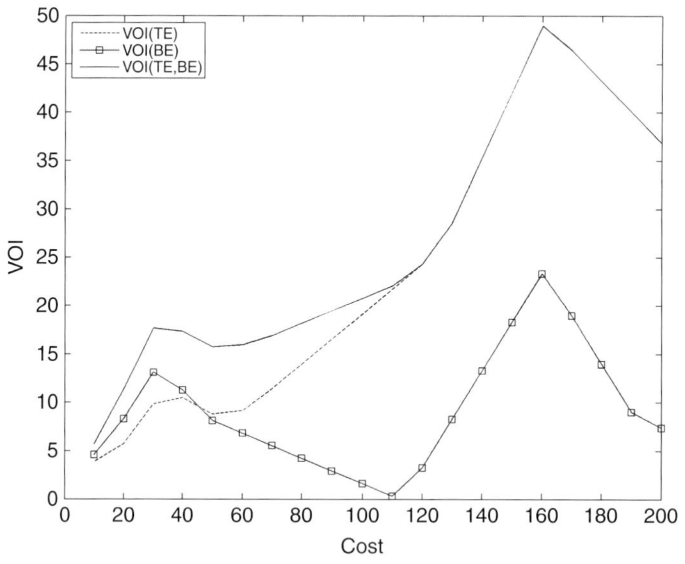

Figure 6.18 Basin modeling example: VOI for a single exploration well at Prospects TE or BE versus exploration wells at both TE and BE prospects.

and VOI for (TE, BE) for large costs. For moderate to small costs, there is added value in knowing both TE and BE since we now get valuable perfect information about two possibly lucrative prospects.

Finally, let us study **sequential information gathering** for this example. The assumption of the free selection of prospects for development remains, but with sequential testing, the petroleum company can drill the second exploration well, if it so desires, after seeing the outcome of the first exploration well. We study this situation only for the two biggest prospects – i.e., (TE, BE). Sequential information gathering was discussed earlier in Section 5.8.

The prior value is the same as in Equation (6.5), while the posterior value calculations are different. As described in Section 5.8, one now performs the first test and only then continues with the second test if its added value is larger than the price of the second experiment. In the situation with perfect information, the posterior value is

$$PoV_{seqtest}\left(x_{\mathbb{K},2} \mid x_{\mathbb{K},1}\right) = \sum_{e_1=1}^{4} \max\left\{\text{Stop}\left(x_{\mathbb{K},1}\right), \text{ContVal}\left(x_{\mathbb{K},1}\right) - P_{\mathbb{K},2}\right\} p\left(x_{\mathbb{K},1} = e_1\right), \quad (6.7)$$

where the two terms within the max term indicate two alternatives: (i) no further testing before making the decision and (ii) continue testing at the other prospect before making the decision. These are given by

$$\text{Stop}\left(x_{\mathbb{K},1}\right) = \sum_{j \in \{\text{TE,TW,BE,BW}\}} \max\left\{0, \sum_{x=1}^{4} \text{Rev}_{j,x}\, p(x_j = x \mid x_{\mathbb{K},1} = e_1) - \text{Cost}\right\}, \quad (6.8)$$

$$\text{ContVal}\left(x_{\mathbb{K},1}\right) = \sum_{e_2=1}^{4} \sum_{j \in \left\{\substack{\text{TE,TW,}\\ \text{BE,BW}}\right\}} \left(\begin{array}{l} \max\left\{0, \sum_{x=1}^{4} \text{Rev}_{j,x}\, p\left(x_j = x \mid x_{\mathbb{K},1} = e_1, x_{\mathbb{K},2} = e_2\right) - \text{Cost}\right\} \\ \cdot p\left(x_{\mathbb{K},2} = e_2 \mid x_{\mathbb{K},1} = e_1\right) \end{array} \right). \quad (6.9)$$

The idea behind sequential testing is that one can choose to stop testing – i.e., Equation (6.8) – for some outcomes of the evidence $x_{\mathbb{K},1} = e_1$, while continuing testing – i.e., Equation (6.9) – for other outcomes. In Equation (6.7), the petroleum company can stop testing if the added value of the second experiment is less than its price, denoted $P_{\mathbb{K},2}$. This flexibility results in a larger VOI than static testing at one or both prospects (the static information-gathering case was outlined in Figure 6.18). The final posterior value of sequential testing is obtained by maximizing over the experimentation sequence – i.e., whether we should start testing at $x_{\mathbb{K},1} = TE$ or $x_{\mathbb{K},1} = BE$ in our case. Note that this VOI analysis cannot be done without knowing the price of testing. The price $P_{\mathbb{K},2}$ must now take part explicitly. When static testing is done, we can first compute the VOI and then compare it with the price of the experiment. The VOI computation can therefore be done without knowing the price – this is no longer possible when studying the VOI of sequential experimentation.

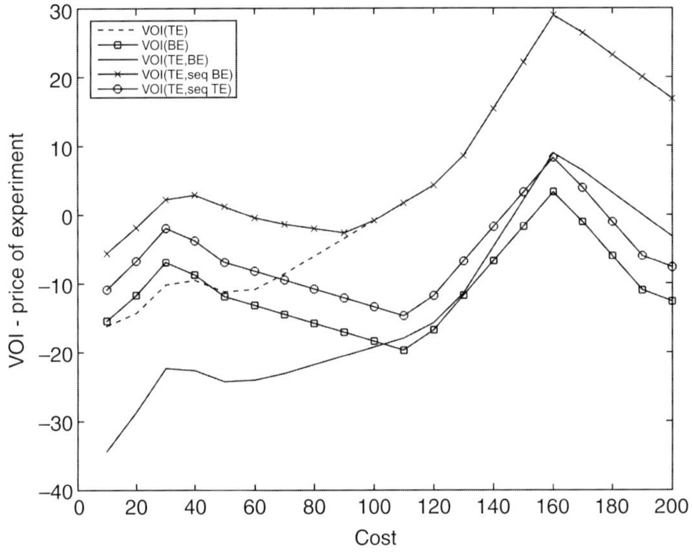

Figure 6.19 Basin modeling example: VOI for exploration well(s) at one or two prospect(s) versus the VOI of sequential testing where the decision maker can drill the second exploration well after observing the result of the first.

Figure 6.19 shows the VOI results of sequential testing where the petroleum company starts exploring at the BE or TE prospects and then can drill a second exploration well after observing the outcome of the first well.

The prices of the experiments (20 MMBOE/well) have been subtracted from the VOI results in all cases to make the plots comparable. Note that there is something to gain by sequential exploration drilling when the costs are quite small. In this case, the decision maker can sometimes avoid drilling the second exploration well because sufficient information is gained from the first well. When the costs are large, it is optimal to drill the exploration well at the large Prospect TE, and the petroleum company need not drill a subsequent exploration well at BE.

6.2.3 *Risky business: petroleum prospect risking example*

Keywords: *risking, basin and petroleum systems modeling, petroleum prospects, exploration wells, empirical modeling, Gaussian model, perfect information, partial information, high decision flexibility, decoupled value function*

Let us analyze hydrocarbon resource data from another part of the North Sea. The decisions are again about petroleum development and production of hydrocarbon prospects, and we study the information obtained from exploration wells. The prospect volumes are generated using computer models for hydrocarbon migration for a number of uncertain geological input variables. The software we build on here was developed by Sylta (2004); see also Sylta (2008) and Manoharan (2014).

In the context of the framework outlined in Chapter 5, the highlights of this example are:

- The **exploration wells** will be assumed to provide **perfect information** about the hydrocarbon profits at the location where they are drilled. Only one exploration well is drilled. Thus, the information gathering gives **partial perfect information**.
- Regarding the three properties of spatial decision situations:

1. The **spatial modeling** contains a combination of geological and statistical modeling to describe the **volumes of hydrocarbons**. The geological understanding is largely based on **basin modeling** with uncertain input parameters.
2. We assume that the decision maker is **free to select profitable prospects** for development, so there is **high decision flexibility**.
3. When we choose to develop prospects, the total value is the sum of the values of the prospects. The value at one prospect only depends on the revenues and costs at that prospect alone; therefore, the **value function decouples**.

Framing the decision situation

We consider 27 petroleum prospects from offshore Norway. The 27 prospects are a subset of a larger geological interpretation of the North Viking Graben. A prospect is profitable if its **volume of hydrocarbon** and the associated revenue exceeds the expected cost of establishing infrastructure and producing and processing the hydrocarbons. The distinctions of interest are the profits $\boldsymbol{x} = (x_1, ..., x_{27})$, defined as the difference between uncertain revenues (derived from uncertain geological risking variables and resulting uncertain hydrocarbon volumes) and fixed costs. Potential development would take place at the individual prospects, and we assume that the petroleum company has a free selection problem, where the decision maker has alternatives $a_i \in \{0,1\}$, $i = 1,\ldots,27$, representing development or not. There is no coupling in value between prospects, and the value function for the alternative of development is $v(x_i, a_i) = x_i$.

Information gathering

The information again involves an exploration well made at a single prospect. An exploration well provides perfect information about the distinctions of interest (volumes or profits in this case) at the prospect where it is drilled. As we have previously discussed, information at one location will also be informative of other prospects when there is statistical dependence. VOI analysis is useful for selecting the exploration site.

Modeling

The main uncertainties in this part of the North Sea are believed to be fault seal parameters, source rock parameters, geothermal gradients, and rock flow parameters. The local fault model was built specifically for this area by Sylta (2008).

We perform risking over these geological variables using Gaussian distributions with mean and variance based on the currently available geological knowledge. The risking results in 524 realizations of geological parameters, which are inputs to the basin modeling (migration) software. The outputs studied here are volumes at the 27 prospects for the 524 realizations. Given our focus on the prospect volumes alone, this can be considered a **non-parametric approach**. The empirical approach uses the "data" by themselves without enforcing parametric assumptions. The "data" are not field experiments but rather the volume outputs of computer experiments using geological concepts from basin modeling.

The profits are defined by the revenues with costs subtracted. We assume a cost of 1 billion to establish infrastructure and produce the hydrocarbons in the reservoir. This is subtracted from revenues corresponding to every realization of hydrocarbon volumes.

We denote these realizations of profits by x_i^b, where the subscript is an index for the prospect number $i = 1,\ldots,27$, while the superscript is the scenario $b = 1,\ldots,524$. We consider each of the 524 risking scenarios to be equally likely. Figure 6.20 shows the mean as well as the approximate 10th and 90th percentiles of profits at all prospects. Prospect 26 is clearly the most lucrative, while Prospects 1–18 and 22 have very small oil accumulations in these basin modeling runs, leading to negative expected profits. Prospects 19–21, 23–25, and 27 are near 0 profits and could potentially benefit from more information.

Figure 6.21 shows a cross-plot of the 524 realizations for Prospects 20 and 25. In Figure 6.21 (left), we show the fitted 70% Gaussian ellipse for the bivariate distribution of these two prospects. This ellipse does not seem to capture the data very well, as it focuses

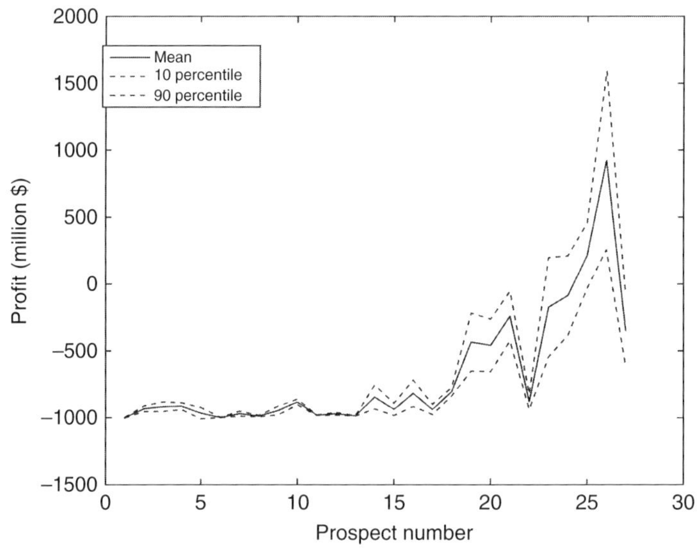

Figure 6.20 Petroleum prospect risking example: mean, 10th, and 90th percentile for profits at the 27 prospects. The volumes and profits are obtained by risking over 524 realizations of geological input parameters.

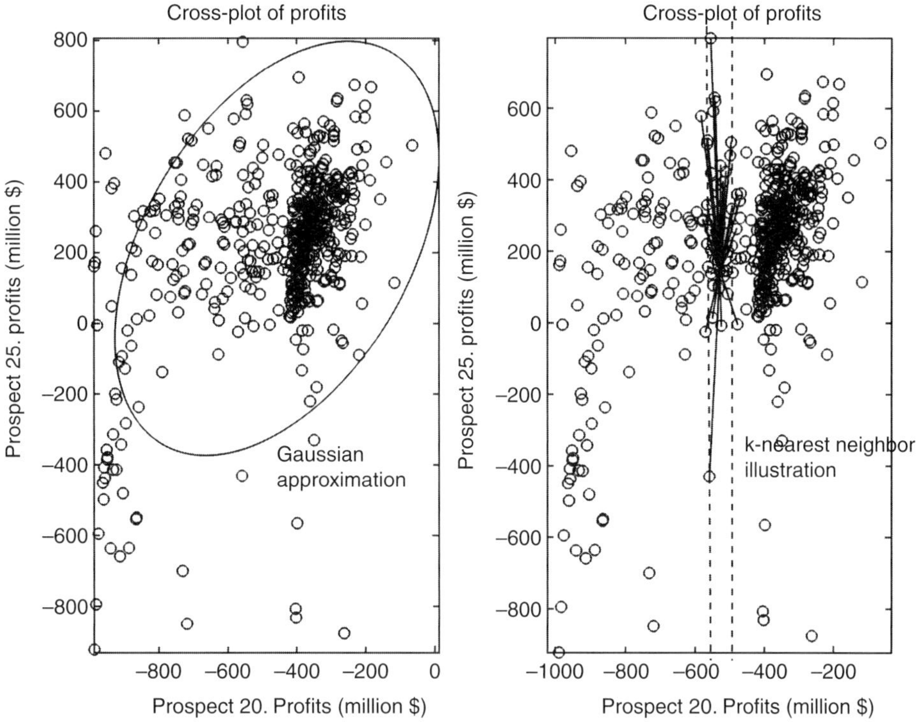

Figure 6.21 Petroleum prospect risking example: realizations of the 524 profits for Prospects 20 (first axis) and 25 (second axis). The left display shows a Gaussian approximation to the bivariate distribution. The right display illustrates the k-nearest neighbors of an observation for Prospect 20.

on the central parts of the data, ignoring a tail to the lower left with low profits. Some other distribution may be more appropriate.

Figure 6.21 (right) shows the 50 nearest neighbors of a data point for Prospect 20. This approach of k-nearest neighbors is used to approximate the conditional distributions in the VOI expression.

VOI analysis

The prior value is determined by making the development decision at every prospect without additional information. We simply use the empirical data to approximate the prior mean for each prospect. We get prior value

$$PV = \sum_{i=1}^{27} \max\left\{0, \frac{1}{524}\sum_{b=1}^{524} x_i^b\right\}. \tag{6.10}$$

For approximating the posterior value and to perform the VOI analysis, we need to condition on the outcomes of possible exploration wells. Within the completely empirical approach, an exploration well would pinpoint one of the 524 scenarios. We apply a

k-nearest neighbor approach to smooth across a few scenarios and use the outcome of profits at the neighboring scenarios to approximate the conditioning.

The posterior value of information at x_j is then computed as follows:

$$PoV(x_j) = \sum_{i=1}^{27} \frac{1}{524} \sum_{b=1}^{524} \max\left\{0, \hat{E}\left(x_i \mid x_j^b\right)\right\}, \quad \hat{E}\left(x_i \mid x_j^b\right) = \frac{1}{k} \sum_{c \in N_{b,k,j}} x_i^c, \tag{6.11}$$

where the $N_{b,k,j}$ notation means the k-nearest neighbors of realization number b, computed in the dimension defined by prospect j. For instance, Figure 6.21 (right) shows the nearest neighbor of a realization for Prospect $j = 20$ with $i = 25$. The dashed lines define the neighborhood region for realization number b. It is not symmetric; it depends on the configuration of the neighbors. The separation between the lines depends on the choice of k. If the true model was a continuous parametric pdf, a conditioning on Prospect 20 like this would cut the pdf along the vertical line defined by the conditioning profit at Prospect 20. Instead, with the non-parametric approach, we first find the k-nearest realizations of the conditioning profit (x_{20}^b in Figure 6.21 right) and then average the prospect's profits over the realizations identified by the neighborhood. From Figure 6.21 (right), this average is $\hat{E}(x_{25} \mid x_{20}^b)$. The maximum of this approximation of the conditional expectation and 0 is then averaged over all 524 realizations (inner sum over b in Equation (6.11)). Finally, we sum over all 27 prospects (outer sum over i in Equation (6.11)) because the value is decoupled. If the company considered drilling two exploration wells, we would condition on the realizations in a similar way but now perform the k-nearest neighbor calculation in the space defined by the two exploration prospects in the conditioning statement. Through double expectation, we could apply the same procedure for the prior value before making the average over j instead of the straightforward sum in Equation (6.10). The suggested approximation of the conditional expectation in Equation (6.11) is just one possible approach here – the approximation can be done in other ways, including kernels and clustering approaches (see the hands-on projects in Section 7.2).

Figure 6.22 (top) shows the VOI approximation (the posterior value minus the prior value) obtained using this k-nearest neighbor approach to conditioning on the realizations from the basin modeling runs, with $k = 50$. The VOI is large for Prospects 23–26 and quite small for the rest. A large part of the VOI is perfect information at the selected prospect, but some value is also due to information propagating to the dependent prospects.

Let us compare this non-parametric VOI result with those based on a parametric Gaussian model. This Gaussian model is simply based on taking the empirical mean and covariance of the 27 prospects from the 524 realizations. As we saw in the bivariate display earlier, this approximation did not capture all details of the distribution, but then again it may not matter since we are interested in making decisions, not in the details of the probability distributions. The highest VOIs are again achieved for Prospects 23–26, with some differences in the actual VOI values. An exploration well at Prospect 26 is now less valuable, while information at Prospect 24 is more valuable compared to the non-parametric calculations. For the prospects with expected profits below 0, the VOI is smaller for the Gaussian approximation. These effects are caused by the approximations imposed by the k-nearest

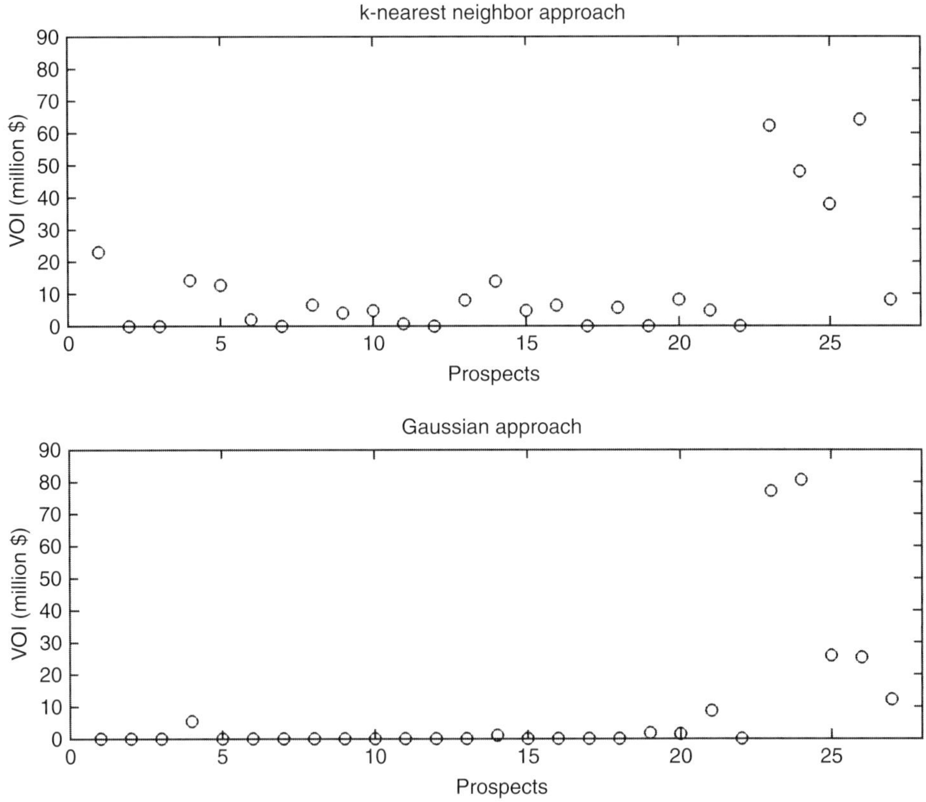

Figure 6.22 Petroleum prospect risking example: the VOI comparison of the *k*-nearest neighbor approach (top) and the results obtained by the Gaussian approximation (bottom).

neighbor method as well as the Gaussian model. For the Gaussian approximation, one may lose important information in the tail, and this is reflected in smaller (or sometimes larger) VOI. This effect holds for both the marginal effects at the prospect where we get perfect information as well as at the other dependent prospects where this information propagates.

6.3 Reservoir characterization from geophysical data

In this section, we zoom in and scrutinize a finer spatial resolution than the large basin-scale examples in the previous section. The scale is now that of a reservoir. A petroleum company must decide whether to drill wells or not at defined reservoir units. There is prior knowledge about the reservoir from previous geological and geophysical data, but there is still uncertainty in the spatial distribution of oil (and other reservoir variables), and this makes it hard to make the drilling decisions. It may be worthwhile to purchase additional geophysical information to better characterize the subsurface before making the decisions. When is it worth acquiring this additional geophysical information? Geophysical data may include seismic travel time and amplitude information, electromagnetic resistivity data, ground-penetrating radar, and other remotely sensed data.

Reflection seismic data are acquired by emitting sound waves at the surface and then recording reflected echoes from the subsurface formations to create a 3-D seismic image of the subsurface. The processing, imaging, interpretation, and inversion of reflection seismic data is a big industry. The reflected amplitudes depend on many factors, including the lithology and pore fluids. Interpretation of seismic amplitudes requires rock physics models to understand the links between the observed seismic data and the reservoir properties (lithology, fluids) of interest. This interpretation is non-unique and uncertain. Often in addition to reflections at normal incidence to the subsurface horizons, it is also useful to get information from reflection at a range of reflection angles, obtained by recording reflections at increasing distances (offset) between the source and the receivers in the seismic experiment. The additional information in amplitude variation with the reflection angle (or offset) can help to reduce the ambiguity of the rock physics interpretations for reservoir characterization. The text by Avseth et al. (2005), among others, describes the use of rock physics models for quantitative seismic interpretation. Many different attributes can be extracted from seismic data and used for reservoir characterization. For the purpose of the examples that we consider here, the decision maker is considering whether it is valuable to integrate seismic amplitude-versus-offset (AVO) data to inform the reservoir well placement decisions. The extraction of seismic AVO information entails the interpretation of amplitudes at different reflection angles from what is called pre-stack seismic data, which are much larger data sets than the post-stack (or partial-stack) data sets used for the interpretation of normal-incidence amplitudes. Seismic AVO analysis is typically more costly than just getting the reflection amplitude data at zero incidence angles (stacked seismic data). In the second example, in addition to seismic AVO data we also consider acquiring information from electromagnetic (EM) surveys. EM geophysical methods use EM waves to image the resistivity of the subsurface, usually at a much lower spatial resolution than reflection seismic imaging. An advantage of EM data is that they are quite sensitive to the presence of oil or gas accumulations since hydrocarbon-saturated rocks are much more resistive than brine-saturated rocks. However, there are uncertainties associated with the EM interpretations, just as there are uncertainties associated with the seismic interpretations. Should we acquire EM data instead of seismic AVO analysis? Perhaps we should acquire only seismic AVO and no EM data? When is it valuable to acquire both?

In both examples, we assume a risk-neutral decision maker, no constraints for the decision situation, and a static decision situation without opportunities for sequential decision making. The spatial modeling involves geostatistical models for discrete and continuous variables, using some of the methods discussed in Chapter 4. Seismic AVO data were discussed and illustrated in Chapters 2 and 4. In Section 6.3.1, the distinction of interest is the presence or absence of oil at the reservoir units, modeled by a discrete Markov random field. We compare the VOI of partial and total seismic tests. In Section 6.3.2, the modeling involves transformed Gaussian random field models for oil saturation and reservoir porosity, but then we approximate the model to work directly on reservoir unit revenues. We conduct VOI analysis for seismic and EM geophysical data used for reservoir characterization.

6.3.1 *Black gold in a white plight: reservoir characterization example*

Keywords: *reservoir characterization, drilling decisions, Markov random fields, seismic data, imperfect information, partial information, total information, high decision flexibility, decoupled value function*

Let us consider a reservoir where the top reservoir horizon has been identified. This is a two-dimensional domain in the east–north plane, but with varying depth since the top reservoir is not flat but instead follows an interpreted horizon in the subsurface. The spatial distribution of reservoir variables in this domain is relevant for drilling decisions at selected units. As relevant background material, we illustrated the Gaussian mixture model for seismic AVO data in Chapter 2 and inversion of seismic AVO data to facies classes for a similar two-dimensional spatial model in Section 4.6. The case is inspired by an oil reservoir in the North Sea, the Glitne field – see, e.g., Avseth et al. (2001), Mukerji et al. (2001), and Eidsvik et al. (2004b). We are interested in analyzing the VOI for two attributes of seismic AVO data processed along the top reservoir. This example was also discussed in Bhattacharjya et al. (2010).

We view the highlights of this example in the context of the framework outlined in Chapter 5:

- The **seismic data** provide **imperfect information** about the reservoir variables. The spatial coverage is usually large but could focus on selected spatial regions. Thus, there could be **partial or total imperfect information**. We consider static information gathering.
- Regarding the three properties of spatial decision situations:

1. We use a discrete Markov random field model to represent the reservoir facies. The focus is on defining the prior pdf for spatial **reservoir variables** and relating these to the **seismic amplitude data** through a likelihood model based on rock physics and wave physics relations between reservoir properties and seismic amplitudes.
2. We assume that the decision maker **freely selects profitable reservoir units** – i.e., there is **high decision flexibility**. The decision is whether to drill a well at a reservoir unit or not.
3. When reservoir units are selected, the total value is the sum of the values of individual reservoir units. Again, this situation is one where the **value function decouples**. The value at each unit depends only on the oil content at that unit, the volume of the unit, the oil price, and the costs of producing the unit.

Framing the decision situation

We assume that the geometry of the reservoir has been mapped out to a sufficient extent, including the cap rock formation and the depth of different reservoir units. In the following, we split the top reservoir into 100 units or cells, allocated on a regular grid of size

20 × 5. The distinction of interest is the reservoir litho-fluid facies at the site, denoted $x_i \in \{1,2,3\}$, $i = 1,\ldots,100$. The three classes are oil sand, brine sand, or shale. The lucrative outcome is an oil-filled sand unit ($x_i = 1$). At each reservoir unit, the decision is whether to drill a production well or not. Thus, we have alternatives $a_i \in \{0,1\}$, $i = 1,\ldots,N$, where $N = 100$ and the alternative $a_i = 1$ indicates the decision to drill. We assume that the decision maker can drill as many units as are profitable. The value of the drilling decision depends on the uncertain facies outcome. There is a large cost of drilling (set to 2 monetary units), but a large revenue (set to 5 monetary units) if the well discovers oil sand. When there is no oil sand at the drilling location, the petroleum company receives no revenues but must still pay the cost of drilling. The decision is based on expected profits, assuming a risk-neutral decision maker. We assume that the value function decouples between sites. As for several examples in Section 5.4 and in Section 6.2, we assume the known cost of production and revenue associated with the lucrative outcome – i.e., $v(x_i, a_i = 1) = \text{Rev} \cdot I(x_i = 1) - \text{Cost}$.

Information gathering

Potential information-gathering schemes include a **partial or total seismic test**, with two different seismic attributes. Seismic tests provide imperfect information about the reservoir facies variables of interest. In this case, the information gathering does not necessarily mean acquiring a new seismic survey; here, the two attributes require different processing and analysis of the seismic data: **post-stack versus pre-stack**. We will compare the use of just normal-incidence reflectivity (from post-stack seismic data) or both normal-incidence reflectivity and amplitude change with angle (the AVO gradient) for providing information that might help with the decision. If AVO attributes are purchased, the decision maker will have to pay for pre-stack seismic AVO processing, inversion, and analysis. VOI analysis can reveal if it is worth the price. We will further study the value of partial testing or processing only a part of the domain and compare this with the total information of the AVO attributes over the entire field.

Figure 6.23 shows the situation with a 20 × 5 grid and the location of partial tests. The grid coordinates are determined by the sailing directions and the resolution of the seismic survey, but for simplicity these coordinates may be interpreted as (east, north).

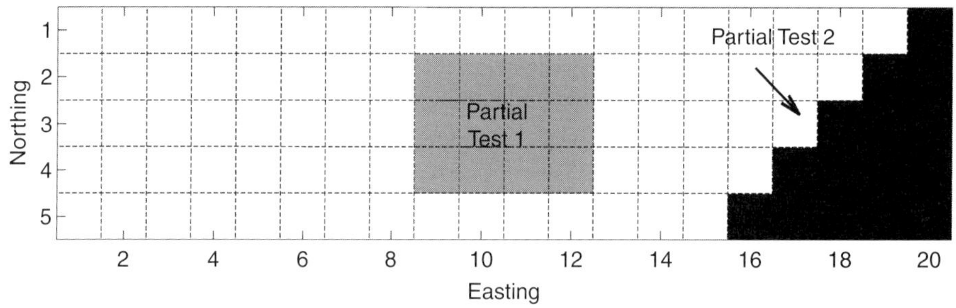

Figure 6.23 Reservoir characterization example: illustration of a grid with 20 x 5 reservoir cells where the decision maker can drill for oil. The gray and black subsets of cells indicate areas where the decision maker is considering purchasing partial seismic amplitude data.

Modeling

The **seismic AVO attributes** are informative of the facies classes of interest. The seismic AVO data at a cell $j = 1,...,m$ is denoted $\boldsymbol{y}_j = (y_{j,1}, y_{j,2})$. Note that the data at each cell can be univariate: when the processing provides only the zero-offset attributes of seismic data, the measurement at site j is $y_{j,1}$. With pre-stack processing of seismic data, both attributes (AVO intercept and gradient) are available at the cell, and we get the bivariate response $\boldsymbol{y}_j$.

To integrate the seismic AVO data for facies prediction, we must define a likelihood model. We considered a particular likelihood model for seismic AVO data in Chapter 2 (and Section 4.6) when we presented Gaussian mixtures. In general, the modeling of seismic amplitudes from facies classes can be complicated and uses elastic wave propagation theory. Reflection amplitudes depend on several aspects of the reservoir, including rock types and pore fluids, and the appropriate rock physics models connecting reservoir variables to the seismic signatures would vary from case to case. In this specific example, we first assume conditional independence, implying that we can model the likelihood unit by unit and that the reflection amplitudes only depend on the facies and fluids at that location. Second, we build the likelihood model using a graphical representation (see Section 2.3) involving porosity, density, and velocity variables in addition to the facies and the seismic AVO data. We note that the seismic amplitudes are directly related to the velocities and density (elastic properties) and only indirectly to the facies variable. However, once we have a graphical relation of variables, we can propagate deterministic physics-based relations (along with uncertainty) in the graph to get a connection between the facies and seismic AVO. Our approach is to build a likelihood model using this graphical representation of reservoir variable connections, established rock physics relations for the North Sea region, and well data in the vicinity of the reservoir.

The likelihood model for seismic AVO given only the facies variable is approximated by a Gaussian distribution where the mean value depends on the facies class and the variance is constant between classes – i.e.,

$$p(\boldsymbol{y}_i \mid x_i = k) = N(\boldsymbol{\mu}_k, \boldsymbol{T}),$$
$$\boldsymbol{\mu}_1 = (0.03, -0.21), \quad \boldsymbol{\mu}_2 = (0.08, -0.15), \quad \boldsymbol{\mu}_3 = (0.02, 0). \tag{6.12}$$

The covariance matrix $\boldsymbol{T}$ has diagonal elements 0.06^2 and 0.17^2 and an off-diagonal term including negative correlation at –0.7. Seismic AVO attributes typically have a negative correlation.

In Figure 6.24, we visualize the marginal pdfs for the zero-offset data $y_{j,1}$ (left) and the angle-dependent $y_{j,2}$ (right). Based on the mean values and the likelihood variances, there is much overlap between the classes, and a seismic AVO measurement cannot discriminate very well between the classes. The pdf for brine sand is a little different from the other two pdfs for the zero-offset seismic attribute. The response distributions for shale and oil sand are very similar for zero-offset data but differ somewhat for the AVO gradient. The oil sand shows a more negative AVO gradient. It appears that when both attributes are available, the data should clearly be more valuable because they can improve the prediction of facies classes. The underlying spatial dependence in the oil saturation will also help discriminate

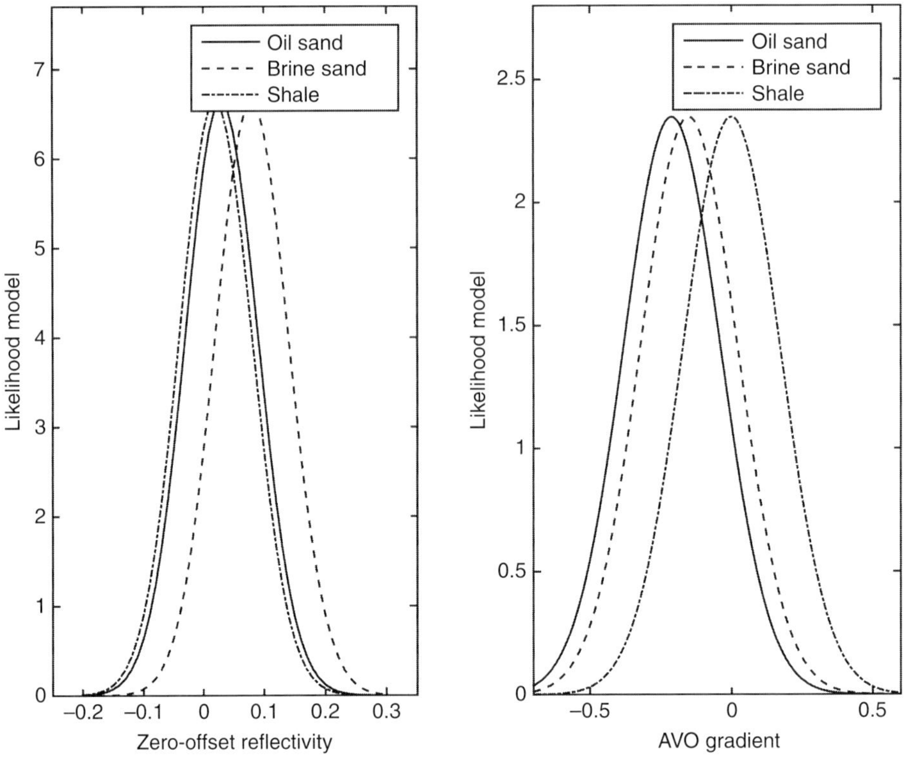

Figure 6.24 Reservoir characterization example: likelihood model for seismic AVO data. The zero-offset reflectivity (left) represents the amplitude at the top reservoir for zero incidence angles. The seismic AVO gradient (right) represents the change of amplitude as a function of the incidence angle. The likelihood model is conditional on three different facies classes (oil sand, brine sand, and shale).

between the classes, because the dependence means that data at one location contain some information of the saturation at all locations, as the facies have spatial dependency. This aspect could make partial spatial information-gathering schemes rather attractive.

The reservoir is heterogeneous, and its interpretation and facies characterization benefit from spatial modeling. Spatial dependence in the **facies variable** at the top reservoir is incorporated through the use of a first-order **Markov random field** (MRF) formulation: the Ising model as described in Chapter 4. This implies that the probability for oil in a given cell, given the outcome in the entire field, depends only on the outcome in the four neighboring cells. The joint prior pdf for the spatial facies variable is then represented by

$$p(\boldsymbol{x}) = \frac{\exp\left(\beta \sum_{i \sim j} I(x_i = x_j) + \sum_{i=1}^{N} \alpha_i(x_i)\right)}{Z}, \tag{6.13}$$

where β is the interaction parameter, while $I(A)$ is an indicator function that equals 1 if A is true and 0 otherwise. The first summation is over all pairs x_i and x_j, which are the nearest neighbors in the grid, and Z is a normalizing constant. The external field or pointwise prior function $\alpha_i(x_i)$ is a function of the outcome at every cell. Without this external field, each class would be equally likely in the prior model. With the external field, we can incorporate some of the geological knowledge about the reservoir. As an example, in this case, oil sands are more likely to occur at the shallower parts of the reservoir zones, as oil is lighter than brine. We achieve this by specifying larger $\alpha_i(x_i)$ potentials for $x_i = 1$ than for $x_i = 2$ and $x_i = 3$ when i represents a spatial location at shallow depth.

We specify β based on an analogue seismic AVO data set. Thus, we assume that the spatial interaction is about the same for similar reservoir settings. The maximum likelihood estimate (MLE) of β is computed by evaluating the marginal likelihood $p(y;\beta)$ for a set of β values. This marginal likelihood requires that we sum out all dependent uncertain variables $x_i, i = 1,\ldots,100$. This is possible by using the forward–backward algorithm (Appendix A.4) for the moderate-sized MRF model. The marginal likelihood is shown in Figure 6.25. The plot shows that the MLE for the interaction parameter is about 0.9.

Only the marginal probabilities are needed for making decisions at the reservoir units. We compute the prior marginal probabilities by the forward–backward algorithm (Appendix A.4) for MRFs, taking the interaction parameter β and the pointwise parameters

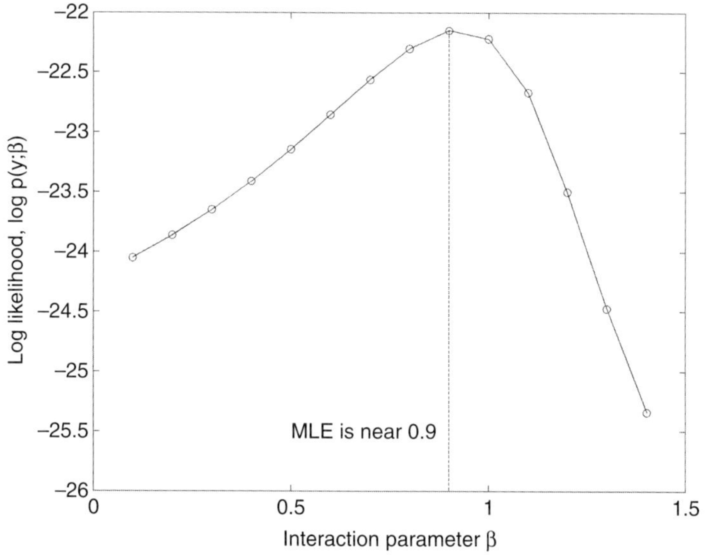

Figure 6.25 Reservoir characterization example: maximum likelihood estimation for the spatial interaction parameter in the Markov random field. The plot shows the marginal likelihood of a currently available seismic amplitude data set as a function of the interaction parameter on the first axis.

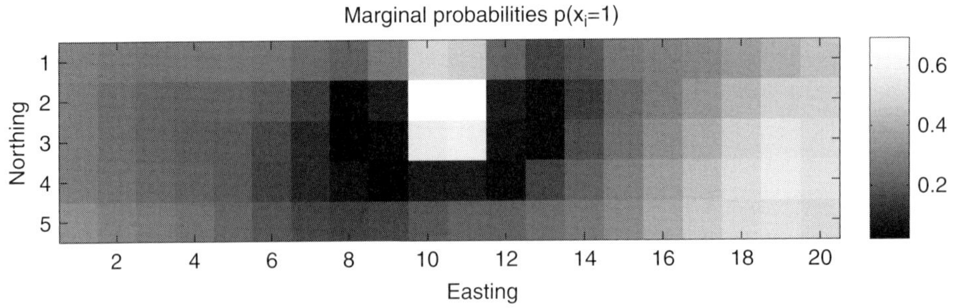

Figure 6.26 Reservoir characterization example: marginal probability of the oil sand facies at all reservoir units on the 20 × 5 grid.

$\alpha_i(x_i)$ as inputs. The prior probability of the oil sand facies is shown in Figure 6.26. We see that the northernmost central part has a high probability of oil but that there is large heterogeneity in the spatial facies classes in the grid cells.

VOI analysis

Let us now evaluate the VOI for various data attributes and test schemes. We assume fixed revenues, Rev = 5, monetary units in the event of oil and fixed drilling costs, Cost = 2, monetary units no matter the outcome. The value under the drilling alternative is the expected value based on the marginal probability of oil sands. The prior value then becomes

$$PV = \sum_{i=1}^{N} \max\left\{0, \text{Rev}\cdot p(x_i = 1) - \text{Cost}\right\}. \tag{6.14}$$

The marginal probabilities are computed by one run of a forward–backward algorithm (Appendix A.4).

If the petroleum company purchases seismic AVO data, it can make informed decisions based on the conditional probability of success at each reservoir unit. This decision may change depending on the observation. The posterior value calculation integrates over all possible data, and we have

$$PoV = \sum_{i=1}^{N} \int \max\left\{0, \text{Rev}\cdot p(x_i = 1 \mid \boldsymbol{y}) - \text{Cost}\right\} p(\boldsymbol{y})\, d\boldsymbol{y}. \tag{6.15}$$

We use Monte Carlo sampling to approximate the multidimensional integral over the data. One data set is sampled by first generating a realization $\boldsymbol{x}^b$ of the discrete facies from the MRF model $p(\boldsymbol{x})$ in Equation (6.13); next, a data set $\boldsymbol{y}^b$ is sampled from the likelihood model $p(\boldsymbol{y} \mid \boldsymbol{x}^b)$ in Equation (6.12) given the facies realization. For the inner calculation, $p(x_i = 1 \mid \boldsymbol{y})$, we use forward–backward computations. The forward–backward algorithm is also used when we sample a realization from the MRF model (Appendix A.4).

Table 6.2. *VOI results of total and partial testing as well as post-stack (zero-offset data) versus pre-stack (both zero-offset data and AVO gradient) seismic amplitude processing*

	Total Test	Partial Test 1	Partial Test 2
Zero-Offset Data	4.9	0.5	2.1
AVO Data	35.9	1.9	10.6

Table 6.2 shows the Monte Carlo approximations of VOI for total and partial tests for seismic attributes. The partial test regions are shown in Figure 6.23.

The results indicate that total testing is more valuable than partial testing but that the second partial test does surprisingly well considering that it only explores 15 out of 100 grid units. Partial Test 1 does not do so well because the test is done in an area with high chances of oil a priori; thus, the prior value is high, and the decision maker would most likely drill here in any event. The area of Partial Test 2 is more uncertain, and the expected profits at cells in that area are close to 0, indicating that seismic processing in this southeastern area could likely help in making better drilling decisions.

The value with seismic AVO attributes is much larger than that of only the zero-offset attributes. Together, the two attributes will discriminate between the facies classes much better than only one of the attributes. The VOI of a total test with both attributes is 36 monetary units. In comparison, total perfect information yields a VOI of 70. This could be important in a case where the company wishes to compare two processing schemes known to have different accuracies of the seismic likelihood. The case of perfect information is representative of seismic processing that is extremely accurate and in which there is no interpretation uncertainty. This is not possible in practice, but one could imagine spending substantial labor and money (expensive processing) to obtain better-quality data, and the VOI analysis could then be used to check whether this extensive processing work is worth the price.

6.3.2 *Reservoir dogs: seismic and electromagnetic data example*

Keywords: *reservoir characterization, drilling decisions, transformed Gaussian variables, seismic data, electromagnetic data, imperfect information, total information, high decision flexibility, decoupled value function, decision regions*

Let us again study drilling decisions at defined reservoir units at a grid of the top reservoir horizon. We are interested in comparing the VOI of seismic AVO data processed along the top reservoir with the VOI of an EM survey. Unlike the previous subsection, we will now model reservoir variables using a continuous sample space. The spatial model is for the saturation and porosity in the reservoir units, and the profits result from producing oil at the reservoir units. The modeling of spatially correlated continuous reservoir variables and VOI analysis of seismic and EM data was studied by Eidsvik et al. (2008) and Rezaie et al. (2014).

In the context of the framework outlined in Chapter 5, the highlights for this example are:

- The **seismic or electromagnetic data** provide **imperfect information** about the reservoir variables. The spatial coverage is usually large, and in this example we have **total imperfect information**. We focus on static information gathering.
- Regarding the three properties of spatial decision situations:

  1. The reservoir heterogeneity is modeled by a continuous random field. The focus is on defining the prior pdf for spatial **reservoir variables** and relating these to the **geophysical data** by a likelihood model using rock physics and wave physics relations.
  2. We assume that the decision maker **freely selects profitable reservoir units** – i.e., there is **high decision flexibility**. The decision is whether to drill a well at a reservoir unit or not.
  3. When reservoir units are selected, the total value is the sum of the values of individual reservoir units. We thus have a situation where the **value function decouples**. The value at each unit depends only on the oil content at that unit, the volume of the unit, the oil price, and the costs of producing the unit.

Framing the decision situation

We assume that the geometry of the reservoir has been mapped out from seismic travel time information. A grid of reservoir units or cells is defined at the top reservoir zone. This grid is of size $n_1 \times n_2$, where $n_1 = 25$ and $n_2 = 25$, and the total number of cells is $n = n_1 n_2 = 625$. The decision maker can freely select drilling sites. At every reservoir unit, the petroleum company decides whether to drill a well or not – i.e., alternatives are $a_i \in \{0,1\}$, $i = 1,\ldots,N$, with $N = n = 625$. The distinctions of interest, $\boldsymbol{x} = (x_1,\ldots,x_n)$, are the profits from a reservoir unit. These profits are related to porosity and oil saturation, as we describe in detail in the spatial modeling later. The value function decouples to the sum of individual values obtained from the reservoir units. At cell i, the value $v(x_i, a_i)$ only depends on the alternative taken at that reservoir unit and the uncertain distinction of interest at that cell.

Information gathering

We will study the value of reprocessing seismic AVO data and that of acquiring and processing EM resistivity data. It is possible to purchase either AVO data, EM data, or both. We will consider total testing here, but it could be possible for a petroleum company to purchase partial tests as well – for instance, EM data only along one east–west or north–south line in the grid.

We assume that prior knowledge about the reservoir includes initial post-stack seismic interpretation. The question is whether to process the angle-dependent information in the seismic amplitude data – i.e., perform **pre-stack seismic AVO** analysis. These amplitude-versus-angle data are informative of the reservoir saturation and could help in making better drilling decisions. The petroleum company can also consider purchasing an EM survey. The **EM resistivity** information is obtained by processing and inversion of the phase and amplitude data measured by sea bed logging sensors along with depth information and overburden resistivity properties. The integration of seismic data and

EM resistivity data is important for reservoir characterization. While interpretation of zero-offset seismic data can be ambiguous because of saturation or porosity effects, the combination of elastic and resistivity data may help resolve this ambiguity in the interpretation of saturation and porosity (Gomez et al. 2008). Adding these sources of information will reduce the uncertainty in the predicted profits. However, as we have mentioned several times now, reducing uncertainty alone does not necessarily make the data valuable to the decision maker.

Modeling

Our approach is based on a Bayesian model for the porosity and saturation along the top reservoir horizon and for the seismic AVO data and EM resistivity data. The saturation is two-phase brine or oil dominated. Brine saturation at a reservoir unit is denoted by $s_i \in (s_{\min}, s_{\max})$, where $s_{\min} = 0.1$ is the irreducible water saturation, while $s_{\max} = 0.9$ is associated with the residual oil saturation. The oil saturation is $1 - s_i$, and the porosity is $\phi_i \in (\phi_{\min}, \phi_{\max}) = (0.15, 0.4)$. Both saturation and porosity are spatial variables with a heterogeneous spatial distribution along the reservoir horizon. The minimum and maximum values are defined by the reservoir situation. Typically, they are calibrated from geological conditions and well and core data. The saturation and porosity variables are represented at every grid cell $i = 1, \ldots, n_1 n_2$. The prior knowledge and the available seismic interpretation are used to build the current reservoir model before considering the inclusion of the seismic AVO information or EM resistivity data.

In Figure 6.27, we summarize the current knowledge of the porosity and saturation reservoir variables, as well as the data that have been used to get to this description. The data include zero-offset seismic amplitudes (top left) and seismic travel time data (top right). The porosity and oil saturation estimates (middle row) come from a quantitative seismic reservoir characterization workflow. They are obtained by Bayesian inversion, applying a logistic Gaussian prior model for the saturation and porosity (Eidsvik et al. 2008), and a rock physics model, similar to the soft sand model used by Bachrach (2006) for relating the reservoir variables to the zero-offset seismic data at the top of the reservoir. In the prior model, the seismic travel time is used to interpret the depth of the horizon. Shorter travel times indicate shallower depths where high oil saturation is a priori more likely since oil is lighter than brine.

The displays show that the zero-offset seismic amplitudes are informative of the porosity in this case. Relatively stronger positive reflections correspond to lower porosity in this reservoir. Uncertainty levels for porosity (bottom left) are rather low as compared with the ones for oil saturation (bottom right). In the following, when we consider the value of purchasing the seismic AVO gradient attribute or EM resistivity data, we will fix the porosity in the reservoir and assume that only the saturation is uncertain. We will further assume that the saturation is directly related to the reservoir profits defined as the difference between the revenues and drilling costs.

The distinctions of interest $\boldsymbol{x} = (x_1, \ldots, x_n)$ are the **profits** associated with drilling wells and producing oil from the n reservoir units. At unit i, profit $x_i = \text{Rev}_i - \text{Cost}$, where Rev_i

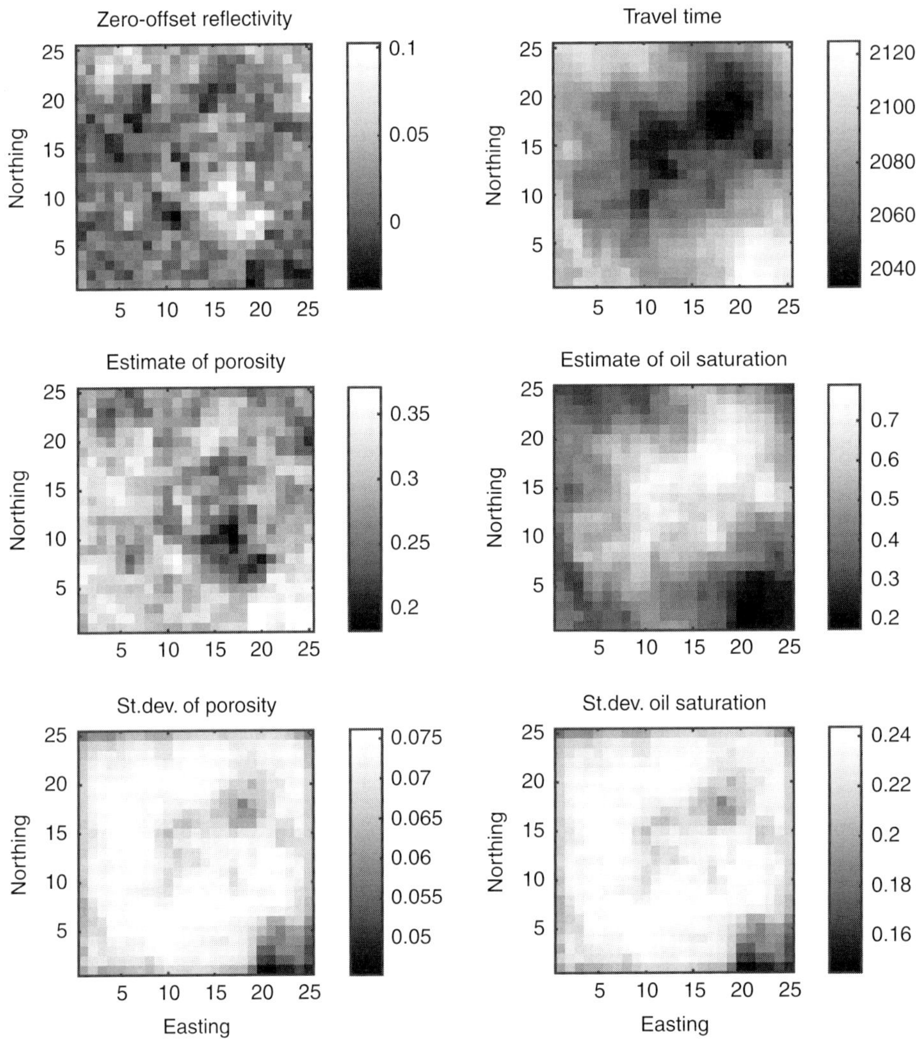

Figure 6.27 Seismic and electromagnetic data example: a summary of prior knowledge about the reservoir variables. The zero-offset seismic data and the seismic travel time data (top) are used to construct the current estimate of porosity and saturation (middle) with associated uncertainties (bottom).

is the uncertain revenue from the reservoir unit and where the subtracted cost of drilling is assumed to be fixed and constant for all units. The revenues are linearly related to the uncertain **saturation** variable: $\text{Rev}_i = \text{Rev}_0\, \varphi_i \left(1 - s_i\right)$, where the constant Rev_0 incorporates factors such as the cell area, the assumed reservoir thickness, the recovery factor, and the oil price. We assume that the profits are Gaussian in the prior model – i.e., $p(\boldsymbol{x}) = N(\boldsymbol{\mu}, \boldsymbol{\Sigma})$.

The mean μ of profits depends on several physical parameters of the model, such as the expected saturation, porosity, and reservoir depth and thickness, as well as on the price of oil and the cost of drilling wells.

Figure 6.28 shows the prior mean used for profits. This is largely based on the predictions of saturation and porosity from the currently available zero-offset seismic data and the seismic travel time information. Note how the saturation influences the profits, making higher profits more likely at some of the central units. The prior variance–covariance $\boldsymbol{\Sigma}$ of profits is defined by the variance terms specified from propagating the uncertainty in the saturation to profits, while the spatial correlation is modeled by an exponential correlation function with a range of 5 or 10 grid cells. We compare the effect of different correlation ranges for the VOI analysis in the following text.

At this stage of building the reservoir model, there is still uncertainty around the spatial distribution of saturation and hence profits. The decision maker has to decide whether to purchase pre-stack seismic AVO attributes, EM resistivity data, or both. Purchasing these sources of information could reduce the uncertainty in the predicted profits. The seismic AVO data and EM resistivity are related to the reservoir saturation by rock and fluid properties described by rock physics relations and by the physics of elastic and electromagnetic waves. These relations, along with the uncertainty in the acquisition and

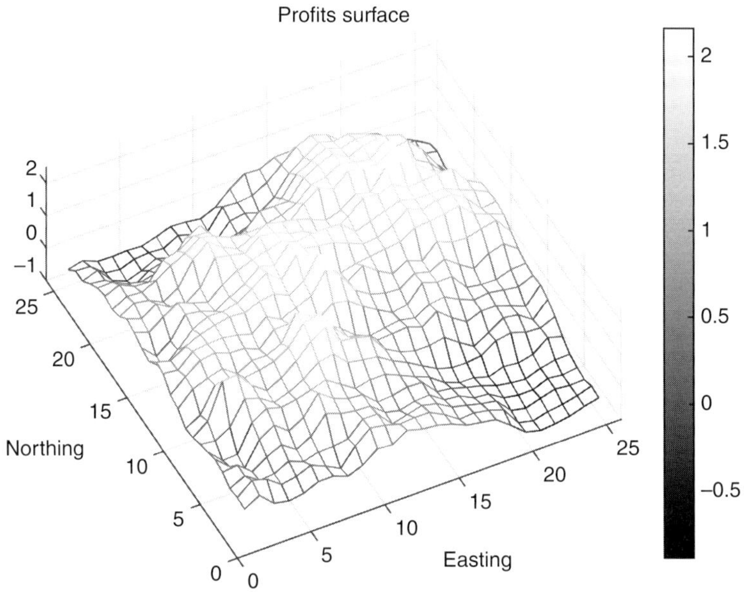

Figure 6.28 Seismic and electromagnetic data example: expected value for profits at all reservoir units. The top reservoir domain is split into 625 reservoir units on a regular grid. The prior model for profits is based on post-stack seismic data.

processing of data, define the likelihood model $p(\boldsymbol{y} \mid \boldsymbol{x})$. In the likelihood model, porosity is used as a fixed explanatory variable, and profits are modeled via saturation.

In Figure 6.29, model curves for the two geophysical attributes as a function of saturation and porosity (dots) are plotted with associated revenues (first axis). The modeling uses well-known rock physics relations and theories described in texts such as Mavko et al. (2009) and Schon (2011). The elastic moduli of the rocks with different porosities and saturations are first computed using a soft sand model (Dvorkin and Nur 1996; Bachrach 2006) tuned to a North Sea well log, along with Gassmann's fluid substitution equations. This is followed by the Aki and Richards approximations for the seismic P-wave reflectivity and AVO gradient. The cap rock shale was assigned fixed elastic properties. Archie's law, with parameters tuned to a North Sea well log, is used to compute the resistivity as a function of porosity and saturation. The seismic AVO gradient (left) varies strongly with porosity but has a weaker trend with saturation, especially for high porosity. Logarithmic resistivity (right) has a clear trend as a function of saturation but is a weaker function of porosity.

Non-spatial inputs include the fixed reservoir thickness, the cell area, the recovery factor, and the oil price, which when multiplied together result in revenue $\text{Rev}_0 = 18$ monetary units. The profits are obtained by subtracting the cost of development, which is set to Cost = 2 monetary units. As expected, Figure 6.29 shows that profits increase with higher porosity and decrease with increasing water saturation.

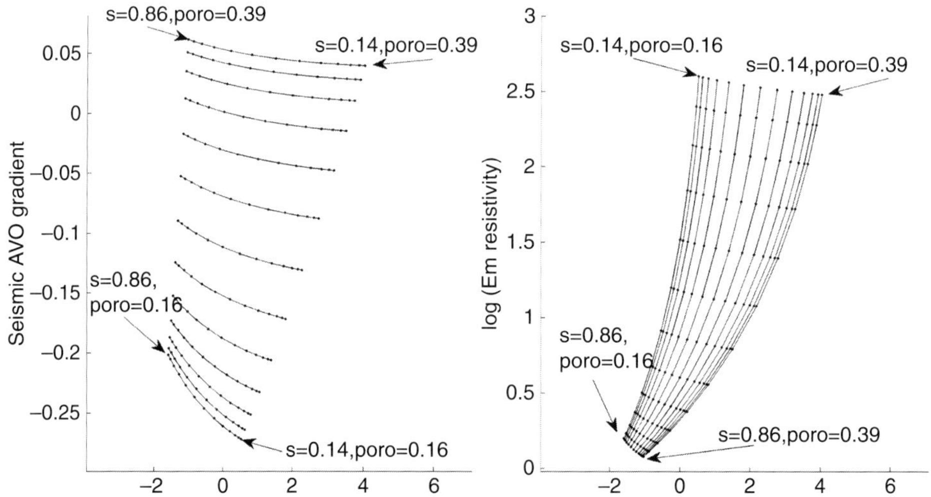

Figure 6.29 Seismic and electromagnetic data example: the geophysical response variables (second axis) depend on the saturation and porosity through rock physics models.

In the likelihood model for the seismic AVO data and the electromagnetic data, we assume conditional independence. This holds for the seismic AVO data, the EM data, and between seismic AVO and EM data. Thus, the likelihood factors into products between the data and between the reservoir units. Let $y_I = (y_{1,1}, \ldots, y_{1,m_1})$ denote the seismic AVO data, where $m_1 = n = 625$ since the seismic AVO attribute is available over the entire domain (total but imperfect information). The individual terms comprising the likelihood are modeled in a linear fashion, resulting in $p(y_{1,i} \mid x_i) = N(f_{1,i} + F_{1,i}(x_i + \text{Cost}), \tau_1^2)$, where the expectation term (the mean of the distribution) is plotted in Figure 6.29. Critically, this model assumes a linear relationship between the revenues x_i+Cost and the expected seismic pre-stack attribute $E(y_{1,i} \mid x_i)$. In this model for the likelihood mean value, we fit the intercept $f_{1,i}$ and slope $F_{1,i}$ using a linearized version of the rock physics modeling described earlier. The fixed terms $f_{1,i}$ and $F_{1,i}$ depend on the spatial porosity as covariates, which are now assumed to be fixed based on the prior post-stack reservoir characterization. The variance τ_1^2 can be estimated from well logs of saturation, porosity, and elastic properties, as well as seismic AVO data. This accuracy of the data would depend on the rock and fluid properties, data quality, and processing; in this example, we have set the noise to $\tau_1^2 = 0.035^2$. A very elaborate processing scheme could have higher accuracy, but of course at a higher price.

Let $\boldsymbol{y}_2 = (y_{2,1}, \ldots, y_{2,m_2=m_1})$ denote the EM data (being the logarithm of electromagnetic resistivity). Note that m_2 here, but does not have to equal m_1. The EM data may have different coverage than the seismic attributes. The individual terms of the EM likelihood are defined by $p(y_{2,i} \mid x_i) = N(f_{2,i} + F_{2,i}(x_i + \text{Cost}), \tau_2^2)$, where the expectation term is based on a linearization of Archie's law with parameters tuned to a North Sea well log. This is plotted in Figure 6.29. Again, the fixed terms $f_{2,i}$ and $F_{2,i}$ depend on porosity. The variance τ_2^2 can be estimated from well logs of saturation, porosity, and resistivity, but the accuracy of the EM data would generally also depend on the processing and inversion. We have set this noise to $\tau_2^2 = 2^2$ in this example.

We group both sets of data together in a vector of length $m_1 + m_2$ such that $\boldsymbol{y} = (\boldsymbol{y}_1, \boldsymbol{y}_2)$ and similarly group the other terms, $\boldsymbol{f} = (\boldsymbol{f}_1, \boldsymbol{f}_2)$ and $\boldsymbol{F} = (\boldsymbol{F}_1, \boldsymbol{F}_2)$ for the forward models and $\boldsymbol{T} = \text{diag}(\tau_1^2 \boldsymbol{I}_{m_1}, \tau_2^2 \boldsymbol{I}_{m_2})$ for the covariance matrix of the seismic AVO and EM data.

Under the Gaussian and linear modeling assumptions connecting revenues to the data, the posterior mean of profits becomes

$$E(\boldsymbol{x} \mid \boldsymbol{y}) = \boldsymbol{\mu} + \boldsymbol{\Sigma} \boldsymbol{F}^t (\boldsymbol{F} \boldsymbol{\Sigma} \boldsymbol{F}^t + \boldsymbol{T})^{-1} (\boldsymbol{y} - \boldsymbol{f} - \boldsymbol{F}\boldsymbol{\mu}) - \text{Cost} \cdot \mathbf{1}, \tag{6.16}$$

where $\text{Cost} \cdot \mathbf{1}$ is a length n vector of the drilling costs. The marginal likelihood of the seismic AVO and EM data is a Gaussian given by $p(\boldsymbol{y}) = N(\boldsymbol{f} + \boldsymbol{F}\boldsymbol{\mu}, \boldsymbol{F}\boldsymbol{\Sigma}\boldsymbol{F}^t + \boldsymbol{T})$.

VOI analysis

The decision about whether to drill or not at a reservoir unit is determined by the expected profit x_i at the reservoir units. Thus, the VOI is defined by

$$VOI(\boldsymbol{y}) = \sum_{i=1}^{N} \int \max\{0, E(x_i \mid \boldsymbol{y})\} p(\boldsymbol{y}) d\boldsymbol{y} - \sum_{i=1}^{N} \max\{0, E(x_i)\}. \tag{6.17}$$

Here, the data $\boldsymbol{y}$ could be seismic AVO data, EM resistivity data, or both.

The prior value, the second term in Equation (6.17), is obtained from the sum of profits from all units that have a positive prior mean based on the assumed Gaussian prior model for profits. For the posterior value, the first term in Equation (6.17), we use the closed-form analytical solution to this integral, which is valid under Gaussian linear modeling assumptions (see Sections 5.3–4 and Appendix A.1).

In Figure 6.30, we plot the VOI of seismic AVO data (SD), the EM resistivity data (ED), and both of these geophysical data (SD + ED) as a function of the cost of drilling. This cost is assumed to be constant for all cells in the grid. Obtaining both data (SD + ED) obviously results in the highest VOI. The VOI for the seismic AVO data is a little less than that of the EM data for our parameter settings. Notably, the VOI of both SD and ED is not the sum of the VOI of SD and the VOI of ED; the additional value of collecting both data types instead of just one is low. All curves peak around Cost = 2.5 monetary units, at which level added information is most valuable. For a very low drilling cost, the drilling decisions are easy to make without further information and the VOI is small. Similarly, at very large drilling costs, it is not necessary to acquire more data before making the decision to avoid drilling and, again, the VOI is small.

Next, we compare these VOI results with the price of the different geophysical experiments. As an example, say the processing required for pre-stack seismic data has a price of 10 monetary units. Figure 6.30 shows that the VOI of the seismic AVO gradient is greater

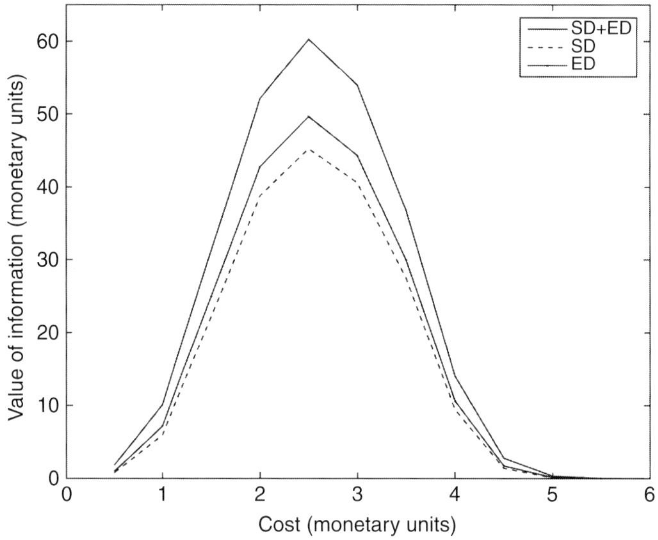

Figure 6.30 Seismic and electromagnetic data example: the VOI of seismic AVO gradient data (SD) and electromagnetic resistivity data (ED) as a function of the fixed drilling cost at the reservoir units.

than this processing price for drilling costs between 1 and 4 monetary units. In this cost range, it would be valuable for the petroleum company to purchase the AVO gradient information. As another example, say that obtaining EM resistivity data at our accuracy has a price equal to 40 monetary units. The VOI of EM resistivity data is greater than this price for drilling costs between 2 and 3.5 monetary units.

By constructing **decision regions,** one can figure out which data are most valuable to purchase. Decision regions are obtained by plotting the optimal choice of the information-gathering scheme with the price of the seismic AVO gradient (P_{SD}) on one axis and the price of the EM resistivity data (P_{ED}) on the other axis. For various price levels, the decision regions are defined by picking the most valuable information relative to its price of the experiment – i.e.,

$$\arg\max\left\{\text{VOI}_{SD+ED} - (P_{ED} + P_{SD}), \text{VOI}_{SD} - P_{SD}, \text{VOI}_{ED} - P_{ED}, 0\right\}. \tag{6.18}$$

The decision maker should purchase the data with the largest difference between the VOI and the associated price, if any of these are positive. For these calculations, the drilling cost is set to Cost = 2.

The decision regions are shown in Figure 6.31. We display the case of a short spatial correlation range (five cells, left) and a long correlation range (10 cells, right). If the price of seismic AVO and EM data is very low, the company benefits from purchasing both. This is shown in the lower left corners in the display. If the price of one or both data types is too high, the value of this information cannot justify purchasing this data. The decision boundaries are computed by equating the values for the different types of data. For instance, the boundary between ED or SD is defined by the line $P_{ED} = \text{VOI}_{ED}\text{-VOI}_{SD} + P_{SD}$.

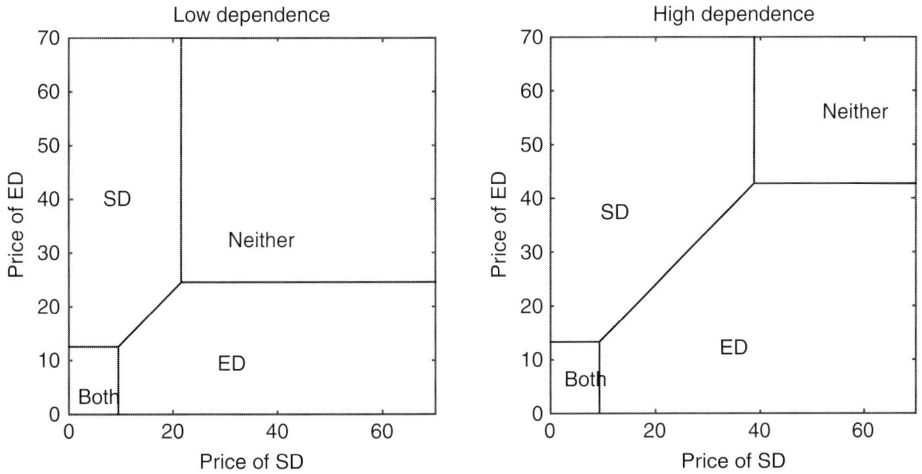

Figure 6.31 Seismic and electromagnetic data example: decision regions for geophysical testing schemes. The left display is indicative of small spatial correlation in the reservoir profits, while the results of the right display are based on large spatial dependence.

For the parameters used here, there is no boundary between purchasing both or neither. Both data are informative about the underlying uncertainty related to saturation and are thus related to the profits, and for a large range of prices it is valuable to obtain at least one of the geophysical data. The spatial dependence in the model clearly matters since the decision regions are different in the left and right displays. When there is high correlation in the profit variables (right display), the regions of only SD and only ED are dominant for a larger range of prices.

6.4 Mine planning and safety

In this section, we study information-gathering options for making improved mining decisions. The evaluation of information criteria in the mining industry has been studied extensively – see, e.g., Rivoirard (1987) and Chiles and Delfiner (2012). There are often large costs involved with the planning and development of mines. Profits are a function of the uncertain ore grade. Mining also involves issues related to the safety and prevention of mine hazards. In this case, the decision maker's value is created from the avoidance of a disastrous event with potential loss of life. In both situations, there are uncertainties, and the decision maker must choose whether it is worthwhile to acquire additional information to aid in decision making under uncertainty.

6.4.1 I love rock and ore: mining oxide grade example

Keywords: *oxide grade example, mining, X-ray data, Gaussian random field, perfect information, imperfect information, partial information, low decision flexibility, decoupled value function, decision regions*

In the mining oxide project considered here, there is a lot of prior geological knowledge and some borehole data that are highly informative about the oxide grade at the locations where they have been drilled. We studied some of the aspects of the currently available data in Chapter 2 and used them for spatial modeling in Section 4.4. Even though current data provide extensive knowledge about the resource, there is still uncertainty about the spatial distribution of the oxide. We study whether it will be worthwhile for the mining company to drill additional boreholes to improve the spatial predictions about the oxide grades and thus make a better decision.

We view the highlights of this section in the context of the framework outlined in Chapter 5:

- The **borehole data** are of two types. One provides **perfect information** about the oxide grade at the locations where they are acquired, while the other provides **imperfect information** about the oxide grade at the locations where they are acquired. We will compare the two measurement types. Data are acquired in boreholes, while our interest is in the entire spatial domain, so this is an example of **partial information gathering**. We focus on static information gathering.

- Regarding the three properties of spatial decision situations:
  1. We use a **Gaussian random field model** for the spatial distribution of the oxide grade.
  2. We assume that the decision maker either **selects the whole mining project or not** – i.e., there is **low decision flexibility**.
  3. If the decision maker decides to start mining, the total value is the sum of individual values from mining blocks. We thus have a situation where the **value function decouples to a sum**. The value at each unit depends on the oxide grade at that unit and the recovery factors and costs of producing the unit.

As in Eidsvik and Ellefmo (2013), we assume a risk-neutral decision maker and no constraints for the decision situation, and we consider the static situation without opportunities for sequential decision making.

Framing the decision situation

The distinction of interest, $x(\mathbf{s}_i) = x_i$, is the **oxide grade** at locations $\mathbf{s}_i$ in the mine. It will be profitable to start mining if the aggregated oxide resources are abundant and if the total revenue exceeds the total operational costs. We consider a simple decision situation with two alternatives: **mining or not**. Going ahead with the mining would involve building infrastructure and extracting the resources in the entire region of interest. We denote the mining alternatives by $a \in \{0,1\}$, where $a = 1$ denotes mining and $a = 0$ denotes not mining. The value function decouples to the sum over all defined blocks or composites in the mine. A preliminary analysis has defined around 8000 blocks in total, each of a fixed volume. The discretization into blocks entails an assumption about the equal oxide grade levels inside every block. The development decision, $a = 1$, involves the company mining all the blocks. Some blocks are, however, known to be largely waste rock, and these would only incur a cost without generating revenues. For the central part of the ore, all blocks are taken to the plant and processed for mineral production. Note that the company could probably gain value by choosing a more complex development strategy with more decision flexibility, such as through sequential decisions, because of the way mining is done from the top (open pit) or via tunneling operations.

Information gathering

It is useful to evaluate different schemes for collecting more exploratory borehole data about the spatial distribution of the oxide grade. Several exploration boreholes have already been drilled, and oxide grade measurements of two types are available from 1871 locations. One data type is obtained from X-ray fluorescence (**XRF**) spectrometry in the laboratory, which is assumed to provide **perfect information** about the oxide grade sample. Another type of data is obtained by portable X-ray meter (**XMET**) acquisition; this kind of information gathering is made on the location with a handheld meter, and it provides **imperfect information** about the oxide grade. The XMET data (imperfect information) are available at 1871 locations. The XRF data (perfect information) are observed at 103 of the 1871 current observation locations. The VOI question is whether more borehole data are needed at this stage, and should the additional measurements be of the XMET type or the XRF type?

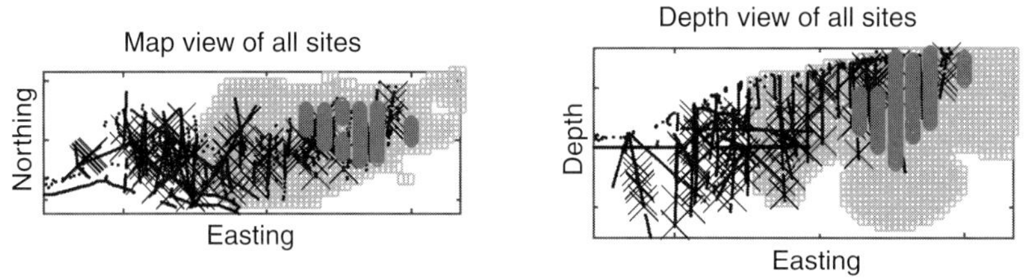

Figure 6.32 Mining oxide grade example: map views of current data locations (black), planned data-gathering sites (gray circles), and mining development block sites (gray squares).

The mining company plans to drill at 265 additional locations to collect oxide grade measurements. However, going through the actual drilling operation has a price, and it may not be worthwhile to do so. VOI analysis is useful for checking if the value from the scheme is greater than the price of data acquisition.

We assume that boreholes, once drilled, are sampled every meter. A core sample of the rock formation is then analyzed and results in a measured value of the oxide grade at the location where the sample was acquired. As in the currently available data, we may collect data of two different accuracies and price: XMET or XRF. Figure 6.32 illustrates the currently available data (XRF as crosses and XMET as small black dots), as well as blocks where mining development will be done and the locations where the mining company could collect more measurements.

The mining development will focus on the spatial blocks at a high altitude and in the eastern zone (gray squares in the display). The planned boreholes (indicated by gray circles in the display) are in the center of this development domain.

Modeling

The example in Section 4.4 illustrates parameter estimation and spatial prediction based on the currently available oxide grade data from this deposit. We recall the modeling aspects briefly here.

We use a Gaussian regression model for the oxide grade and estimate parameters in this model from available geological knowledge about the mining ore formation in addition to the available XRF data and XMET data at the observation sites. The preliminary geological analysis has revealed three dominant rock classes. We use this rock class as a covariate in a Gaussian regression model for the oxide grade. Thus, the marginal distribution for the oxide grade at location $\mathbf{s}_i$ is defined by $p(x(\mathbf{s}_i)) = N(\beta_0 + \beta_1 h(\mathbf{s}_i), \sigma^2)$, where the mean of the oxide grade varies in the spatial coordinate along with the variations of the geological class covariate $h(\mathbf{s}_i) \in \{1,2,3\}$. The class represents the mineralization grade that has been constructed from a smooth interpretation of the preliminary data by experts in the local geology.

Spatial correlation is modeled by $Cov(x(\mathbf{s}_i), x(\mathbf{s}_j)) = \sigma^2 \exp(-\eta t_{ij})$, where t_{ij} is the Euclidean distance between the two sites $\mathbf{s}_i$ and $\mathbf{s}_j$ and η is the correlation decay parameter. The imperfect XMET data are assumed to be unbiased estimates of the oxide grade at the locations where the data are acquired, with an independent Gaussian noise term with variance τ^2. Model parameters are estimated by maximum likelihood, as described in Section 4.4. The regression parameter $\beta_1 = 1.32$ for the slope with a mineralization class clearly indicates a significant trend of an increased oxide grade from Class 1 to Classes 2 and 3.

The suggested model for oxide grade may seem implausible because grade is positive and bounded upward. Nevertheless, the Gaussian modeling approach appears to provide a reasonable description when accounting for the mineralization covariate. This is indicated by the quantile–quantile (QQ) plot in Section 4.4 as well as the analysis by Eidsvik and Ellefmo (2013).

VOI analysis

The VOI is computed using the expected revenues and costs from the mine. For computing the prior and posterior values, we make certain assumptions about the mining processes that will be executed if the mine is developed. (Various extensions are possible here.) First, the block values have been calculated assuming that the blocks have already been uncovered and that they will be mined. Second, we assume known costs for mining 1 tonne of oxide ore. In addition, the blocks with a currently predicted grade above cut-off are processed after mining, whereas the waste rock (with a predicted grade below cut-off) is not processed any further. This entails that only a fraction of the blocks contain possible revenues but that there are costs associated with all blocks. The cut-off is fixed in our application.

Let Rev_l be the revenue factor associated with resource block $l = 1,\ldots,7936$, which provides the revenue once it is multiplied with the expected grade. We set Rev_l to 0 outside the ore, since these blocks are not processed further. The revenue factor is a product of several elements: the price of the commodity extracted from the oxide, block volume, density, and certain processing parameters. Let $\mathbf{Rev} = (\text{Rev}_1,\ldots,\text{Rev}_{7936})$ be the (column) vector of revenue levels and $\mathbf{Cost} = \left(\text{Cost}_1,\ldots,\text{Cost}_{7936}\right)$ the costs in the set of resource blocks. The cost includes both the cost of mining and processing a block, except if a block is waste rock, then the cost of processing is not included in the cost for that block. The prior value is calculated from the expected profits (Prof) based on the currently available data – i.e.,

$$\text{PV} = \max\left\{0, E\left(\text{Prof}\right)\right\}, \qquad E\left(\text{Prof}\right) = \mathbf{Rev}^t \ \boldsymbol{\mu}_{\bar{x}} - \mathbf{Cost}^t \ \mathbf{1}, \tag{6.19}$$

where $\boldsymbol{\mu}_{\bar{x}}$ is the vector of the mean oxide block grades given the current data. This is computed from the Gaussian regression model, similar to Kriging (see Chapter 4 and Appendix A.1).

Consider what happens when more data $\boldsymbol{y}$ are collected. The new data are either XMET or XRF in the planned boreholes. The posterior value is

$$\begin{aligned} \text{PoV}\left(\boldsymbol{y}\right) &= \int \max\left\{0, E\left(\text{Prof} \mid \boldsymbol{y}\right)\right\} p(\boldsymbol{y})d\boldsymbol{y}, \\ E\left(\text{Prof} \mid \boldsymbol{y}\right) &= \mathbf{Rev}^t \boldsymbol{\mu}_{\bar{x}\mid \boldsymbol{y}} - \ \mathbf{Cost}^t \mathbf{1}, \end{aligned} \tag{6.20}$$

where $\boldsymbol{\mu}_{\bar{x}\mid \boldsymbol{y}}$ is the prediction of the block grades given the current data and planned data $\boldsymbol{y}$. We need to integrate over all possible prospective data outcomes for the posterior value computation. The pdf of the planned data is Gaussian and denoted $p(\mathbf{y}) = N(\boldsymbol{\mu}_y, \boldsymbol{\Sigma}_y)$. The conditional mean of the oxide grade at resource blocks given the planned data $\boldsymbol{\mu}_{\bar{x}|y}$ is computed from the Gaussian regression model, similar to Kriging. The mean $\boldsymbol{\mu}_y$ and covariance $\boldsymbol{\Sigma}_y$ of the planned data are also computed from the known equations for Gaussian variables (see Chapter 4 and Appendix A.1). As discussed in Sections 5.3–4, the Gaussian linear formulation enjoys closed-form analytical solutions for the posterior value and the VOI (see also Appendix A.1). We can hence compute the VOI analytically in this situation.

The mining company should decide to purchase the data if the VOI is larger than the price of data acquisition. The XRF data is perfect information, and the VOI of the XRF data is always larger than the VOI of the imperfect XMET data. However, the XRF data are more expensive than the imperfect XMET data. Figure 6.33 shows the decision regions for the XMET data and XRF data. The company should choose to purchase XRF data for a rather small window of XRF and XMET prices. When the price of XRF exceeds \$0.75 million and the XMET price exceeds \$0.5 million, the VOI results show that neither data type should be acquired.

VOI analysis can also help the mining company design the data-gathering scheme – i.e., choose which boreholes to drill. We can compare the VOI of the XMET data at all boreholes with the VOI of the XRF data at a subset of boreholes. Figure 6.34 shows the decision regions for the XRF data in a subset consisting of the two longest boreholes, versus the XMET data in all boreholes, or nothing. The decision regions are computed as a function of the drilling price and the price of laboratory experiments. When the price of laboratory

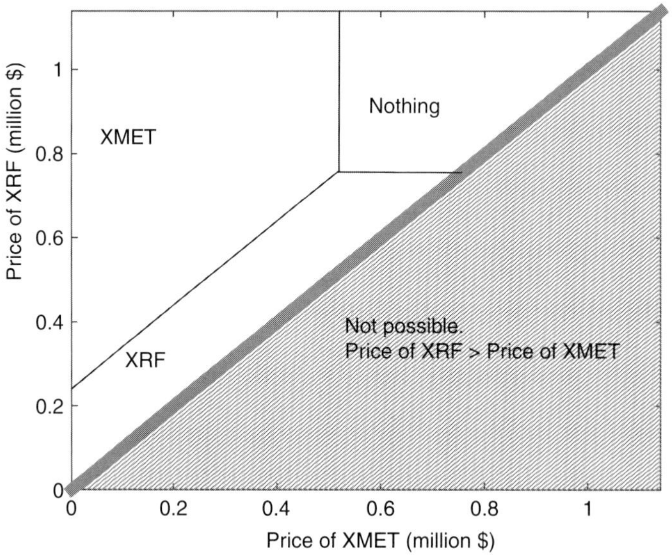

Figure 6.33 Mining oxide grade example: decision regions for full sampling of XMET and XRF data. The price of drilling is the same, but the processing of the XRF data is more expensive.

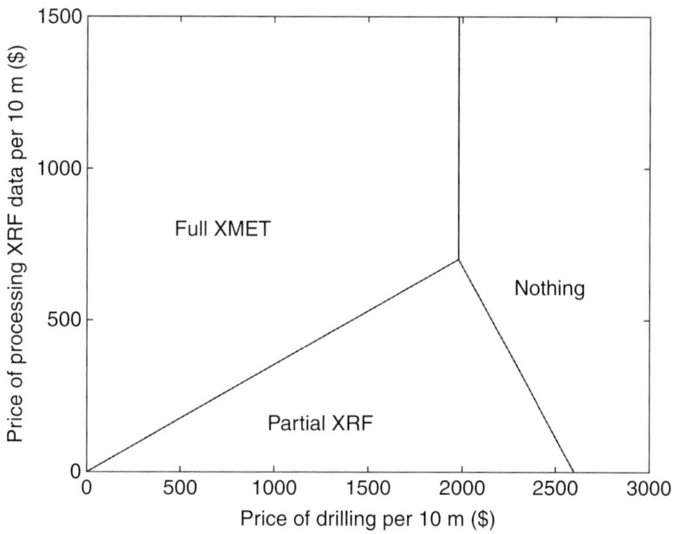

Figure 6.34 Mining oxide grade example: decision regions for partial sampling of XRF data or full sampling of XMET data. The axes represent the price of drilling per 10 m and the price of acquiring and processing XRF data in the laboratory.

experiments is low and the price of drilling is not too high, the partial XRF testing is preferred. When laboratory prices get higher, one should acquire total tests of XMET unless the price of drilling gets too high.

The decision theoretic notion of VOI is useful because it evaluates the information content in the context of a decision situation and is in monetary units. However, as we discussed in Chapter 5, many other measures can be applied to evaluate designs for data acquisition. In the following analysis, we will compare the VOI results with the integrated prediction variance and the entropy. Other measures used in the mining industry are discussed in Rivoirard (1987), Vann et al. (2003), Chiles and Delfiner (2012), and Eidsvik and Ellefmo (2013).

In Table 6.3, we summarize the relative VOI, prediction variance, and entropy of the partial sampling schemes. It is not always appropriate to compare these measures in general, as they are based on fundamentally different principles, but we note that the entropy criterion tends to prefer perfect information more than the others. For the entropy measure, it is more informative to collect perfect information in the 10 longest boreholes than to collect imperfect information in all boreholes.

6.4.2 We will rock you: rock hazard example

Keywords: *rock hazard example, mining, joint frequency data, spatial generalized linear model, Laplace approximation, imperfect information, partial information, high decision flexibility, decoupled value function, decision regions*

Table 6.3. *Relative comparison of partial or total XRF testing with total XMET*

	VOI	Kriging Variance	Entropy
XRF Partial (5)/XMET Total	0.39	0.76	0.69
XRF Partial (10)/XMET Total	0.94	0.92	1.21
XRF Total/XMET Total	1.45	1.06	1.68

In this example, we consider decisions for rock support in mining and tunneling operations. It is necessary to make assessments of the rock strength to avoid rock fall. The decision is whether to increase support at selected tunnel locations to avoid uncertain future rock falls. One may acquire data from boreholes to improve the prediction of variables for rock fall analysis, such as the spatial intensity of joints within the rock mass. Joints are opening mode fractures in rocks, and zones that are heavily fractured might be weaker and more prone to rock falls.

We view the highlights of this section in the context of the framework outlined in Chapter 5:

- **Borehole data** are informative of **joint frequencies**, which are uncertain and considered **imperfect information**. Data are acquired in boreholes, while the key areas of interest are the selected tunnel locations where support may be increased, so this is an example of **partial information gathering**. We focus on static information gathering.
- Regarding the three properties of spatial decision situations:
  1. We use a **Gaussian random field model with a Poisson likelihood** for the spatial distribution of joint frequencies.
  2. The decision maker chooses to **increase the rock support** at a selected site if this is expected to be the best alternative. The decision maker is free to select as many sites as are profitable – i.e., there is **high decision flexibility**. Here, the notion of "profit" is essentially avoiding losses arising from the hazard and associated damages.
  3. The decision maker's choice (increasing the rock support or not at a site) only depends on that particular location in the mine. We thus have a situation where the **value function decouples**. The value at any unit depends on the cost of rock fall and the cost of adding support to increase stability.

We assume a risk-neutral decision maker and no constraints for the decision situation, and we consider the static situation without opportunities for sequential decision making. We studied a joint frequency data set in Section 4.5. In this section, we conduct VOI analysis for joint frequency data (see also Evangelou and Eidsvik 2015).

Framing the decision situation

The distinction of interest is the intensity of joints, denoted $x(\mathbf{s}_i) = x_i$, at locations $\mathbf{s}_i$ in the mine. A couple of mining tunnel locations at a depth of 250 m have experienced rock fall, and the underlying decision is whether the mining company should increase the support along the tunnels at this depth. A set of $N = 52$ critical tunneling locations near a depth of 250 m has been selected. The decision to add support at any of these locations comes at the cost of bolting equipment and labor.

If the decision is to **add support** (at a cost), we assume that there will be no future rock fall. We denote this alternative by $a_j = 1, j = 1,\dots,N$. When no added support is provided ($a_j = 0$), rock fall may occur, and the cost of this event depends on the uncertain size of the fall via the joint intensity at the locations. This decision is slightly different from the previously considered decision situations in the sense that here we consider a minimization problem where the decision maker is interested in minimizing costs rather than maximizing profits.

Information gathering

Information about the predicted joint intensity at spatial locations could help the engineers make better decisions about stability. By acquiring **joint frequency data** in boreholes, one obtains information about the spatial intensity of joints. We consider a baseline set of possible boreholes to drill. This data set has size $m = 1615$, where data are sampled every 4 m in boreholes. This baseline setup is next split into subsets for various partial tests. The first partial testing scheme is to acquire and analyze only half of the boreholes, resulting in $m = 768$ measurement locations. Another potential scheme is to analyze only every second location (i.e., every 8 m) in half of the boreholes, with $m = 383$. The last scheme is to acquire data only in a quarter of the boreholes, with $m = 383$. Note that the data size in the last two data-gathering options is the same, but the price of experimentation may be different. VOI is used to compare the different information-gathering schemes for different price ranges.

This analysis will be done retrospectively, meaning that for this mining location, we have all $m = 1615$ data from all boreholes, but now we study this case in hindsight, evaluating whether other information-gathering schemes would have been just as good. Figure 6.35 shows the situation with the boreholes and the 52 sites identified for adding rock support. The 52 sites are plotted along the plane at a depth of 250 m.

Modeling

The hierarchical model formulation has a **Poisson likelihood model** for the joint counts at each core sample location and a **Gaussian spatial process** for the latent logarithmic intensity of joints. This model is hence a spatial generalized linear model (GLM), as was discussed in Section 4.5. The data are denoted $\mathbf{y} = (y_1,\dots,y_m)$ at the m observation sites. Conditional on the spatially varying intensity, the Poisson likelihood is $p(y(\mathbf{s}_i) \mid x(\mathbf{s}_i)) = \text{Poisson}(V_i \exp(x(\mathbf{s}_i)))$, $i = 1,\dots,m$, for fixed volume V_i (fixed diameter and 4-m cores). We assume a Gaussian distribution for the log intensity $x(\mathbf{s}_i)$ with a constant mean (μ) and variance structure defined by a nugget effect (τ^2) and a correlated effect with an exponential anisotropic covariance function with variance σ^2. There is more spatial correlation along the geological strike of the formations and less correlation perpendicular to the strike. When we discretize the spatial process $x(\mathbf{s}_i)$ on a grid of spatial locations,

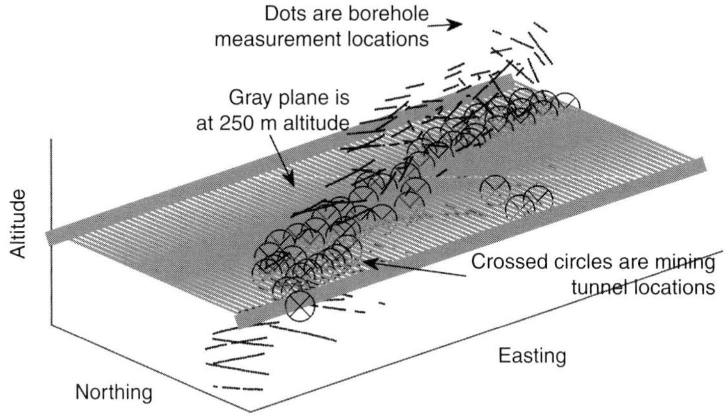

Figure 6.35 Rock hazard example: illustration of joint frequency data and the 52 alternative locations concerning rock stability. The 52 sites are at an altitude of 250 m, and their locations are indicated across the shaded plane.

we can write $p(\boldsymbol{x}) = N(\boldsymbol{\mu}, \boldsymbol{\Sigma})$, where $\boldsymbol{\mu} = \mu\mathbf{1}$ and the covariance matrix has entries $\Sigma_{ij} = \tau^2 I(i=j) + \sigma^2 \exp(-\eta_s t_{ij,s} - \eta_p t_{ij,p})$. Here, the lag distances $t_{ij,s}$ and $t_{ij,p}$, as well as the spatial correlation parameters $\boldsymbol{\eta} = (\eta_s, \eta_p)$, are defined in the strike direction (subscript s) and the direction perpendicular to the geological strike (subscript p).

We specify the statistical modeling parameters from the available data in this mine. The estimation procedure is based on optimizing an approximation of the marginal likelihood $p(\boldsymbol{y}) = p(\boldsymbol{y};\sigma,\tau,\boldsymbol{\eta},\mu)$ of the joint count data. For details, please see Section 4.5 and Appendix A.2. The maximum MLEs of the parameter values are as follows (with standard errors in parentheses): mean 1.55 (0.06), partial sill 0.13 (0.014), and nugget 0.04 (0.003), and the in-strike correlation range, $3/\eta_s$, is about 300 (41) m. Thus, the marginal variance in the latent intensity process is $\sigma^2 + \tau^2 = 0.17$ at all sites. The correlation range perpendicular to the strike is set to a quarter of the in-strike correlation range – i.e., 75 m. For the initial analysis, we fix the parameters, but we later analyze the parameter values according to the uncertainty ranges to study the effect of the mean, variance, and correlation range on the VOI. The conditional pdf of joint frequencies given the data is $p(\boldsymbol{x} \mid \boldsymbol{y})$. This posterior does not have a closed-form solution, but we can approximate it by a Gaussian approximation, and this will be useful for performing analytical approximations for the VOI expressions.

VOI analysis

The prior value for this decision situation is

$$PV = \sum_{j=1}^{52} \max\left\{-\text{Cost}, -\text{Rev}\cdot E\left(\exp\left(\zeta x(\mathbf{s}_j)\right)\right)\right\}, \tag{6.21}$$

where Cost is the cost of added support at a location $\mathbf{s}_j$, while Rev$\cdot E\left(\exp\left(\zeta x\left(\mathbf{s}_j\right)\right)\right)$ is the expected cost associated with rock fall at that location, with $x\left(\mathbf{s}_j\right)$ being the logarithm of the joint intensity at that location. The decision maker should choose the alternative that minimizes the expected costs at each of the locations. Note that in general, Cost, Rev, and ζ (a scale factor) depend on rock mechanics, fluid composition, geometric considerations, cost of rock mass transport, and other engineering inputs. Here, the input variables have been set to Cost = 20000, $\zeta = 3$, and Rev = 100 monetary units.

Let $\boldsymbol{y}$ denote a generic joint count data set acquired according to a specific spatial design. The posterior value is

$$PoV\left(\boldsymbol{y}\right)=\sum_{j=1}^{52}\max\left\{-\text{Cost},-\text{Rev}\cdot E\left(\exp\left(\zeta x\left(\mathbf{s}_j\right)\right)\mid \boldsymbol{y}\right)\right\}p\left(\boldsymbol{y}\right). \tag{6.22}$$

The VOI is defined as the difference between the posterior and prior values (for a risk-neutral decision maker). The computation of the prior value in Equation (6.21) is analytically available under the Gaussian modeling assumptions for $x\left(\mathbf{s}_j\right)$ (see Appendix A.2). The posterior value is not straightforward in this case: first, the conditional distribution of the joint intensity has no closed-form expression. Second, the marginal distribution of the data is unknown. We briefly outline an analytical approximation to the VOI for this setting. This analytical solution relies on a Gaussian approximation and the Laplace approximation for integral expressions. The details for the approximate GLM calculations are provided in Appendix A.2.

The idea is to first construct a Gaussian approximation to the posterior. Let

$$\hat{\boldsymbol{x}}=\arg\max_{\boldsymbol{x}}\left\{p\left(\boldsymbol{y}\mid\boldsymbol{x}\right)p\left(\boldsymbol{x}\right)\right\}, \tag{6.23}$$

and let $\boldsymbol{\Sigma}^{-1}+\hat{\boldsymbol{D}}$ be the negative inverse Hessian of $\log p\left(\boldsymbol{x}\mid\boldsymbol{y}\right)$ evaluated at $\hat{\boldsymbol{x}}$. The matrix $\hat{\boldsymbol{D}}$ is diagonal with the i-th element given by $V_i\exp\left(\hat{x}\left(\mathbf{s}_j\right)\right)$, where V_i again equals the volume of the 4-m cores. Then, an approximation to the conditional mean and variance is

$$E\left(\boldsymbol{x}\mid\boldsymbol{y}\right)\approx\hat{\boldsymbol{x}},\ Var\left(\boldsymbol{x}\mid\boldsymbol{y}\right)\approx\left(\boldsymbol{\Sigma}^{-1}+\hat{\boldsymbol{D}}\right)^{-1}. \tag{6.24}$$

Critically, the posterior mean and variance depend on $\boldsymbol{y}$ only through $\hat{\boldsymbol{x}}$. We can approximate the distribution of $\hat{\boldsymbol{x}}$ by a multivariate normal with mean $\boldsymbol{\mu}$ and known covariance. The integral expressions required in Equation (6.22) can then be approximated in a similar manner to the method for the Gaussian model in Sections 5.3 and 5.4.

Let us now interpret the results. First, we note that the prior mean for the joint intensity is quite high in this case, and the prior decision is to add support at all locations. Additional data will either push the decision more clearly toward added support or toward avoiding support when the observations indicate that this is not necessary. The VOI depends, of course, on the spatial acquisition design. When data are acquired near one of the 52 prediction sites, the information gain is large for that location. If data are far from a prediction site, there is hardly any added value at this location. In addition to evaluating the full design (entire set of 1615 data), we split the data into various subsets and compute the VOI for each acquisition design. The results are as follows:

- All boreholes (data size $m = 1615$): VOI is 216 000 monetary units.
- Half of the boreholes (data size $m = 768$): VOI is 165 000 monetary units.
- Every second observation in half of the boreholes (data size $m = 383$): VOI is 159 000 monetary units.
- Quarter of the boreholes (data size $m = 383$): VOI is 96 000 monetary units.

The VOI clearly decreases when we collect less data. However, the decrease is slower than one would expect from a fractional splitting of the data. Moreover, the reduction depends a lot on the borehole design, since there is spatial correlation in the model. In particular, we note that the data size is the same for the last two options, but the strategy with more boreholes and coarser core samples of joint counts has a much higher VOI. The mining company benefits from such VOI evaluation when planning the data acquisition.

These VOI results must be compared with the price levels of the different data acquisition schemes. Based on the various fractions of the borehole samples and the associated price of data acquisition, we next construct decision regions. Here, we compare the scheme defined by a quarter of the boreholes (Quarter) with the analysis of every second observation in half of the boreholes (Half–Half). Note that the total number of data is the same. Thus, the processing of joint count data is assumed to be equal for the two schemes, but the cost of drilling is larger for the Half–Half option. Like in the previous sections, the notion of decision regions for data collection relies on selecting the best data-gathering scheme as follows:

$$\arg\max\left\{VOI_{\text{Half-Half}} - P_{\text{Half-Half}}, VOI_{\text{Quarter}} - P_{\text{Quarter}}, 0\right\}, \tag{6.25}$$

where the price of the data gathering is denoted P_*, with the subscripts denoting the acquisition scheme. Figure 6.36 shows the decision regions as a function of the price of the drilling parameter (first axis) and the price of processing per sample (second axis).

The Quarter scheme drills about $383 \cdot 4 \approx 1500$ m, while the Half–Half option drills about $383 \cdot 8 \approx 3000$ m. When the drilling cost is small, the more valuable Half–Half scheme is only a little more expensive than drilling a quarter of the boreholes. The mining company should then decide to drill half of the boreholes and analyze samples every 8 m. The price of processing per sample – i.e., counting the joints in a core sample – is the same for both data acquisition schemes. If the prices are too high, the "Nothing" region indicates that the optimal alternative is to purchase no data.

Next, we study sensitivity to the statistical model parameters. This analysis is done by varying each parameter at a time from its reference level. The low and high cases are determined by ±2 standard deviations from the reference levels, which are set from the joint count data described earlier in this section.

The results of the sensitivity analysis are shown in Table 6.4 for the full spatial design, with a reference VOI of 216 000 monetary units. The VOI is much lower in both the low

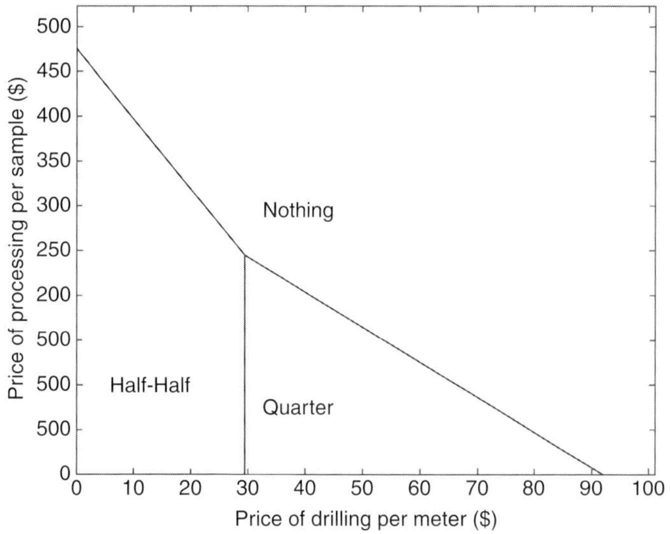

Figure 6.36 Rock hazard example: decision regions for Half–Half or Quarter data-gathering schemes. The price of drilling (per meter) is on the first axis, while the price of processing a borehole data sample is on the second axis.

Table 6.4. *VOI results for different configurations of statistical model parameters*

VOI	Mean	Correlation	Partial Sill	Nugget
Low	131 000	164 000	218 000	228 000
High	113 000	252 000	202 000	205 000

and high cases for the mean, indicating large sensitivity to the prior mean value. In fact, the VOI is highest for values near the prior mean in this case, at which the decision maker is most indifferent about the alternatives a priori, and therefore the data become very valuable. When the mean value is low, there tends to be few joints, and the company should decide not to provide extra support. Additional data are unlikely to change this decision, and the VOI is low. Similarly, for a large mean, data are unlikely to change the decisions about adding support, and the VOI is low.

The VOI increases with longer spatial correlation because the spatial data carry more information to the 52 prediction sites with stronger dependence. The VOI decreases with a higher partial sill and nugget. This comes as a surprise since large prior uncertainty tends to make data more valuable. This counterintuitive result occurs because the nugget and partial sill also impact the expected number of joint counts in the Poisson distribution, not just the variability of the joint intensity.

We believe that the Poisson distribution with a Gaussian process model for the intensity is a suitable model for such data (see also Ellefmo and Eidsvik 2009). The approximate analytical results make it possible to perform VOI analysis routinely for different input conditions.

6.5 Groundwater management

In this section, we study an example concerning groundwater management. There are important decisions to be made in this domain regarding maintaining the quality of water. This has an impact on the quality of life through its effects on agriculture, availability of drinking water, etc. Geophysical measurements such as those conducted by electromagnetic methods may provide information about properties such as the salinity of the water, and this information may help in making better decisions.

6.5.1 Salt water wells in my eyes: groundwater management example

Keywords: *groundwater characterization, geological scenario, multiple-point statistics, electromagnetic data, perfect information, imperfect information, low spatial decision flexibility, complex value function*

This example concerns decisions about potentially recharging groundwater with freshwater. The subsurface distribution of the sand and shales, which impact the groundwater flow and recharge in the aquifer, is uncertain. Geophysical measurements conducted by a transient or time-domain electromagnetic method (TEM) may be valuable for better characterizing the subsurface and thus helping to make informed decisions about the recharge and selection of possible recharge locations. This example, inspired by a real situation on the California coast, is related to groundwater use for agriculture and the decision to mitigate salt water intrusion. The example is described in Trainor-Guitton et al. (2011) as well as in the chapter on VOI in Caers (2011).

In the context of the framework outlined in Chapter 5, the highlights are:

- **The electromagnetic data** provide **imperfect information** about the **spatial variables and the geologic scenario**. Data cover the entire spatial domain of interest, and this is an example of **total imperfect information**. We focus on static information gathering.
- Regarding the three properties of spatial decision situations:
    1. We use a model with **geological scenarios defined by channel directions** and then propagate spatial uncertainty given the scenario.
    2. The decision maker chooses either **no recharge or recharge from the center, north, or south locations.** The four alternatives can easily be enumerated, and this is a setting with **low decision flexibility**.
    3. The value function associated with the different alternatives involves complex spatial modeling and subsurface fluid flow to give a scalar response. We thus have a situation with a **coupled value function**.

We assume a risk-neutral decision maker and no constraints for the decision situation, and we consider the static situation without opportunities for making sequential decisions.

Framing the decision situation

Groundwater is pumped out from a coastal fluvial aquifer and is used for agriculture. Excessive groundwater pumping causes seawater intrusion, leading to increased water salinity and decrease in usability for crops. Artificial groundwater recharge by pumping freshwater into the subsurface is considered a way to mitigate salt water intrusion.

We assume that the main uncertainty pertains to the geological scenarios denoted x, where $x \in \{1,2,3\}$ represents three geological scenarios to be defined in the spatial modeling later.

The four alternatives $a \in \{1,2,3,4\}$ considered are: (i) no recharge, (ii) recharge at a central location, (iii) recharge at a northern location, and (iv) recharge at a southern location. Figure 6.37 illustrates the situation. Value comes from the use of fresh groundwater to grow cash crops sold at profit. The value function is highly nonlinear – it cannot be decoupled to a sum over some spatial variables represented at geographical units.

Information gathering

Geophysical surveying by the time domain TEM provides information about the subsurface geological uncertainty. TEM can be a land-based or airborne method. In this example, we consider the **airborne TEM** (see Chapter 7 for further hands-on projects with land-based data). Pulsed electrical currents from a transmitter coil induce electrical currents and fields in the subsurface. The time-varying induced electrical field gives rise

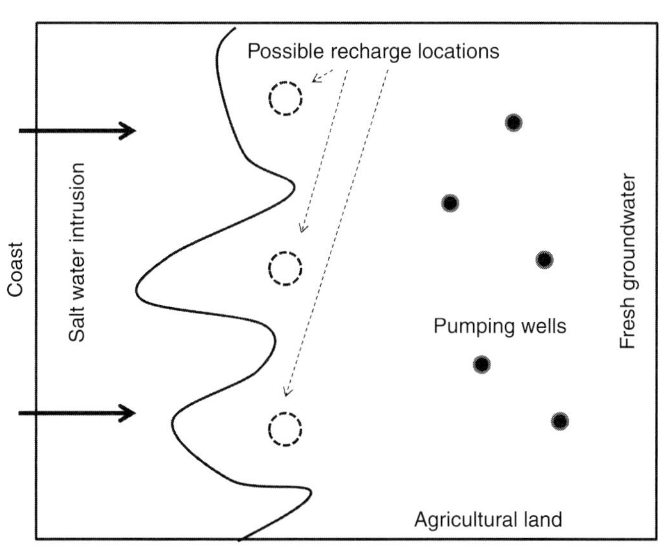

Figure 6.37 Groundwater management example: locations for performing groundwater recharge.

to a magnetic field that is recorded by a receiver coil. The basic data are the magnetic field response as a function of time. The basic data are inverted for the subsurface resistivity that can then be interpreted in terms of subsurface lithology patterns, since different lithologies have different resistivities.

Modeling

The spatial modeling relies on embedding the groundwater reservoir variables in a few discrete classes or geological scenarios. The heterogeneous subsurface consists of high-porosity, high-permeability sand channels within low-permeability background shales. The spatial distributions and orientations of these channels are uncertain. As a result, the subsurface flow and the impact of recharge are uncertain. The prior geological uncertainty is modeled as being one of three different scenarios of sand channel orientations: dominantly northeast, dominantly southeast, and a mix of both. A priori, the geological scenarios have equal probabilities – i.e., 1/3 for each.

The spatial modeling starts with these geological scenarios. Given a scenario, we can generate multiple realizations of the spatial variables of interest using Monte Carlo sampling. Using a multiple-point geostatistical technique (see Section 4.7), 50 facies realizations were generated for each geological scenario. All of these facies realizations were populated with porosities and permeabilities appropriate for sand and shales. The value function is computed from these realizations as inputs.

VOI analysis

To assess the expected value under the three uncertain geological scenarios and the four alternatives, we rely on the Monte Carlo simulations of spatial variables. Flow simulations are run for each of the simulations from every geologic scenario and for all four alternatives. This simulates 10 years of groundwater flow. The volume of fresh groundwater (salinity less than 150 ppm of chloride) available after 10 years is converted to a monetary value using fixed numbers for the amount of crops grown per unit volume of fresh water and profit per ton of crop. The expected values are obtained by averaging over the 50 realizations for each geological scenario under each alternative. For the alternatives where recharge is performed, the cost of recharge is subtracted. Thus, this requires a total of $50 \cdot 3 \cdot 4 = 600$ computational flow simulations.

The values (in units of million $) are as shown in Table 6.5. Again, we denote the values by $v(x,a)$, where the first index denotes the geologic scenario and the second index denotes the alternative. The prior value is then

$$\begin{aligned} PV &= \max_{a \in A} \left\{ E\left(v(x,a)\right)\right\} = \max_{a \in A} \left\{ \sum_x v(x,a)\, p(x) \right\}, \\ &= \max \left\{ \sum_x v(x,1)\, p(x), \sum_x v(x,2)\, p(x), \sum_x v(x,3)\, p(x), \sum_x v(x,4)\, p(x) \right\}. \end{aligned} \tag{6.26}$$

Table 6.5. *Expected values for all geological scenarios under all alternatives for groundwater recharge*

Channels	Recharge	Value
Northeast channels	No recharge	13.66
Northeast channels	Central recharge	13.8
Northeast channels	North recharge	13.9
Northeast channels	South recharge	12.8
Mixed channels	No recharge	12.15
Mixed channels	Central recharge	11.1
Mixed channels	North recharge	12.9
Mixed channels	South recharge	11.3
Southeast channels	No recharge	11.6
Southeast channels	Central recharge	9.89
Southeast channels	North recharge	9.66
Southeast channels	South recharge	9.44

Filling in the values $v(x,a)$ from the table, and using equal prior probability of 1/3 for each geological scenario, we get

$$PV = \max\left\{\begin{bmatrix}\frac{1}{3}(12.15+13.66+11.6), \frac{1}{3}(11.1+13.8+9.89), \\ \frac{1}{3}(12.9+13.9+9.66), \frac{1}{3}(11.3+12.8+9.44)\end{bmatrix}\right\}. \qquad (6.27)$$
$$= \max\{12.47, 11.59, 12.15, 11.18\} = 12.47$$

The optimal decision is not to recharge. A decision tree representation of the problem is shown in Figure 6.38.

The value with perfect information is calculated by interchanging the expectation and the max operation in the equation for prior value to give

$$PoV(x) = E\left(\max_{a\in A}\{v(x,a)\}\right) = \sum_x \max_{a\in A}\{v(x,a)\}\, p(x). \qquad (6.28)$$

Using the numbers for $v(x,a)$ in this example,

$$PoV(x) = \frac{1}{3}(12.9) + \frac{1}{3}(13.9) + \frac{1}{3}(11.6) = 12.8. \qquad (6.29)$$

The situation with perfect information (clairvoyance) is represented by the flipped tree in Figure 6.39.

The value of perfect information is the difference between the posterior value with perfect information and the prior value:

$$VOI(x) = PoV(x) - PV = 12.8 - 12.47 = 0.33. \qquad (6.30)$$

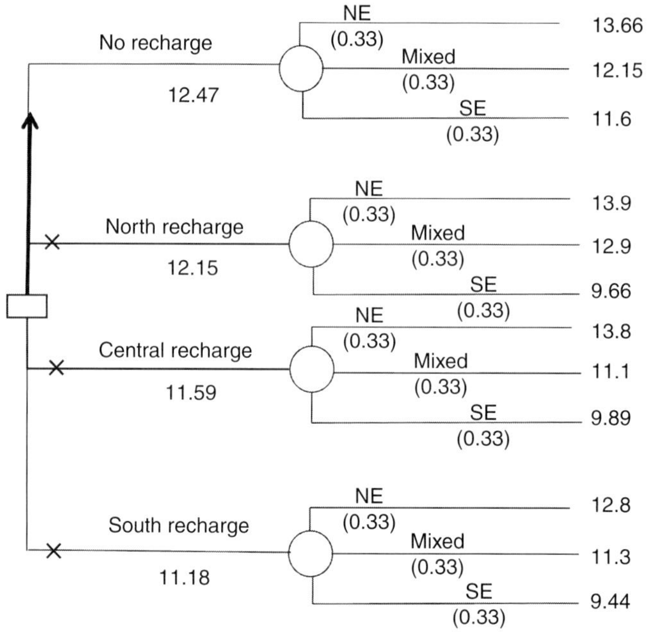

Figure 6.38 Groundwater management example: decision tree representing the prior situation showing the alternatives, the geological scenario uncertainty, and the associated values.

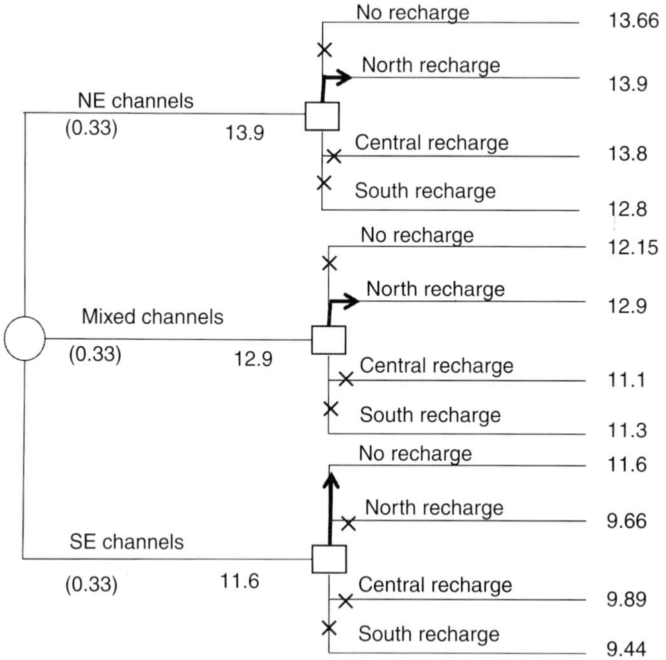

Figure 6.39 Groundwater management example: flipped decision tree representing perfect information about the geological scenario uncertainty before the alternatives and the associated values.

This represents an upper bound on how much a risk-neutral decision maker should be willing to pay to get relevant information about the uncertainty – in this case, about the subsurface conditions.

We will continue this example using influence diagrams. While this is a fairly simple example that can be solved using either decision trees or influence diagrams, the use of influence diagrams allows us to potentially handle more complex VOI problems than can be handled by decision trees.

We illustrate how this situation can be solved by influence diagrams using Netica as the software. Chapter 7 has a hands-on example. First, represent chance nodes for the geologic uncertainty with the three states ("MixedChannels," "NEChannels," and "SEChannels"). Next, add a decision node with four states ("NoRecharge," "CentralRecharge," "NorthRecharge," and "SouthRecharge"). Finally, define a value node. Figure 6.40 shows the setup.

Next, we fill in the tables in this display with the probabilities and values defined in Table 6.5. The prior value (or expected utility) can then be obtained by selecting "compile" or similar from a menu bar. Netica solves the network by internally creating a junction tree for efficient inference. The expected utility is reported to be 12.47, and the optimal decision is "NoRecharge." The value with perfect information can be calculated by first adding an arc from the "Channels" node to the decision node. This arc indicates that the uncertain factor will be known before the decision. As we have shown, compiling will give 12.8 as the expected utility, and the value of perfect information is $12.8 - 12.47 = 0.33$. Note that the optimal decision (from the "optimize decisions" menu) is to recharge at the north for both the mixed or northeast channel orientations and no recharge for the southeast channel orientation.

The value of perfect information is an upper bound. The value of imperfect information about the subsurface geological uncertainty obtained from geophysical surveys such as TEM cannot exceed this bound. The basic TEM data are the magnetic field response as a

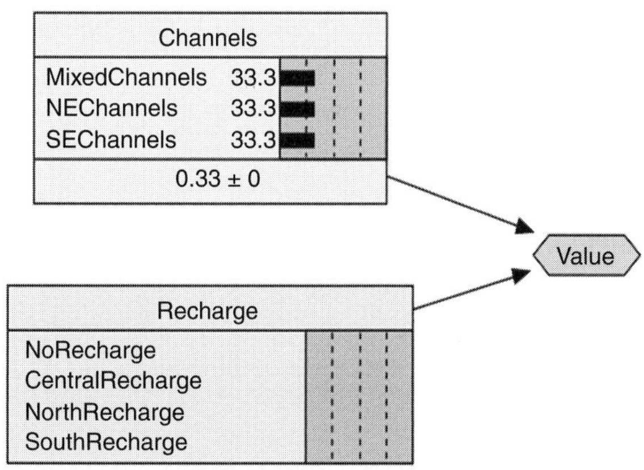

Figure 6.40 Groundwater management example: basic influence diagram for the groundwater recharge example.

function of time. These data have to be inverted for the subsurface resistivity that can be an indicator of subsurface lithologies. Clay-rich shales typically have lower resistivity (less than 30 ohm-m) compared to water-saturated sands (~80 omh-m).

Thus, TEM data after inversion and interpretation can help identify the subsurface channel heterogeneity (e.g., Auken et al. 2008). In order to compute the VOI for this geophysical data, we use the data reliability given by the conditional probability *p(data interpretation|true geological scenario)*. Note that this involves not just how reliably the receiver coil records the magnetic field (instrumental error) but also the inversion and the interpretation of the inversion in terms of the underlying geologic scenario. In other engineering applications, it is possible to get numbers for the reliability of tests by conducting repeated experiments under well-controlled laboratory conditions. However, in geoscience applications it is much more challenging, as geology is much more varied from site to site and less well controlled; in addition, here the value is derived not just from the test result itself but from the geological interpretation of the test. Assessments from experts who have worked with these kinds of data in similar geological environments provide one way of obtaining the required conditional probabilities. Another potential approach is that of Monte Carlo simulations, as described in Trainor-Guitton et al. (2011).

As mentioned earlier, using a multiple-point geostatistical technique, 50 facies realizations were generated for each geological scenario. All of these facies realizations were populated with resistivity appropriate for sand and shales, and TEM data were forward simulated. These were then inverted, and finally the inverted resistivity images were interpreted in terms of channel orientations. The interpretation was done by a local spatial autocorrelation to find the direction of maximum spatial correlation, which was taken to be the interpreted orientation of the subsurface channels. The interpretations were compared with the true channel orientations, and the counts of correct versus incorrect interpretations were used to obtain the conditional probability table, as shown in Table 6.6.

The posterior value of TEM must be computed as an average over all TEM interpretations, in each case making the optimal decision from Bayes' rule for the conditional probabilities and conditional expected values. We illustrate how this situation can be solved by influence diagrams. Let us add a chance or uncertainty node for TEM data to our original influence diagram. This network is shown in Figure 6.41.

Table 6.6. *The reliability measures represented as a conditional probability table for TEM interpretation (columns) given the true geological scenario (rows)*

Channels	Interpretation of Northeast	Interpretation of Mixed	Interpretation of Southeast
Northeast channels	0.98	0	0.02
Mixed channels	0	0.96	0.04
Southeast channels	0	0	1

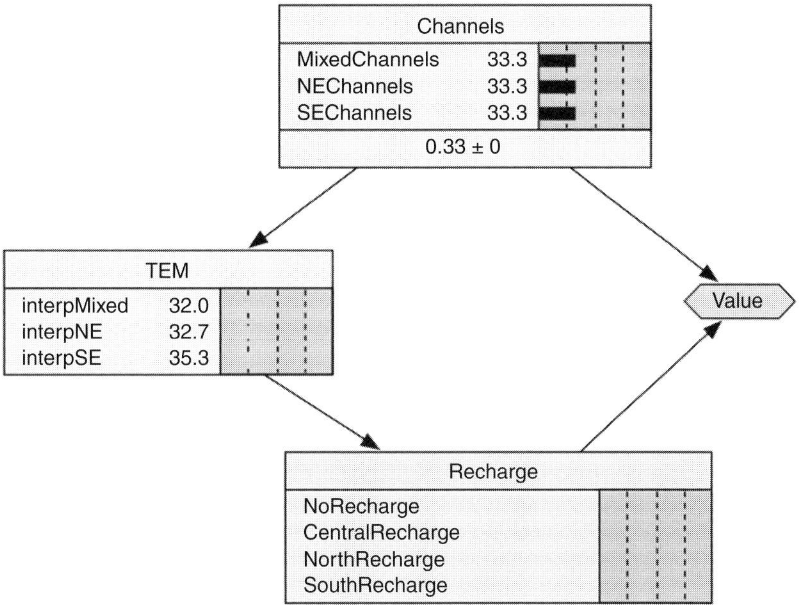

Figure 6.41 Groundwater management example: influence diagram for the groundwater recharge example with a node for geophysical (TEM) data.

To solve the influence diagram, we select "compile" from the menu bar in Netica for solving the influence diagram by the junction tree algorithm. The expected utility given in the messages window (12.78) is close to the value with perfect information (12.8). This is not surprising, since the data are very reliable, though not perfect. Perfect information would be represented by "1" on the diagonal and "0" elsewhere. The value of this imperfect TEM information is 12.78 – 12.47 = 0.31. The TEM survey should then be acquired and interpreted if the price of the data is less than this value. The optimal decisions (from the "optimize decisions" menu) show that the optimal decision is to recharge at the north for both cases of TEM data interpreted as mixed or northeast channel orientations and no recharge when the TEM interpretation is southeast channel orientation. This is similar to the case with perfect information.

6.6 Bibliographic notes

VOI analysis is popular in several domains of applications. For instance, there are abundant health and medical applications – see, e.g., Yokota and Thompson (2004b), Willan and Pinto (2005), and Welton et al. (2008) for a partial overview of this vast field or Braithwaite and Scotch (2013) for a recent paper advocating the use of VOI in clinical management and testing. Another domain is that of product engineering – see, e.g., Keisler and Brodfuehrer (2009), who apply VOI analysis to automobile manufacturing, or Panchal et al. (2009), who apply VOI analysis to improved product design for data center cooling systems. Please see the bibliographic notes in Chapter 3 for further references on VOI applications in general. Our focus in this section is on literature specifically related to VOI in the Earth sciences. Several relevant articles have been grouped into the following categories. Interested readers may pursue the mentioned papers or references therein to find more literature on the particular Earth sciences topic.

Petroleum

The petroleum industry has a long history of VOI analysis dating back to at least Grayson (1960). See Bratvold et al. (2009) for a recent overview of VOI in the industry. Despite the interest in VOI analysis, most efforts have been at a high level without fully integrating spatial modeling and spatial decision making. Bickel and Smith (2006) and Bickel et al. (2008b) study decision making for reservoir development, with dependent petroleum prospects. Their focus is on the best sequential strategy and not on VOI analysis. Martinelli et al. (2011) calculate the VOI of perfect information from exploration wells at segments, modeled by a Bayesian network, as in Section 6.2.1. Martinelli et al. (2013b) use basin modeling to learn a network and then conduct VOI analysis, similar to Section 6.2.2. Manoharan (2014) conducts VOI analysis for petroleum prospect risking, not unlike what we described in Section 6.2.3.

We list a number of papers that value geophysical data and, to some extent, contain elements of spatial or multivariate modeling: Houck and Pavlov (2006) calculate the VOI of electromagnetic data using a threshold method and a sensitivity map for spatially distributed anomalies. Kolbjørnsen et al. (2012) evaluate the VOI of electromagnetic data using a linear fluid factor model. Bickel et al. (2008a) consider a multivariate reservoir model and conduct VOI analysis for seismic data with constraints on the well budget. Houck (2007) and Pinto et al. (2011) study the VOI of time-lapse seismic data. Blangy et al. (2014) evaluate the value of time-lapse seismic data at a more qualitative level. Bhattacharjya et al. (2010) present VOI analysis for seismic amplitude data for making improved spatial decisions about drilling at units where discrete facies variables are the distinctions of interest using a MRF model as in Section 6.3.1. Eidsvik et al. (2008) apply a logistic model to saturation and porosity and analyze the VOI of seismic amplitude data and EM data for reservoir drilling decisions, similar to Section 6.3.2. Rezaie et al. (2014) extend this to a spatial closed-skew Gaussian representation. Barros et al. (2014) describe a VOI workflow for history matching in closed-loop reservoir management using creative computational techniques (see Section 5.10).

A topic related to VOI of geophysical data is that of constructing spatial acquisition designs for surveys in the petroleum industry. Examples include Vermeer (2012), who outlines the many aspects of seismic survey design; Curtis (2004), who provides a tutorial on traditional statistical design of experiments for geophysical data acquisition; and Horesh et al. (2010), who look at inverse problems, design, and uncertainty quantification. Unlike VOI analysis, these approaches for constructing useful experiments are not based on decision theoretic principles.

Mining planning and safety

The mining industry has a long history of evaluating various kinds of spatial information – see, e.g., the book by Chiles and Delfiner (2012) or the various methods presented by Rivoirard (1987) for measuring the information content, such as the Kriging variance, slope, correlation, or weight of the mean. These measures are, however, not directly based on any decision situation. Froyland et al. (2004) compute the VOI of infill drilling decisions (scheduling). Phillips et al. (2009) apply VOI analysis for the selection of ore grade scanning data, which is applied to an iron ore mine in Sweden where there is uncertainty regarding categorical ore types at spatial sections. Eidsvik and Ellefmo (2013) conduct VOI analysis of perfect and imperfect oxide grade measurements, assuming a Gaussian spatial regression model for the oxide grade in a mining project, as shown in Section 6.4.1.

Karam et al. (2007) outline principles for decision situations and information-gathering schemes in mining and tunneling operations. Zetterlund et al. (2011) conduct VOI analysis of rock measurements for decisions about tunneling projects. They model classes of deformation as spatial variables along with a beta distribution for the probability of sealing fractures. The case has low decision flexibility with only two alternatives. Evangelou and Eidsvik (2015) use the Poisson regression model for joints and study a decision situation of high decision flexibility, similar to Section 6.4.2.

Groundwater, hydrology, and geothermal resources

Wagner et al. (1992) introduce VOI analysis for groundwater management in their stochastic optimization for pumping and recharge costs. Their approach is simulation and optimization based but includes spatial correlation in the model for hydraulic conductivity. Borisova et al. (2005) present VOI analysis for the management and control of agricultural nitrogen pollution in the Susquehanna River Basin. They model the economic uncertainties as well as the uncertainties of pollutant transport in the watershed. Nowak et al. (2012) use measures related to VOI for studying the value of hydrological field campaigns in the situation of a contaminant source in an aquifer. Trainor-Guitton et al. (2011, 2013) study the VOI of EM data for groundwater sustainability decisions in an aquifer. They use extensive spatial modeling to estimate the value function of the different spatial alternatives for recharge, similar to our example in Section 6.5. Trainor-Guitton et al. (2014) study an example of a hidden geothermal resource where magnetotelluric data could be useful for guiding the spatial drilling alternatives.

Environmental applications

Borsuk et al. (2001) describe VOI analysis for management decisions regarding the water quality of a river estuary. They use a Bayesian network for the various model components, where edges and conditional probabilities are guided by (spatial) physical processes. De Bruin et al. (2001) apply VOI for evaluating geographical information related to the volumes of sand at a port in the Netherlands. Bouma et al. (2009) present an application of VOI analysis of satellite observations for monitoring algal bloom in the North Sea.

Bates et al. (2014) conduct VOI analysis for a spatial decision situation with several alternatives regarding contaminant management in a fjord. The value calculation builds on several factors, including the environmental (spatial) impact. Ballari et al. (2012) present VOI for mobile wireless sensor networks for environmentally connected decisions. The mobility of sensors offers substantial opportunities for spatial or spatiotemporal information gathering.

Biological conservation, forestry, and fishing

Polasky and Solow (2001) study information measures for the problem of selecting biological conservation sites. Williams et al. (2011) study the VOI of monitoring exercises for management alternatives about the conservation of a rare bird species. Kangas (2010) describes a situation with a forest owner who can harvest immediately or after 10 years and studies the VOI in this context. Several data-gathering options are discussed in the paper, such as small test area sampling and remote sensing. Mantyniemi et al. (2009) study the VOI of control measurements for fish management, applied to a herring population with uncertain dynamic models. Hansen and Jones (2008) review VOI approaches in fisheries.

Agriculture and climate forecasting

Byerlee and Anderson (1982) present VOI analysis for agricultural applications. Adams et al. (1995) discuss the VOI for weather information for agricultural management, in particular the value of El Niño Southern Oscillation forecasts for an enterprise. An example with farming policies, with or without climate information, is shown in Cabrera et al. (2006). Meza et al. (2008) review evidence for the economic value of seasonal climate forecasts for agriculture.

7

Problems and projects

Figure 7.1 concepts, equations seep in and stay for good, if you do it yourself

Our aim in this chapter is to make it easier for the reader to get started with value of information (VOI) analysis. The problems and projects provide some training in statistics, decision analysis, and spatial modeling, but the main goal is to allow the reader to learn VOI analysis by solving problems and completing projects. We have included a variety of questions – there are smaller constructed examples as well as more real-world cases. Some of the problems can be solved by pen and paper, while larger projects require more hands-on computer implementation that comes with models or data sets available for download from the book website (srb.stanford.edu/voi). This website also provides MATLAB and Netica files and other auxiliary informational documents that we hope will be useful for the problems and projects.

Section 7.1 mainly contains problem or project versions of examples used in the book, such as those used to illustrate VOI analysis for spatial decision making in Chapter 5. The real data examples in Sections 7.2–5 follow the same section structure as in Chapter 6. Readers can modify various parameters of the examples and study the effect on the VOI. Note that the model descriptions in Sections 7.2–5 are very short. Before starting a hands-on project, we advise the reader to review the associated example(s) with much more explanation and discussion in Sections 6.2–5 and to consult the book website. If the decision maker's utility function is not explicitly specified for a problem or a project, the reader should assume that the decision maker is risk neutral – i.e., has a linear utility function.

7.1 Problems and tutorial hands-on projects

The first few problems cover concepts from Chapters 2 and 3. They typically start with some modeling assumptions and methodological aspects, and then progress to VOI analysis. For each exercise, we indicate the most relevant chapter in parentheses, where the topic of the exercise is discussed. Some exercises are necessarily more complex – for instance, those involving spatial modeling and spatial decision situations (Chapters 4 and 5), and we present these as hands-on projects. Note that several models are described in some detail in the appendix.

7.1.1 Problem sets

1. Consider the power law pdf, $p(x)=\frac{\beta-1}{\alpha}\left(\frac{\alpha}{x}\right)^{\beta}$, for $x>\alpha>0$ and $\beta>1$.

   a. (2) Show that the cumulative distribution function is $F(x)=1-\left(\frac{\alpha}{x}\right)^{\beta-1}$.

   b. (2) Show that the expected value is $E(x)=\frac{\beta-1}{\beta-2}\alpha$.

   c. (2) One can generate a random variable x from this power law distribution by inversion. Set $\beta=3$, $\alpha=1$. Generate a uniform random variable $u=U(0,1)$. Set $u=1-F(x)$, and solve for the sampling variable x. Repeat the sampling $B=1000$ times, display the realizations in various ways, and calculate the mean to compare with the result from b.

   d. (3) Suppose that the power law distribution mentioned earlier describes the distribution of revenue from a project that a decision maker is considering investing in. He or she either invests or does not invest. The decision maker invests only if the expected revenue is larger than Cost. Calculate the prior value as a function of Cost and the statistical model parameters. Assume that $\beta>2$.

e. (3) Assume that one can purchase perfect information about the project revenue. Calculate the posterior value and the VOI. Compute and plot the VOI for a variety of input parameters: Cost, $\beta > 2$, α.

f. (3) Use the sampling scheme in c. to compute a Monte Carlo approximation of the VOI. Compare this with the analytical result in e.

2. Assume that the profit of a project has a univariate Gaussian pdf

$$p(x) = \frac{1}{\sqrt{2\pi}}\frac{1}{\sigma}\exp\left(-\frac{(x-\mu)^2}{2\sigma^2}\right), \quad -\infty < x < \infty.$$

A decision maker will invest in a project that has a positive expected value.

a. (2) Use symmetry arguments to show that the mean μ of the Gaussian satisfies $\int (x-\mu)\, p(x)\, dx = E(x) - \mu = 0$.

b. (2) Use a transformation of variables: $z = g(x) = \frac{x-\mu}{\sigma}$, $x = g^{-1}(z) = \mu + \sigma z$, and the transformation formula $p(z) = \left|\frac{dg^{-1}}{dz}\right| p\left(g^{-1}(z)\right)$ to show that the variable z is standard normal distributed.

c. (2) Use integration by parts – i.e., $\left(\int u \cdot v' = u \cdot v \,| - \int u' \cdot v\right)$ to show that $\int_a^b z\phi(z)\, dz = \int_a^b 1 \cdot z\phi(z)\, dz = \phi(a) - \phi(b)$, where the standard normal pdf is $\phi(z) =$ $\phi(z) = \frac{1}{\sqrt{2\pi}}\exp\left(-\frac{z^2}{2}\right)$.

d. (3) Suppose that a decision maker is considering getting perfect information about the profit. The posterior value of perfect information is $PoV(x) = \int \max\{0, x\}\, p(x)\, dx = \int_0^\infty x p(x)\, dx$. Use the results from b. and c. along with the symmetry properties of the standard normal pdf $\phi(z)$ and its cumulative distribution $\Phi(z)$ to show that $PoV(x) = \sigma\phi\left(\frac{\mu}{\sigma}\right) + \mu\Phi\left(\frac{\mu}{\sigma}\right)$.

e. (3) Plot the analytical VOI solution as a function of the mean in the range $\mu \in (-2, 2)$ and standard deviation $\sigma \in (0.1, 2)$. Draw realizations from the Gaussian, and use Monte Carlo integration to check the analytical solution for the VOI.

3. Consider a univariate skew-normal model, as described in Section 4.5, with pdf $p(x) = 2\Phi(\lambda(x-\mu))N(\mu, \sigma^2)$, with standard normal cumulative distribution $\Phi(z)$. This is a special case of a closed skew normal with parameters $q = 1$, $\varsigma = 0$, and $\Delta = 1$.

a. (2) Plot this skew-normal pdf for a grid of x values and for different values of the location $\mu \in (-2, 2)$, scale $\sigma \in (0.1, 1)$, and skewness $\lambda \in (-10, 10)$ parameters.

b. (2) Draw samples from this skew-normal distribution by rejection sampling using proposal density $q(x) = N(\mu, \sigma^2)$. Study the acceptance rate.
c. (3) Assume that x represents an uncertain profit of a project. Compute the prior value and the value of perfect information using Monte Carlo sampling (rejection) from the skew normal.
d. (3) Assume instead that the decision maker can purchase imperfect information according to the Gaussian likelihood $p(y \mid x) = N(x, \tau^2)$. This is a special case of the closed skew-normal model, and there is a closed-form solution for the posterior mean $E(x \mid y)$; see Dominguez-Molina et al. (2003). Use this result and Monte Carlo sampling over data y to compute the VOI for this decision situation. Study the VOI for different measurement standard deviations $\tau^2 \in (0.1, 1)$ and compare with the result in c.

4. Consider the bivariate Gaussian pdf $p(\boldsymbol{x}) = N(\boldsymbol{0}, \boldsymbol{\Sigma})$ for profits $\boldsymbol{x} = (x_1, x_2)$ at two projects, just like the "For whom the bell tolls" examples in Chapters 2 and 3. Assume variance 1 for both projects and correlation $-1 < \rho < 1$. The decision maker selects a project (alternative $a_i = 1, i = 1,2$) if its expected profit is positive and otherwise avoids investment (alternative $a_i = 0$, $i = 1,2$). We consider a decision situation with a free selection of projects without constraints (see the example in Section 3.4). Assume that the decision maker can purchase total imperfect information, $y_j = x_j + N(0, \tau^2)$, $j = 1,2$ – i.e., likelihood $p(\boldsymbol{y} \mid \boldsymbol{x}) = N(\boldsymbol{x}, \boldsymbol{T})$, $\boldsymbol{T} = \tau^2 \boldsymbol{I}$.

a. (2) Show that the conditional distribution of profits, given the imperfect information, is Gaussian with covariance $\boldsymbol{\Sigma}_{x|y} = (\boldsymbol{\Sigma}^{-1} + \boldsymbol{T}^{-1})^{-1}$ and mean $\boldsymbol{\mu}_{x|y} = \boldsymbol{\Sigma}_{x|y} \boldsymbol{T}^{-1} \boldsymbol{y}$. Here, use the relation $p(\boldsymbol{x} \mid \boldsymbol{y}) \propto p(\boldsymbol{x}) p(\boldsymbol{y} \mid \boldsymbol{x})$, and complete the quadratic forms for $\boldsymbol{x}$ in the exponent of the Gaussian pdfs.
b. (2) Use the formula for linear combinations of Gaussian variables to show that $p(\boldsymbol{y}) = N(\boldsymbol{0}, \boldsymbol{\Sigma} + \boldsymbol{T})$ and $p(\boldsymbol{\mu}_{x|y}) = N(\boldsymbol{0}, \boldsymbol{\Sigma}_{x|y} \boldsymbol{T}^{-1} (\boldsymbol{\Sigma} + \boldsymbol{T}) \boldsymbol{T}^{-T} \boldsymbol{\Sigma}_{x|y})$.
c. (3) Use the result from b. and from Exercise 2 to compute the posterior value of total imperfect information: $PoV(\boldsymbol{y}) = \sum_{j=1}^{2} \int \max\{0, \mu_{x|y,j}\} p(\boldsymbol{y}) d\boldsymbol{y}$. (Hint: note that the only part of the data that is informative of the decision is $\boldsymbol{\mu}_{x|y}$, so the expression simplifies to: $PoV(\boldsymbol{y}) = \sum_{j=1}^{2} \int \max\{0, \mu_{x|y,j}\} p(\mu_{x|y,j}) d\mu_{x|y,j}$.)
d. (3) Cross-plot the VOI results as a function of the correlation ρ and for different test accuracies τ of the imperfect test.
e. (3) Assume next that the decision maker can gather information about only one of the projects (partial testing) say, $y_1 = x_1 + N(0, \tau^2) = \boldsymbol{F}\boldsymbol{x} + N(0, \tau^2)$, where $\boldsymbol{F} = (1, 0)$. Calculate the posterior value of this partial test.
f. (3) Let the price of a test be P_j for Project $j = 1,2$. The decision maker should purchase the test that has the largest VOI compared with the test price if it is positive.

Table 7.1. *The joint pdf for the two petroleum prospects and their marginal probabilities*

	Failure Prospect 2	**Success Prospect 2**	**Marginal Probability**
Failure Prospect 1	0.85	0.05	0.9
Success Prospect 1	0.05	0.05	0.1
Marginal Probability	0.9	0.1	1

The price of testing both projects is $P_1 + P_2$. Make a diagram of the decision regions as a function of price P_1 on the first axis and P_2 on the second axis. Draw this diagram for a couple of correlation parameters ρ and test accuracies τ.

5. Consider a binary bivariate example concerning the selection of petroleum Prospects 1 and/or 2 for development. The joint probability density function (pdf) is defined by Table 7.1. The petroleum company aims to develop infrastructure and produce hydrocarbons. The decision about selecting prospects for development depends on the probabilities of failure and success as well as the revenues associated with the successful outcome and the cost of selection. In the event of a success at Prospect 1, the company receives revenue $\text{Rev}_1 = \text{Rev} = 3$, and the same at Prospect 2, $\text{Rev}_2 = \text{Rev} = 3$. The fixed cost of selection, no matter the outcome of the prospect, is set to $\text{Cost}_1 = \text{Cost}_2 = \text{Cost}$.

a. (2) Suppose that the company can obtain perfect information at one of the prospects. Use Bayes' rule to show that the conditional probabilities are:

$$p(x_1 = 1 \mid x_2 = 1) = p(x_2 = 1 \mid x_1 = 1) = 0.5,$$

$$p(x_1 = 1 \mid x_2 = 0) = p(x_2 = 1 \mid x_1 = 0) = 0.055.$$

b. (3) Assume that the company can freely select prospects when they are expected to be profitable. Show that the prior value is $PV = 2\max\{0, 0.3 - \text{Cost}\}$.

c. (3) Show that the posterior value of total perfect information at both Prospects 1 and 2 is $PoV(\boldsymbol{x}) = 0.2\max\{0, 3 - \text{Cost}\}$. Plot the posterior value, with the prior value in b., as a function of Cost. Interpret the results.

d. (3) Compute the posterior value of perfect partial information at Prospect 1:

$$PoV(x_1) = \max\{0, 3 - \text{Cost}\}\, p(x_1 = 1) + \sum_{z=0}^{1} \max\{0, 3 \cdot p(x_2 = 1 \mid x_1 = z) - \text{Cost}\}\, p(x_1 = z)$$

e. (3) Plot the posterior value in d. with the one from c. and the prior value in b. as a function of Cost. Interpret the results.

6. Let us expand the same binary bivariate example, concerning the selection of petroleum Prospects 1 and/or 2 for development, with imperfect information. We denote the imperfect test(s) by y_1 and/or y_2 at the two prospects. This could, for instance, be seismic test(s) at the petroleum prospect(s). We assume that these are binary: positive or negative – i.e., $y_j \in \{0,1\}$ for $j = 1,2$. In the likelihood model, we assume conditional independence and symmetry – i.e., $p(y_j = k \mid x_j = k) = \gamma$ for $k = 0,1$ and for $j = 1,2$.

   a. (2) Compute the marginal bivariate distribution for the test results by marginalizing over $\boldsymbol{x}$.
   b. (3) The posterior value of imperfect information at Prospect 1 is

$$PoV(y_1) = \sum_{z=0}^{1} \begin{pmatrix} \max\{0, 3 \cdot p(x_1 = 1 \mid y_1 = z) - \text{Cost}\} \\ + \max\{0, 3 \cdot p(x_2 = 1 \mid y_1 = z) - \text{Cost}\} \end{pmatrix} p(y_1 = z).$$

   Set likelihood reliability $\gamma = 0.9$ and compute this posterior value as a function of Cost. Plot the results for various Cost with the prior value and the posterior values of perfect information in the previous exercise. Interpret the results.
   c. (3) Compute the posterior value of total imperfect information $PoV(y_1, y_2)$, assuming reliability $\gamma = 0.9$. Plot the result as a function of Cost and compare with the other information-gathering schemes.

7. The following game motivated Bernoulli to introduce the logarithmic utility function in 1738. This is known as the St. Petersburg's paradox. Consider a lottery where a coin is tossed repeatedly until it lands heads, and the owner of the lottery is rewarded \$ $2^{n_{\text{trial}}}$, where n_{trial} is the total number of coin tosses. Assume that heads and tails are equally likely outcomes and that the coin tosses are independent.

   a. (2) What is the pdf for the number of tosses until the coin lands heads? What is the pdf for the reward?
   b. (3) How much would you be willing to pay to own this lottery? (Thinking about this carefully may make the following parts more enjoyable.)
   c. (3) What would a risk-neutral person be willing to pay to own the lottery? Consider the solution obtained and list some reasons why you think that this game is referred to as a paradox.
   d. (3) Suppose that the lottery terminates after a maximum of n_{trial} coin tosses. Plot the expected value of the lottery as a function of n_{trial} from 1 to 50.
   e. (3) Consider a person with an initial wealth of \$1000 and a logarithmic utility function $u(x) = \log(x)$, where x is the person's total wealth. How much would this person be willing to pay to own the lottery for maximum coin tosses $n_{\text{trial}} = 10$? (Hint: Use the same approach that was used for the general definition of VOI as a personal indifference price – i.e., by equating the expected utilities for the situations with and without the lottery.) Plot this amount as a function of n_{trial} from 1 to 50. Then, plot this amount as a function of initial wealth \$100 to \$100 000.

f. (3) Draw an influence diagram for someone who is deciding whether to play this lottery or not. Try this first with a small n_{trial} initially – say, $n_{\text{trial}} = 3$ – and then try to generalize the diagram for a larger n_{trial}. For $n_{\text{trial}} = 3$, also construct the conditional probability tables and the value table for the influence diagram.

8. In this problem, we will build decision trees for a situation related to CO_2 sequestration, where there is a risk of leakage (Konishi, 2014). The decision maker can proceed with CO_2 injection or suspend sequestration. The latter incurs a tax of 80 monetary units. The former only has a cost of injection equal to 30 monetary units, but the injected CO_2 may leak. If leakage occurs, there will be a fine of 60 monetary units (i.e., a cost of 90 in total). The probability of leakage is $p(x=1)=0.3$, while a safe or sealing formation has probability $p(x=0)=0.7$.

   a. (3) Draw a decision tree for this situation. What alternative does the maximum expected value (or minimum expected cost) have?
   b. (3) What if the decision maker could get perfect information about the leaking or sealing formation? Draw a decision tree for this situation with perfect information. What is the maximum expected value (or minimum expected cost)?
   c. (3) Assume next that the decision maker can purchase a geophysical experiment with a binary outcome, indicating whether the formation is leaking or not. The reliability of the test is defined by $p(y=1 \mid x=1)=0.9$ and $p(y=0 \mid x=0)=0.95$. Draw the decision tree for this situation with imperfect information and compute the maximum expected value.

9. Consider the oil wildcatter example described in Chapter 3. Figure 3.9 shows the influence diagram for the oil wildcatter's decision situation where he or she is considering whether to purchase a seismic test for more information about the amount of oil.

   a. (2) Compute the pre-posterior probability for the seismic test result and the posterior probability of the amount of oil given a seismic test result using Bayes' rule.
   b. (3) Draw the decision tree corresponding to the influence diagram.
   c. (3) Solve the decision tree for a risk-neutral decision maker – i.e., find the optimal strategy and the expected value.
   d. (3) Solve the decision tree for a decision maker with an exponential utility function, with risk aversion coefficient $\gamma = 0.002$ (M$)^(-1) – i.e., find the optimal strategy and the certain equivalent.

10. Consider the "MacKenna's gold" example described in Chapter 3, where a person owns an oil field. Figure 3.11 shows the influence diagram, and Figure 3.12 shows the probability tree that was used to compute the certain equivalent. Make the same assumptions about the person's utility function as was made earlier. Suppose that the person's financial advisor suggests a lottery/deal through a futures contract where the payoff is –$50 million when the price of gold is high and $100 million when the price of gold is low.

a. (3) Suppose that the person does not own the oil field and only owns the futures contract (as determined by the price of gold). Update the influence diagram in Figure 3.11 and the probability tree in Figure 3.12 with this new situation. Find the certain equivalent for this new situation.
b. (3) Now suppose that the person owns both lotteries (the oil field and the futures contract). Update the influence diagram in Figure 3.11 to also include the new lottery (along with the older lottery of the oil field) by adding another value node and a corresponding value table.
c. (3) Edit the influence diagram mentioned earlier by merging the value nodes together into one node. What are the parents of this value node? Form the value table corresponding to this new value node by merging the previous two value tables. (Hint: the new value table will grow in size.)
d. (3) Using the formulation in c. mentioned earlier, update the probability tree in Figure 3.12 and compute the new certain equivalent.
e. (3) Compare the certain equivalents of the three situations where the person (i) owns only the oil field, (ii) owns only the futures contract, and (iii) owns both.

11. This problem is similar to the "Never break the chain" examples. We consider the following Markov chain models, differing in Markov transition probabilities $P(k,l)$:

(i) Independent case: $P(0,0) = P(1,1) = 0.5$, $p(x_1 = 1) = 0.5$.
(ii) Equal dependence: $P(0,0) = P(1,1) = 0.9$, $p(x_1 = 1) = 0.5$.
(iii) Absorbing: $P(0,0) = 1$, $P(1,1) = 0.9$, $p(x_1 = 1) = 0.5$.

a. (2) Simulate Markov chains realizations of length $n = 100$ from each of these models and compare the results. (The simulation is done by simulating uniform random variables $U(0,1)$. If the number is larger than 0.5, the initial state $x_1 = 1$; otherwise, it is $x_1 = 0$. At the subsequent step, we draw new independent uniform variables and set the correct state according to the conditional probability defined by the Markov transition rule.)
b. (2) Compute the marginal probabilities for x_i, $i = 1,\ldots,n$, for the Markov chain with an absorbing state. Compare with realizations.
c. (2) Assume that we know that $x_{20} = 1$ and compute the conditional probabilities for all states x_i, $i = 1,\ldots,n$. (Note that the computation is straightforward for the subsequent states but more difficult for the states prior to the observation site.)
d. (2) Compute similarly the probabilities when we know that $x_{20} = 0$.
e. (3) Assume that the nodes represent petroleum prospects. The decision situation is that of free selection of sites. Let us compare the VOI of information at single nodes and for the different Markov chain models. We assign costs associated with the selection of petroleum prospects and revenues associated with successful prospects. In this example, assume fixed revenues and costs: Rev = 3 and Cost =1, identical for all sites. Compute and plot the VOI of the single observation nodes. Discuss and interpret the results. For which node is it better to collect data, and what should the price of the experiment be to make it worthwhile?

12. Let us consider the hidden Markov model version of the "Never break the chain" examples, which adds a hierarchical level of uncertainty at every node in the Markov chain model.

    a. (2) Simulate a Markov chain of length $n = 250$ with equal dependence: $P(0,0) = P(1,1) = p$, $p(x_1 = 1) = 0.5$. We set $p = 0.9$. For each node or time step, generate conditionally independent data $y_j = x_j + N(0, \tau^2)$, $j = 1,...,n$. We set $\tau^2 = 0.4^2$. Plot the data $\boldsymbol{y} = (y_1,...,y_n)$ and interpret.
    b. (2) Describe the forward recursion for evaluating the marginal likelihood $p(\boldsymbol{y})$. (This is based on marginalizing over one of the x_j variables at a time to get $p(y_j \mid y_{j-1},...,y_1)$.) Compute the marginal likelihood model for a grid of values for p and τ^2, given the data from a., and find the maximum likelihood estimate (MLE). Compare with the values used in the simulation.
    c. (2) Use the forward–backward recursions to find the marginal probabilities $p(x_j = 1 \mid \boldsymbol{y})$ for all $j = 1,...,n$. Use similarly the forward–backward recursions to sample realizations $\boldsymbol{x}^b$ from $p(\boldsymbol{x} \mid \boldsymbol{y})$, $b = 1,...,100$.
    d. (3) Suppose that the nodes represent petroleum prospects. The decision situation is that of free selection of nodes or sites. In addition to the current data $\boldsymbol{y}$, the decision maker considers purchasing perfect information at a single node. Like in the previous exercise, there are costs associated with the selection of prospects and revenues associated with successful prospects: Rev = 3 and Cost = 1, identical for all sites. Compute and plot the VOI of the single observation nodes. Discuss the results. (Note that you will get different results every time because the current data $\boldsymbol{y}$ vary from simulation to simulation.)
    e. (3) Suppose instead that the decision maker can only select disjoint blocks of size 10 – i.e., 25 blocks in total. A block is selected if the sum of expected revenues for that block is larger than the sum of costs for the block. Compute the prior value of this decision situation. Compute the VOI of observations at a single node for this decision situation.

13. Consider the logistic integral that is common for modeling success probabilities in generalized linear models (GLM):

$$I = \int_{-\infty}^{\infty} \frac{\exp(x)}{1+\exp(x)} \frac{1}{\sqrt{2\pi\sigma^2}} \exp\left(-\frac{(x-\mu)^2}{2\sigma^2}\right) dx.$$

Set parameter $\sigma^2 = 0.1^2$, and let μ vary around $\pm 4\sigma$.

    a. (2) Approximate the integral by Monte Carlo sampling from the Gaussian with mean μ and variance σ^2. Plot the results as a function of μ.
    b. (2) Approximate the integral by the Laplace approximation. This entails expanding the exponent in the integrand to a quadratic form. We have $\frac{\exp(x)}{1+\exp(x)} = \frac{1}{1+\exp(-x)} = \exp(-\log(1+\exp(-x)))$. The method further relies on

maximizing the exponent and finding the curvature at the mode from the second derivative of the exponent. Plot the approximation as a function of μ.

c. (2) Approximate the integral using an alternative function in the integrand (Demidenko 2004). This approach relies on the fact that the logistic expression is rather similar to the cumulative distribution of the standard Gaussian:

$$\frac{\exp(x)}{1+\exp(x)} \approx \int_{-\infty}^{\alpha x} \frac{1}{\sqrt{2\pi}} \exp\left(-\frac{z^2}{2}\right) dz, \quad \alpha = \frac{16\sqrt{3}}{15\pi} \approx 1.7,$$

where the seemingly arbitrary number for α fits the logistic function for a wide range of parameters (Demidenko 2004). The expected value of the cumulative Gaussian is now analytically tractable as another cumulative, and we get the estimate:

$$\hat{I} = \int_{-\infty}^{\frac{\alpha\mu}{\sqrt{1+\alpha^2\sigma^2}}} \frac{1}{\sqrt{2\pi}} \exp\left(-\frac{z^2}{2}\right) dz = \Phi\left(\frac{\alpha\mu}{\sqrt{1+\alpha^2\sigma^2}}\right).$$

Plot this integral approximation as a function of μ.

d. (3) Assume that the logistic model represents the probability of success, awarded with a revenue defined by Rev = 3, while there is Cost = 1 in either outcome (success and failure). For this decision situation, the prior value is $\max\{\text{Rev}\cdot I - \text{Cost}, 0\}$, while the posterior value with perfect information is

$$I_{post} = \int_{-\infty}^{\infty} \max\left\{0, \text{Rev}\cdot\left(\frac{\exp(x)}{1+\exp(x)}\right) - \text{Cost}\right\} \frac{1}{\sqrt{2\pi\sigma^2}} \exp\left(-\frac{(x-\mu)^2}{2\sigma^2}\right) dx.$$

Approximate the posterior value by Monte Carlo sampling from $N(\mu, \sigma^2)$.

By introducing $a = \log\left(\frac{\text{Cost} / \text{Rev}}{1 - \text{Cost} / \text{Rev}}\right)$, the posterior value equals

$$I_{post} = \text{Rev}\int_{a}^{\infty}\left(\frac{\exp(x)}{1+\exp(x)}\right) \frac{1}{\sqrt{2\pi\sigma^2}} \exp\left(-\frac{(x-\mu)^2}{2\sigma^2}\right) dx - \text{Cost}\left(1-\Phi(a)\right).$$

The first term can again be approximated by the cumulative normal approximation to the first part of the integrand. This method results in

$$\hat{I}_{post} = \text{Rev}\left(\Phi\left(\frac{\mu-a}{\sigma}\right) - \Phi_2\left(\frac{\mu-a}{\sigma}, \frac{-\alpha\mu}{\sqrt{1+\alpha^2\sigma^2}}; \frac{-\alpha\sigma}{\sqrt{1+\alpha^2\sigma^2}}\right)\right) - \text{Cost}\left(1-\Phi(a)\right),$$

where $\Phi_2(z_1, z_2; \rho)$ is the cumulative distribution function of bivariate standard Gaussian variables with correlation ρ (function "mvncdf" in MATLAB).

Compare this approximate posterior value and the VOI solution with the Monte Carlo approximation.

7.1.2 Hands-on projects

We now present some larger-sized projects from Chapters 3–5.

Time after time: time-lapse seismic example

This example is related to time-lapse seismic monitoring of producing reservoirs and was used in Chapter 3 to illustrate concepts from decision analysis, influence diagrams, and VOI analysis. Readers not familiar with the software Netica or Kevin Murphy's MATLAB Bayes Net Toolbox (BNT) might want to start with a simpler influence diagram project, as described in Section 7.5, before proceeding to this one.

1. (3) The influence diagram model shown in Figure 3.14 is available on the website as a Netica project. Open the influence diagram and see how the different conditional probability distribution (CPD) tables and the value function tables are assigned.
2. (3) Recreate the results for the three scenarios described in Table 3.2.
3. (3) Explore the sensitivity of the results to changes in various reservoir factors such as depth, rock and fluid properties, and spatial heterogeneity.
4. (3) If you are familiar with Kevin Murphy's BNT toolbox (see also Sections 7.2 and 7.5), you can construct the influence diagram in MATLAB and test the various scenarios. A MATLAB file is provided on the website to build and run the model.
5. (3) The value of time-lapse seismic monitoring comes from being able to monitor unexpected changes. In this influence diagram model, unexpected changes come from unexpected breakthroughs (with states "true" or "false") or bypassed oil (with states "high," "medium," or "low"). These in turn depend on the level of spatial heterogeneity, "high," or "low." In general, high spatial heterogeneity will lead to more chances of high bypassed oil and unexpected breakthrough behavior. The CPD tables for the nodes "bypassed oil" and "breakthrough" have been elicited from experts, but the numbers are quite uncertain. Let α and β denote the probability of bypassed oil being "high" and the probability of breakthrough being "true" when spatial heterogeneity is high, nominally set to 20% each. These probabilities are lower when spatial heterogeneity is low – say, $\alpha / 2$ and $\beta / 2$, respectively. Explore the sensitivity of the outputs to these parameters. How sensitive is the ratio of the value if time-lapse seismic monitoring is performed to the value if it is not performed –i.e., value(perform_seismic = true)/ value(perform_seismic = false), to α and β? Vary α and β from 0.02 to 0.4, and make a contour plot of the value ratio as a function of α and β. You will probably find the MATLAB model using the BNT toolbox useful for this exercise, as you can run it in a for-loop with a range of inputs.

Norwegian wood: forestry example

This project builds on the forestry example that was presented in Sections 4.4 and 5.3–4.

Part I: parameter estimation and Kriging

Here we will study the simulation of Gaussian processes or random fields. We simulate spatial data and perform parameter estimation and prediction (Kriging). Consider an exponential covariance model for the spatial distinction of interest $x(\mathbf{s})$.

$$Cov\left(x(\mathbf{s}_i), x(\mathbf{s}_j)\right) = \Sigma_{ij}\left(|\mathbf{t}_{ij}|\right) = \sigma^2 \exp\left(-\eta|\mathbf{t}_{ij}|\right),$$
$$y(\mathbf{s}_j) = x(\mathbf{s}_j) + N(0, \tau^2), \quad j = 1, \ldots, m.$$

where $|\mathbf{t}_{ij}|$ is the Euclidean distance between the two sites $\mathbf{s}_i$ and $\mathbf{s}_j$, and the measurements $y_j = y(\mathbf{s}_j)$ are assumed to be conditionally independent. The unit square is the domain of interest. We assume a mean increasing with east and north coordinates as follows: $\mu_i = \alpha\left((s_{i1} - 0.5) + (s_{i2} - 0.5)\right)$ for site $\mathbf{s}_i = (s_{i1}, s_{i2})$ on the unit square.

1. (4) Simulate $m = 100$ random data sites within the unit square. This is done by simulating uniform numbers along each axis. Plot the data sites. Set parameter values $\sigma^2 = 1$, $\eta = 10$, and $\tau^2 = 0.1^2$. Form the spatial covariance entries from the exponential covariance function to build a $m \times m$ covariance matrix for the data y_j, $j = 1, \ldots, m$. Take its Cholesky factorization and simulate dependent zero-mean Gaussian data variables, then add the mean using $\alpha = 1$.
2. (4) Use the data to estimate the model parameters $\alpha, \sigma^2, \tau^2, \eta$. Do this by maximum likelihood, iterating between an update for the mean parameter, and updating the covariance parameters. Monitor the likelihood function at each step of the algorithm to check for convergence.
3. (4) Use the Gaussian model with the estimated parameters to perform Kriging – i.e., predict variables $x(\mathbf{s})$, where sites are on a regular grid of size 25 × 25 for the unit square. Visualize the Kriging surface and the prediction standard error.
4. (4) Generate conditional realizations at the grid nodes given the data. This can be done in different ways. One alternative is to simulate unconditional realizations of the regular lattice variables, simulate a perturbed version of the data set at the $m = 100$ observation sites, and then add together the unconditional simulation and the spatially weighted (Kriged) perturbed data variable. Compare the simulations with the Kriged surface in Step 3.

Part II: VOI analysis and spatial design

Let us use variance, entropy, and VOI to compare three spatial designs of size $m = 49$. The goal is to make better decisions at all sites in the 25 × 25 regular grid for the unit square ($n = 625$ sites in total). The mean and covariance are the same as those in Part I of this project.

1. (5) Construct design a. based on random sampling – i.e., picking m random cells (the MATLAB command is *randsample(n,m)*). Construct design b. along the center lines (as in Section 4.4). Construct design c. by focused sampling near the northwest and southeast diagonal: 25 samples on the diagonal, and four series of six sites every three

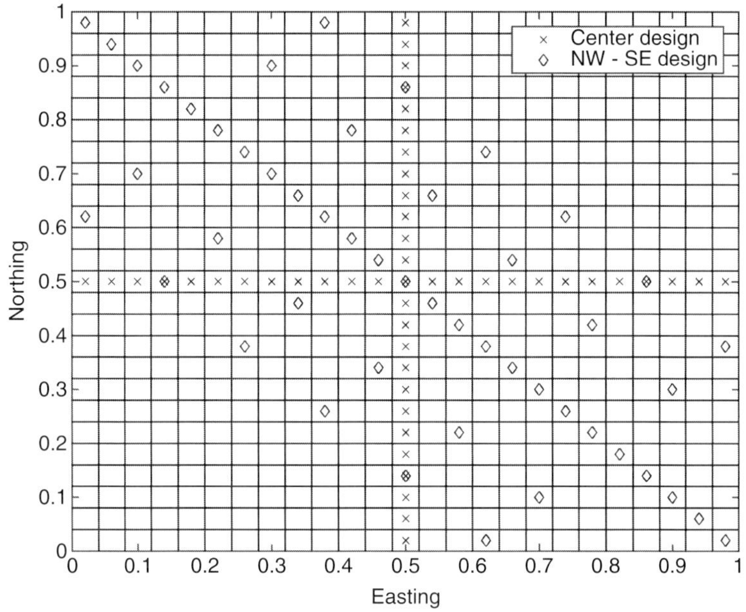

Figure 7.2 Illustration of the spatial designs of experiment. The center design has survey locations along the middle north–south column and the middle east–west row in the grid (crosses). The northwest–southeast design has survey lines near this diagonal (diamonds).

columns and rows away from this diagonal (this is a total of m sampling sites). Plot the three designs. Designs b. and c. are shown in Figure 7.2.

2. (5) Compute the total (sum over all prediction cells) reduction in the variance for variables $x(\mathbf{s})$ on the regular grid. Compare the reduction obtained by the three designs. Similarly, compute the reduction in the entropy obtained by the three designs. (The result should not depend on the mean parameter α.)
3. (5) Assume that the Gaussian variables $x(\mathbf{s})$ represent uncertain profits at the cells on the regular grid. The decision maker would select all the cells if the total profit (sum over all prediction cells) is positive. This is a decision situation with low decision flexibility. Compute the prior value and the VOI for designs a.–c. Does the result depend on α?
4. (5) Assume instead that the decision maker is free to select individual cells that are profitable. This is a decision situation with high decision flexibility. What is the prior value and the VOI of all three designs? How does the result depend on α? Compare with the situation with less decision flexibility.

The tree amigos: conservation biology example

We study the fitting of parameters in a Markov random field (MRF) (Section 4.6) and compare designs for information gathering with the goal of making better decisions about biological conservation, similar to the example in Section 5.4. We consider a regular grid of size $n_1 \times n_1$ and two possible classes (presence or absence) at every cell. We will use the forward–backward algorithm for grids for computations (see the code on the book website).

Part I: parameter specification

1. (4) Generate a realization on a size 10 × 10 grid from an Ising model with interaction parameter $\beta = 0.75$.
2. (4) Fit a neighborhood probability model based on this realization. The fitting will be of increasing complexity, cell-by-cell, center cell given nearest neighbors (northeast, south, and west), given 8 nearest neighbors and given 12 nearest neighbors. We refer to these as zero-, first-, second-, and third-order neighborhoods on the lattice. The conditional probabilities are obtained by counting the number of 0s and 1s at the center node for various configurations of neighbors. In the situation with the third-order neighborhood, there are a total of 4096 possible configurations for the neighbors.
3. (4) Generate 10 new realizations from the Ising model. For each realization, evaluate the probability p of success at the center nodes, given the neighboring configurations for all zero-, first-, second-, and third-order neighborhoods. Compare this prediction with the actual realization at the center node. Use a score function to favor observations x at the center node that are consistent with high probability p. The Brier score is defined by $\text{Score} = \left(I(x=0)-(1-p)\right)^2 + \left(I(x=1)-p\right)^2$. Study how the score varies with increasing neighborhood order. Relate what you see to overfitting the model to the data.

Part II: value of information

For the VOI analysis, the setting is as in Section 5.4 with a grid of size 3 × 3.

1. (5) Assume that the decision maker can freely select cells on the grid for conservation. If the species is present at a cell, a reward of Rev = 2 is obtained. A site selection has Cost = 1. Compute the VOI for total imperfect data gathering. Assume imperfect testing with accuracy $\gamma = 0.9$ in the likelihood model, as in Section 5.4. Computing can be done exactly with the forward algorithm for the likelihood $p(\boldsymbol{y})$ and backward calculation of $p(x_i = 1 \mid \boldsymbol{y})$; see Appendix A.3.
2. (5) Consider now partial tests along the various vertical lines (data size $m = 3$) and a diamond-like design (data size $m = 4$). Compare the VOI of these partial tests with the VOI for total testing from 1. (Note that the marginal probability of the partial test can be computed by marginalizing over the $p(\boldsymbol{y})$. A trick for the forward–backward recursion is to model the data at unobserved sites as completely uninformative data – i.e., set $\gamma_{i,\text{unobserved}} = 0.5$ at the unobserved sites.)

Go with the flow: petroleum simulation example

This hands-on project builds on the petroleum reservoir simulation example that was described in Section 5.5. The VOI is computed for seismic amplitude data, which is informative of the subsurface geology.

You can complete most of the exercise by downloading geostatistical realizations of the reservoir and flow response values from the website, or you may create your own realizations and flow simulations. For the latter, you will need to run software for flow in porous media. Commercial software for this purpose includes Eclipse, which is commonly used in the petroleum industry. The website contains code based on the freely available MATLAB reservoir simulation toolbox (MRST) developed by Lie et al. (2012), which can be downloaded from www.sintef.no/Projectweb/MRST.

Part I: downloading realization outputs

Consider the two geological scenarios and their associated production realizations in Section 5.5. For this exercise, you need to download the production profiles.

1. (5) Compute the prior value and posterior value of the drill or do not drill decision situation. Assume a range of reliability parameters for the scenario interpretation and compare the VOI.

Part II: generate realizations from geologic scenario

1. (4) Define a spatial grid model and a drilling strategy. (You may use what was described in Section 5.5 and available on the website, or you may define your own. A corner design is commonly used for illustration.) Using geostatistical algorithms, generate multiple realizations of facies, porosity, and permeability for two (or more) geologic scenarios. Training images of facies for the two geological scenarios used in Section 5.5 are provided on the website. (Alternatively, you can download the multiple realizations for porosity and permeability from the website.)
2. (5) Run the reservoir simulator for all realizations of porosity and permeability (for each geologic scenario) to create realizations of the production profile over the future years. (The simulator requires more input variables such as initial saturation, production rates, rock compositions, type of flow, etc. You can check the website for these inputs.) Use the output production profiles and profits to approximate the prior value, assuming a cost of development.
3. (5) Assume a rock physics model for seismic acoustic impedance, which takes porosity as input and returns the expected seismic impedance conditional on a realization of the porosity field. (Check the website for the modeling codes.) Create synthetic seismic acoustic impedance for all realizations and visualize the result.
4. (5) Compute a summary statistic of the data for each realization. The variogram of impedance data is one possibility. Classify the realizations into the two (or more) geologic scenario classes according to this summary statistic. This defines the reliability measure represented as a confusion matrix. Use the reliability measure to approximate the VOI of an imperfect seismic interpretation.
5. (5) Form the reliability measure for the seismic data based on clustering techniques. For instance, you may use multidimensional scaling (Scheidt and Caers 2009). Use the results to approximate the VOI of an imperfect seismic interpretation.

Part III: VOI analysis using approximate Bayesian computing

1. (4) Define a spatial grid model and a drilling strategy. (You may use what was described in Section 5.5 and available on the website, or you may define your own. A corner design is commonly used for illustration.) Using geostatistical algorithms, generate multiple realizations of facies, porosity, and permeability from the prior model. Training images for facies for the two geological scenarios used in Section 5.5 are provided on the website. (Alternatively, you can download the multiple realizations for porosity and permeability from the website.)
2. (5) Run the reservoir simulator for all realizations of porosity and permeability to create realizations of the production profile over the future years. (The simulator requires more input variables such as initial saturation, production rates, rock compositions, type of flow, etc. You can check the website for these inputs.) Use the output production profiles and profits to approximate the prior value, assuming a cost of development.
3. (5) Assume a rock physics model for seismic acoustic impedance, which takes porosity as input and returns the expected seismic impedance conditional on a realization of the porosity field. (Check the website for the modeling code.) Create synthetic seismic acoustic impedance for all realizations and visualize the result.
4. (5) Define a summary statistic from the seismic data and generate the loss matrix for each of the realizations. You may, for instance, choose loss as the sum of square distance between synthetic seismic impedances, variogram properties of the synthetic seismic impedances, or some other low-dimensional summary statistic. Use the approximate Bayesian computing algorithm presented in Section 5.6 to approximate the VOI. Use different values of the threshold parameter and discuss the results.

Frozen: hydropower example

The project builds on the hydropower example presented in Section 5.6. The goal is to evaluate snow depth data-gathering schemes for optimization of dam levels.

1. (4) Define a spatial grid model and a system of dams with their catchments in a river basin. Also define high and low (or more) water levels for each dam. (You may use what was described in Section 5.6 and available on the website or define your own.) Generate multiple realizations of snow depth for each grid cell. You can define your own spatial stochastic model or generate a transformed Gaussian random field of some kind (Section 5.6). See the code on the website for one candidate model.
2. (5) Use the geometry of catchments to calculate the value (water) running into the dams for the realizations mentioned earlier. Use the aggregated water in catchments to study what the situation will be for all low–high alternatives for all realizations. Study in particular the situation for the all-low and all-high situations. What is the mean value, and what is the variance over the realizations? Approximate the prior value. Note: you will get different answers every time because of the random snow depths in Step 1.

3. (5) Define an acquisition design for snow measurements. (You may use the one in Section 5.6 or on the website, or you may create your own.) Define further a summary statistic for the snow data at the acquisition sites and compute the loss matrix for each of the realizations. You may choose any summary statistic and corresponding loss function, such as the squared difference in the mean snow thickness at the acquisition sites. Use the approximate Bayesian computing-based algorithm in Section 5.6 to compute the posterior value and conduct VOI analysis when there is an opportunity to perform snow measurements at the indicated locations.
4. (5) Suppose that the decision situation has less decision flexibility: the decision maker is only allowed to adjust the uppermost dams in the river system, while the bottom ones have to be lowered at this time (say, due to flood avoidance systems or for agricultural use). Compute the prior value and the VOI of this decision situation.

7.2 Hands on: exploration of petroleum prospects

The reader is advised to go through Section 6.2 before starting on these hands-on projects, as most exercises are extensions of what was discussed in that section.

7.2.1 Gotta get myself connected: Bayesian network example

For this project, you will need to download Kevin Murphy's Bayesian Network Toolbox (BNT) for MATLAB. This is available at http://code.google.com/p/bnt/. Additional code using BNT is provided on the book website. Readers not familiar with BNT may wish to start with a simpler BNT network project described in Section 7.5 before proceeding to this project.

Part I: small network

Consider the small network shown in Figure 7.3 with 12 nodes, of which 6 are segment nodes, where a petroleum company is considering development. This network is used in Martinelli et al. (2013a). The probability model is provided on the book website. The marginal probabilities for the three discrete classes (dry, gas, and oil) are displayed in Table 7.2, along with the revenues (costs) for the dry, gas, and oil outcomes.

1. Use the code to compute the marginal probabilities and the expected values (intrinsic values) shown in Table 7.2.
2. Assume that the decision maker drills the most profitable Segment 3 and finds oil. Compute the conditional probabilities at all Segments 1, 2, 4, 5, and 6. What is now the (second) most profitable segment? A myopic drilling strategy is based on sequential alternatives using forward updating of conditional probabilities in this way. Assume that the second well at the (conditional) second-best segment is dry. What is the next alternative? Should the decision maker continue drilling?
3. Assume free selection of segments, and compute $VOI(x_{\mathbb{K}})$ for a single exploration well. Which segment is most valuable?

Table 7.2. *Marginal probabilities and expected revenues for the six segments in the small Bayesian network*

	Segment 1	Segment 2	Segment 3	Segment 4	Segment 5	Segment 6
p(Dry)	0.2	0.1	0.8	0.3	0.15	0.34
p(Gas)	0.52	0.72	0.01	0.02	0.68	0.52
p(Oil)	0.28	0.18	0.19	0.68	0.17	0.14
Rev(Dry)	−20	−25	−1	−15	−22	−8
Rev(Gas)	6	3	9	0	4	5
Rev(Oil)	3	1	6	7	2	1
E(Value)	−0.04	−0.16	0.43	0.15	−0.25	0.05

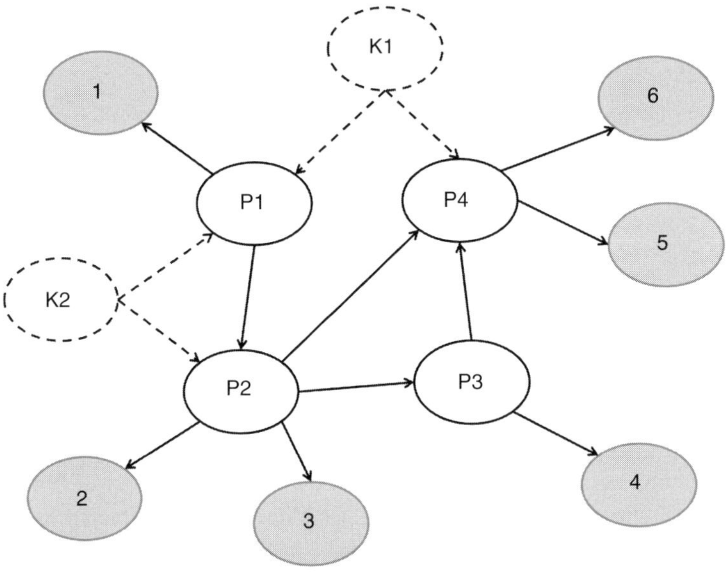

Figure 7.3 A Bayesian network model with 12 nodes. The nodes marked 1–6 (gray) are the observable nodes. They may represent petroleum fields.

Part II: North Sea network with 25 segments

1. Download and run the North Sea network. Compute the VOI of perfect information at single-segment exploration wells and verify the results displayed in Figures 6.3–6, Section 6.2.1.
2. Repeat the same exercise as in Step 1, but this time with uncertainty in the kitchen variables. Assume that the kitchens generate gas with a probability of 0.8 and are otherwise dry.

3. Assume that one can only acquire imperfect information about the charge variable – for example, from seismic data or analysis by basin and petroleum systems modeling. Let the accuracy of the imperfect test be $p(y_j = k \mid x_j = k) = 0.7$. Compute the VOI of partial imperfect information at single segments and compare with the results in Steps 1 and 2.

7.2.2 *Basin street blues: basin modeling example*

The petroleum exploration example with four Prospects TE, BE, TW, and BW in the North Sea is presented in Section 6.2.2. It relies on basin and petroleum systems modeling (BPSM) outputs to obtain the parameters of the network. If you have your own case with BPSM runs, you can analyze that instead.

For most of this project, you will need to download Kevin Murphy's BNT for MATLAB. This is available at http://code.google.com/p/bnt/.

1. Study the simulated 24 runs of BPSM with the different input conditions and accumulations in Prospects TE and BE (Chapter 2, Table 2.1, or download this from the website). Plot the results as a function of the various input parameters and use the design of experiment and analysis of variance to study the effects of the input variables on each output.
2. The learning of the network is based on clustering. Run the code (on the website) for this model specification. You may try to modify and change the conditional probability tables of the network and study the effect on the subsequent VOI analysis.
3. Use the established BN to reproduce the marginal and conditional probabilities, as well as the VOI of perfect information at single prospects (Figures 6.14–17).
4. Assume that only imperfect information will be available at single prospects. Compare the VOI of exploration at single wells when the accuracy of the imperfect test is $p(y_j = k \mid x_j = k) = 0.7$ for a test. Compare with the VOI of perfect information. Assume instead that the accuracy depends on the expected prospect size – i.e., the volumes for the success outcomes displayed in Table 6.1. Larger accumulations are easier to detect geophysically than smaller accumulations. One way to model this is to set accuracy to $p(y_j = k \mid x_j = k) = \dfrac{\exp\left(\mathrm{Vol}_j / 100\right)}{1 + \exp\left(\mathrm{Vol}_j / 100\right)}$.

   Compute the VOI of such imperfect tests at single prospects for different cost ranges.
5. Suppose that an exploration well provides perfect information about the reservoir, trap, and source attributes at single prospects. This will be more informative than just knowing the success, partial success, partial failure, or failure outcome at the leaf nodes. Compute the VOI of this sort of perfect information at the reservoir, trap, and source parent nodes to the leaf prospect variable. Compare the results.

7.2.3 Risky business: petroleum prospect risking example

This exercise builds on the 524 volume scenarios at 27 prospects in the Viking Graben, presented in Section 6.2.3. The data are based on basin modeling runs kindly provided to us by Øyvind Sylta and Migris (www.migris.no). For this project, you must download the results from risking. If you have your own risked volumes over multiple prospects, you may of course try a similar analysis for that case.

1. Consider Prospects 20 and 25, which were analyzed in Figure 6.21. Construct a kernel density estimate for these bivariate data. (See the m-files for this example.) The kernel density estimate can be represented on the grid of prospect volumes (or profits). Visualize the density estimate.
2. Compute the VOI of single-prospect exploration as was done in Figure 6.22. Vary the number of neighbors k and study the effect on the VOI.
3. Compute the VOI of single-prospect exploration wells using the kernel density estimate for all pairs of variables. This approach builds on the results in Step 1, where it is possible to approximate the conditional distributions by summing over the discrete grid of profits at two prospects. Hence, each conditional expectation required for the posterior value is now computed from the discretized conditional pdf on the cells used for the kernel density estimate. Compare the results with that of the k-nearest neighbors.
4. Compute the VOI of two exploration wells using the k-nearest neighbors for the two prospects in the conditioning statement. Compare the VOI of the two exploration wells with the VOI of single exploration wells.

7.3 Hands on: reservoir characterization from geophysical data

The hands-on projects in this section build on the examples in Section 6.3. The reader is advised to go through that section before starting on these hands-on projects.

7.3.1 Black gold in a white plight: reservoir characterization example

The petroleum example with a MRF model and seismic amplitude–versus-offset (AVO) data information was presented in Section 6.3.1. We start by modeling the seismic attributes from facies and predicting lithofacies from the current seismic data. Next, we conduct VOI analysis for processed seismic AVO data. You will need to download the data set and the MATLAB files from the website. If you have your own example, you may try this workflow for that case instead.

Part I: modeling and prediction from seismic amplitude versus offset data

1. Download the rock physics modeling associated with the hierarchical model in Figure 7.4. Fix the lithofacies variable (facies and saturation variables) at the three

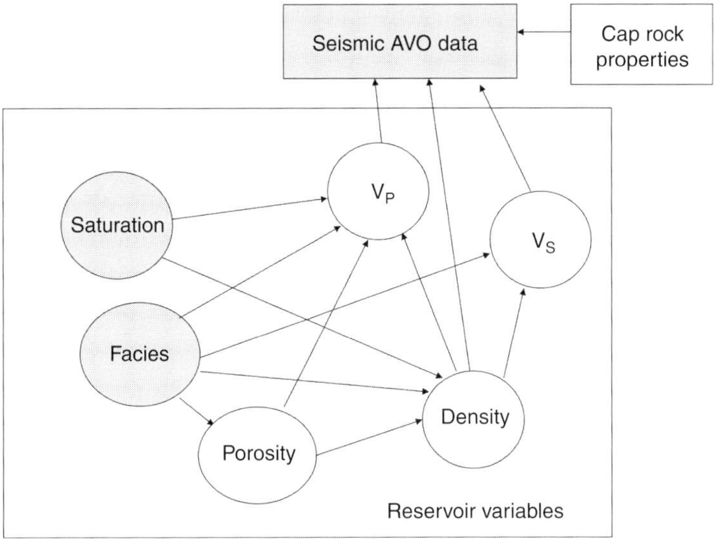

Figure 7.4 A graphical model for reservoir variables and seismic reflection data. The facies and saturation variables of main interest are connected to the seismic reflectivity data via rock physics models represented by edges in the graph.

classes (oil sand, brine sand, and shale); generate realizations of porosity conditional on the facies; generate elastic properties (P-wave velocity, S-wave velocity, and density) given the facies, saturation, and porosity; and finally generate seismic AVO data given the elastic properties. Study the propagated uncertainty for the seismic AVO data within each facies class. Plot the fitted Gaussian pdf from the multiple realizations. Cross-plot the propagated synthetic seismic with the observed seismic AVO data (see data file on website).

2. Consider the available seismic AVO data on the size 75 × 75 grid. (A part of this was shown in Section 4.6.) Implement a Markov chain Monte Carlo sampler for posterior assessment of the model for lithofacies, given the seismic AVO data. Use the likelihood model established from Step 1, as presented in Section 6.3.1. Use further a MRF prior model for the lithofacies with spatial interaction $\beta = 0.75$ for equal colors (oil sand, brine sand, or shale). Study the convergence of the Markov chain Monte Carlo sampler by plotting the fraction of different colors ($x_i = 1,2,3$) in the realizations. Classify the lithofacies at every cell based on the color that occurs most often at that cell.
3. Repeat the exercise in Step 2, but this time condition only on the zero-offset attribute of the seismic amplitude data. Compare this classification with both data.
4. Linear discriminant analysis: $\arg\max_k \left\{ \left(\boldsymbol{y}_i - \boldsymbol{\mu}_k \right)^t \boldsymbol{\Sigma}^{-1} \left(\boldsymbol{y}_i - \boldsymbol{\mu}_k \right) \right\}$ is used to construct the pointwise classifier at a site i into one of the classes $k = 1,2,3$. Calculate and draw the classification boundaries in the data space $\boldsymbol{y}_i = \left(y_{i1}, y_{i2} \right)$ of zero-offset

reflectivity and AVO gradient. Apply this method to construct a pointwise classifier that only integrates the data at that particular cell. This is equal to a model with $\beta = 0$. Compare classifications using both seismic AVO attributes.

Part II: VOI analysis for seismic amplitude versus offset data

Consider the grid of size 20 × 5 that was studied in Section 6.3.1. The MRF prior model is an Ising model with spatial interaction parameter $\beta = 0.9$ and an external field $\alpha_i(x_i)$, $i = 1,\ldots,100$, $x_i = 1,2,3$. Use the forward–backward algorithm for Monte Carlo sampling for the outer loop over data $\boldsymbol{y}$ as well as for analytical calculation of the marginal success probability given data $p(x_i = 1 \mid \boldsymbol{y})$. The code is available at the website. (Note that this VOI analysis takes some time to run on the computer.)

1. Reproduce the marginal prior probabilities in Figure 6.26 and the VOI results in Table 6.2 with a total seismic AVO test with both attributes. Compare the VOI results with the one assuming independence – i.e., $\beta = 0$ in the MRF prior.
2. Assume that the accuracy of the seismic data is much better so that the standard deviations of the likelihood are half. Compute the VOI of this situation, with $\beta = 0.9$ and $\beta = 0$. Discuss the results.

7.3.2 *Reservoir dogs: seismic and electromagnetic data example*

The example with a Gaussian model for profits and seismic AVO data or electromagnetic (EM) resistivity data was presented in Section 6.3.2. We start by rock physics modeling for the seismic attributes and the EM resistivity given the saturation and porosity variables. We then turn to VOI analysis. The project requires MATLAB and data files that are downloadable from the website. If you have your own example, you may try a similar workflow for that case.

Part I: rock physics modeling

1. The rock physics model for the seismic data is based on a soft sand rock physics model with parameters tuned from well logs and Gassmann's fluid substitution. The EM resistivity model is based on Archie's law. Run the MATLAB forward models provided on the website to study the expected data response as a function of saturation. Make plots for different porosities.
2. Suppose that the porosity is fixed from seismic post-stack data, as in Section 6.3.2. Run the code available online for this example to plot the expected seismic and resistivity response as a function of the reservoir unit revenues (and profits). One version of this plot is shown in Figure 6.29. Study the associated derivatives numerically and compare the linearized model with the assumed nonlinear rock physics models.

Part II: VOI analysis for seismic and electromagnetic data

1. Compute the VOI of seismic AVO and EM resistivity data for fixed parameter settings as indicated in the example and the associated code. Plot the VOI as a function of drilling cost, but also vary the accuracy of the AVO data and the EM resistivity data and study the effects.
2. Suppose that the EM resistivity data are only collected along the two center lines. Compute the VOI of this partial information-gathering scheme and compare with total testing.
3. What if it is not possible to freely select units in the grid? Instead, the decision maker can only select blocks of variables. The grid is split into 25 disjoint blocks of size 5×5. The decision maker will select a block if the sum of expected revenues from the 25 cells in a block is larger than the sum of the costs. We first assume additive revenues and costs. Compute the VOI of total seismic AVO and EM testing.
4. Decisions are made at blocks, as in Step 3, but now there are shared costs. If the block is selected, there is only one common cost. Decrease this cost from 25 Cost to 0. Plot the VOI results of total seismic AVO and EM testing.

7.4 Hands on: mine planning and safety

These projects pertain to the decision situations for mining and tunneling presented in Section 6.4. The reader is advised to go through that section before starting on the exercises.

7.4.1 I love rock and ore: mining oxide grade example

These exercises require downloading a data set from the website and running and modifying MATLAB files. Note that the data set has aspects similar to the real data but has been modified for confidentiality reasons. Of course, if you have similar data of your own, you may try the same exercises.

Part I: data analysis

1. Cross-plot the available X-ray fluorescence (XRF) and X-ray meter (XMET) data and plot each XRF and XMET along with the mineralization covariate. Tune model parameters for the regression parameter and the nugget effect from the XMET and XRF data. (Note that the numbers here are different from the ones in Section 6.4.1 because we have modified the original data set.)
2. Estimate the variogram from the XMET data. Assume an isotropic covariance model.
3. Use your established model from Steps 1 and 2, as well as the locations of data, to predict (by Kriging) the oxide grade at the selected blocks (download from the website). Predict also the sum of oxide grades at these blocks and the associated variance.

Part II: VOI analysis

1. All blocks will be utilized if the aggregated value from all blocks exceeds the cost of development. Use the Gaussian model to compute the analytical posterior value of XMET data in all boreholes. Compute the VOI.
2. The planned data consist of 10 boreholes (of equal length and sampling intensity). Suppose that we can collect XMET data in only one borehole. Which one is most valuable from a VOI perspective?
3. Do Exercises 1 and 2 with XRF data instead of XMET data and compare the results.

7.4.2 We will rock you: rock hazard example

The exercises are done in MATLAB, with the code available at the website. Note that the data have been slightly modified for confidentiality reasons.

Part I: data analysis

1. Make a histogram of the joint frequency data. Compute the average of the data. Compare the empirical results with the fitted Poisson distribution. without any hierarchical spatial modeling. The Poisson distribution is $p(y_i = k) = \frac{\lambda^k}{k!}\exp(-\lambda)$, $k = 0, 1, \ldots,$ and parameter $\hat{\lambda} = \bar{y}$ is the sample mean.
2. The marginal likelihood for covariance parameters $\boldsymbol{\theta}$ can be written as follows: $p(\boldsymbol{y};\boldsymbol{\theta}) = \frac{p(\boldsymbol{x};\boldsymbol{\theta})\,p(\boldsymbol{y} \mid \boldsymbol{x})}{p(\boldsymbol{x} \mid \boldsymbol{y};\boldsymbol{\theta})}$. Expand the numerator in an exponential quadratic form and show how an iterative optimization of the posterior would proceed. Use the MATLAB code to optimize the Laplace approximation of the marginal likelihood.

Part II: VOI analysis

1. Use the analytic forms for the prior and posterior values to conduct VOI analysis for full samples and subsample the joint frequency borehole data to only every second or fourth sample. Discuss the results.
2. Perform VOI analysis for individual boreholes. Assume a fixed price for drilling per meter. Study which borehole is the most valuable for acquiring joint frequency data.
3. Instead of the joint counts, one can imagine getting direct measurements of the log joint intensity $x(\mathbf{s})$ at selected sites, possibly through geophysical testing. Assume imperfect measurements of the log intensity at all rock hazard sites of interest, $y(\mathbf{s}_j) = x(\mathbf{s}_j) + N(0, \tau^2)$, $j = 1, \ldots, N$. Study the effect of accuracy τ and compare with the joint counts in Steps 1 and 2.

7.5 Hands on: groundwater management

This example considers decisions about groundwater recharge and was described in Section 6.5. The reader is advised to go through that section before starting on these exercises. We split the hands-on project into two parts: the first uses Netica software; the second uses the BNT.

7.5.1 Part I: salt water wells in my eyes – groundwater monitoring in Netica

We will describe how to use the software Netica (obtainable from www.norsys.com) to build and solve the influence diagrams. The free download version running in the limited mode will suffice for this example. After downloading and extracting, double-click on the Netica icon to start. From the File menu, start a new network. (Alternatively, you can download the pre-built network from the website for the book.) In Netica, nodes representing uncertainty or chance nodes are called "nature" nodes. In addition, there are decision nodes, utility (or value) nodes, and deterministic or constant nodes. Add a nature node to the new network representing the geological uncertainty.

By right-clicking and going to the properties of this nature node, fill out the name ("Channels," say) and the three states ("MixedChannels," "NEChannels," and "SEChannels"). Do not assign any numeric vales to the three discrete states. In the table for this node (right-click), assign the prior probabilities for each geological scenario. Add a decision node ("Recharge") with four states ("NoRecharge," "CentralRecharge," "NorthRecharge," and "SouthRecharge") and finally a utility ("Value") node. Connect the "Channels" node and "Recharge" node to the "Value" node by arrows as shown in Figure 6.40. Fill in the table for the "Value" node as shown in Table 6.5.

1. The prior value or expected utility is obtained by selecting "compile" and "optimize decisions" from the "Network" menu at the top. If "compile" is grayed out, it indicates that the network has already been compiled. Netica solves the network by internally building a junction tree for fast inference. The expected utility is reported in the messages window to be 12.47, and the optimal decision is "NoRecharge." You may need to bring up the message window to see the computed expected utility. The value with perfect information can be calculated by first adding a link from the "Channels" node to the decision node. This link indicates that the uncertain factor will be known before the decision. Optimizing the decision will result in 12.8 as the expected utility, representing the posterior value with perfect information.
2. Let us add a node (chance or "nature" node in Netica language) for the transient electromagnetic method (TEM) data to our original influence diagram. This network is shown in Figure 6.41. Fill in the reliabilities. Select "optimize decisions" from the "Network" menu at the top. The expected utility (12.78) is close to the value with perfect information (12.8) obtained earlier. If we change the table to represent perfect information ("1" on the diagonal and "0" elsewhere), we get back 12.8 as the expected utility. The value *with* imperfect information is 12.78, so the value *of* this imperfect information is 12.78 – 12.47 = 0.31.
3. A different mode of electromagnetic survey, a land-based survey, has coarser spacing and hence lower resolution. The reliability is as shown in Table 7.3.

Table 7.3. *Reliability measure for the land-based electromagnetic survey*

Channels	Interpretation of Northeast	Interpretation of Mixed	Interpretation of Southeast
Northeast Channels	0.6	0.35	0.05
Mixed Channels	0.2	0.6	0.2
Southeast Channels	0.05	0.35	0.6

On average, the three different channel scenarios are correctly identified about 60% of the time. The northeast channels are not very likely to be misinterpreted as southeast channels (only 5% of the time), and vice versa, but the northeast and southeast channels can be misinterpreted as mixed channels 35% of the time. This survey is somewhat cheaper, but is this any good? How much should the decision maker be willing to pay to purchase this information?

4. Compute the VOI for the groundwater example when the three different channel scenarios have unequal prior probabilities: [0.15, 0.15, 0.70]. Compute both the value of perfect information and the value of imperfect EM data with different reliability.

7.5.2 Part II: salt water wells in my eyes – groundwater monitoring in BNT

In this example, we will use influence diagrams in MATLAB to solve the same groundwater recharge case as in Section 7.5.1. You will need to download the BNT for MATLAB written by Kevin Murphy. This is available at http://code.google.com/p/bnt/.

1. After installing BNT, use the help documentation and familiarize yourself with the basic BNT functions for creating simple Bayes' nets and influence diagrams.
2. Create a three-node graph with nodes "geology," "recharge" and "value," where both "geology" and "recharge" are parents of "value." Define the number of states for each node: three states for "geology," four for "recharge," and one for "value." Use the description from the previous exercise or from Section 6.5 to name the states. You can also follow along with the associated *m*-file on the book website. Using this graph as an input to the "mk_limid" BNT function, create an influence diagram with "geology" as a "chance" node, "recharge" as a "decision" node, and "value" as a "utility" node. Assign the probability tables for the geology node using "tabular_CPD." Start with equal probabilities for all three geological scenarios. Assign the table for the value node using "tabular_utility_node." Be careful about the ordering of the value numbers from Table 6.5. The values for node numbered 1 ("geology") cycles uncertain outcomes first, followed by the different recharge alternatives. Thus, the ordering of the values in the table correspond to: (geology_scenario_1, recharge_1), (geology_scenario_2, recharge_1), … ,(geology_scenario_n, recharge_1), (geology_scenario_1, recharge_2), and so on. Assign uniform initial policies using "tabular_decision_node."

Next, set up the junction tree inference engine ("jtree_limid_inf_engine"), and solve the influence diagram ("solve_limid") to get the prior value (maximum expected utility) and the optimal strategy ("strategy" function). Is this consistent with the calculations using the Netica influence diagram and the decision tree in Section 6.5?

3. Compute the value with perfect information. Add an arc from "geology" to "recharge," and re-solve the influence diagram.
4. Compute the value with imperfect EM information. Add another "chance" node, "tem," to the graph. (In the MATLAB *m*-file, the "tem" node is numbered 2). This represents the imperfect EM data and will have three states ("interpMixed," "interpNE," and "interpSE"). Add arcs from "geology" to "tem" and from "tem" to "recharge." Assign the CPD tables for the "tem" node using information given in Section 6.5. Solve the influence diagram for the posterior value and, finally, get the value of imperfect information. Compute the VOI for both high-resolution and low-resolution (land-based) EM data.
5. Assign unequal prior probabilities to the geological scenarios with more probability for the "SEchannels" scenario, and study the trade-offs between these prior geological probabilities and the reliability of the EM data.

Appendix: selected statistical models and sampling methods

The first five sections (Appendices A.1–5) provide further mathematical descriptions of statistical models that are used in the book. The final section (Appendix B) presents a variety of sampling methods.

Appendix A.1: Gaussian distribution

The multivariate Gaussian model is fundamental in statistics. We present some properties of the Gaussian distribution; consult, e.g., Mardia et al. (1980), Anderson (2003), or Johnson and Wichern (2012) for more detailed discussions.

A.1.1 Definition and properties

The Gaussian pdf for random variable $\boldsymbol{x} = (x_1,...,x_n)$, viewed as an $n \times 1$ vector, with model parameters $\boldsymbol{\mu}$ and $\boldsymbol{\Sigma}$ is

$$p(\boldsymbol{x}) = N(\boldsymbol{\mu},\boldsymbol{\Sigma}) = \frac{1}{(2\pi)^{n/2}} |\boldsymbol{\Sigma}|^{-1/2} \exp\left(-\frac{1}{2}(\boldsymbol{x}-\boldsymbol{\mu})^t \boldsymbol{\Sigma}^{-1}(\boldsymbol{x}-\boldsymbol{\mu})\right), \tag{A.1}$$

with continuous sample space $-\infty < x_i < \infty$, $i = 1,...,n$. The distribution is recognized by the exponent of a quadratic form in $\boldsymbol{x}$. The normalizing constant in front of the exponent ensures that $\int p(\boldsymbol{x})d\boldsymbol{x} = 1$.

Joint and marginal probabilities

For the Gaussian distribution:

$$E(\boldsymbol{x}) = \boldsymbol{\mu} = \begin{pmatrix} \mu_1 \\ \mu_2 \\ \cdots \\ \mu_n \end{pmatrix}, \quad Var(\boldsymbol{x}) = \boldsymbol{\Sigma} = \begin{bmatrix} \Sigma_{11} & \Sigma_{12} & \cdots & \Sigma_{1n} \\ \Sigma_{21} & \Sigma_{22} & \cdots & \Sigma_{2n} \\ \cdots & \cdots & \cdots & \cdots \\ \Sigma_{n1} & \Sigma_{n2} & \cdots & \Sigma_{nn} \end{bmatrix}. \tag{A.2}$$

The marginal pdf $p(x_i) = N(\mu_i, \Sigma_{ii})$ and $Cov(x_i, x_j) = \Sigma_{ij}$.

Conditioning

Let $\mathbb{K}$ be a subset of the indices $\{1,...,n\}$. The marginal pdf of block variable $\boldsymbol{x}_{\mathbb{K}}$ is $p(\boldsymbol{x}_{\mathbb{K}}) = N(\boldsymbol{\mu}_{\mathbb{K}}, \boldsymbol{\Sigma}_{\mathbb{K}})$, where we pick entries identified by $\mathbb{K}$ in the mean vector and in the rows and columns of the covariance matrix.

If we split the random vector into two blocks, $\boldsymbol{x} = \begin{pmatrix} \boldsymbol{x}_{\mathbb{L}} \\ \boldsymbol{x}_{\mathbb{K}} \end{pmatrix}$, with mean $\boldsymbol{\mu} = \begin{pmatrix} \boldsymbol{\mu}_{\mathbb{L}} \\ \boldsymbol{\mu}_{\mathbb{K}} \end{pmatrix}$ and covariance structure $\boldsymbol{\Sigma} = \begin{pmatrix} \boldsymbol{\Sigma}_{\mathbb{L}} & \boldsymbol{\Sigma}_{\mathbb{LK}} \\ \boldsymbol{\Sigma}_{\mathbb{KL}} & \boldsymbol{\Sigma}_{\mathbb{K}} \end{pmatrix}$, the conditional pdf of $\boldsymbol{x}_{\mathbb{L}}$ given $\boldsymbol{x}_{\mathbb{K}}$ is also Gaussian and has the mean and covariance matrix

$$\begin{aligned} \boldsymbol{\mu}_{\mathbb{L}|\mathbb{K}} &= \boldsymbol{\mu}_{\mathbb{L}} + \boldsymbol{\Sigma}_{\mathbb{LK}} \boldsymbol{\Sigma}_{\mathbb{K}}^{-1} (\boldsymbol{x}_{\mathbb{K}} - \boldsymbol{\mu}_{\mathbb{K}}), \\ \boldsymbol{\Sigma}_{\mathbb{L}|\mathbb{K}} &= \boldsymbol{\Sigma}_{\mathbb{L}} - \boldsymbol{\Sigma}_{\mathbb{LK}} \boldsymbol{\Sigma}_{\mathbb{K}}^{-1} \boldsymbol{\Sigma}_{\mathbb{KL}}. \end{aligned} \tag{A.3}$$

The conditional covariance structure can be recognized from the inverse covariance matrix $\boldsymbol{Q} = \boldsymbol{\Sigma}^{-1}$, sometimes called the precision matrix. Assuming the same block structure $\boldsymbol{Q} = \begin{pmatrix} \boldsymbol{Q}_{\mathbb{L}} & \boldsymbol{Q}_{\mathbb{LK}} \\ \boldsymbol{Q}_{\mathbb{KL}} & \boldsymbol{Q}_{\mathbb{K}} \end{pmatrix}$, then

$$\begin{aligned} p(\boldsymbol{x}_{\mathbb{L}} \mid \boldsymbol{x}_{\mathbb{K}}) &\propto p(\boldsymbol{x}) \propto \exp\left(-\frac{1}{2}(\boldsymbol{x}-\boldsymbol{\mu})^t \boldsymbol{Q}(\boldsymbol{x}-\boldsymbol{\mu})\right) \\ &\propto \exp\left(-\frac{1}{2}(\boldsymbol{x}_{\mathbb{L}}-\boldsymbol{\mu}_{\mathbb{L}})^t \boldsymbol{Q}_{\mathbb{L}}(\boldsymbol{x}_{\mathbb{L}}-\boldsymbol{\mu}_{\mathbb{L}}) - (\boldsymbol{x}_{\mathbb{L}}-\boldsymbol{\mu}_{\mathbb{L}})^t \boldsymbol{Q}_{\mathbb{LK}}(\boldsymbol{x}_{\mathbb{K}}-\boldsymbol{\mu}_{\mathbb{K}})\right). \end{aligned} \tag{A.4}$$

From the quadratic form in the exponent, $p(\boldsymbol{x}_{\mathbb{L}} \mid \boldsymbol{x}_{\mathbb{K}}) = \mathrm{N}\left(\boldsymbol{\mu}_{\mathbb{L}} - \boldsymbol{Q}_{\mathbb{L}}^{-1}\boldsymbol{Q}_{\mathbb{LK}}(\boldsymbol{x}_{\mathbb{K}} - \boldsymbol{\mu}_{\mathbb{K}}), \boldsymbol{Q}_{\mathbb{L}}^{-1}\right)$. The block entries in the precision matrix are defined via $\boldsymbol{Q\Sigma} = \boldsymbol{I}$ – i.e.,

$$\begin{aligned} &\boldsymbol{Q}_{\mathbb{L}}\boldsymbol{\Sigma}_{\mathbb{L}} + \boldsymbol{Q}_{\mathbb{LK}}\boldsymbol{\Sigma}_{\mathbb{KL}} = \boldsymbol{I}, \quad -\boldsymbol{Q}_{\mathbb{L}}\boldsymbol{\Sigma}_{\mathbb{LK}}\boldsymbol{\Sigma}_{\mathbb{K}}^{-1} = \boldsymbol{Q}_{\mathbb{LK}}, \\ &\boldsymbol{Q}_{\mathbb{L}}\boldsymbol{\Sigma}_{\mathbb{LK}} + \boldsymbol{Q}_{\mathbb{LK}}\boldsymbol{\Sigma}_{\mathbb{K}} = \boldsymbol{0}, \quad \boldsymbol{Q}_{\mathbb{L}}\left(\boldsymbol{\Sigma}_{\mathbb{L}} - \boldsymbol{\Sigma}_{\mathbb{LK}}\boldsymbol{\Sigma}_{\mathbb{K}}^{-1}\boldsymbol{\Sigma}_{\mathbb{KL}}\right) = \boldsymbol{I}, \end{aligned} \tag{A.5}$$

and we can identify terms in Equation (A.3).

For a single variable, this means that the full conditional mean and variance equal

$$E(x_i \mid \boldsymbol{x}_{-i}) = \mu_i - \frac{1}{Q_{ii}} \sum_{j \neq i} Q_{ij}(x_j - \mu_j), \qquad Var(x_i \mid \boldsymbol{x}_{-i}) = \frac{1}{Q_{ii}}. \tag{A.6}$$

These full conditional calculations can sometimes be simplified by imposing conditional independence. This is represented by 0 entries in the precision matrix $\boldsymbol{Q}$, giving a sparse structure. Instead of summing over all $j \neq i$, only the modeled non-zero entries take part in Equation (A.6). This sparse structure of the precision matrix relies on conditional independencies modeled by the Markov property. This is critical for graphical representations, where sparseness facilitates modeling and interpretation – see, e.g., Whittaker (1990) and Meinshausen and Buhlmann (2006). Special cases include time series models (Brockwell and Davis 2009) and Gaussian Markov random fields (MRFs) (Rue and Held 2005). The literature on sparse matrix computation is enormous because it plays a prominent role in numerical analysis. For instance, a finite difference approximation to a differential equation involves solving sparse linear systems $\boldsymbol{Qx} = \boldsymbol{\mu}$. Having a sparse structure is also key to the success of iterative solvers of large linear systems such as conjugate gradients – see, e.g., Golub and van Loan (1996).

Linear transformations

A transformation $\boldsymbol{y} = \boldsymbol{F}\boldsymbol{x} + \boldsymbol{b}$, for a size $m \times n$ fixed matrix $\boldsymbol{F}$ and known length m vector $\boldsymbol{b}$, has Gaussian pdf $p(\boldsymbol{y}) = N(\boldsymbol{F}\boldsymbol{\mu}+\boldsymbol{b}, \boldsymbol{F}\boldsymbol{\Sigma}\boldsymbol{F}^t)$. In particular, a Gaussian variable can always be transformed to independent zero-mean and unit-variance variables $z = (z_1, \ldots, z_n)$ by setting $\boldsymbol{F} = \boldsymbol{L}^{-1}$ and $\boldsymbol{b} = -\boldsymbol{L}^{-1}\boldsymbol{\mu}$ as follows:

$$z = \boldsymbol{y} = \boldsymbol{L}^{-1}(\boldsymbol{x} - \boldsymbol{\mu}), \qquad \boldsymbol{x} = \boldsymbol{\mu} + \boldsymbol{L}z, \quad \boldsymbol{L}\boldsymbol{L}^t = \boldsymbol{\Sigma}. \tag{A.7}$$

The result can be derived from the transformation formula of random variables:

$$\begin{aligned} &\boldsymbol{x} = \boldsymbol{x}(z) = \boldsymbol{\mu} + \boldsymbol{L}z, \qquad J(\boldsymbol{x}(z)) = |\boldsymbol{L}| = |\boldsymbol{\Sigma}|^{1/2}, \\ &p(z) = |J(\boldsymbol{x}(z))|\, p_x(\boldsymbol{x}(z)) = \frac{1}{(2\pi)^{n/2}} \exp\left(-\frac{1}{2} z^t z\right) = \prod_{i=1}^{n} \frac{1}{\sqrt{2\pi}} \exp\left(-\frac{z_i^2}{2}\right). \end{aligned} \tag{A.8}$$

The matrix $\boldsymbol{L}$ is called the Cholesky factorization of the covariance matrix Σ. It is a lower triangular matrix – i.e., it has 0 entries in the upper right triangle.

This important transformation in Equation (A.7) ensures that many computations for dependent normal distributed random variables can build on known results about independent univariate variables. As an example, consider the probability that the random vector falls within distance unit 1 from the mean, as defined by the covariance ellipsoid – i.e.,

$$p\left(\left|(\boldsymbol{x} - \boldsymbol{\mu})^t \boldsymbol{\Sigma}^{-1} (\boldsymbol{x} - \boldsymbol{\mu})\right| < 1\right) = \int_{\left|(x-\mu)^t \Sigma^{-1} (x-\mu)\right| < 1} p(\boldsymbol{x})\, d\boldsymbol{x}. \tag{A.9}$$

The computation is straightforward once we transform to independent variables. It translates to the sum of squared independent standard normal variables $\sum_{i=1}^{n} z_i^2 < 1$, and this is directly available from the chi-square distribution since $\sum_{i=1}^{n} z_i^2$ is chi-square distributed with n degrees of freedom. Note that the probability of outcomes within the circle decreases quickly as a function of the dimension n (Table A.1). The chance of a sample close to the

Table A.1. *Probability of having standard normal samples within the unit circle and the unit square displayed as a function of the variable dimension*

Dimension of Variable	1	2	3	4	5	10	20
$p\left(\left\|(\boldsymbol{x}-\boldsymbol{\mu})^t \boldsymbol{\Sigma}^{-1}(\boldsymbol{x}-\boldsymbol{\mu})\right\| \le 1\right)$	0.68	0.39	0.20	0.09	0.04	0.0002	2e-10
$\prod_{i=1}^{n} p(-1 < z_i < 1)$	0.68	0.46	0.31	0.21	0.14	0.02	4e-4

mean gets extremely small in high dimensions as it is always in the tail of the distribution. The probability of all independent variables occurring within the unit square is also displayed in Table A.1.

Consider also the linear transformation $\boldsymbol{y} = \boldsymbol{F}\boldsymbol{x} + \boldsymbol{e}$, where $p(\boldsymbol{e}) = N(\boldsymbol{0}, \boldsymbol{T})$ and $\boldsymbol{x}$ and $\boldsymbol{e}$ are independent. The pdf of $\boldsymbol{y}$ is $p(\boldsymbol{y}) = N(\boldsymbol{F}\boldsymbol{\mu}, \boldsymbol{F}\boldsymbol{\Sigma}\boldsymbol{F}^T + \boldsymbol{T})$. The joint distribution of $\boldsymbol{x}$ and $\boldsymbol{y}$ is Gaussian with mean $\begin{pmatrix} \boldsymbol{\mu} \\ \boldsymbol{F}\boldsymbol{\mu} \end{pmatrix}$ and covariance $\begin{pmatrix} \boldsymbol{\Sigma} & \boldsymbol{\Sigma}\boldsymbol{F}^T \\ \boldsymbol{F}\boldsymbol{\Sigma} & \boldsymbol{F}\boldsymbol{\Sigma}\boldsymbol{F}^T + \boldsymbol{T} \end{pmatrix}$. The conditional distribution of $\boldsymbol{x}$ given $\boldsymbol{y}$ is also Gaussian with mean and variance:

$$\begin{aligned} \boldsymbol{\mu}_{x|y} &= \boldsymbol{\mu} + \boldsymbol{\Sigma}\boldsymbol{F}^T \left(\boldsymbol{F}\boldsymbol{\Sigma}\boldsymbol{F}^T + \boldsymbol{T}\right)^{-1} \left(\boldsymbol{y} - \boldsymbol{F}\boldsymbol{\mu}\right), \\ \boldsymbol{\Sigma}_{x|y} &= \boldsymbol{\Sigma} - \boldsymbol{\Sigma}\boldsymbol{F}^T \left(\boldsymbol{F}\boldsymbol{\Sigma}\boldsymbol{F}^T + \boldsymbol{T}\right)^{-1} \boldsymbol{F}\boldsymbol{\Sigma}. \end{aligned} \tag{A.10}$$

Parameter estimation

For multiple data sets of $\boldsymbol{x}^1, \ldots, \boldsymbol{x}^B$, the empirical mean and covariance estimates are

$$\hat{\boldsymbol{\mu}} = \frac{1}{B}\sum_{b=1}^{B} \boldsymbol{x}^b, \quad \hat{\boldsymbol{\Sigma}} = \frac{1}{B}\sum_{b=1}^{B} \left(\boldsymbol{x}^b - \hat{\boldsymbol{\mu}}\right)\left(\boldsymbol{x}^b - \hat{\boldsymbol{\mu}}\right)^t. \tag{A.11}$$

Suppose instead that the mean is modeled by $\boldsymbol{\mu} = \boldsymbol{H}\boldsymbol{\beta}$, where the size $n \times k$ matrix $\boldsymbol{H}$ consists of known explanatory variables and that $\boldsymbol{\beta}$ is a length k vector of fixed but unknown regression parameters. If the data contain total perfect information $\boldsymbol{y} = \boldsymbol{x}$ and the prior covariance $\boldsymbol{\Sigma}$ is known, the maximum likelihood estimator (MLE) of the regression parameter $\boldsymbol{\beta}$ equals that of the weighted least-squares estimate:

$$\tilde{\boldsymbol{\beta}} = \left(\boldsymbol{H}^t \boldsymbol{\Sigma}^{-1} \mathbf{H}\right)^{-1} \boldsymbol{H}^t \boldsymbol{\Sigma}^{-1} \boldsymbol{x}. \tag{A.12}$$

If the covariance matrix is $\boldsymbol{\Sigma} = \sigma^2 \boldsymbol{I}$, this simplifies to the ordinary least-squares $\tilde{\boldsymbol{\beta}} = \left(\boldsymbol{H}^t \mathbf{H}\right)^{-1} \boldsymbol{H}^t \boldsymbol{x}$. An estimate of the noise level is $\hat{\sigma}^2 = \frac{1}{n}\left(\boldsymbol{x} - \boldsymbol{H}\tilde{\boldsymbol{\beta}}\right)^t \left(\boldsymbol{x} - \boldsymbol{H}\tilde{\boldsymbol{\beta}}\right)$. This estimate is asymptotically unbiased, but for a finite sample we may use $n - k$ in the denominator instead to achieve an unbiased variance estimate.

Assume that we have imperfect information $\boldsymbol{y} = (y_1, \ldots, y_m)$ and the model is

$$p(\boldsymbol{x}) = N(\boldsymbol{H}\boldsymbol{\beta}, \boldsymbol{\Sigma}), \qquad p(\boldsymbol{y} \mid \boldsymbol{x}) = N(\boldsymbol{F}\boldsymbol{x}, \boldsymbol{T}). \tag{A.13}$$

The marginal likelihood of the data is $p(\boldsymbol{y}) = N(\boldsymbol{G}\boldsymbol{\beta}, \boldsymbol{C})$ for $\boldsymbol{G} = \boldsymbol{F}\boldsymbol{H}$ and $\boldsymbol{C} = \boldsymbol{F}\boldsymbol{\Sigma}\boldsymbol{F}^T + \boldsymbol{T}$. The log-likelihood as a function of $\boldsymbol{\beta}$ and unknown fixed nuisance parameters $\boldsymbol{\theta}$ in the prior covariance matrix $\boldsymbol{\Sigma} = \boldsymbol{\Sigma}(\boldsymbol{\theta})$, and/or the likelihood noise matrix $\boldsymbol{T} = \boldsymbol{T}(\boldsymbol{\theta})$, becomes

$$l(\boldsymbol{\theta}, \boldsymbol{\beta}) = -\frac{m}{2}\log(2\pi) - \frac{1}{2}\log|\mathbf{C}| - \frac{1}{2}(\boldsymbol{y} - \boldsymbol{G}\boldsymbol{\beta})^t \mathbf{C}^{-1} (\boldsymbol{y} - \boldsymbol{G}\boldsymbol{\beta}). \tag{A.14}$$

The MLEs of $\boldsymbol{\beta}$ and $\boldsymbol{\theta}$ are obtained by

$$\left(\hat{\boldsymbol{\beta}}, \hat{\boldsymbol{\theta}}\right) = \text{argmax} \left\{l(\boldsymbol{\theta}, \boldsymbol{\beta})\right\}. \tag{A.15}$$

For fixed $\boldsymbol{\theta}$, the MLE of $\boldsymbol{\beta}$ is analytically available. We have

$$\frac{dl}{d\beta} = \boldsymbol{G}^t\mathbf{C}^{-1}\boldsymbol{y} - \boldsymbol{G}^t\mathbf{C}^{-1}\boldsymbol{G}\beta = \mathbf{0}, \quad \hat{\beta} = \hat{\beta}(\boldsymbol{y};\boldsymbol{\theta}) = \left(\boldsymbol{G}^t\mathbf{C}^{-1}\boldsymbol{G}\right)^{-1}\boldsymbol{G}^t\mathbf{C}^{-1}\boldsymbol{y}. \tag{A.16}$$

A direct calculation shows that $Var(\hat{\beta}) = \left(\boldsymbol{G}^t\mathbf{C}^{-1}\boldsymbol{G}\right)^{-1}$.

Treating the regression parameter $\boldsymbol{\beta}$ as fixed, the MLE of nuisance parameters $\boldsymbol{\theta}$ can be obtained by numerical maximization. Set $z = \boldsymbol{y} - \boldsymbol{G}\beta$, and let $\mathbf{Q} = \mathbf{C}^{-1}$. Denote a component of $\boldsymbol{\theta}$ by θ_r, $r = 1,\ldots,d$. (In our setting of Chapter 4, the components are prior variance σ^2, correlation decay η, nugget τ^2, and $d = 3$.) The first derivative (score) of the log-likelihood in Equation (A.14) with respect to element θ_r becomes

$$\frac{dl}{d\theta_r} = -\frac{1}{2}\operatorname{trace}\left(\mathbf{Q}\frac{d\mathbf{C}}{d\theta_r}\right) + \frac{1}{2}z^t\mathbf{Q}\frac{d\mathbf{C}}{d\theta_r}\mathbf{Q}z, \tag{A.17}$$

where we have used $\frac{d\log|\mathbf{C}|}{d\theta_r} = \operatorname{trace}\left(\mathbf{Q}\frac{d\mathbf{C}}{d\theta_r}\right), \frac{dz^t\mathbf{C}^{-1}z}{d\theta_r} = -z^t\left(\mathbf{Q}\frac{d\mathbf{C}}{d\theta_r}\mathbf{Q}\right)z.$

The expected Hessian is

$$E\left(\frac{d^2l}{d\theta_r d\theta_{r'}}\right) = -\frac{1}{2}\operatorname{trace}\left(\mathbf{Q}\frac{d\mathbf{C}}{d\theta_{r'}}\mathbf{Q}\frac{d\mathbf{C}}{d\theta_r}\right), \tag{A.18}$$

where we used $E\left(z^t\boldsymbol{\Sigma}z\right) = \operatorname{trace}\left(\boldsymbol{\Sigma}Var\left(z\right)\right)$, assuming $E\left(z^t\right) = \mathbf{0}$.

Algorithm: the iterative Fisher scoring algorithm for obtaining the MLE of β and θ

Initiate β^0, $\boldsymbol{\theta}^0$ by least-squares estimation and empirical variograms or other approaches.

Iterate for $b = 0,1,\ldots,$ until convergence:

$\mathbf{C} = \mathbf{C}\left(\boldsymbol{\theta}^b\right)$

$\beta^{b+1} = \beta^{b+1}(\boldsymbol{y};\boldsymbol{\theta}) = [\boldsymbol{G}^t\mathbf{C}^{-1}\boldsymbol{G}]^{-1}\boldsymbol{G}^t\mathbf{C}^{-1}\boldsymbol{y},$

$z = \boldsymbol{y} - \boldsymbol{G}\beta^{b+1}$

$\mathbf{Q} = \mathbf{C}^{-1}, \mathbf{C}_r^* = \frac{d\mathbf{C}\left(\boldsymbol{\theta}^b\right)}{d\theta_r}, r = 1,\ldots,d$

$u_r = \frac{dl}{d\theta_r} = -\frac{1}{2}\operatorname{trace}\left(\mathbf{Q}\mathbf{C}_r^*\right) + \frac{1}{2}z^t\mathbf{Q}\mathbf{C}_r^*\mathbf{Q}z, \qquad r = 1,\ldots,d$

$V_{\{rr'\}} = E\left(\frac{d^2l}{d\theta_r d\theta_{r'}}\right) = -\frac{1}{2}\operatorname{trace}\left(\mathbf{Q}\mathbf{C}_r^*\mathbf{Q}\mathbf{C}_{r'}^*\right), \qquad r,r' = 1,\ldots,d.$

$\boldsymbol{\theta}^{b+1} = \boldsymbol{\theta}^b + \mathbf{V}^{-1}\mathbf{u}.$

$b = b+1.$

It is convenient to parameterize the model such that θ_r is on the real line. Asymptotically, as the number of data $n \to \infty$, the MLEs are unbiased and Gaussian distributed with variance obtained from the inverse Hessian $E\left(\frac{d^2 l}{d\boldsymbol{\theta}^2}\right)^{-1}$ and $E\left(\frac{d^2 l}{d\beta^2}\right)^{-1} = \left(\boldsymbol{G}^t \mathbf{C}^{-1} \boldsymbol{G}\right)^{-1}$. The speed of convergence to this asymptotic result also depends on the design of experiment as well as the true model parameters.

Note that these derivations are for fixed parametric covariance structure such as in spatial covariance models. In a graphical model with an unknown conditional independence structure, there have been several attempts trying to specify the conditional independence structure from data. This is more difficult than estimating the parametric covariance matrix. If the non-zero structure cannot be specified by auxiliary data or expert opinion, it is possible to try novel techniques for sparse estimation – see, e.g., Meinshausen and Buhlmann (2006), Friedman et al. (2009), and Khare et al. (2015).

Sampling

There are several ways to sample random Gaussian vectors. The preferred approach depends on the situation, especially on the structure of the covariance matrix. We present a common method based on Cholesky factorization of the covariance matrix.

Algorithm: sampling from $p(x) = N(\mu, \Sigma)$

1. Find the Cholesky factorization $\boldsymbol{\Sigma} = \boldsymbol{L}\boldsymbol{L}^t$ of the variance–covariance matrix.
2. Sample independent standard Gaussian variables z_i from $N(0,1), i = 1,\ldots,n$.
3. Set $\boldsymbol{x} = \boldsymbol{\mu} + \boldsymbol{L}\boldsymbol{z}$.

The method relies on sampling independent standard Gaussian variables, which is a built-in routine in most software (see Appendix B).

Alternative approaches rely on different decompositions of the covariance structure – for instance, the eigendecomposition of $\boldsymbol{\Sigma}$ – see, e.g., Gray (2006). If the covariance matrix can be embedded in a circulant matrix, methods from the spectral domain are immediately applicable. In spatial applications, this can often be achieved by using a regular grid and wrapping the grid around to form a torus. Under stationarity assumptions, the covariance matrix is then fully defined by the covariance from one point to all the others (length n, not $n(n-1)/2$). The discrete Fourier transform of the resulting matrix and the inverse Fourier transform of the equal size matrix of independent variables give a realization of the random field with the specified covariance structure in $\boldsymbol{\Sigma}$ – see, e.g., Chan and Wood (1997). The posterior is likely not stationary, but it may be possible to re-use unconditional samples.

If the precision matrix is sparse, it is more efficient to use the Cholesky factorization of $\boldsymbol{Q} = \boldsymbol{\Sigma}^{-1}$ instead of the covariance matrix $\boldsymbol{\Sigma}$. The Cholesky factor of the precision matrix is not as sparse as the precision matrix, but the fill-in is rather small for many models (Rue and Held 2005).

Assume again the Gaussian hierarchical model for imperfect information – i.e., a prior pdf $p(\boldsymbol{x}) = N(\boldsymbol{\mu},\boldsymbol{\Sigma})$ and a likelihood model $p(\boldsymbol{y} \mid \boldsymbol{x}) = N(\boldsymbol{Fx},\boldsymbol{T})$. Conditional samples can be drawn directly based on the Cholesky factor of the conditional covariance matrix in Equation (A.10). Alternatively, samples can be generated as follows, re-using the unconditional samples as in the following algorithm.

Algorithm: sampling conditional Gaussian samples given linear Gaussian data *y*

1. Draw an unconditional sample $\boldsymbol{x}$ from $p(\boldsymbol{x}) = N(\boldsymbol{\mu},\boldsymbol{\Sigma})$.
2. Sample a noise term $\boldsymbol{e}$ from $p(\boldsymbol{e}) = N(\boldsymbol{0},\boldsymbol{T})$.
3. Set a conditional sample $\boldsymbol{x} = \boldsymbol{x} + \boldsymbol{\Sigma F}^t\left(\boldsymbol{F\Sigma F}^t + \boldsymbol{T}\right)^{-1}\left(\boldsymbol{y} + \boldsymbol{e} - \boldsymbol{Fx}\right)$.

A direct calculation shows that the mean and covariance of the expression in Step 3 match those in Equation (A.10). This conditional sampling method is particularly useful if the dimension of the data m is small relative to the dimension n of the distinction of interest. If the size of data is large, the inverse in Step 3 may be time consuming. Again, it is possible to factorize the matrices in different ways.

In the context of geostatistical simulations, drawing from a multivariate Gaussian distribution can be done sequentially using the chain rule of probability distributions. This is the so-called sequential Gaussian simulation (SGSIM) algorithm described in Chapter 4.

A.1.2 Decision analysis and VOI results

We show analytic results for computing the posterior value for Gaussian linear models. The following result is a known property of the Gaussian distribution, see e.g. Schlaiffer (1959) and Bickel (2008). For a Gaussian variable w with mean m and variance r^2:

$$\begin{aligned} E\left(\max\{0,w\}\right) &= \int \max\{w,0\}\, p(w)\,dw = \int_0^\infty w p(w)\,dw = \int_{-m/r}^{\infty} (m+rz)\phi(z)\,dz \\ &= m\int_{-m/r}^{\infty} \phi(z)\,dz + r\int_{-m/r}^{\infty} z\phi(z)\,dz = m\left(1-\Phi\left(-m/r\right)\right) + r\phi\left(-m/r\right) \qquad \text{(A.19)} \\ &= m\Phi\left(m/r\right) + r\phi\left(m/r\right), \end{aligned}$$

where the cumulative distribution of the standard Gaussian is $\Phi(z) = \int_{-\infty}^{z} \frac{1}{\sqrt{2\pi}} e^{-\frac{x^2}{2}}\,dx$, and its pdf is $\phi(z) = \frac{1}{\sqrt{2\pi}} e^{-\frac{z^2}{2}}$. The solution uses a transformation of variables $z = \frac{(w-m)}{r}$ to get standard normal integral expressions. It further uses $\int z\exp\left(-\frac{z^2}{2}\right)dz = -\exp\left(-\frac{z^2}{2}\right) + const$ as well as symmetry properties of $\phi(z)$ and $\Phi(z)$.

Similar results hold for an exponential utility function – see, e.g., Bickel (2008). Consider utility function $u(w) = 1 - \exp(-\gamma w)$. Then

$$\begin{aligned}
&\int \exp(-\gamma w)\, p(w)\, dw = \int \exp(-\gamma w) \frac{1}{\sqrt{2\pi}r} \exp\left(-\frac{(w-m)^2}{2r^2}\right) dw \\
&= \int \frac{1}{\sqrt{2\pi}r} \exp\left(-\frac{w^2}{2r^2} + \frac{w(m-\gamma r^2)}{r^2} - \frac{m^2}{2r^2}\right) dw \\
&= \exp\left(\frac{(m-\gamma r^2)^2}{2r^2} - \frac{m^2}{2r^2}\right) \int \frac{1}{\sqrt{2\pi}r} \exp\left(-\frac{w^2}{2r^2} + \frac{w(m-\gamma r^2)}{r^2} - \frac{(m-\gamma r^2)^2}{2r^2}\right) dw \\
&= \exp\left(-m\gamma + \frac{\gamma^2 r^2}{2}\right),
\end{aligned} \tag{A.20}$$

where we expanded the Gaussian quadratic form to get the integral over the $N(m-\gamma r^2, r^2)$ pdf. The certain equivalents in Section 3.2 are computed by the inverse utility function of this integral solution – i.e., $CE = \frac{-1}{\gamma} \log\left(1 - 1 + \exp\left(-\gamma m + \frac{\gamma^2 r^2}{2}\right)\right) = m - \frac{\gamma r^2}{2}$. For VOI computations, related expressions exist, but the integrals are incomplete – i.e.,

$$\int_{\alpha}^{\infty} \exp(-\gamma w)\, p(w)\, dw = \exp\left(-m\gamma + \frac{\gamma^2 r^2}{2}\right)\left(1 - \Phi\left(\frac{\alpha - (m-\gamma r^2)}{r}\right)\right). \tag{A.21}$$

See Bickel (2008) for details regarding VOI calculations for exponential utility functions.

Assuming again a risk-neutral decision maker, we next show analytical expressions for the two-action decision situation and decoupling of the value function. We assume that the distinction of interest $\boldsymbol{x}$ are profits with prior pdf $p(\boldsymbol{x}) = N(\boldsymbol{\mu}, \boldsymbol{\Sigma})$ and that information is obtained by a linear model $\boldsymbol{y} = \boldsymbol{F}\boldsymbol{x} + \boldsymbol{e}$, $p(\boldsymbol{e}) = N(\boldsymbol{0}, \boldsymbol{T})$. The marginal distribution of data is $p(\boldsymbol{y}) = N(\boldsymbol{F}\boldsymbol{\mu}, \boldsymbol{F}\boldsymbol{\Sigma}\boldsymbol{F}^T + \boldsymbol{T})$.

The conditional mean $E(\boldsymbol{x} \mid \boldsymbol{y}) = \boldsymbol{\mu}_{x|y} = \boldsymbol{\mu} + \boldsymbol{\Sigma}\boldsymbol{F}^T(\boldsymbol{F}\boldsymbol{\Sigma}\boldsymbol{F}^T + \boldsymbol{T})^{-1}(\boldsymbol{y} - \boldsymbol{F}\boldsymbol{\mu})$ depends linearly on the random data. The distribution of $\boldsymbol{\mu}_{x|y}$ is thus Gaussian with

$$\begin{aligned}
&E(\boldsymbol{\mu}_{x|y}) = \boldsymbol{\mu}, \\
&Var(\boldsymbol{\mu}_{x|y}) = \boldsymbol{\Sigma}\boldsymbol{F}^T(\boldsymbol{F}\boldsymbol{\Sigma}\boldsymbol{F}^T + \boldsymbol{T})^{-1}\boldsymbol{F}\boldsymbol{\Sigma} = \boldsymbol{R}.
\end{aligned} \tag{A.22}$$

In the situation with low decision flexibility (select all sites or none), we have the posterior value

$$PoV(\boldsymbol{y}) = \int \max\left\{0, E\left(\sum_{i=1}^{n} x_i \mid \boldsymbol{y}\right)\right\} p(\boldsymbol{y})\, d\boldsymbol{y} = \int \max\{0, w(\boldsymbol{y})\}\, p(\boldsymbol{y})\, d\boldsymbol{y}, \tag{A.23}$$

where the expected profits $w = w(\boldsymbol{y}) = E\left(\sum_{i=1}^{n} x_i \mid \boldsymbol{y}\right)$ has a pdf derived from Equation (A.22):

$$p(w) = N\left(\mu_w, r_w^2\right), \quad \mu_w = \sum_{i=1}^{n} \mu_i, \quad r_w^2 = \sum_{k=1}^{n} \sum_{l=1}^{n} R_{kl}. \tag{A.24}$$

The integral in Equation (A.23) is multivariate, but the only variable that matters is the linear combination of the data – i.e., $w = E\left(\sum_{i=1}^{n} x_i \mid \boldsymbol{y}\right)$, which takes part within the maximum. As a consequence, we can integrate over the univariate distribution of w. Mathematically, we do a transformation of variables and integrate over $\left(w, \boldsymbol{w}_{\_}\right)$ instead of $\boldsymbol{y}$. The second component $\boldsymbol{w}_{\_}$ contains $m-1$ dimensions of the data that are independent of w. Since the variable $\boldsymbol{w}_{\_}$ is not involved in the maximization, it integrates out. We get

$$PoV(\boldsymbol{y}) = \int \max\left\{0, E\left(\sum_{i=1}^{n} x_i \mid \boldsymbol{y}\right)\right\} p(\boldsymbol{y}) d\boldsymbol{y} = \int \max\{0, w\} p(w) dw, \tag{A.25}$$

and the solution is similar to that in Equation (A.19). From Equation (A.24) we get

$$PoV(\boldsymbol{y}) = \int \max\left\{0, E\left(\sum_{i=1}^{n} x_i \mid \boldsymbol{y}\right)\right\} p(\boldsymbol{y}) d\boldsymbol{y} = \mu_w \Phi\left(\mu_w / r_w\right) + r_w \phi\left(\mu_w / r_w\right). \tag{A.26}$$

For the case with high spatial flexibility (select any site) and a decoupled value function,

$$PoV(\boldsymbol{y}) = \sum_{i=1}^{n} \int \max\{0, E(x_i \mid \boldsymbol{y})\} p(\boldsymbol{y}) d\boldsymbol{y} = \sum_{i=1}^{n} \int \max\{0, w_i(\boldsymbol{y})\} p(\boldsymbol{y}) d\boldsymbol{y}, \tag{A.27}$$

where the expected profits are $E\left(x_i \mid \boldsymbol{y}\right) = w_i\left(\boldsymbol{y}\right) = w_i$, with Gaussian pdf $p(w_i) = N\left(\mu_i, r_i^2\right)$, where $r_i^2 = R_{ii}$ are the diagonal elements of matrix $\boldsymbol{R}$ in Equation (A.22). Again, the integrals are multivariate over the entire sample space of data $\boldsymbol{y}$, but in each integral only the linear combination of data is relevant for the maximization. We get

$$PoV(\boldsymbol{y}) = \sum_{i=1}^{n} \int \max\{0, E(x_i \mid \boldsymbol{y})\} p(\boldsymbol{y}) d\boldsymbol{y} = \sum_{i=1}^{n} \int \max\{0, w_i\} p(w_i) dw_i, \tag{A.28}$$

and using the closed-form solution in Equation (A.19) and the parameters in Equation (A.22):

$$PoV(\boldsymbol{y}) = \sum_{i=1}^{n} \int \max\{0, E(x_i \mid \boldsymbol{y})\} p(\boldsymbol{y}) d\boldsymbol{y} = \sum_{i=1}^{n} \left(\mu_i \Phi\left(\mu_i / r_i\right) + r_i \phi\left(\mu_i / r_i\right)\right). \tag{A.29}$$

Appendix A.2: Generalized linear models

A.2.1 Definition and properties

Generalized linear models (GLMs) are very popular in statistics, and we list some properties of GLMs that are directly useful for this book. Excellent textbooks on GLMs include McCullagh and Nelder (1989) and Dobson and Barnett (2008). Extensive content related to these models is available in Demidenko (2004).

Table A.2. *The pdf of the binomial with n_{trial} trials and success probability p and the Poisson distribution with intensity λ. For the binomial distribution, the probability of success depends on the latent variable via a logistic transform. For the Poisson distribution, the intensity depends on the latent variable via a log transform*

	Probability Mass Function (pdf)	Link
Binomial	$p(y=k)=\frac{n_{trial}!}{(n_{trial}-k)!k!}p^k(1-p)^{n_{trial}-k}, \quad k=0,1,..,n_{trial}$	$p=\frac{\exp(x)}{1+\exp(x)}$
Poisson	$p(y=k)=\frac{\lambda^k}{k!}\exp(-\lambda), \quad k=0,1,..$	$\lambda=\exp(x)$

Table A.2 gives the univariate pdfs of the Poisson and binomial distributions. They are the two most common distributions for discrete response GLMs. The rightmost column shows the link functions, where the variable x is on the real line. Here, the inverse link functions are logistic and log variables.

Joint model

Assume that we have data $\boldsymbol{y}=(y_1,...,y_n)$ and possibly some known covariates, organized in matrix $\boldsymbol{H}$, as in Appendix A.1. In a hierarchical framework, we regress the latent variable $\boldsymbol{x}=(x_1,...,x_n)$ on the covariates. We set $p(\boldsymbol{x})=N(\boldsymbol{\mu},\boldsymbol{\Sigma})$, where $\boldsymbol{\mu}=\boldsymbol{H\beta}$ and $\boldsymbol{\Sigma}=\boldsymbol{\Sigma}(\boldsymbol{\theta})$ as in Appendix A.1. We assume that the data are conditionally independent $p(y_i \mid \mathbf{y}_{-i},\boldsymbol{x})=p(y_i \mid x_i)$, $i=1,...,n$.

A GLM is straightforward to simulate based on the Gaussian model and the likelihood.

Algorithm: sampling from a GLM

1. Draw a random Gaussian sample $\boldsymbol{x}$ from pdf $p(\boldsymbol{x})=N(\boldsymbol{\mu},\boldsymbol{\Sigma})$.
2. For each $i=1,...,n$, sample y_i using the conditional independence likelihood $p(y_i \mid x_i)$.

Conditioning

The posterior pdf $p(\boldsymbol{x}\mid\boldsymbol{y})$ is not available in closed form but can be approximated in various ways. Maximization of the posterior is identical to

$$\hat{\boldsymbol{x}}=\arg\max_{\boldsymbol{x}}\{p(\boldsymbol{y}\mid\boldsymbol{x})p(\boldsymbol{x})\}. \quad (A.30)$$

Denote by $\boldsymbol{\Sigma}^{-1}+\hat{\boldsymbol{D}}$ the negative Hessian (second derivatives) of $\log\left(p(\boldsymbol{y}\mid\boldsymbol{x})p(\boldsymbol{x})\right)$ evaluated at the posterior mode $\hat{\boldsymbol{x}}$. Because of the conditional independence, the matrix $\hat{\boldsymbol{D}}$ is diagonal. Straightforward differentiation of the form in Table A.2 gives the i-th element $\hat{D}_{i,i}=\exp(\hat{x}_i)$ for the Poisson model. The binomial exponent is available from Table A.2; $\log\left(\frac{n_{trial}!}{(n_{trial}-k)!k!}\right)+kx-n_{trial}\log\left(1+\exp(x)\right)$, and differentiation results in $\hat{D}_{i,i}=-\frac{n_{trial}\exp(\hat{x}_i)}{1+\exp(\hat{x}_i)}$.

A Gaussian approximation to the posterior has the mean and covariance matrix

$$E(\boldsymbol{x}\mid\boldsymbol{y})\approx\hat{\boldsymbol{x}},\ Var(\boldsymbol{x}\mid\boldsymbol{y})\approx\left(\boldsymbol{\Sigma}^{-1}+\hat{\boldsymbol{D}}\right)^{-1}. \tag{A.31}$$

Assume that the goal is to predict $x_j=x(\mathbf{s}_{0,j})$ at a site $\mathbf{s}_{0,j}$. We let $E(x_j)=\mu_j$,and $Var(x_j)=\sigma_j^2$, and $\boldsymbol{\Sigma}_{j,*}$ denotes the covariance between x_j and $\boldsymbol{x}$. By the Gaussian properties from Equation (A.3),

$$\begin{aligned}\kappa_j&=E(x_j\mid\boldsymbol{x})=\mu_j+\boldsymbol{\Sigma}_{j,*}\boldsymbol{\Sigma}^{-1}(\boldsymbol{x}-\boldsymbol{\mu}),\\ \xi_j^2&=Var(x_j\mid\boldsymbol{x})=\sigma_j^2-\boldsymbol{\Sigma}_{j,*}\boldsymbol{\Sigma}^{-1}\boldsymbol{\Sigma}_{j,*}^t.\end{aligned} \tag{A.32}$$

Moreover, we can approximate the conditional distribution given data $\boldsymbol{y}$ by the Laplace approximation – see, e.g., Evangelou and Zhu (2012):

$$\begin{aligned}\nu_j&=E(x_j\mid\boldsymbol{y})\approx\mu_j+\boldsymbol{\Sigma}_{j,*}\boldsymbol{\Sigma}^{-1}(\hat{\boldsymbol{x}}-\boldsymbol{\mu}),\\ \zeta_j^2&=Var(x_j\mid\boldsymbol{y})\approx\sigma_j^2-\boldsymbol{\Sigma}_{j,*}\left(\boldsymbol{\Sigma}+\hat{\boldsymbol{D}}^{-1}\right)^{-1}\boldsymbol{\Sigma}_{j,*}^t.\end{aligned} \tag{A.33}$$

Note the similarity with the Gaussian case in Equation (A.10) and the differences in the matrix inverse expressions.

Parameter estimation

The estimation of unknown regression parameters $\boldsymbol{\beta}$ and covariance parameters $\boldsymbol{\theta}$ can be done via the Laplace approximation of the marginal likelihood. From Bayes' formula and the Gaussian approximation, this is defined by

$$\hat{p}(\boldsymbol{y};\boldsymbol{\beta},\boldsymbol{\theta})=\hat{p}(\boldsymbol{y})=\frac{p(\boldsymbol{y}\mid\hat{\boldsymbol{x}})p(\hat{\boldsymbol{x}})}{\hat{p}(\hat{\boldsymbol{x}}\mid\boldsymbol{y})}=\frac{p(\boldsymbol{y}\mid\hat{\boldsymbol{x}})p(\hat{\boldsymbol{x}})}{N\left(\hat{\boldsymbol{x}},\left(\boldsymbol{\Sigma}^{-1}+\hat{\boldsymbol{D}}\right)^{-1}\right)}, \tag{A.34}$$

where the denominator is the Gaussian approximation to the posterior defined by Equation (A.31), while the numerator is defined by the model.

The maximum likelihood estimate (MLE) must be computed numerically.

Algorithm: maximum likelihood estimation by the Laplace approximation

Initiate β^0, $\boldsymbol{\theta}^0$ by least-squares estimation and empirical variograms or other approaches.
Iterate for $b = 0,1,\ldots$ until convergence:

1. Compute $\log \hat{p}\left(y;\boldsymbol{\beta}^b,\boldsymbol{\theta}^b\right)$.
2. Approximate the derivatives $\mathbf{u}$ and the Hessian $\mathbf{V}$ of the approximate log-likelihood.
3. $\left(\boldsymbol{\beta}^{b+1},\boldsymbol{\theta}^{b+1}\right) = \left(\boldsymbol{\beta}^b,\boldsymbol{\theta}^b\right) + \mathbf{V}^{-1}\mathbf{u}$.
4. $b = b+1$.

The algorithm may overshoot, and it could be wise to apply more robust updating methods, especially in the beginning of the iterative optimization scheme.

A.2.2 Decision analysis and VOI results

We derive the approximate VOI computations for the spatial GLM with Poisson likelihood. Suppose that we have the situation with a free selection of sites $j = 1,\ldots,N$ with two alternatives at each site. Using generic notation with a log intensity x_j, revenues denoted Rev per random event at a site, and fixed costs denoted Cost, the prior value can be written

$$PV = \sum_{j=1}^{N} \max\left\{0, \mathrm{Rev}\cdot E\left(\exp\left(x_j\right)\right) - \mathrm{Cost}\right\}. \tag{A.35}$$

The setting is the same for minimizing expenses (maximizing negative expenses) and maximizing profits. For the situation described in Section 6.4.2, the value of Rev is tied to expenses associated with random rock fall when we choose the alternative of not adding rock support. Rock fall is again connected with the random number of joints with log intensity x_j. (Here, we assume that unit volume $V_j = 1$; otherwise, the expected counts become $V_j E(\exp(x_j))$.) The value of Cost is the expenses associated with the alternative of adding support with a non-random outcome. If we subtract the fixed Cost for both alternatives, we get the formula in Equation (A.35).

From the Gaussian modeling assumptions,

$$\begin{aligned} E\left(\exp\left(x_j\right)\right) &= \int \frac{1}{\sqrt{2\pi\sigma_j^2}} \exp\left(-\frac{\left(x_j-\mu_j\right)^2}{2\sigma_j^2} + \frac{x_j\sigma_j^2}{\sigma_j^2}\right) dx_j \\ &= \exp\left(\mu_j + \frac{\sigma_j^2}{2}\right) \int \frac{1}{\sqrt{2\pi\sigma_j^2}} \exp\left(-\frac{\left(x_j-\left(\mu_j+\sigma_j^2\right)\right)^2}{2\sigma_j^2}\right) dx_j = \exp\left(\mu_j + \frac{\sigma_j^2}{2}\right), \end{aligned} \tag{A.36}$$

which can be inserted in Equation (A.35).

For the posterior value, we can have partial or total sampling given data $\boldsymbol{y}$, which are Poisson distributed, given the intensity of counts $\boldsymbol{x}$. Recall that the mean and variance in Equation (A.33) depend on $\boldsymbol{y}$ only through $\hat{\boldsymbol{x}}$. The asymptotic distribution of $\hat{\boldsymbol{x}}$ is the Gaussian with mean $\boldsymbol{\mu}$ and variance

$$\mathbf{R} = \boldsymbol{\Sigma} E_{\hat{x}} \left(\boldsymbol{\Sigma} + \hat{\boldsymbol{D}}^{-1} \right)^{-1} \boldsymbol{\Sigma}. \tag{A.37}$$

The diagonal entries of this matrix are denoted r_j^2, $j = 1,...,N$. Notice the similarity between this result and that for the Gaussian models in Equation (A.22). For the Poisson model $\hat{D}_{jj}^{-1} = e^{-\hat{x}_j}$, so $E_{\hat{x}} \hat{D}_{jj}^{-1} = e^{-\mu_j + r_j^2/2}$, which depends on $\mathbf{R}$ itself. We can use iterative updates to get $\mathbf{R}$ (Evangelou and Eidsvik 2015).

The posterior value is

$$\begin{aligned} PoV(\boldsymbol{y}) &= \sum_{j=1}^{N} \max\left\{0, \text{Rev} \cdot E\left(\exp\left(x_j\right) \mid \boldsymbol{y}\right) - \text{Cost}\right\} p(\boldsymbol{y}) d\boldsymbol{y} \\ &\approx \sum_{j=1}^{N} \int \max\left\{0, \text{Rev} \cdot \exp\left(\hat{x}_j + \frac{\xi_j^2}{2}\right) - \text{Cost}\right\} p\left(\hat{x}_j\right) d\hat{x}_j \\ &\approx \sum_{j=1}^{N} \exp\left(\frac{\xi_j^2}{2}\right) \text{Rev} \int_{e_j}^{\infty} \exp\left(\hat{x}_j\right) p\left(\hat{x}_j\right) d\hat{x}_j - \text{Cost} \int_{e_j}^{\infty} p\left(\hat{x}_j\right) d\hat{x}_j \end{aligned} \tag{A.38}$$

$$\approx \text{Rev} \cdot \exp\left(\mu_j + \xi_j^2/2 + r_j^2/2\right) \Phi\left(\frac{e_j - \mu_j - r_j^2}{r_j}\right) - Cost \cdot \Phi\left(\frac{e_j - \mu_j}{r_j}\right),$$

where $e_j = (\log(\text{Cost} / \text{Rev}) - \xi_j^2 / 2)$, with ξ_j defined in Equation (A.32).

The VOI is the difference between the posterior and prior values. Because of the approximations done, this is not necessarily always positive. To ensure positive VOI, we can apply double expectation $E\left(E\left(\exp(x) \mid \boldsymbol{y}\right)\right)$ and use the Laplace approximation on the expression for the prior value as well. This gives

$$PV = \sum_{j=1}^{N} \max\left\{0, \text{Rev} \cdot \exp\left(\mu_j + \frac{\xi_j^2}{2} + \frac{r_j^2}{2}\right) - Cost\right\}. \tag{A.39}$$

With these established analytical results, it is possible to conduct VOI analysis without Monte Carlo approximations.

Appendix A.3: Markov chains and hidden Markov models

A.3.1 Definition and properties

There is an enormous body of literature on Markov chains and stochastic processes – see, e.g., Taylor and Karlin (1994) or Lawler (2006) – and on hidden Markov models – see, e.g.,

MacDonald and Zucchini (1997). Again, our focus is on providing more background for the results used in this book.

Denote the random variable by $x_i \in \{1,\ldots,d\}$, $i = 1,\ldots,n$. In the simplest case, there are only two discrete states: $d = 2$. The modeling is based on an ordering in the variables so that x_{i-1} is before x_i in some sense. Sometimes this might be an obvious temporal ordering. A Markov chain is defined by conditional probabilities and by reducing the conditioning from all previous variables to only a subset of the nearest or most recent variables. A lag r, or order r, Markov chain is defined by

$$p(x_i = l \mid x_{i-1},\ldots,x_1) = p(x_i = l \mid x_{i-1},\ldots,x_{i-r}), \quad l = 1,\ldots,d. \tag{A.40}$$

For a first-order Markov chain

$$p(x_i = l \mid x_{i-1},\ldots,x_1) = p(x_i = l \mid x_{i-1}), \quad l = 1,\ldots,d. \tag{A.41}$$

We focus on this first-order Markov chain, but the models and methods easily generalize to the r-th order.

Joint and marginal probabilities

The joint pdf for all variables $\boldsymbol{x} = (x_1, x_2, \ldots, x_n)$ in a first-order Markov chain is

$$p(\boldsymbol{x}) = p(x_1)\prod_{i=2}^{n} p(x_i \mid x_{i-1}), \tag{A.42}$$

where the initial state has pdf $p(x_1)$.

Marginal probabilities are obtained by recursively summing out the previous variables

$$p(x_{i+1} = l) = \sum_{k=1}^{d} p(x_i = k)\, p(x_{i+1} = l \mid x_i = k), \quad l = 1,\ldots,d, \quad i = 1,\ldots,n-1. \tag{A.43}$$

Time invariance or time homogeneity entails that the conditional probabilities do not depend on the location or time step i – i.e., $p(x_{i+1} = l \mid x_i = k) = P(k,l)$. These one-step conditional probabilities can be organized in a size $d \times d$ Markov transition matrix

$$\boldsymbol{P} = \begin{pmatrix} P(1,1) & P(1,2) & \ldots & P(1,d) \\ P(2,1) & P(2,2) & \ldots & P(2,d) \\ \ldots & \ldots & \ldots & \\ P(d,1) & \ldots & \ldots & P(d,d) \end{pmatrix}, \tag{A.44}$$

where $\sum_{l=1}^{d} P(k,l) = 1$ for all $k = 1,\ldots,d$. Time-invariant transition probabilities are convenient because of easy interpretation and the parsimonious structure with just a few unknown parameters to specify.

In this situation, the conditional distribution over many time steps is defined by multiplying the transition matrix many times – i.e., $\boldsymbol{P}^{i-j} = \boldsymbol{P} \cdot \ldots \cdot \boldsymbol{P}$ holds the $i-j$ step conditional probabilities $p(x_i = k \mid x_j = l)$, $k,l = 1,...,d$. Regular Markov chains are characterized by the possibility to reach every state, so all entries in $\boldsymbol{P}^{i-j}$ are strictly positive. This is not obvious in general since some states may be absorbing states. For a formal discussion, see, e.g., Taylor and Karlin (1994) or one of the other textbooks.

If the distribution for the initial state $p(x_1)$ is defined directly from the distribution associated with the transition matrix $\boldsymbol{P}$, the marginal probabilities in Equation (A.43) are the same for every location or time step – i.e., $p(x_{i+1} = l) = p(x_i = l)$ for all $l = 1,...,d$. This stationary distribution satisfies $p(x_{i+1} = l) = \sum_{k=1}^{d} p(x_i = k)P(k,l)$ for any l. In vector matrix form $\boldsymbol{p} = \boldsymbol{pP}$, $\boldsymbol{p} = (p(x_i = 1),\ldots,p(x_i = d))$, and $\sum_{k=1}^{d} p(x_i = k) = 1$. The stationary distribution is an eigenvector of the transition matrix $\boldsymbol{P}$ with an eigenvalue of 1. The stationary distribution will coincide with the limiting distribution defined from $\lim_{i\to\infty} \boldsymbol{P}^i$ under certain regularity conditions (see, e.g., Gamerman and Lopes (2006)). This means that no matter where we initiate the Markov chain, it will converge to the stationary distribution.

Sampling from a Markov chain

To sample a realization $\boldsymbol{x}$ from a first-order Markov chain, we directly use the model.

Algorithm: sampling a first-order Markov chain

1. Draw an initial state x_1 from (stationary distribution) $p(x_1)$.
2. For $i = 1,...,n-1$, sample x_{i+1} from $p(x_{i+1} \mid x_i)$.

The class is picked by drawing uniform numbers and comparing the uniform random variables with the transition probabilities (see Appendix B).

Conditioning

As we indicated earlier, the forward conditioning over multiple time steps is defined by $\boldsymbol{P}^{i-j} = \boldsymbol{P} \cdot \ldots \cdot \boldsymbol{P}$, which equals the following recursive calculation for $i = j+1,...,n$:

$$p(x_i = k \mid x_j = l) = \sum_{q=1}^{d} p(x_i = k, x_{i-1} = q \mid x_j = l) = \sum_{q=1}^{d} P(q,k)\,p(x_{i-1} = q \mid x_j = l). \tag{A.45}$$

The full conditional distribution of x_i is

$$p(x_i = k \mid \boldsymbol{x}_{-i}) = p(x_i = k \mid x_{i-1} = q, x_{i+1} = l) \propto P(q,k)P(k,l), \tag{A.46}$$

and normalization gives

$$p(x_i = k \mid x_{i-1} = q, x_{i+1} = l) = \frac{P(q,k)P(k,l)}{\sum_{k=1}^{d} P(q,k)P(k,l)}. \tag{A.47}$$

The backward transition probability of a first-order Markov chain is defined by

$$p(x_i = k \mid x_{i+1},\ldots,x_n) = p(x_i = k \mid x_{i+1} = l) = \frac{p(x_i = k, x_{i+1} = l)}{p(x_{i+1} = l)} = \frac{P(k,l)\,p(x_i = k)}{p(x_{i+1} = l)}. \tag{A.48}$$

The backward computation over multiple time steps can be computed recursively, $i = j-1,\ldots,1$, and

$$p(x_i = k \mid x_j = l) = \sum_{q=1}^{d} p(x_i = k \mid x_{i+1} = q)\,p(x_{i+1} = q \mid x_j = l). \tag{A.49}$$

Conditional backward sampling proceeds by recursively drawing variables x_i, $i = j-1,\ldots,1$, from the backward transition probabilities in Equation (A.48) rather than summing over the subsequent state x_{i+1} as in Equation (A.49).

Hidden Markov model

The hidden Markov model (HMM) represents a latent distinction of interest $\boldsymbol{x} = (x_1,\ldots,x_n)$ by a Markov chain, while the measurements $\boldsymbol{y} = (y_1,\ldots,y_m)$ provide imperfect information about $\boldsymbol{x}$. Here we assume data size $m = n$. The most common setting is that of a parsimonious model with a first-order Markov chain with time-invariant transition probabilities $p(x_{i+1} = l \mid x_i = k) = P(k,l)$ and a likelihood model involving conditional independence $p(y_i \mid \boldsymbol{y}_{-i}, \boldsymbol{x}) = p(y_i \mid x_i)$. The data could be either discrete or continuous. For the discrete data situation, the likelihood model $p(y_i = k \mid x_i = k)$ indicates the accuracy of the measurement. With a continuous model for the data, some of the parameters of the likelihood $p(y_i \mid x_i = k)$ depend on the latent class $k \in \{1,\ldots,d\}$. The joint distribution is

$$p(\boldsymbol{x},\boldsymbol{y}) = p(x_1)\left(\prod_{i=2}^{n} p(x_i \mid x_{i-1})\right)\left(\prod_{i=1}^{n} p(y_i \mid x_i)\right). \tag{A.50}$$

Sampling data $\boldsymbol{y}$ from the marginal distribution $p(\boldsymbol{y})$ of an HMM is done by first sampling a Markov chain $p(\boldsymbol{x})$ as shown earlier and then sampling y_i from $p(y_i \mid x_i)$ for all $i = 1,\ldots,n$.

The most common goals of analyzing an HMM are (i) marginal likelihood calculation $p(\boldsymbol{y};\boldsymbol{\theta}) = p(\boldsymbol{y})$ for parameter estimation, (ii) evaluation of posterior marginal probabilities $p(x_i \mid \boldsymbol{y})$, $i = 1,\ldots,n$, and (iii) sampling a realization from the posterior $p(\boldsymbol{x} \mid \boldsymbol{y})$. We illustrate the forward–backward algorithm for these tasks. For more background, see Scott (2002). The algorithm is based on recursively summing out one latent variable at a time. The first step is a forward calculation that defines the marginal likelihood. Some expressions from this forward recursion must be stored in memory, as they will be useful in the backward recursion.

Algorithm: the forward recursion involves prediction, updating, and marginalization

$$\begin{aligned} p(x_i = k \mid y_1,\ldots,y_{i-1}) &= \sum_{j=1}^{d} p(x_{i-1} = j, x_i = k \mid y_1,\ldots,y_{i-1}) \\ &= \sum_{j=1}^{d} p(x_i = k \mid x_{i-1} = j)\, p(x_{i-1} = j \mid y_1,\ldots,y_{i-1}), \end{aligned} \tag{A.51}$$

$$\begin{aligned} p(x_i = k \mid y_1,\ldots,y_i) &= \frac{p(x_i = k, y_i \mid y_1,\ldots,y_{i-1})}{p(y_i \mid y_1,\ldots,y_{i-1})} \\ &= \frac{p(y_i \mid x_i = k)\, p(x_i = k \mid y_1,\ldots,y_{i-1})}{p(y_i \mid y_1,\ldots,y_{i-1})}, \end{aligned} \tag{A.52}$$

$$p(y_i \mid y_1,\ldots,y_{i-1}) = \sum_{k=1}^{d} p(y_i \mid x_i = k)\, p(x_i = k \mid y_1,\ldots,y_{i-1}). \tag{A.53}$$

Initiate by $i = 1$ and a predictive distribution $p(x_1)$; proceed using Equations (A.52) and (A.53); set $i = 2$; calculate Equations (A.51), (A.52), and (A.53): set $i = 3$, etc., until the last time step $i = n$.

Note how the conditional independence assumptions simplify the expressions, i.e. $p(x_i = k \mid x_{i-1} = j, y_1,\ldots,y_{i-1}) = p(x_i = k \mid x_{i-1} = j)$ and $p(y_i \mid x_i = k, y_1,\ldots,y_{i-1}) = p(y_i \mid x_i = k)$.

The product of all normalizing constants $p(y_i \mid y_1,\ldots,y_{i-1})$ in Equation (A.53) is the marginal likelihood

$$p(\mathbf{y}) = \prod_{i=1}^{n} p(y_i \mid y_1,\ldots,y_{i-1}). \tag{A.54}$$

Algorithm: the backward recursion for posterior marginal probabilities

1. Start at n with $p(x_n \mid y_1,\ldots,y_n)$ – i.e., the result of the final forward step.
2. Step backward for $i = n-1,\ldots,1$ using

$$p(x_i = k \mid y_1,\ldots,y_n) = \sum_{l=1}^{d} p(x_i = k \mid y_1,\ldots,y_{i+1}, x_{i+1} = l)\, p(x_{i+1} = l \mid y_1,\ldots,y_n). \tag{A.55}$$

The backward steps use $p(x_i = k \mid \boldsymbol{y}, x_{i+1}) = p(x_i = k \mid y_1,\ldots,y_i,x_{i+1}) = p(x_i = k \mid y_1,\ldots, y_{i+1},x_{i+1})$ from the conditional independence assumptions. The additional conditioning on y_{i+1} allows more efficient re-use of the computations from the forward recursion. The first term in Equation (A.55) can be derived from the forward recursion by noticing that

$$\begin{aligned} p(x_i = k \mid y_1,\ldots,y_{i+1},x_{i+1}) &= \frac{p(x_i = k, x_{i+1}, y_{i+1} \mid y_1,\ldots,y_i)}{p(y_{i+1} \mid x_{i+1}, y_1,\ldots,y_i)\, p(x_{i+1} \mid y_1,\ldots,y_i)} \\ &= \frac{p(x_{i+1} \mid x_i = k)\, p(x_i = k \mid y_1,\ldots,y_i)}{p(x_{i+1} \mid y_1,\ldots,y_i)}. \end{aligned} \tag{A.56}$$

All components are given either from the Markov transition model, or the forward calculation. The resulting $p(x_i = k \mid \boldsymbol{y})$ is the posterior given all data, which is also called the smoothing distribution in this context of the (dynamic) HMM.

Sampling from an HMM is done by backwards sampling.

Algorithm: sampling from the posterior of a hidden Markov model

1. Draw x_n from $p(x_n \mid y_1,\ldots,y_n)$ derived in the forward recursion in Equation (A.52).
2. For $i = n-1,\ldots,1$, draw x_i from $p(x_i \mid y_1,\ldots,y_i,x_{i+1})$ using Equation (A.56).

A.3.2 Decision analysis and VOI results

The decision situation we have used in association with such discrete distinctions of interest involves revenues in the success outcome and costs in the failure outcome(s). This means that the prior value is

$$PV = \sum_{j=1}^{N} \max\{0, \text{Rev} \cdot p(x_j = 1) - \text{Cost}\}, \tag{A.57}$$

when we have a high decision flexibility – i.e., a free selection of sites, $N = n$, and two alternatives at each site. Note that the prior may include current data – for instance, modeled by an HMM.

For the posterior value, we can have total or partial imperfect or perfect data $\boldsymbol{y}$. If data (perfect or imperfect) are discrete and only at a few sites, we can directly compute the posterior value as a sum over the data variables – i.e.,

$$PoV(\boldsymbol{y}) = \sum_{\boldsymbol{y}} \sum_{j=1}^{N} \max\{0, \text{Rev} \cdot p(x_j = 1 \mid \boldsymbol{y}) - \text{Cost}\} p(\boldsymbol{y}). \tag{A.58}$$

The forward–backward algorithm is used for the inner probabilities in Equation (A.58). The VOI is the difference between this posterior value and the prior value in Equation (A.57).

If the data are located at many sites, the outer sum involves a lot of terms. Moreover, the probabilities for partial tests $\boldsymbol{y}_{\mathbb{K}}$, where $\mathbb{K} \subset \{1,...,n\}$, may require specific marginalization techniques similar to Equation (A.53). It might then be easier to use Monte Carlo sampling for the outer sum while maintaining the forward–backward recursion to solve for the inner probabilities. For partial testing, we can sample the entire vector $\boldsymbol{y}$ and discard the non-observed parts to get $\boldsymbol{y}_{\mathbb{K}}$. The same holds for continuous data when the outer integral cannot be solved numerically.

Algorithm: Monte Carlo approximation of the posterior value for a Markov chain model or HMM

1. Sample data variables $\boldsymbol{y}^b$, $b = 1,...,B$ from $p(\boldsymbol{y})$.
2. Compute the posterior probability $p(x_j = 1 \mid \boldsymbol{y}^b)$, $j = 1,...,N$, and $b = 1,...,B$ by the forward–backward algorithm.
3. Average the values $PoV(\boldsymbol{y}) \approx \frac{1}{B}\sum_{b=1}^{B}\sum_{j=1}^{N} \max\{0, \text{Rev} \cdot p(x_j = 1 \mid \boldsymbol{y}^b) - \text{Cost}\}$.

Appendix A.4: Categorical Markov random fields

The categorical Markov random field (MRF) model extends the one-dimensional Markov chains and hidden Markov models to the two- and three-dimensional spatial domains. There is no longer a natural ordering of the variables, and dependence is modeled either by the conditional structure or by potentials over subsets of variables.

A.4.1 Definition and properties

The most common setting represents variables on a regular grid. In two dimensions, the grid size could be $n_1 \times n_2$, where the total number of cells is $n = n_1 n_2$. Let the random variable $x_i \in \{1,\ldots,d\}$, $i = 1,\ldots,n$, and in the simplest case there are only two discrete states $d = 2$. Similar to the Markov chain, the MRF model reduces the conditioning to the nearest neighbors, but in the spatial model there is no natural ordering of the variables of interest $\boldsymbol{x} = (x_1,...,x_n)$. One variable does not naturally come before another.

Conditional model formulation

The Markov assumption means that the full conditional distribution reduces to the conditional distribution given variables in the neighborhood – i.e.,

$$p(x_i \mid \boldsymbol{x}_{-i}) = p(x_i \mid x_j; j \in N_i), \tag{A.59}$$

where the neighborhood N_i can be of various orders. The MRF on a regular grid has the following neighborhoods:

First order: N_i consists of grid cells to the north, east, south, and west of the current cell;
Second order: N_i is north, northeast, east, southeast, south, southwest, west, and northwest.
And so on.

The Ising model is the most popular MRF model. The conditional probabilities are then

$$p\left(x_i \mid x_j; j \in N_i\right) \propto \exp\left(\beta \sum_{j \in N_i} I\left(x_i = x_j\right)\right). \tag{A.60}$$

for $x_i \in \{1,2\}$ and where β is the spatial interaction parameter to first order neighbors. If this interaction parameter is zero, the variables are independent. It is usually set to a positive value, imposing similarity in neighborhood colors.

The model in Equation (A.60) does not favor one color over the other. We can increase the chance of colors at selected cells by adding an external field in the form of single-node potentials in the exponent as follows:

$$p\left(x_i \mid x_j; j \in N_i\right) \propto \exp\left(\beta \sum_{j \in N_i} I\left(x_i = x_j\right) + \alpha_i\left(x_i\right)\right). \tag{A.61}$$

Note, however, that the potentials $\alpha_i\left(x_i\right)$ do not have a one-to-one relation with the marginal probability $p_i\left(x_i\right) = p\left(x_i\right)$ at cell i on the grid.

The extension of the Ising model to more than two colors is sometimes called the Potts model.

Joint model formulation for Markov random fields

The associated joint distribution of the MRF, called the Gibbs formulation, is given by

$$p(\boldsymbol{x}) = \frac{\exp\left(-\sum_{k \in C} \psi_k\left(\boldsymbol{x}_k\right)\right)}{Z} \propto \exp\left(-\sum_{k \in C} \psi_k\left(\boldsymbol{x}_k\right)\right), \tag{A.62}$$

where $\boldsymbol{x}_k$ is a subset of the variable $\boldsymbol{x}$, and C is the set of all cliques. The normalizing constant Z results from summing over the entire sample space of $x_1, \ldots, x_n$, and it does not depend on $\boldsymbol{x}$. The cliques are representative of the interaction or dependence structure of the MRF, and the clique potentials $\psi_k\left(\boldsymbol{x}_k\right)$ assign a penalty (or reward) to configurations $\boldsymbol{x}_k$ within a clique type k. The potentials can be specified to make it more likely to get patches of similar color in the grid. For a first order neighborhood on the regular grid, there are just two types of cliques: single-cell cliques and vertical and horizontal cell-pair cliques. For the Ising model, the positive clique potentials are β for equal colored cell pairs (north–south) and (east–west) on the grid. The zero potential for cliques with black and white cells then penalize differences in colors. For a second-order neighborhood, the cliques are of size 2 × 2. When the cliques get larger, there is additional flexibility for incorporating spatial patterns via low or high clique potentials for various configurations of $\boldsymbol{x}_k$ – see, e.g., Tjelmeland and Besag (1996).

It is not obvious that the definition of the MRF by conditional probabilities in Equation (A.59) results in a unique joint statistical model. However, the Hammersley–Clifford theorem (Besag 1974) states that there is such a one-to-one correspondence between the conditional model formulation over neighborhoods in Equation (A.59) and the joint formulation over cliques in Equation (A.62). It is relatively easy to show that this holds one way, while the other is harder. For the easier approach, given a joint distribution over cliques, we show that the full conditional depends only on the neighboring variables. We have

$$p(x_i \mid \boldsymbol{x}_{-i}) \propto p(\boldsymbol{x}) \propto \exp\left(-\sum_{k \in C} \psi_k(\boldsymbol{x}_k)\right) \propto \exp\left(-\sum_{k \in C \cap i} \psi_k(\boldsymbol{x}_k)\right) = p(x_i \mid x_j; j \in N_i), \quad (A.63)$$

since the cliques where x_i is involved are the ones with x_i and its interactions with the neighbors of cell i. None of the variables outside the neighborhood interact with x_i in the joint distribution, since there are no cliques associated with such pairs.

Hidden Markov random fields

Assume an MRF for the latent variable $\boldsymbol{x}$. Imperfect data $\boldsymbol{y} = (y_1, \ldots, y_m)$ can be gathered, and the likelihood is $p(\boldsymbol{y} \mid \boldsymbol{x})$. The joint distribution for latent variables and data is

$$p(\boldsymbol{x}, \boldsymbol{y}) = p(\boldsymbol{x}) p(\boldsymbol{y} \mid \boldsymbol{x}). \quad (A.64)$$

Again, we impose the customary assumption of conditionally independent data – i.e., $p(y_i \mid x_i), i = 1, \ldots, m$. For location $i = 1, \ldots, n$, the full conditional distribution then becomes

$$p(x_i \mid \boldsymbol{x}_{-i}, \boldsymbol{y}) = p(x_i \mid y_i, x_j; j \in N_i). \quad (A.65)$$

Assuming an Ising model, we have full conditional distribution

$$p(x_i \mid \boldsymbol{x}_{-i}, \boldsymbol{y}) = p(x_i \mid y_i, x_j; j \in N_i) \propto \exp\left(\beta \sum_{j \in N_i} I(x_i = x_j) + \log p(y_i \mid x_i)\right). \quad (A.66)$$

Note that the structure of the full conditional in Equation (A.66) is the same as for the Ising model with an external prior field in Equation (A.61).

Calculations for Markov random fields and hidden Markov random fields

We discuss marginalization and conditioning for prior and posterior pdfs together. If the data are conditionally independent, the calculations for the posterior model (with data) follow the same steps as those in the prior (without data) because the likelihood terms enter as an external field; see Equation (A.66).

For small-sized grids and moderate-sized neighborhoods, it is possible to perform exact marginalization and conditioning for MRFs. The idea is similar to the forward–backward algorithm used for the hidden Markov model in Appendix A.3. For large-sized MRFs, the solution requires approximations. The most popular approach has been Markov chain Monte Carlo (MCMC) sampling.

Consider the Ising model with an external field (which could be a conditionally independent likelihood term as in Equation (A.66)). The model can be written

$$p(\boldsymbol{x}) = \frac{\exp\left(\beta \sum_{i \sim j} I(x_i = x_j) + \sum_i \alpha_i(x_i)\right)}{Z} = \frac{h(\boldsymbol{x})}{Z}, \tag{A.67}$$

where the short notation $i \sim j$ denotes a clique – i.e., all pairs of north–south or east–west nodes on the grid. The normalizing constant Z is given by

$$Z = \sum_{x_1=1}^{d} \ldots \sum_{x_n=1}^{d} h(\boldsymbol{x}). \tag{A.68}$$

This normalizing constant is very hard to compute because it is a sum over all variables. The evaluation of the joint, marginal, and conditional distributions is then computationally demanding.

Suppose that the grid nodes are numbered sequentially from northwest to southeast. Then, x_1 is the categorical value in the northwest corner, x_{n_1} the one in the southwest corner, $x_{n_1(n_2-1)+1}$ in the northeast corner, and x_n in the southeast corner. For the Ising model, with $d = 2$, the neighbors of an interior node i are defined by $i+1$, $i-1$, $i-n_1$, and $i+n_1$. If node i is on the edge or a corner node, some of these neighbors vanish. Assume $n_1 \le n_2$, and n_1 is quite small (say, 10 or 20). It is then possible to explore the probability model analytically using the forward–backward algorithm. If the dimensions are larger, the combinatorial increase in $d^{\min(n_1,n_2)}$ becomes too demanding.

To understand the forward–backward algorithm in this context, write the probability function for $\boldsymbol{x}$ in a sequential manner as follows:

$$\begin{aligned} p(\boldsymbol{x}) &= p(x_1 \mid x_2,\ldots,x_n)\, p(x_2 \mid x_3,\ldots,x_n) \ldots p(x_{n-1} \mid x_n)\, p(x_n) \\ &= p(x_1 \mid x_2,\ldots,x_{1+n_1})\, p(x_2 \mid x_3,\ldots,x_{2+n_1}) \ldots p(x_{n-1} \mid x_n)\, p(x_n) \\ &= \frac{h(x_1 \mid x_2,\ldots,x_{1+n_1})\, h(x_2 \mid x_3,\ldots,x_{2+n_1}) \ldots h(x_{n-1} \mid x_n)\, h(x_n)}{Z}, \end{aligned} \tag{A.69}$$

where we use the Markov property but choose to condition on all buffer variables in the sequential lineup. The buffer is of length n_1 in this case, indicating the length of the north–south column in the grid. It gets shorter for the last (easternmost) column. The terms in this expression are defined by $h(x_1 \mid x_2,\ldots,x_{1+n_1}) = \exp\left(\beta\left(I(x_1 = x_2) + I(x_1 = x_{1+n_1})\right) + \alpha_1(x_1)\right)$ for the first location, then goes on like this for x_2, x_3, and so on until $h(x_{n-1} \mid x_n) = \exp\left(\beta I(x_{n-1} = x_n) + \alpha_{n-1}(x_{n-1})\right)$, $h(x_n) = \exp(\alpha_n(x_n))$.

Algorithm: forward calculation for small-sized MRFs

1. Start with

$$Z_1\left(x_2,\ldots,x_{n_1+1}\right)=\sum_{x_1=1}^{d}h\left(x_1 \mid x_2,\ldots,x_{1+n_1}\right), \tag{A.70}$$

2. For general $i \le n-n_1$

$$Z_i\left(x_{i+1},\ldots,x_{i+n_1}\right)=\sum_{x_i=1}^{d}h\left(x_i \mid x_{i+1},\ldots,x_{i+n_1}\right)Z_{i-1}\left(x_i,\ldots,x_{i+n_1-1}\right). \tag{A.71}$$

3. At the final step n, calculate

$$Z=Z_n=\sum_{x_n=1}^{d}h\left(x_n\right)Z_{n-1}\left(x_n\right), \tag{A.72}$$

which is the normalizing constant in Equation (A.67).

Here, for the intermediate step, $(x_{i+1},\ldots,x_{i+n_1})$ has d^{n_1} possible values of buffer configurations. As the buffer length gradually decreases in the last (easternmost) column, the number of possible configurations gets smaller.

Algorithm: backward recursion scheme for the marginal probabilities

1. Start with the marginal for x_n:

$$p\left(x_n\right)=\frac{\sum_{x_1=1}^{d}\cdots\sum_{x_{n-1}=1}^{d}h\left(\boldsymbol{x}\right)}{Z}=\frac{h\left(x_n\right)Z_{n-1}\left(x_n\right)}{Z},\quad x_n\in\{1,\ldots,d\}. \tag{A.73}$$

2. For $x_{n-1}\in\{1,\ldots,d\}$, first arrange the joint density $p(x_{n-1},x_n)$ and then sum out x_n:

$$p_{n-1}(x_{n-1})=\frac{\sum_{x_1=1}^{d}\cdots\sum_{x_{n-2}=1}^{d}\sum_{x_n=1}^{d}h(\boldsymbol{x})}{Z}=\frac{\sum_{x_n=1}^{d}h(x_n)h(x_{n-1}\mid x_n)Z_{n-2}(x_n,x_{n-1})}{Z}. \tag{A.74}$$

3. For a general node i, construct the joint probability for the buffer of length n_1, and then the marginal $p_i(x_i)$ for node i is obtained by summing over all buffer configurations $x_{i+1},\ldots,x_{i+n_1}$ for each possible value $x_i\in\{1,\ldots,d\}$.

The conditional situation given data $\boldsymbol{y}$ is not dissimilar. The likelihood term can be regarded as an external field, where $\alpha_i(x_i) = \log p(y_i \mid x_i)$, $x_i \in \{1,\ldots,d\}$. The marginal pdf of main interest is then the posterior marginal $p(x_i \mid \boldsymbol{y})$. The normalizing constant depends on both the model parameters and the data.

Note that for numerical stability, it is preferred to work on the logarithmic scale and to perform robust summing operations. This entails

$$\sum_{x=1}^{d} h(x) = \sum_{x=1}^{d} \exp(\log h(x)) = \exp(lh_{\max}) \sum_{x=1}^{d} \exp(\log h(x) - lh_{\max}), \qquad \text{(A.75)}$$

where $lh_{\max} = \max\{\log h(x)\}$.

Sampling

We demonstrate a recursive backward sampling algorithm for drawing a variable $\boldsymbol{x}$ from the joint pdf $p(\boldsymbol{x})$. A similar approach would work for conditional sampling.

Algorithm: sampling for small-sized MRFs

1. Run the forward recursion for sequential normalizing constants in Equation (A.71).
2. Sample the value of x_n from probability vector

$$p_n(x_n) = \frac{1}{Z} \sum_{x_1=1}^{d} \cdots \sum_{x_{n-1}=1}^{d} h(\boldsymbol{x}) = \frac{1}{Z} h(x_n) Z_{n-1}(x_n), \qquad x_n \in \{1,\ldots,d\} \qquad \text{(A.76)}$$

where the sequential normalizing constants in the forward recursion are re-used.

3. Continue backward, generating x_{n-1} conditional on the sample of x_n from probability vector

$$p(x_{n-1} \mid x_n) = \frac{h(x_{n-1} \mid x_n) Z_{n-2}(x_n, x_{n-1})}{Z_{n-1}(x_n)}, \qquad x_{n-1} \in \{1,\ldots,d\}, \qquad \text{(A.77)}$$

and so on for all $i = n-2,\ldots,1$.

Parameter estimation

There are unknown interaction parameters in the MRF that must be specified. The MRF parameters are the clique potentials that reduce to the pairwise interaction term β for the Ising model. We may also have unknown parameters in the likelihood model, but we focus on the interaction parameter in this discussion.

In many models, it is easier to maximize the likelihood with perfect information $\boldsymbol{x}$. But in this MRF case it is still complicated to compute $p(\boldsymbol{x})$ because of the normalization constant. By the forward–backward recursion it is possible to compute the likelihood and perform maximum likelihood estimation (MLE) for small grids. For larger grids, one often maximizes the pseudo-likelihood to specify the parameters. The pseudo-likelihood for general clique potentials ψ is defined via the full conditionals, $cl(\psi) = \prod_i p(x_i \mid x_j; j \in N_i)$. Since we have the Markovian structure, each term is computed by counting the number of different neighborhood clique configurations (Besag 1986).

If the latent variable is not directly observed but only imperfectly informed by data $\mathbf{y}$, we can estimate parameters from the marginal likelihood model $p(\mathbf{y}) = p(\mathbf{y};\beta)$. For a large-sized grid, MCMC algorithms are the preferred approximation. For small-sized grids, we can use the forward–backward algorithm. The marginal likelihood is the ratio of normalizing constants:

$$p(\mathbf{y}) = \frac{p(\mathbf{x})p(\mathbf{y} \mid \mathbf{x})}{p(\mathbf{x} \mid \mathbf{y})} = \frac{h_x(\mathbf{x})p(\mathbf{y} \mid \mathbf{x})Z_{x|y}}{Z_x h_{x|y}(\mathbf{x} \mid \mathbf{y})} = \frac{Z_{x|y}}{Z_x}, \tag{A.78}$$

because the functions in $h_{x|y}(\mathbf{x} \mid \mathbf{y})$ are defined by the function in the prior model $h_x(\mathbf{x})$ and the terms in the likelihood model $p(\mathbf{y} \mid \mathbf{x})$.

Algorithm: likelihood evaluation for small-sized MRFs

1. Run the forward recursion without data to get the normalizing constant Z_x in the prior model $p(\mathbf{x})$.
2. Run the forward recursion with data to get the normalizing constant $Z_{x|y}$ in the posterior model $p(\mathbf{x} \mid \mathbf{y})$.
3. Compute the ratio of the normalizing constants $p(\mathbf{y}) = \frac{Z_{x|y}}{Z_x}$ as in Equation (A.78).

A.4.2 Decision analysis and VOI results

The decision situation we have used in association with MRFs is similar to that for the Markov chain in Appendix A.3. For the prior value,

$$PV = \sum_{j=1}^{N} \max\left\{0, \text{Rev}\cdot p(x_j = 1) - \text{Cost}\right\}, \tag{A.79}$$

when we have a free selection of sites, $N = n$, and two alternatives at each site. The marginal pdf in the prior must now be computed by forward–backward recursions or by sophisticated MCMC sampling.

For the posterior value,

$$PoV(\mathbf{y}) = \sum_{\mathbf{y}} \sum_{j=1}^{N} \max\left\{0, \text{Rev}\cdot p(x_j = 1 \mid \mathbf{y}) - \text{Cost}\right\} p(\mathbf{y}). \tag{A.80}$$

For small-sized grids, the forward–backward algorithm solves for the inner probabilities in Equation (A.80). The outer loop can be summed out exactly for partial testing at only a small set of sites. Alternatively, for larger data sizes, samples from $p(\mathbf{y})$ can be used to get a Monte Carlo approximation like we described in Appendix A.3.2. Samples of $\mathbf{y}^b$,

$b = 1,\dots,B$ can be obtained by first sampling from $p(\boldsymbol{x})$ and then drawing conditionally independent samples for $\boldsymbol{y}$ from $p(y_i \mid x_i)$ given the realization for the latent field.

Appendix A.5: Discrete graphs and Bayesian networks

We show the main modeling and computational elements for discrete graphical models. In some sense, Markov chains and Markov random fields are special cases of this situation. For details regarding a more formal description, please consult a reference such as Pearl (1988), Shachter (1988), Jensen (1996), Jensen and Nielsen (2007), Darwiche (2009), or Koller and Friedman (2009).

A.5.1 Definition and properties

Bayesian networks (BNs), also called belief networks, are graphical models consisting of nodes and edges. We assume n nodes. The edges connect the nodes together, and they are directed. An edge that indicates direction from one node to another is an arc or, equivalently, an arrow. The arrows indicate the (causal) conditional dependence structure of the model. The graph is acyclic, meaning that there are no directed paths leading back to a node. These graphs are termed directed acyclic graphs (DAGs).

The variables of interest are $\boldsymbol{x} = (x_1,\dots,x_n)$. Data $\boldsymbol{y}$ could either be included in some of these nodes or be represented by additional nodes. The BN model for $\boldsymbol{x}$ is defined from conditional statements,

$$p(\boldsymbol{x}) = \prod_{i=1}^{n} p\left(x_i \mid x_j; j \in pa(i)\right), \tag{A.81}$$

where $pa(i)$ indicates the parent nodes of i. Top nodes (root nodes) in the network have empty parent sets. For the discrete situation that we consider here, the probabilities in Equation (A.81) can be collected in conditional probability tables.

If there is a directed path from node i to node j, we say that i is an ancestor of j and that j is a descendent of i.

Marginal and conditional probabilities

Several independence properties can be proven from the semantics of the network. The local Markov independence property of a belief network is the property that any node in the network is independent of its non-descendents given its parents. Other global independence properties can also be derived based on a graphical analysis of the network.

The full conditional probability of x_i is

$$p(x_i \mid \boldsymbol{x}_{-i}) = p\left(x_i \mid x_j; j \in pa(i), j \in ch(i), j \in op\left(ch(i)\right)\right), \tag{A.82}$$

where the child nodes are denoted $ch(i)$ and the other parents of such a child node j are denoted $op(j)$. The latter nodes, sometimes called wife or husband nodes, are required because all parents are needed in the modeled probability of the child. This set of conditioning nodes in Equation (A.82), which makes a variable conditionally independent of the rest, is sometimes called the Markov blanket. A related concept is d-separation: if two nodes i and j are separated by a node (or a collection of nodes) s, then the conditional probability $p(x_i \mid x_j, x_s) = p(x_i \mid x_s)$. This kind of d-separation can be useful for conditional independence statements not directly enforced by the model.

The marginal probability at a node $i \in \{1,...,n\}$ is obtained by marginalization over the parents, as follows:

$$p(x_i) = \sum_{x_{pa(i)}} p(x_i \mid x_j; j \in pa(i))\, p(\boldsymbol{x}_{pa(i)}). \tag{A.83}$$

Note that this requires the joint pdf of the parent nodes, which is not always straightforward to obtain.

Assume that we can get perfect or imperfect data denoted by $\boldsymbol{y}$. The conditional probability of a node given information $\boldsymbol{y}$ is given by Bayes' formula:

$$p(x_i \mid \boldsymbol{y}) = \frac{p(x_i, \boldsymbol{y})}{p(\boldsymbol{y})} = \frac{p(x_i, \boldsymbol{y})}{\sum_{x_i} p(x_i, \boldsymbol{y})}, \tag{A.84}$$

which in this general form requires the joint distribution over the information (denominator) and the joint distribution of the variable of interest and the data (numerator).

Unlike Equations (A.81) and (A.82), the computations in Equations (A.83) and (A.84) are not directly defined from the model. We need to marginalize over variables for these computations. There are many ways of computing these marginal (and conditional) probabilities, and the method of choice depends on the situation.

Parameter estimation

Suppose that the edge structure of the BN is known; the specification of unknown parameters $\boldsymbol{\theta}$ is typically done by maximum likelihood estimation (MLE) or fully Bayesian methods. If we have total perfect information in the form of data replicates $\boldsymbol{x}^1,...,\boldsymbol{x}^B$, the conditional probability tables can be computed by counting. All conditional probabilities are then estimated directly by

$$\hat{p}(x_i = l \mid x_j = k_j; j \in pa(i)) = \frac{\sum_{b=1}^{B} I(x_i^b = l, x_j^b = k_j; j \in pa(i))}{\sum_{b=1}^{B} I(x_j^b = k_j; j \in pa(i))}. \tag{A.85}$$

If there are few replicates, it is possible to borrow structure within the network structure. It is also possible to assign Dirichlet prior pdfs to the network probabilities $\boldsymbol{\theta}$ in a fully Bayesian procedure. If the data $\boldsymbol{y}$ are partial or imperfect, we can estimate parameters $\boldsymbol{\theta}$ using the marginal likelihood $p(\boldsymbol{y};\boldsymbol{\theta})$, again possibly with a Dirichlet prior model $p(\boldsymbol{\theta})$.

In many applications, the graph structure is assigned from expert opinion. If the graph structure is unknown, we must specify the edge structure as well as the parameters $\boldsymbol{\theta}$. This is more difficult than the parameter estimation (or learning) problem for a fixed structure – see, e.g., Chickering (1996).

Junction tree algorithm

There are several algorithms for efficient computation of BNs (see the discussions in Chapters 2.3 and 3.3).

A very successful approach for efficient marginalization in BNs is the junction tree algorithm (JTA). We present the idea only briefly since the full explanation of the JTA requires a great deal of set notation. The original presentation of the algorithm was provided by Lauritzen and Spiegelhalter (1988); see also Cowell et al. (2007) for a complete description. Rather than knowing all the computational aspects of this famous algorithm, one can of course just run it using numerously available software. In our analysis in this book, we used the Bayesian Network Toolbox (BNT) software implemented by K. Murphy.

A core idea of the JTA algorithm is to reform the BN to another network of clusters of nodes, sometimes called supernodes. The resulting graph consists of clusters that are formed in a special way that simplifies marginalization. This graph is built from the structure of the BN, but unlike the shape of a BN model, it is singly connected – i.e., it does not have branches that meet. The graph is also undirected. This is achieved by creating a triangulated graph – i.e., by adding edges between parent nodes that are not already connected – i.e., the so-called "moralizing" operation. A triangulated graph has no cycles of size larger than 3.

The junction graph has cluster nodes that define the cliques of the model, and the junction tree is built by utilizing the role of separators in the triangulated structure. There is no unique junction tree for a given BN model, and the optimal junction tree depends on the situation. There are a number of heuristic methods for constructing junction trees.

Once the junction tree is established, it is relatively straightforward to compute marginal and conditional probabilities using recursive computations (belief or evidence propagation) along the tree. These use operations on the conditional probability tables for the clique configurations. The procedure is often called query or message passing in the BN literature.

To compute several probabilities at the same time, the JTA may utilize the clustering nodes and then use a local message passing algorithm (Pearl 1988) on the established tree. This avoids repeated variable elimination and redundant computation. Message passing on the junction tree can be thought of as a generalization of the forward–backward algorithm for a hidden Markov model to BNs (Murphy 2012).

A.5.2 Decision analysis and VOI results

The decision situation we have used for BNs is similar to that for the Markov chain and Markov random fields in Appendices A.3–4. The prior value is

$$PV = \sum_{j=1}^{N} \max\left\{0, \text{Rev} \cdot p\left(x_j = 1\right) - \text{Cost}\right\}, \tag{A.86}$$

when we have a free selection of sites, $N = n$, and two alternatives at each site. The marginal pdf in the prior must now be computed by efficient marginalization.

For the posterior value,

$$PoV(\mathbf{y}) = \sum_{y}\sum_{j=1}^{N} \max\left\{0, \text{Rev}\cdot p\left(x_j = 1 \mid \mathbf{y}\right) - \text{Cost}\right\} p(\mathbf{y}). \tag{A.87}$$

The JTA is useful for computing both $p(\mathbf{y})$ and $p(x_j = 1 \mid \mathbf{y})$ for various kinds of information $\mathbf{y}$. If the information set is small, the outer sum can be solved exactly. Otherwise, it can be approximated by Monte Carlo sampling, as presented in Appendices A.3–4. Re-using samples or computations may be possible.

Appendix B: Sampling methods

The starting point of random number generation on the computer is the generation of **independent uniform variables**. These algorithms generate numbers that mimic (uniform) randomness and independence. Consider a simple algorithm that specifies a seed (start value, 10 here) and continues for $b = 0, 1, \ldots$ as follows:

$$10, u^0 = 10/c, u^b = \text{mod}\left(a \cdot c \cdot u^{b-1}, c\right)/c, \quad a = 5, \; c = 21. \tag{A.88}$$

This gives the following sequence of uniform numbers u^b: 0.47, 0.38, 0.91, 0.52, 0.62, 0.09, which may appear to be a random and independent sequence at first sight. We can perform tests to check whether a sequence gives independent uniform numbers. Obviously, some more inspection shows that the choice in Equation (A.88) is not very intelligent. In fact, the next number in the sequence is $\text{mod}(0.09 \cdot 21 \cdot 5, 21)/21 = 0.47$ – i.e., exactly the same as the first number, and the process will just repeat itself. The very basic-level research on Monte Carlo sampling is in constructing deterministic sequences of "uniformly distributed" numbers. For a long time, MATLAB used a congruential simulator algorithm similar to the one mentioned earlier with $a = 7^5 = 16{,}807$ and $c = 2^{31} - 1 = 2{,}147{,}483{,}647$. The resulting sequence repeats itself after c periods and covers the uniform domain very well.

Given the uniform variables, two direct methods for simulation are readily available. They are called **inversion** and **transformation**. We illustrate them by examples. Inversion works as follows: (i) draw a uniform random number u between 0 and 1, denoted by $U(0,1)$, and (ii) set $x = F^{-1}(u)$, where F^{-1} is the inverse cumulative distribution function of x. Note that F is a monotone increasing function, and the inverse then exists. We can prove correctness by

$$p\left(F^{-1}(u) < x\right) = p\left(u < F(x)\right) = F(x). \tag{A.89}$$

As an example, suppose we want to sample a binomial variable with $n_{trial} = 5$ trials and success probability $p = 0.4$. The pdf is $p(x) = \frac{n_{trial}!}{(n_{trial} - x)!\,x!} p^x (1-p)^{n_{trial}-x}$, $x = 0, \ldots, n_{trial}$, and the associated cumulative distribution function $F(x) = \sum_{y=0}^{x} p(y)$ is displayed in Table B.1.

Suppose that we generate a uniform random number $u = 0.05$, then $F(0) > u$, and the inverse function sets $x = 0$. Naturally, the chance of ending up in $x = 0$ is the same as that

Table B.1. *Probability distribution for the binomial with success probability p = 0.4 and five trials*

$x=0$	$x=1$	$x=2$	$x=3$	$x=4$	$x=5$
$p(0)=0.08$	$p(1)=0.26$	$p(2)=0.34$	$p(3)=0.23$	$p(4)=0.08$	$p(5)=0.01$
$F(0)=0.08$	$F(1)=0.34$	$F(2)=0.68$	$F(3)=0.91$	$F(4)=0.99$	$F(5)=1$

of a uniform random number being smaller than 0.08, which is the same as $F(0)$. Similarly, we get $x = 1$ if the generated uniform random number is between 0.08 and 0.34, and so on. Outcome $x = 5$ only happens if $u > 0.99$.

Inversion is efficient in many situations, but the cumulative (or its inverse) function is not always available in a closed form. For instance, the Gaussian cumulative distribution is defined by an integral expression, and one has to store the cumulative distribution for a dense set of values in the interval (0, 1) to use inversion. Using such a look-up table is not always achievable.

The common way of generating a Gaussian variable, and many other distributions, is to use transformations. We illustrate an algorithm for simulating independent bivariate Gaussians from two uniform variables. Suppose that we have generated two independent $U(0,1)$ variables u_1 and u_2. Then the outputs x_1 and x_2 from the following algorithm

$$\theta = 2\pi u_1,\ r = -2\log u_2,\ x_1 = r\cos(\theta),\ x_2 = r\sin(\theta), \tag{A.90}$$

are two independent standard Gaussian variables. Here, the idea is to use polar coordinates – i.e., simulate a random angle θ and a random positive distance r, and then take the sine and cosine part of the (distance, angle) vector. By the transformation formula, we have $p(x_1, x_2) = |J|\, p(u_1) p(u_2) = |J|$. The inverse transformation is given by $u_1 = \frac{\arctan(x_2/x_1)}{2\pi}$, $u_2 = \exp\left(-\frac{x_1^2+x_2^2}{2}\right)$, and the Jacobian of this inverse transformation is

$$J = \begin{bmatrix} \frac{1}{2\pi}\frac{d\arctan\left(\frac{x_2}{x_1}\right)}{dx_1} & \frac{d\exp\left(-\frac{x_1^2+x_2^2}{2}\right)}{dx_1} \\ \frac{1}{2\pi}\frac{d\arctan\left(\frac{x_2}{x_1}\right)}{dx_2} & \frac{d\exp\left(-\frac{x_1^2+x_2^2}{2}\right)}{dx_2} \end{bmatrix} = \begin{bmatrix} \frac{-\left(x_2/x_1^2\right)}{2\pi}\frac{x_1^2}{x_1^2+x_2^2} & -x_1\exp\left(-\frac{x_1^2+x_2^2}{2}\right) \\ \frac{\left(1/x_1\right)}{2\pi}\frac{x_1^2}{x_1^2+x_2^2} & -x_2\exp\left(-\frac{x_1^2+x_2^2}{2}\right) \end{bmatrix}$$

$$= \frac{1}{2\pi}\frac{x_1^2}{x_1^2+x_2^2}\exp\left(-\frac{x_1^2+x_2^2}{2}\right)\left(\frac{x_2^2}{x_1^2}+1\right) = \frac{1}{2\pi}\exp\left(-\frac{x_1^2+x_2^2}{2}\right),$$

where we use the kernel rule and $\frac{d\arctan x}{dx} = \frac{1}{1+x^2}$. This Jabobian is recognized as the pdf of the bivariate Gaussian distribution.

Correlated Gaussian can be generated in various ways, such as the Cholesky factorization described in Appendix A.1. Most computational softwares such as MATLAB and R have algorithms for generating random numbers from many different distributions.

Rejection sampling

In many situations, we can write $p(\boldsymbol{x}) = \frac{p_0(\boldsymbol{x})}{Z}$, where Z is a normalizing constant, not depending on $\boldsymbol{x}$, while $p_0(\boldsymbol{x})$ is a function that can be evaluated. As an example, this is typically the case in posterior models in the Bayesian context, where $p(\boldsymbol{x})$ is the prior model and the likelihood of data $\boldsymbol{y}$ is defined by $p(\boldsymbol{y} \mid \boldsymbol{x})$. The posterior distribution is

$$p(\boldsymbol{x} \mid \boldsymbol{y}) \propto p(\boldsymbol{y} \mid \boldsymbol{x}) p(\boldsymbol{x}). \tag{A.91}$$

Here, the proportionality sign is used instead of the normalizing constant, which is (the often intractable) $p(\boldsymbol{y}) = \sum_{x} p(\boldsymbol{x}, \boldsymbol{y})$. For several modeling assumptions, we can evaluate $p_0(\boldsymbol{x} \mid \boldsymbol{y}) = p(\boldsymbol{y} \mid \boldsymbol{x}) p(\boldsymbol{x})$, but we cannot sample directly from $p(\boldsymbol{x} \mid \boldsymbol{y})$. We present a reject/accept method for generating realizations from such $p(\boldsymbol{x}) = \frac{p_0(\boldsymbol{x})}{Z}$. This method relies on an alternative pdf $q(\boldsymbol{x})$ that is easy to draw from.

Algorithm: rejection sampling

1. Draw a sample $\boldsymbol{x}$ from $q(\boldsymbol{x})$.
2. Evaluate $r \geq \max_x \left(\frac{p_0(\boldsymbol{x})}{q(\boldsymbol{x})} \right)$.
3. Accept sample $\boldsymbol{x}$ if $\frac{p_0(\boldsymbol{x})}{rq(\boldsymbol{x})} > U(0,1)$.

The factor r defines an envelope $rq(\boldsymbol{x})$ that is always larger or equal to $p_0(\boldsymbol{x})$. We see that the normalizing constant is not needed since it cancels from the acceptance rate in Step 3.

The mean number of accepted samples becomes $\frac{1}{r}$. Thus, if the maximum ratio of the two densities is close to 1, we accept most proposals. If the maximum ratio is high, the acceptance rate goes down. Typically, we must make the tails of the proposal distribution $q(\boldsymbol{x})$ at least as heavy as that of $p(\boldsymbol{x})$; otherwise, $r \to \infty$ in the tails, and we would accept no samples. If we can find a natural proposal distribution that makes r computable, this method performs very well. However, in high dimensions, it seems very hard to find attractive proposal distributions that give small r, and the acceptance rate tends to decline with increasing variable dimensions.

Approximate Bayesian computing

Approximate Bayesian computing (ABC) is similar to rejection sampling in its accept/reject step. It has been limited to posterior sampling in Bayesian contexts. Notably, for this approach, the expected relation or forward model $E(\boldsymbol{y} \mid \boldsymbol{x}) = \boldsymbol{f}(\boldsymbol{x})$ between the distinction of interest and the data is known, but no further assumptions are made about the likelihood model. For this reason, ABC is often referred to as likelihood-free sampling. Instead of a formal likelihood description, a similarity measure is used in the conditioning to the data.

Algorithm: approximate Bayesian computing

Repeat for samples $b=1,...,B$.

1. Sample $\boldsymbol{x}^b$ from $p(\boldsymbol{x})$.
2. Evaluate the forward model $\boldsymbol{f}(\boldsymbol{x}^b)$
3. Accept within a tolerance of data – i.e., if the sample has dissimilarity: $d(\boldsymbol{y}, \boldsymbol{f}(\boldsymbol{x}^b)) < \varepsilon$

In Step 3 of the algorithm, a summary statistic and distance measure must be selected, as well as a threshold for acceptance ε. The approach has become popular in several application domains, and it has been intensely studied in the statistics literature (Marin et al. 2012).

Markov chain Monte Carlo algorithms

The basis of Markov chain Monte Carlo (MCMC) sampling is to simulate a Markov chain over the state space of the variable $\boldsymbol{x}$. The Markov chain converges to its limiting distribution under mild regularity conditions. And by construction of the MCMC algorithm, this limiting distribution is equal to the distribution $p(\boldsymbol{x})$, which we aim to sample from.

The Markov chain must be irreducible so that it is possible to reach every state from any other in a number of Markov chain iterations. If this is not the case, one risks sampling only a subset of the sample space. Naturally, Markov chains with absorbing states are not allowed. The sampler is initiated in $\boldsymbol{x}^0$ and then each transition $b = 0,1,...$ moves according to a transition distribution or kernel $P(\boldsymbol{x}^b, \boldsymbol{x})$. After a transient phase, often called the burn-in, samples are assumed to be from the desired pdf. Note that if the initial sample is from the correct pdf, the Markov chain will stay within that (stationary) distribution, and any random sample is an exact sample from the correct distribution.

Subsequent MCMC samples will be dependent, but the sample average will still converge to the right value under mild regularity assumptions. The key to successful implementation is to construct a Markov chain that mixes well – i.e., produces samples that are not too dependent. Such a Markov chain will traverse the probability space in fewer iterations than a poorly mixing Markov chain, and one can run the chain for a shorter time before extracting relevant statistics.

The Gibbs sampler is an MCMC method using the full conditional pdfs in the Markov chain transitions. Considering a single-site updating scheme, the Gibbs sampler changes only one variable per iteration:

Algorithm: Gibbs sampling (single site)

Initiate $\boldsymbol{x}^0 = \left(x_1^0,\ldots,x_n^0\right)$. Iterate as follows for $b = 1,\ldots,B$:

1. Pick a component $i \in \{1,\ldots,n\}$. Sample x_i^b from the full conditional distribution:

$$p\left(x_i \mid x_1^{b-1},\ldots,x_{i-1}^{b-1},x_{i+1}^{b-1},\ldots,x_n^{b-1}\right) \propto p\left(\boldsymbol{x}_{-i}^{b-1},x_i\right) \tag{A.92}$$

2. Set $x_j^b = x_j^{b-1}, \quad j \neq i$.

The approach relies on the ability to sample from full conditionals. This is straightforward for many hierarchical models, where the full conditionals only depend on some neighboring variables. Gibbs sampling is then an efficient way to divide and conquer. For the MRF model, this Gibbs sampler would update the color of one cell at a time from the conditional distribution at the selected cell, given the current state of the neighboring cells.

The Gibbs sampler can also update block variables at every time step. The full conditional distributions for blocks are often more complicated to derive, but the sampler gets better mixing properties in the Markov chain, and the block sampler does not have to run as long as a single-site sampler.

Metropolis–Hastings (MH) sampling is based on accepting or rejecting a proposed sample at every iteration:

Algorithm: Metropolis–Hastings sampling

Initiate $\boldsymbol{x}^0 = \left(x_1^0,\ldots,x_n^0\right)$. Iterate as follows for $b = 1,\ldots,B$:

1. Draw a proposal from $q\left(\boldsymbol{x}^* \mid \boldsymbol{x}^{b-1}\right)$.
2. Set $\boldsymbol{x}^b = \boldsymbol{x}^*$ if the acceptance probability

$$\min\left(1,\frac{p\left(\boldsymbol{x}^*\right)q\left(\boldsymbol{x}^{b-1} \mid \boldsymbol{x}^*\right)}{p\left(\boldsymbol{x}^{b-1}\right)q\left(\boldsymbol{x}^* \mid \boldsymbol{x}^{b-1}\right)}\right) > U(0,1), \text{ otherwise set } \boldsymbol{x}^b = \boldsymbol{x}^{b-1}. \tag{A.93}$$

There are a multitude of MH algorithm variants. They all evaluate the acceptance rate in Step 2, which includes the proposal distribution and the target pdf $p\left(\boldsymbol{x}^*\right)$, but they differ in the proposal distribution $q\left(\boldsymbol{x}^* \mid \boldsymbol{x}^{b-1}\right)$. The flexibility in choosing the proposal gives opportunities for creating efficient MH samplers, and there are only few regularity conditions on the proposal distribution $q\left(\boldsymbol{x}^* \mid \boldsymbol{x}^{b-1}\right)$. On the other hand, selecting a useful proposal distribution could be a difficult task since some choices induce much poorer mixing properties than others. We discuss some proposal mechanisms that have been common in applications.

For the MRF model, a single-site MH sampler attempts to randomly change the color at site i and accepts the change with probability $\min\left(1, \frac{p(\boldsymbol{x}^*)}{p(\boldsymbol{x}^{b-1})}\right)$. Then, $x_j^* = x_j^{b-1}$ for all sites $j \neq i$, and the acceptance ratio simplifies to $\frac{p(\boldsymbol{x}^*)}{p(\boldsymbol{x}^{b-1})} = \exp\left(-\sum_{k \in C \cap \{i\}} \left(\psi_k\left(\boldsymbol{x}_k^*\right) + \psi_k\left(\boldsymbol{x}_k^{b-1}\right)\right)\right)$, which only contains cliques where cell i is involved.

The random walk MH sampling defines proposal $q\left(\boldsymbol{x}^* \mid \boldsymbol{x}^{b-1}\right) = N\left(\boldsymbol{x}^{b-1}, h^2\boldsymbol{I}\right)$ – i.e., symmetric around the current state $\boldsymbol{x}^{b-1}$. This approach is easy to implement and has shown to be useful for several continuous models. For this random walk MH algorithm, the mixing properties are (asymptotically) optimal with an average acceptance rate of about 0.25. More sophisticated MH proposals can be constructed by using derivative information at the current state. The Langevin–Metropolis algorithm defines proposal $q\left(\boldsymbol{x}^* \mid \boldsymbol{x}^{b-1}\right) = N\left(\boldsymbol{x}^{b-1} + h\nabla \log p\left(\boldsymbol{x}^{b-1}\right), h^2\boldsymbol{I}\right)$. Since the proposal uses more of the specific model in the proposal, it tends to traverse the distribution function more quickly. The samples are less dependent, and the mixing properties are usually very good. The mixing properties are (asymptotically) optimal under an average acceptance rate of about 0.5. Note that the faster mixing comes at the cost of evaluating the gradient.

Slice sampling is one of many MCMC methods that generate variables by augmentation. The idea is to sample an auxiliary variable u in addition to $\boldsymbol{x}$ and in this way improve the mixing properties of the Markov chain. The slice sampler generates the auxiliary variable by cutting or slicing the pdf horizontally and then moving to an updated variable with a pdf above this slice. The rationale behind cutting is not dissimilar to that of inversion.

Algorithm: slice sampling

Initiate $\boldsymbol{x}^0 = \left(x_1^0, \ldots, x_n^0\right)$. Iterate as follows for $b = 1, \ldots, B$:

1. Draw an augmented variable u from $U\left(0, p\left(\boldsymbol{x}^{b-1}\right)\right)$.
2. Sample $\boldsymbol{x}^b$ from uniform pdf $U\left(C\left(\boldsymbol{x} \mid u\right)\right)$, with constrained region $C\left(\boldsymbol{x} \mid u\right) = \left\{p\left(\boldsymbol{x}\right) > u\right\}$.

The constraint in the second step may be hard to calculate in practice – i.e., the region of valid proposals is not easily identified. Suggested solutions construct a valid region using gradual stepping-out schemes.

References

Adams, R. M., Bryant, K. J., McCarl, B. A., Legler, D. M., O'Brien, J., Solow, A., and Weiher, R., (1995). Value of improved long-range weather information. *Contemporary Economic Policy*, 13, 10–19.

Alemu, E. T., Palmer, R. N., Polebitski, A., and Meaker, B. (2011). Decision support system for optimizing reservoir operations using ensemble streamflow predictions. *Journal of Water Resources Planning and Management*, 137, 72–82.

Alkhatib, A., and King, P. (2014). An approximate dynamic programming approach to decision making in the presence of uncertainty for surfactant-polymer flooding. *Computational Geosciences*, 18, 243–263.

Allard, D., and Naveau, P. (2007). Simulating and analyzing spatial skew normal random fields. *Communications in Statistics*, 36, 1821–1834.

Anderson, T. W. (2003). *An Introduction to Multivariate Statistical Analysis*. New York: Wiley.

Armstrong, M., Galli, A. G., Beucher, H., Le Loc'h, G., Renard, D., Dogliez, B., Eschard, R., and Geffroy, F. (2011). *Plurigaussian Simulations in Geosciences*, 2nd edn. Berlin: Springer-Verlag.

Arpat, B., and Caers, J. (2007). Conditional simulations with patterns. *Mathematical Geology*, 39, 177–203.

Ash, R. B. (1965). *Information Theory*. New York: Dover Publications.

Auken, E., Chistiansen, A., Jacobsen, L., and Sørensen, K. (2008). A resolution study of buried valleys using laterally constrained inversion of TEM data. *Journal of Applied Geophysics*, 65, 10–20.

Avseth, P., Mukerji, T., Jørstad, A., Mavko, G., and Veggeland, T. (2001). Seismic reservoir mapping from 3-D AVO in a North Sea turbidite system. *Geophysics*, 66, 1157–1176.

Avseth, P., Mukerji, T., and Mavko, G. (2005). *Quantitative Seismic Interpretation*. Cambridge University Press.

Bachrach, R. (2006). Joint estimation of porosity and saturation using stochastic rock-physics modelling. *Geophysics*, 71, O53–O63.

Ballari, D., de Bruin, S., and Bregt, A. K. (2012). Value of information and mobility constraints for sampling with mobile sensors. *Computers & Geosciences*, 49, 102–111.

Banerjee, S., Gelfand, A. E., Finley, A. O., and Sang, H. (2008). Gaussian predictive process models for large spatial data sets. *Journal of the Royal Statistical Society, Series B*, 70, 825–848.

Banerjee, S., Gelfand, A., and Carlin, B. (2004). *Hierachical Modeling and Analysis for Spatial Data*. Boca Raton, FL: Chapman & Hall/CRC Press.

Bardossy, A., and Li, J. (2008). Geostatistical interpolation using copulas. *Water Resources Research*, 44, W07412.

Barros, E. G. D., Jansen, J. D., and van den Hof, P. M. J. (2014). Value of information in closed loop reservoir management. Extended abstract, *14th European Conference on the Mathematics of Oil Recovery (ECMOR)*, Catania, Italy.

Bates, M. E., Sparrevik, M., de Lichy, N., and Linkov, I. (2014). The value of information for managing contaminated sediments. *Environmental Science and Technology*, 48, 9478–9485.

Beaumont, M., Zhang, W., and Balding, D. (2002). Approximate Bayesian computation in population genetics. *Genetics*, 162, 2025–2032.

Bergseng, E., Ørka, H. O., Næsset, E., and Gobakken, T. (2015). Assessing forest inventory information obtained from different inventory approaches and remote sensing data sources. *Annals of Forest Science*, 72, 33–45.

Bertsekas, D. P. (2012). *Dynamic Programming and Optimal Control*. Vol. II of Approximate Dynamic Programming, 4th edn. Athena Scientific.

Besag, J. (1974). Spatial interaction and the statistical analysis of lattice systems. *Journal of the Royal Statistical Society, Series B*, 36, 192–236.

(1986). On the statistical analysis of dirty pictures. *Journal of the Royal Statistical Society, Series B*, 48, 259–302.

Bhattacharjya, D., and Deleris, L. (2014). The value of information in some variations of the stopping problem. *Decision Analysis*, 11, 189–203.

Bhattacharjya, D., Eidsvik, J., and Mukerji, T. (2010). The value of information in spatial decision making. *Mathematical Geosciences*, 42, 141–163.

(2013). The value of information in portfolio problems with dependent projects. *Decision Analysis*, 10, 341–351.

Bhattacharjya, D., and Mukerji, T. (2006). Using influence diagrams to analyze decisions in 4D seismic reservoir monitoring. *The Leading Edge*, 25, 1236–1239.

Bhattacharjya, D., and Shachter, R. (2007). Evaluating influence diagrams with decision circuits. In *Proceedings of the 23rd Conference on Uncertainty in Artificial Intelligence (UAI)*, AUAI Press, pp. 9–16.

(2008). Sensitivity analysis in decision circuits. In *Proceedings of the 24th Conference on Uncertainty in Artificial Intelligence (UAI)*, AUAI Press, pp. 34–42.

(2010). Three new sensitivity analysis methods for influence diagrams. In *Proceedings of the 24th Conference on Uncertainty in Artificial Intelligence (UAI)*, AUAI Press, pp. 56–64.

Bickel, J. E. (2008). The relationship between perfect and imperfect information in a two-action risk-sensitive problem. *Decision Analysis*, 5, 116–128.

Bickel, J. E., Gibson, R. L., McVay, D. A., Pickering, S., and Waggoner, J. (2008a). Quantifying 3D land seismic reliability and value. *Society of Petroleum Engineers: Reservoir Evaluation and Engineering*, 11, 832–841.

Bickel, J. E., and Smith, J. E. (2006). Optimal sequential exploration: a binary learning model. *Decision Analysis*, 3, 16–32.

Bickel, J. E., Smith, J. E., and Meyer, J. L. (2008b). Modeling dependence among geological risks in sequential exploration decisions. *Society of Petroleum Engineers: Reservoir Evaluation & Engineering*, 11, 233–251.

Bielza, C., Gomez, M., and Shenoy, P. (2010). Modeling challenges with influence diagrams: constructing probability and utility models. *Decision Support Systems*, 49, 354–364.

Bielza, C., and Shenoy, P. (1999). A comparison of graphical techniques for asymmetric decision problems. *Management Science*, 45, 1552–1569.

Blackwell, D. (1953). Equivalent comparisons of experiments. *Annals of Mathematical Statistics*, 24, 265–272.
Blangy, J. P., Schiott, C., Vejbaek, O., and Maguire, D. (2014). The value of 4D seismic: has the promise been fulfilled? *Society of Exploration Geophysics Annual Meeting*, 2552–2557.
Borg, I., and Groenen, P. J. F. (2005). *Modern Multidimensional Scaling: Theory and Applications*, 2nd edn. New York: Springer-Verlag.
Borgault, G. (1997). Using non-Gaussian distributions in geostatistical simulation. *Mathematical Geology*, 29, 315–334.
Borgos, H. G., Omre, H., and Townsend, C. (2002). Size distribution of geological faults: model choice and parameter estimation. *Statistical Modeling*, 2, 217–234.
Borisova, T., Shortle, J., Horan, R. D., and Abler, D. (2005). Value of information for water quality management. *Water Resources Research*, 41, W06004.
Borsuk, M., Clemen, R., Maguire, L., and Reckhow, K. (2001). Stakeholder values and scientific modeling in the Neuse River watershed. *Group Decision and Negotiation*, 10, 355–373.
Bouma, J. A., van der Woerd, H. J., and Kuik, O. J. (2009). Assessing the value of information for water quality management in the North Sea. *Journal of Environmental Management*, 90, 1280–1288.
Braithwaite, R. S., and Scotch, M. (2013). Using value of information to guide evaluation of decision support for differential diagnosis: is it time for a new look? *BMS Medical Informatics and Decision Making*, 13, 105.
Bratvold, R. B., and Begg, S. H. (2010). *Making Good Decisions*. Richardson, TX: Society of Petroleum Engineers.
Bratvold, R. B., Bickel, J. E., Lohne, H. P. (2009). Value of information in the oil and gas industry: Past, present, and future. *Society of Petroleum Engineers: Reservoir Evaluation & Engineering*, 12, 630–638.
Breslow, N. E., and Clayton, D. G. (1993). Approximate inference in generalized linear mixed models. *Journal of the American Statistical Association*, 88, 9–25.
Brockwell, P. J., and Davis, R. A. (2009). *Time Series: Theory and Methods*. New York: Springer-Verlag.
Brown, D. B and Smith, J. E. (2013). Optimal sequential exploration: bandits, clairvoyants, and wildcats. *Operations Research*, 60, 262–274.
Bruland, O., Færevåg, Å., Steinsland, I., Listen, G. E., and Sand, K. (2015). Weather SDM: estimating snow density with high precision using snow depth and local climate. *Hydrology Research*.
Byerlee, D., and Anderson, J. R. (1982). Risk, utility and the value of information in farmer decision making. *Review of Marketing and Agricultural Economics*, 50, 231–246.
Cabrera, V. E., Letson, D., and Podesta, G. (2007). The value of climate information when farm programs matter. *Agricultural Systems*, 93, 25–42.
Caers, J. (2005). *Petroleum Geostatistics*. Richardson, TX: Society of Petroleum Engineers.
(2011). *Modeling Uncertainty in the Earth Sciences*. Chichester, UK: Wiley-Blackwell.
Carlin, B. P., and Louis, T. A. (2000). *Bayes and Empirical Bayes Methods for Data Analysis*. Boca Raton, FL: Chapman & Hall/CRC Press.
Chan, G., and Wood, A. T. A. (1997). An algorithm for simulating stationary Gaussian random fields. *Applied Statistics*, 46, 171–181.
Chickering, M. (1996). Learning Bayesian networks is NP-complete. In D. Fisher and H. Lenz, eds., *Learning from Data: Artificial Intelligence and Statistics V*, New York: Springer-Verlag, pp. 121–130.

Chiles, J. P., and Delfiner, P. (2012). *Geostatistics: Modeling Spatial Uncertainty*, 2nd edn. New York: Wiley.

Christie, M. A., and Blunt, M. J. (2001). Tenth SPE comparative solution project: a comparison of upscaling techniques. *Society of Petroleum Engineers: Reservoir Engineering & Evaluation*, 4, 308–317.

Clemen, R., and Reilly, T. (1999). *Making Hard Decisions with Decision Tools Suite*. Pacific Grove, CA: Duxbury Press.

Clemen, R., and Winkler, R. (1985). Limits for the precision and value of information from dependent sources. *Operations Research*, 33, 427–442.

Covaliu, Z., and Oliver, R. M. (1995). Representation and solution of decision problems using sequential decision diagrams. *Management Science*, 41, 1860–1881.

Cover, T., and Thomas, J. (2006). *Elements of Information Theory*, 2nd edn. New York: Wiley-Interscience.

Cowell, R. G., Dawid, A. P., Lauritzen, S. L., and Spiegelhalter, D. J. (2007). *Probabilistic Networks and Expert Systems*. New York: Springer-Verlag.

Cox, T. F., and Cox, M. A. A. (2001). *Multidimensional Scaling*, 2nd edn. Boca Raton, FL: Chapman and Hall/CRC.

Cressie, N. (1993). *Statistics for Spatial Data*. New York: Wiley.

Cressie, N., and Johannesson, G. (2008). Fixed rank Kriging for very large spatial data sets. *Journal of the Royal Statistical Society, Series B*, 70, 209–226.

Cressie, N., and Wikle, C. K. (2011). *Statistics for Spatio-Temporal Data*. New York: Wiley.

Curtis, A. (2004). Theory of model-based geophysical survey and experimental design: Part A – Linear problems. *The Leading Edge*, 23, 997–1004.

Daly, C., and Caers, J. (2010). Multi-point geostatistics – An introductory overview. *First Break*, 28, 39–47.

Darwiche, A. (2009). *Modeling and Reasoning with Bayesian Networks*. Cambridge University Press.

de Bruin, S., Bregt, A., and van de Ven, M. (2001). Assessing fitness for use: the expected value of spatial data sets. *International Journal of Geographical Information Science*, 15, 457–471.

Demidenko, E. (2004). *Mixed Models: Theory and Applications*. New York: Wiley.

Deutsch, C., and Journel, A. G. (1992). *GSLIB: Geostatistical Software Library and User's Guide*. New York: Oxford University Press.

Diggle, P., and Lophaven, S. (2006). Bayesian geostatistical design. *Scandinavian Journal of Statistics*, 33, 53–64.

Diggle, P. J., Tawn, J. A., and Moyeed, R. A. (1998). Model-based geostatistics. *Journal of the Royal Statistical Society, Series C*, 47, 299–350.

Dobbie, M. J., Henderson, B. L., and Stevens, D. L. (2008). Sparse sampling: spatial design for monitoring stream networks. *Statistics Surveys*, 2, 113–153.

Dobson, A. J., and Barnett, A. (2008). *An Introduction to Generalized Linear Models*. Boca Raton, FL: Chapman & Hall/CRC Press.

Dominguez-Molina, J., Gonzalez-Farias, G., and Gupta, A. (2003). *The Multivariate Closed Skew Normal Distribution*, Technical report 03-12, Department of Mathematics and Statistics, Bowling Green University.

Doucet, A., de Freitas, N., and Gordon, N. (2001). *Sequential Monte Carlo Methods in Practice*. New York: Springer-Verlag.

Dvorkin, J., and Nur, A. (1996). Elasticity of high-porosity sandstones: theory for two North Sea datasets. *Geophysics*, 61, 1363–1370.

Dyer, J. S., and Sarin, R. (1982). Relative risk aversion. *Management Science*, 28, 875–886.

Edwards, W, Miles, R., and von Winterfeldt, D. (2007). Introduction. In W. Edwards, R. Miles, and D. von Winterfeldt, eds., *Advances in Decision Analysis: From Foundations to Applications*. Cambridge University Press, pp. 1–12.

Efron, B., and Tibshirani, R. J. (1993). *An Introduction to the Bootstrap*. Boca Raton, FL: Chapman & Hall/CRC Press.

Efros, A. A., and Freeman, W. T. (2001). Image quilting for texture synthesis and transfer. *In Proceedings of the ACM SIGGRAPH Conference on Computer Graphics*, pp. 341–346.

Eidsvik, J., Avseth, P., Omre, H., Mukerji, T., and Mavko, G. (2004a). Stochastic reservoir characterization using prestack seismic data. *Geophysics*, 69, 978–993.

Eidsvik, J., Bhattacharjya, D., and Mukerji, T. (2008). Value of information of seismic amplitude and CSEM resistivity. *Geophysics*, 73, R59–R69.

Eidsvik, J., and Ellefmo, S. L. (2013). The value of information in mineral exploration within a multi-Gaussian framework. *Mathematical Geosciences*, 45, 777–798.

Eidsvik, J., Mukerji, T., and Switzer, P. (2004b). Estimation of geological attributes from a well log: an application of hidden Markov chains. *Mathematical Geology*, 36, 379–397.

Ellefmo, S. L., and Eidsvik, J. (2009). Local and spatial joint frequency uncertainty and its application to rock mass characterisation. *Rock Mechanics and Rock Engineering*, 42, 667–688.

Evangelou, E., and Eidsvik, J. (2015). *The Value of Information for Correlated GLMs*, Technical report 2/2015, Department of Mathematical Sciences, Norwegian University of Science and Technology (NTNU).

Evangelou, E., and Zhu, Z. (2012). Optimal predictive design augmentation for spatial generalised linear mixed models. *Journal of Statistical Planning and Inference*, 142, 3242–3253.

Evensen, G. (2009). *Data Assimilation: The Ensemble Kalman Filter*. Berlin: Springer-Verlag.

Farquhar, P. (1984). State of the art – utility assessment methods. *Management Science*, 30, 1283–1300.

Ferreira, M. A. R., and Lee, H. K. H. (2007). *Multiscale Modeling: A Bayesian Perspective*. New York: Springer-Verlag.

Forsberg, O. I., and Guttormsen, A. G. (2006). The value of information in salmon farming: harvesting the right fish at the right time. *Aquaculture Economics and Management*, 10, 183–200.

Frazier, P. I., and Powell, W. B. (2010). Paradoxes in learning and the marginal value of information. *Decision Analysis*, 7, 378–403.

Friedman, J., Hastie, T., and Tibshirani, R. (2009). Sparse inverse covariance estimation with the graphical lasso. *Biostatistics*, 9, 432–441.

Froidevaux, R. (1992). Probability field simulation. In A. Soares, ed., *Geostatistics Tróia, Proceedings of the 4th International Geostatistical Congress*, Kluwer Academic Publishers.

Froyland, G., Menabde, M., Stone, P., and Hodson, D. (2004). The value of additional drilling to open pit mining projects. In *Proceedings of Orebody Modelling and Strategie Mine Planning – Uncertainty and Risk Management*, Perth, Australia, pp. 169–176.

Fuentes, M., Chaudhuri, A., and Holland, D. M. (2007). Bayesian entropy for spatial sampling design of environmental data. *Environmental and Ecological Statistics*, 14, 323–340.

Gaetan, C., and Guyon, X. (2010). *Spatial Statistics and Modelling*. New York: Springer-Verlag.

Gamerman, D., and Lopes, H. F. (2006). *Markov Chain Monte Carlo*. Boca Raton, FL: Chapman & Hall/CRC Press.

Gelfand, A. E., Kottas, A., and MacEachern, S. N. (2005). Bayesian nonparametric spatial modeling with Dirichlet mixing. *Journal of the American Statistical Association*, 100, 1021–1035.

Geman, S., and Geman D. (1984). Stochastic relaxation, Gibbs distributions, and the Bayesian restoration of images. *IEEE Transactions on Pattern Analysis and Machine Intelligence*, 6, 721–741.

Genz, A., and Bretz, F. (2009). *Computation of Multivariate Normal and T Probabilities: Lecture Notes in Statistics*. Berlin: Springer-Verlag.

Ginsbourger, D., Rosspopoff, B., Pirot, G., Durrande, N., and Renard, P. (2013). Distance-based Kriging relying on proxy simulations for inverse conditioning. *Advances in Water Resources*, 52, 275–291.

Gneiting, T., and Raftery, A. E. (2007). Strictly proper scoring rules, prediction, and estimation. *Journal of the American Statistical Association*, 102, 359–378.

Golub, G. H., and van Loan, C. F. (1996). *Matrix Computations*. Baltimore, MD: John Hopkins University Press.

Gomez, C. T., Dvorkin, J., and Mavko, G. (2008). Estimating the hydrocarbon volume from elastic and resistivity data: a concept. *The Leading Edge*, 27, 710–718.

Goovaerts, P. (1997). *Geostatistics for Natural Resources Evaluation*. New York: Oxford University Press.

Gramacy, R. B., and Apley, D. W. (2015). Local Gaussian process approximation for large computer experiments. *Journal of Computational and Graphical Statistics*.

Gray, R. M. (2006). Toeplitz and circulant matrices: a review. *Foundations and Trends in Communication and Information Theory*, 2, 155–239.

Grayson, C. J., Jr. (1960). *Decisions Under Uncertainty: Drilling Decisions by Oil and Gas Operators*. Cambridge, MA: Harvard University Press.

Green, P. J., Hjort, N. L., and Richardson, S. (eds.) (2003). *Highly Structured Stochastic Systems*. New York: Oxford University Press.

Guardiano, F., and Srivastava, R. (1993). Multivariate geostatistics: beyond bivariate moments. In A. Soares, ed., *Geostatistics Tróia, Proceedings of the 4th International Geostatistical Congress*, Kluwer Academic Publishers, pp. 133–144.

Hansen, G. J. A., and Jones, M. L. (2008). The value of information in fishery management. *Fisheries*, 33, 340–348.

Hantschel, T., and Kauerauf, A. I. (2009). *Fundamentals of Basin and Petroleum Systems Modelling*. Berlin: Springer-Verlag.

Hastie, T., Tibshirani, R., and Friedman, J. (2009). *The Elements of Statistical Learning*. New York: Springer-Verlag.

Heckerman, D., Horvitz, E., and Nathwani, B. (1989). Update on the pathfinder project. *In Proceedings of the 13th Symposium on Computer Applications in Medical Care*, IEEE Computer Society Press, pp. 203–207.

Heckerman, D., and Shachter, R. (1995). Decision-theoretic foundations for causal reasoning. *Journal of Artificial Intelligence Research*, 3, 405–430.

Hilton, R. (1981). The determinants of information value: synthesizing some general results. *Management Science*, 27, 57–64.

Horesh, L., Haber, E., and Tenorio, L. (2010). Optimal experimental design for the large-scale nonlinear ill-posed problem of impedance imaging. In L. Biegler G. Biros, O. Ghattas, M. Heinkenschloss, D. Keyes, B. Mallick, L. Tenorio, B. van Bloemen Waanders, K. Willcox, and Y. Marzouk, eds., *Large-Scale Inverse Problems and Quantification of Uncertainty*. Wiley, pp. 273–290.

Houck, R. T. (2007). Time-lapse seismic repeatability – how much is enough? *The Leading Edge*, 26, 828–834.

Houck, R. T., and Pavlov, D. A. (2006). Evaluating reconnaissance CSEM survey designs using detection theory. *The Leading Edge*, 25, 994–1004.

Howard, R. (1964). Decision analysis: applied decision theory. In *Proceedings of the 4th International Conference on Operational Research*, Wiley-Interscience, pp. 55–71.

(1966). Information value theory. *IEEE Transactions on Systems Science and Cybernetics*, 2, 22–26.

(1967). Value of information lotteries. *IEEE Transactions on Systems Science and Cybernetics*, 3, 54–60.

(1971). Proximal decision analysis. *Management Science*, 17, 507–541.

(2007). The foundations of decision analysis revisited. In W. Edwards, R. Miles, and D. von Winterfeldt, eds., *Advances in Decision Analysis: From Foundations to Applications*. Cambridge University Press, pp. 32–56.

Howard, R. A., and Abbas, A. (2015). *Foundations of Decision Analysis*. London: Pearson Education.

Howard, R., and Matheson, J. (1984). Influence diagrams. In R. Howard and J. Matheson, eds., *The Principles and Applications of Decision Analysis*, Vol. II. Strategic Decisions Group, pp. 721–762.

Illian, J., Penttinen, A., Stoyan, H., and Stoyan, D. (2008). *Statistical Analysis and Modelling of Spatial Point Patterns*. Chichester, UK: Wiley.

Isaaks, E. H., and Srivastava, R. M. (1989). *An Introduction to Applied Geostatistics*. Oxford: Oxford University Press.

Izenman, A. J. (2008). *Modern Multivariate Statistical Techniques*. New York: Springer-Verlag.

Jaakkola, T. S. (2000). Tutorial on variational approximation methods. In M. Opper and D. Saad, eds., *Advanced Mean Field Methods*. MIT Press, pp. 129–159.

Jensen, F., Jensen, F. V., and Dittmer, S. (1994). From influence diagrams to junction trees. In R. Lopez de Mantaras and D. Poole, eds., *Proceedings of the 10th Conference on Uncertainty in Artificial Intelligence (UAI)*, Morgan Kaufmann, pp. 367–373.

Jensen, F. V. (1996). *An Introduction to Bayesian Networks*. London: UCL Press.

Jensen, F. V., and Nielsen, T. D. (2007). *Bayesian Networks and Decision Graphs*, 2nd edn. New York: Springer-Verlag.

Jensen, J. L., Lake, L. W., Corbett, P. W. M., and Goggin, D. J. (2000). *Statistics for Petroleum Engineers and Geoscientists*, 2nd edn. Amsterdam: Elsevier.

Joe, H. (2014). *Dependence Modeling with Copulas*. Boca Raton, FL: Chapman & Hall/ CRC Press.

Johnson, N. L., Kotz, S., and Balakrishnan, N. (1994). *Continuous Univariate Distributions*. Hoboken, NJ: Wiley.

(1997). *Discrete Multivariate Distributions*. Hoboken, NJ: Wiley.

Johnson, R. A., and Wichern, D. W. (2012). *Applied Multivariate Statistical Analysis*. Delhi: Phi Learning Private Limited.

Journel, A. G. (1983). Nonparametric estimation of spatial distributions. *Mathematical Geology*, 15, 445–468.

Journel, A. G., and Huijbregts, C. J. (1978). *Mining Geostatistics*. Academic Press. Reprint by Blackburn Press, Caldwell, NJ, 2004.

Kangas, A. S. (2010). The value of forest information. *European Journal of Forest Research*, 129, 863–874.

Karam, K. S., Karam, J. S., and Einstein, H. H. (2007). Decision analysis applied to tunnel exploration planning I: principles and case study. *Journal of Construction Engineering and Management*, 133, 344–353.

Kaufman, G. (1993). Statistical issues in the assessment of undiscovered oil and gas resources. *The Energy Journal*, 14, 183–215.

Kazianka, H., and Pilz, J. (2011). Bayesian spatial modeling and interpolation using copulas. *Computers & Geosciences*, 37, 310–319.

Keeney, R. (2007). Developing objectives and attributes. In W. Edwards, R. Miles, and D. von Winterfeldt, eds., *Advances in Decision Analysis: From Foundations to Applications*. Cambridge University Press, pp. 104–128.

Keeney, R., and Raiffa, H. (1976). *Decisions with Multiple Objectives: Preferences and Value Tradeoffs*. Hoboken, NJ: Wiley.

Keisler, J. M. (2004). Value of information in portfolio decision analysis. *Decision Analysis*, 1, 177–189.

(2005). Additivity of information in two-act linear loss decisions with normal priors. *Risk Analysis*, 25, 351–359.

Keisler, J. M., and Brodfuehrer, M. (2009). An application of value-of-information to decision process reengineering. *Engineering Economist*, 54, 197–221.

Keisler, J. M., Collier, Z., Chu, E., Sinatra, N., and Linkov, I. (2014). Value of information analysis: the state of application. *Environment Systems and Decisions*, 34, 3–23.

Kelkar, M., and Perez, G. (2002). *Applied Geostatistics for Reservoir Characterization*. Richardson, TX: Society of Petroleum Engineers.

Khare, K., Oh, S. Y., and Rajaratnam, B. (2015). A convex pseudo-likelihood framework for high dimensional partial correlation estimation with convergence guarantees. *Journal of the Royal Statistical Society, Series B*.

Kim, H. M., and Mallick, B. K. (2004). A Bayesian prediction using the skew Gaussian distribution. *Journal of Statistical Planning and Inference*, 120, 85–101.

Kirkwood, C. (1993). An algebraic approach to formulating and solving large models for sequential decisions under uncertainty. *Management Science*, 39, 900–913.

(2004). Approximating risk aversion in decision analysis applications. *Decision Analysis*, 1, 51–67.

Kolbjørnsen, O., Hauge, R., Drange-Espeland, M., and Buland, A. (2012). Model-based fluid factor for controlled source electromagnetic data. *Geophysics*, 77, E21–E31.

Koller, D., and Friedman, N. (2009). *Probabilistic Graphical Models: Principles and Techniques*. Cambridge, MA: MIT Press.

Konishi, C. (2014). Evaluation of uncertainty and risk of CO_2 sequestration with stochastic models conditioned by seismic and well data. MSc thesis, Stanford University.

Kontoghiorghes, E. J. (ed.) (2005). *Handbook of Parallel Computing and Statistics*. Boca Raton, FL: Chapman & Hall/CRC Press.

Kotz, S., Johnson, N. L., and Balakrishnan, N. (2000). *Continuous Multivariate Distributions*. Hoboken, NJ: Wiley.

Krause, A., and Guestrin, C. (2007). Nonmyopic active learning of Gaussian processes: an exploration-exploitation approach. In *International Conference on Machine Learning*. Omnipress, pp. 449–456.

(2009). Optimal value of information in graphical models. *Journal of Artificial Intelligence Research*, 35, 557–591.

Lantuejoul, C. (2002). *Geostatistical Simulations: Models and Algorithms*. Berlin: Springer-Verlag.

Lauritzen, S. L., and Spiegelhalter, D. J. (1988). Local computations with probabilities on graphical structures and their applications to expert systems. *Journal of the Royal Statistical Society, Series B*, 50, 157–224.

Lawler, G. F. (2006). *Introduction to Stochastic Processes*. Boca Raton, FL: Chapman & Hall/CRC Press.

Le, N. D., and Zidek, J. V. (2006). *Statistical Analysis of Environmental Space-Time Processes*. New York: Springer-Verlag.

Lichtenstein, S., and Slovic, P. (2006). *The Construction of Preference*. Cambridge University Press.

Lie, K. A., Krogstad, S., Ligaarden, I. S., Natvig, J. R., Nilsen, H. M., and Skaflestad, B. (2012). Open-source MATLAB implementation of consistent discretisations on complex grids. *Computational Geosciences*, 16, 297–322.

Lilleborge, M., Hauge, R., and Eidsvik, J. (in press). Information gathering in Bayesian networks applied to petroleum prospecting. *Mathematical Geosciences*.

Lindberg, D. V., and Lee, H. K. H. (2015). Optimization under constraints by applying an asymmetric entropy measure. *Journal of Computational and Graphical Statistics*, 24, 379–393.

Lindgren, F., Rue H., and Lindstrøm J. (2011). An explicit link between Gaussian fields and Gaussian Markov random fields: the stochastic partial differential equation approach. *Journal of the Royal Statistical Society, Series B*, 73, 423–498.

Liu, J. L. (2001). *Monte Carlo Strategies in Scientific Computing*. New York: Springer-Verlag.

Lumley, D. (2001). Time-lapse seismic reservoir monitoring. *Geophysics*, 66, 50–53.

Lumley, D., Behrens, R. A., and Wang, Z. (1997). Assessing the technical risk of a 4-D seismic project. *The Leading Edge*, 16, 1287–1292.

Lundberg, A., Granlund, N., and Gustafsson, D. (2010). Towards automated 'Ground truth' snow measurements – a review of operational and new measurement methods for Sweden, Norway, and Finland. *Hydrological Processes*, 24, 1955–1970.

MacDonald, I. L., and Zucchini, W. (1997). *Hidden Markov and Other Models for Discrete-Valued Time Series*. Boca Raton, FL: Chapman & Hall/CRC Press.

Manoharan, N. (2014). K-nearest neighbour methods for value of information in petroleum decision making. MSc thesis, Norwegian University of Science and Technology (NTNU).

Mantyniemi, S., Kuikka, S., Rahikainen, M., Kell, L. T., and Kaitala, V. (2009). The value of information in fisheries management: North Sea herring as an example. *ICES Journal of Marine Science*, 66, 2278–2283.

Mardia, K. V., Kent, J. T., and Bibby, J. M. (1980). *Multivariate Analysis*. London: Academic Press.

Mariethoz, G., and Caers, J. (2015). *Multiple-Point Geostatistics: Stochastic Modeling with Training Images*. Hoboken, NJ: Wiley & Sons.

Mariethoz, G., Renard, P., and Straubhaar, J. (2010). The direct sampling method to perform multiple point geostatistical simulations. *Water Resources Research*, 46, W11536.

Marin, J. M., Pudlo, P., Robert, C. P., and Ryder, R. J. (2012). Approximate Bayesian computational methods. *Statistics and Computing*, 22, 1167–1180.

Martinelli, G., and Eidsvik, J. (2014). Dynamic exploration designs for graphical models using clustering with applications to petroleum exploration. *Knosys*, 58, 113–126.

Martinelli, G., Eidsvik, J., and Hauge, R. (2013a). Dynamic decision making for graphical models applied to oil exploration. *European Journal of Operational Research*, 230, 688–702.

Martinelli, G., Eidsvik, J., Hauge, R., and Førland, M. D. (2011). Bayesian networks for prospect analysis in the North Sea. *AAPG Bulletin*, 95, 1423–1442.

Martinelli, G., Eidsvik, J., Hokstad, K., and Hauge, R. (2014). Strategies for petroleum exploration based on Bayesian networks: a case study. *Society of Petroleum Engineers Journal*, 19, 564–575.

Martinelli, G., Eidsvik, J., Sinding-Larsen, R., Rekstad, S., and Mukerji, T. (2013b). Building Bayesian networks from basin modeling scenarios for improved geological decision making. *Petroleum Geoscience*, 19, 289–304.

Matheson, J., and Howard, R. (1968). An introduction to decision analysis. In R. Howard and J. Matheson, eds., *The Principles and Applications of Decision Analysis*, Vol. I. Strategic Decisions Group, pp. 17–55.

Matheson, J. E. (1990). Using influence diagrams to value information and control. In R. M. Oliver, J. Q. Smith, eds., *Influence Diagrams, Belief Nets and Decision Analysis*. Wiley & Sons, pp. 25–48.

Mavko, G., Mukerji, T., and Dvorkin, J. (2009). *The Rock Physics Handbook: Tools for Seismic Analysis of Porous Media*, 2nd edn. Cambridge University Press.

McCullagh, P., and Nelder, J. A. (1989). *Generalized Linear Models*. Boca Raton, FL: Chapman & Hall/CRC Press.

McLachlan, G., and Krishnan, T. (2008). *The EM Algorithm and Extensions*. Hoboken, NJ: Wiley.

Meinshausen, N., and Buhlmann, P. (2006). High-dimensional graphs with the lasso. *Annals of Statistics*, 34, 1436–1462.

Merkhofer, M. W. (1977). The value of information given decision flexibility. *Management Science*, 23, 716–727.

Meza, F. J., Hansen, J. W., and Osgood, D. (2008). Economic value of seasonal climate forecasts for agriculture: review of ex-ante assessments and recommendations for future research. *Journal of Applied Meteorology and Climatology*, 47, 1269–1286.

Miles, R. (2007). The emergence of decision analysis. In W. Edwards, R. Miles, and D. von Winterfeldt, eds., *Advances in Decision Analysis: From Foundations to Applications*. Cambridge University Press, pp. 13–31.

Miller, A. C. (1975). The value of sequential information. *Management Science*, 22, 1–11.

Mukerji, T., Avseth, P., Mavko, G., Takahashi, I., and Gonzalez, E. (2001). Statistical rock physics: combining rock physics, information theory, and geostatistics to reduce uncertainty in seismic reservoir characterization. *The Leading Edge*, 20, 313–319.

Muller, W. (2007). *Collecting Spatial Data*. Berlin: Springer-Verlag.

Murphy, K. P. (2012). *Machine Learning: A Probabilistic Perspective*. Cambridge, MA: MIT Press.

Newendorp, P. D., and Schuyler, J. R. (2013). *Decision Analysis for Petroleum Exploration*, 3rd edn. Aurora, CO: Planning Press.

Newman, M. E. J. (2005). Power laws, Pareto distributions and Zipf's law. *Contemporary Physics*, 46, 323–351.

Nowak, W., Rubin, Y., and de Barros, F. P. J. (2012). A hypothesis-driven approach to optimize field campaigns. *Water Resources Research*, 48, W06509.

Panchal, J. H., Paredis, C. J. J., Allen, J. K., and Mistree, F. (2009). Managing design-process complexity: a value-of-information based approach for scale and decision decoupling. *Journal of Computing and Information Science in Engineering*, 9, 021005.

Pearl, J. (1988). *Probabilistic Reasoning in Intelligent Systems*. San Francisco, CA: Morgan Kaufmann.

Peyrard, N., Sabbadin, R., Spring, D., Brook, B., and MacNally, R. (2013). Model-based adaptive spatial sampling for occurrence map construction. *Statistics and Computing*, 23, 29–42.

Phillips, J., Newman, A. M., and Walls, M. R. (2009). Utilizing a value of information framework to inprove ore collection and classification procedures. *The Engineering Economist*, 54, 50–74.

Pinto, J. R., de Agular, J. C., and Moraes, F. S. (2011). The value of information from time-lapse seismic data. *The Leading Edge*, 30, 572–576.

Polasky, R., and Solow, A. R. (2001). The value of information in reserve site selection. *Biodiversity and Conservation*, 10, 1051–1058.

Powell, W. B. (2011). *Approximate Dynamic Programming: Solving the Curses of Dimensionality*, 2nd edn. Hoboken, NJ: Wiley.

Puterman, M. L. (2005). *Markov Decision Processes: Discrete Stochastic Dynamic Programming*. Hoboken, NJ: Wiley.

Pyrcz, M. J., and Deutsch, C. V. (2014). *Geostatistical Reservoir Modeling*. Oxford: Oxford University Press.

Raiffa, H. (1968). *Decision Analysis: Introductory Lectures on Choices under Uncertainty*. Boston: Addison-Wesley.

Rasmussen, C. E., and Williams, C. (2006). *Gaussian Processes for Machine Learning*. Cambridge, MA: MIT Press.

Reeves, R., and Pettitt, A. N. (2004). Efficient recursions for general factorisable models. *Biometrika*, 91, 751–757.

Reich, B. J., and Fuentes, M. (2007). A multivariate semiparametric Bayesian spatial modeling framework for hurricane surface wind fields. *Annals of Applied Statistics*, 1, 249–264.

Remy, N., Boucher, A., and Wu, J. (2008). *Applied Geostatistics with SGeMS*. Cambridge University Press.

Rezaie, J., Eidsvik, J., and Mukerji, T. (2014). Value of information analysis and Bayesian inversion for closed skew-normal distributions: applications to seismic amplitude variation with offset data. *Geophysics*, 79, R151–R163.

Rimstad, K., Avseth, P., and Omre, H. (2012). Hierarchical Bayesian lithology/fluid prediction: a North Sea case study. *Geophysics*, 77, B69–B85.

Rivoirard, J. (1987). Two key parameters when choosing the Kriging neighborhood. *Mathematical Geology*, 19, 851–856.

Royle, J. A. (2002). Exchange algorithm for constructing large spatial designs. *Journal of Statistical Planning and Inference*, 100, 121–134.

Royle, J. A., and Nychka, D. (1998). An algorithm for the construction of spatial coverage designs with implementation in SPLUS. *Computers & Geosciences*, 24, 479–488.

Rubinstein, R. Y., and Kroese, D. P. (2007). *Simulation and the Monte Carlo Method*. Hoboken, NJ: Wiley.

Rue, H., and Held, L. (2005). *Gaussian Markov Random Fields: Theory and Applications*. Boca Raton, FL: Chapman & Hall/CRC Press.

Rue, H., Martino, S., and Chopin, N. (2009). Approximate Bayesian inference for latent Gaussian models using integrated nested Laplace approximations. *Journal of the Royal Statistical Society, B*, 71, 319–392.

Sagan, C. (1994). *Pale Blue Dot: A Vision of the Human Future in Space*. New York: Random House.

Santner, T. J., Williams, B. J., and Notz, W. I. (2003). *Design and Analysis of Computer Experiments*. New York: Springer-Verlag.

Schabenberger, O., and Gotway, C. A. (2009). *Statistical Methods for Spatial Data Analysis*. Boca Raton, FL: Chapman & Hall/CRC Press.

Scheidt, C., and Caers, J. (2009). Uncertainty quantification in reservoir performance using distances and kernel methods – application to a West Africa deepwater turbidite reservoir. *Society of Petroleum Engineers Journal*, 14, 680–692.

Schlaiffer, R. (1959). *Probability and Statistics for Business Decisions*. New York: McGraw-Hill.

Schon, J. H. (2011). *Physical Properties of Rocks: A Workbook*. Elsevier.

Scott, S. L. (2002). Bayesian methods for hidden Markov models: recursive computing in the 21st century. *Journal of the American Statistical Association*, 97, 337–351.

Shachter, R. (1986). Evaluating influence diagrams. *Operations Research*, 34, 871–882.

(1988). Probabilistic inference and influence diagrams. *Operations Research*, 36, 589–605.

(1999). Efficient value of information computation. In *Proceedings of the 15th Conference on Uncertainty in Artificial Intelligence (UAI)*, Morgan Kaufmann, pp. 594–601.

(2007). Model building with belief networks and influence diagrams. In W. Edwards, R. Miles, and D. von Winterfeldt, eds., *Advances in Decision Analysis: From Foundations to Applications*. Cambridge University Press, pp. 177–201.

Shachter, R., and Peot, M. (1992). Decision making using probabilistic inference methods. In D. Dubois, M. P. Wellman, B. D'Ambrosio, and P. Smets, eds., *Proceedings of the 8th Conference on Uncertainty in Artificial Intelligence (UAI)*, Morgan Kaufmann, pp. 276–283.

Shannon, C. E. (1948). A mathematical theory of communication. *Bell System Technical Journal*, 27, 379–324, reprinted in C. E. Shannon and W. Weaver, *The Mathematical Theory of Communication*, University of Illinois Press, 1949, 1998.

Shenoy, P. (1992). Valuation-based systems for Bayesian decision analysis. *Operations Research*, 40, 463–484.

(1998). Game trees for decision analysis. *Theory and Decision*, 44, 149–171.

Shewry, M. C., and Wynn, H. P. (1987). Maximum entropy sampling. *Journal of Applied Statistics*, 14, 165–170.

Silverman, B. W. (1988). *Density Estimation for Statistics and Data Analysis*. Boca Raton, FL: Chapman & Hall/CRC Press.

Srivastava, R. M. (1994). An overview of stochastic methods for reservoir characterization. In J. M. Yarus and R. L. Chambers, eds., *Stochastic Modeling and Geostatistics: Principles, Methods, and Case Studies*. American Association of Petroleum Geologists, pp. 3–16.

Stein, M. L. (1999). *Statistical Interpolation of Spatial Data: Some Theory for Kriging*. New York: Springer-Verlag.

Stein, M. L., Chi, Z., and Welty, L. J. (2004). Approximating likelihood for large spatial data sets. *Journal of the Royal Statistical Society, Series B*, 66, 275–296.

Stien, M., and Kolbjørnsen, O. (2011). Facies modeling using a Markov mesh model specification. *Mathematical Geosciences*, 43, 611–624.

Strebelle, S. (2000). Sequential simulation drawing structures from training images. PhD thesis, Stanford University.

(2002). Conditional simulation of complex geological structures using multiple-point statistics. *Mathematical Geology*, 34, 1–21.

Strebelle, S. B., and Journel A. G. (2001). Reservoir modeling using multiple-point statistics. *Society of Petroleum Engineers Journal*, 71324.

Sucar, L. E., Morales, E. F., and Hoey, J. (2012). *Decision Theory Models for Applications in Artificial Intelligence*. Hershey, PA: IGI Global.
Sylta, Ø. (2004). Hydrocarbon migration modelling and exploration risk. PhD thesis, Norwegian University of Science and Technology (NTNU).
(2008). Analysing exploration uncertainties by tight integration of seismic and hydrocarbon mifration modelling. *Petroleum Geoscience*, 14, 281–289.
Tahmasebi, P., Hezarkhani, A., and Sahimi, M. (2012). Multiple-point geostatistical modeling based on the cross-correlation function. *Computational Geosciences*, 16, 779–797.
Tatman, J., and Shachter, R. (1990). Dynamic programming and influence diagrams. *IEEE Transactions on Systems, Man, and Cybernetics*, 20, 365–379.
Taylor, H. M., and Karlin, S. (1994). *An Introduction to Stochastic Modeling*. London: Academic Press.
Tjelmeland, H., and Austad, H. M. (2012). Exact and approximate recursive calculations for binary Markov random fields defined on graphs. *Journal of Graphical and Computational Statistics*, 21, 758–780.
Tjelmeland, H., and Besag, J. (1996). Markov random fields with higher-order interactions. *Scandinavian Journal of Statistics*, 25, 415–433.
Trainor-Guitton, W. J., Caers, J., and Mukerji, T. (2011). A methodology for establishing a data reliability measure for value of spatial information problems. *Mathematical Geosciences*, 43, 929–949.
Trainor-Guitton, W. J. Hoversten, G. M., Ramirez, A., Roberts, J., Juliusson, E., Key, K., and Mellors, R. (2014). The value of spatial information for determining well placement: a geothermal example. *Geophysics*, 79, W27–W41.
Trainor-Guitton, W. J., Mukerji, T., and Knight, R. (2013). A methodology for quantifying the value of spatial information for dynamic Earth problems. *Stochastic Environmental Research and Risk Assessment*, 27, 969–983.
Tversky, A., and Kahneman, D. (1974). Judgment under uncertainty: heuristics and biases. *Science*, 185, 1124–1131.
Tviberg, S. (2011). To assess the petroleum net present value and accumulation process in a controlled Petromod environment. MSc thesis, Norwegian University of Science and Technology (NTNU).
Ulvmoen, M., Omre, H., and Buland, A. (2010). Improved resolution in Bayesian lithology/fluid inversion from prestack seismic data and well observations: Part 2 – real case study, *Geophysics*, 75, B73–B82.
Vann, J., Jackson, S., and Bertoli, O. (2003). Quantitative Kriging neighborhood analysis for the mining geologist – a description of the method with worked case examples. In *Proceedings of the Fifth International Mining Conference*, The Australasian Institute of Mining and Metallurgy (The AusIMM), pp. 215–223.
van Wees, J., Mijnlie, H., Lutgert, J., Breunese, J., Bos, C., Rosenkranz, P., and Neele, F. (2008). A Bayesian belief network approach for assessing the impact of exploration prospect interdependency: an application to predict gas discoveries in the Netherlands. *AAPG Bulletin*, 92, 1315–1336.
Varin, C., Reid, N., and Firth, D. (2011). An overview of composite likelihood methods. *Statistica Sinica*, 21, 5–42.
Vermeer, G. J. O. (2012). *3D Seismic Survey Design*. Society of Exploration Geophysicists.
von Neumann, J., and Morgenstern, O. (1947). *Theory of Games and Economic Behavior*, 2nd edn. Princeton University Press.

von Winterfeldt, D., and Edwards, W. (1986). *Decision Analysis and Behavioral Research*. Cambridge University Press.

Wackernagel, H. (2003). *Multivariate Geostatistics*, 3rd edn. Berlin: Springer-Verlag.

Wagner, J. M., Shamir, U., and Nemati, H. R. (1992). Groundwater quality management under uncertainty: stochastic programming approaches and the value of information. *Water Resources Research*, 28, 1233–1246.

Welton, N. J., Ades, A. E., Caldwell, D. M., and Peters, T. J. (2008). Research prioritization based on expected value of partial perfect information: a case-study on interventions to increase uptake of breast cancer screening. *Journal of the Royal Statistical Society, Series A*, 171, 807–841.

Whittaker, J. (1990). *Graphical Models in Applied Multivariate Statistics*. Hoboken, NJ: Wiley.

Wiles, L. J. (2004). Economics of weed management: principles and practices. *Weed Technology*, 18, 1403–1407.

Willan, A. R., and Pinto, E. M. (2005). The value of information and optimal clinical trial design. *Statistics in Medicine*, 24, 1791–1806.

Williams, B. K., Eaton, M. J., and Breininger, D. R. (2011). Adaptive resource management and the value of information. *Ecological Modeling*, 222, 3429–3436.

Yokota, F., and Thompson, K. (2004a). Value of information literature analysis: a review of applications in health risk management. *Medical Decision Making*, 24, 287–298.

(2004b). Value of information analysis in environmental health risk management decisions: past, present, and future. *Risk Analysis*, 24, 635–650.

Zan, K., and Bickel, J. E. (2013). Components of portfolio value of information. *Decision Analysis*, 10, 171–183.

Zetterlund, M., Norberg, T., Ericsson, L. O., and Rosen, L. (2011). Framework for value of information analysis in rock mass characterization for grouting purposes. *Journal of Construction Engineering and Management*, 137, 486–497.

Zhang, T., Switzer, P., and Journel, A. (2006). Filter-based classification of training image patterns for spatial simulation. *Mathematical Geology*, 38, 63–80.

Zimmerman, D. L. (2006). Optimal network design for spatial prediction, covariance estimation and empirical prediction. *Environmetrics*, 17, 635–652.

Index